Understanding Engineering Mathematics

Understanding Engineering Mathematics

Bill Cox

Newnes

OXFORD AUCKLAND BOSTON JOHANNESBURG MELBOURNE NEW DELHI

Newnes
An imprint of Butterworth-Heinemann
Linacre House, Jordan Hill, Oxford OX2 8DP
225 Wildwood Avenue, Woburn, MA 01801-2041
A division of Reed Educational and Professional Publishing Ltd

 A member of the Reed Elsevier plc group

First published 2001

© Bill Cox 2001

All rights reserved. No part of this publication may be
reproduced in any material form (including photocopying
or storing in any medium by electronic means and whether
or not transiently or incidentally to some other use of
this publication) without the written permission of the copyright
holder except in accordance with the provisions of the
Copyright, Designs and Patents Act 1988 or under
the terms of a licence issued by the Copyright Licensing
Agency Ltd, 90 Tottenham Court Road, London, England W1P 0LP.
Applications for the copyright holder's written permission
to reproduce any part of this publication should be
addressed to the publishers.

British Library Cataloguing in Publication Data
A catalogue record for this book is available from the British Library

ISBN 0 7506 5098 2

Typeset by Laser Words Private Limited, Chennai, India
Printed in Malta by Gutenberg Press Ltd.

Contents

Preface		**ix**
To the Student		**xi**
1	**Number and Arithmetic**	**1**
	1.1 Review	2
	1.2 Revision	5
	1.3 Reinforcement	27
	1.4 Applications	31
	Answers to reinforcement exercises	32
2	**Algebra**	**37**
	2.1 Review	38
	2.2 Revision	40
	2.3 Reinforcement	73
	2.4 Applications	79
	Answers to reinforcement exercises	81
3	**Functions and Series**	**87**
	3.1 Review	88
	3.2 Revision	90
	3.3 Reinforcement	107
	3.4 Applications	110
	Answers to reinforcement exercises	112
4	**Exponential and Logarithm Functions**	**118**
	4.1 Review	119
	4.2 Revision	120
	4.3 Reinforcement	136
	4.4 Applications	138
	Answers to reinforcement exercises	139
5	**Geometry of Lines, Triangles and Circles**	**142**
	5.1 Review	143
	5.2 Revision	147
	5.3 Reinforcement	160
	5.4 Applications	165
	Answers to reinforcement exercises	167

6 Trigonometry — 170

- 6.1 Review — 171
- 6.2 Revision — 173
- 6.3 Reinforcement — 194
- 6.4 Applications — 197
- Answers to reinforcement exercises — 198

7 Coordinate Geometry — 203

- 7.1 Review — 204
- 7.2 Revision — 205
- 7.3 Reinforcement — 220
- 7.4 Applications — 223
- Answers to reinforcement exercises — 224

8 Techniques of Differentiation — 227

- 8.1 Review — 228
- 8.2 Revision — 230
- 8.3 Reinforcement — 243
- 8.4 Applications — 245
- Answers to reinforcement exercises — 247

9 Techniques of Integration — 250

- 9.1 Review — 251
- 9.2 Revision — 253
- 9.3 Reinforcement — 280
- 9.4 Applications — 285
- Answers to reinforcement exercises — 286

10 Applications of Differentiation and Integration — 290

- 10.1 Review — 291
- 10.2 Revision — 292
- 10.3 Reinforcement — 309
- 10.4 Applications — 311
- Answers to reinforcement exercises — 314

11 Vectors — 317

- 11.1 Introduction – representation of a vector quantity — 318
- 11.2 Vectors as arrows — 319
- 11.3 Addition and subtraction of vectors — 321
- 11.4 Rectangular Cartesian coordinates in three dimensions — 323
- 11.5 Distance in Cartesian coordinates — 324
- 11.6 Direction cosines and ratios — 325
- 11.7 Angle between two lines through the origin — 327
- 11.8 Basis vectors — 328
- 11.9 Properties of vectors — 330
- 11.10 The scalar product of two vectors — 333
- 11.11 The vector product of two vectors — 336

11.12	Vector functions		339
11.13	Differentiation of vector functions		340
11.14	Reinforcement		344
11.15	Applications		347
11.16	Answers to reinforcement exercises		348

12 Complex Numbers — 351

12.1	What are complex numbers?	352
12.2	The algebra of complex numbers	353
12.3	Complex variables and the Argand plane	355
12.4	Multiplication in polar form	357
12.5	Division in polar form	360
12.6	Exponential form of a complex number	361
12.7	De Moivre's theorem for integer powers	362
12.8	De Moivre's theorem for fractional powers	363
12.9	Reinforcement	366
12.10	Applications	370
12.11	Answers to reinforcement exercises	373

13 Matrices and Determinants — 377

13.1	An overview of matrices and determinants	378
13.2	Definition of a matrix and its elements	378
13.3	Adding and multiplying matrices	381
13.4	Determinants	386
13.5	Cramer's rule for solving a system of linear equations	391
13.6	The inverse matrix	393
13.7	Eigenvalues and eigenvectors	397
13.8	Reinforcement	400
13.9	Applications	403
13.10	Answers to reinforcement exercises	405

14 Analysis for Engineers – Limits, Sequences, Iteration, Series and All That — 409

14.1	Continuity and irrational numbers	410
14.2	Limits	412
14.3	Some important limits	416
14.4	Continuity	418
14.5	The slope of a curve	421
14.6	Introduction to infinite series	422
14.7	Infinite sequences	424
14.8	Iteration	426
14.9	Infinite series	428
14.10	Tests for convergence	430
14.11	Infinite power series	434
14.12	Reinforcement	438
14.13	Applications	441
14.14	Answers to reinforcement exercises	442

15 Ordinary Differential Equations — 445

- 15.1 Introduction — 446
- 15.2 Definitions — 448
- 15.3 First order equations – direct integration and separation of variables — 452
- 15.4 Linear equations and integrating factors — 458
- 15.5 Second order linear homogeneous differential equations — 462
- 15.6 The inhomogeneous equation — 468
- 15.7 Reinforcement — 475
- 15.8 Applications — 476
- 15.9 Answers to reinforcement exercises — 480

16 Functions of More than One Variable – Partial Differentiation — 483

- 16.1 Introduction — 484
- 16.2 Function of two variables — 484
- 16.3 Partial differentiation — 487
- 16.4 Higher order derivatives — 489
- 16.5 The total differential — 490
- 16.6 Reinforcement — 494
- 16.7 Applications — 495
- 16.8 Answers to reinforcement exercises — 496

17 An Appreciation of Transform Methods — 500

- 17.1 Introduction — 500
- 17.2 The Laplace transform — 501
- 17.3 Laplace transforms of the elementary functions — 504
- 17.4 Properties of the Laplace transform — 509
- 17.5 The inverse Laplace transform — 512
- 17.6 Solution of initial value problems by Laplace transform — 513
- 17.7 Linear systems and the principle of superposition — 515
- 17.8 Orthogonality relations for trigonometric functions — 516
- 17.9 The Fourier series expansion — 517
- 17.10 The Fourier coefficients — 520
- 17.11 Reinforcement — 523
- 17.12 Applications — 524
- 17.13 Answers to reinforcement exercises — 527

Index — 529

Preface

This book contains most of the material covered in a typical first year mathematics course in an engineering or science programme. It devotes Chapters 1–10 to consolidating the foundations of basic algebra, elementary functions and calculus. Chapters 11–17 cover the range of more advanced topics that are normally treated in the first year, such as vectors and matrices, differential equations, partial differentiation and transform methods.

With widening participation in higher education, broader school curricula and the wide range of engineering programmes available, the challenges for both teachers and learners in engineering mathematics are now considerable. As a result, a substantial part of many first year engineering programmes is dedicated to consolidation of the basic mathematics material covered at pre-university level. However, individual students have widely varying backgrounds in mathematics and it is difficult for a single mathematics course to address everyone's needs. This book is designed to help with this by covering the basics in a way that enables students and teachers to quickly identify the strengths and weaknesses of individual students and 'top up' where necessary. The structure of the book is therefore somewhat different to the conventional textbook, and 'To the student' provides some suggestions on how to use it.

Throughout, emphasis is on the key mathematical techniques, covered largely in isolation from the applications to avoid cluttering up the explanations. When you teach someone to drive it is best to find a quiet road somewhere for them to learn the basic techniques before launching them out onto the High Street! In this book the mathematical techniques are motivated by explaining where you may need them, and each chapter has a short section giving typical applications. More motivational material will also be available on the book web-site. Rigorous proof for its own sake is avoided, but most things are explained sufficiently to give an understanding that the educated engineer should appreciate. Even though you may use mathematics as a tool, it usually helps to have an idea of how and why the tool works.

As the book progresses through the more advanced first year material there is an increasing expectation on the student to learn independently and 'fill in the gaps' for themselves – possibly with the teacher's help. This is designed to help the student to develop a mature, self-disciplined approach as they move from the supportive environment of pre-university to the more independent university environment. In addition the book web-site (www.bh.com/companions/0750650982) will provide a developing resource to supplement the book and to focus on specific engineering disciplines where appropriate.

In the years that this book has been in development I have benefited from advice and help from too many people to list. The following deserve special mention however. Dave Hatter for having faith in the original idea and combining drink and incisive comment well mixed in the local pub. Peter Jack for many useful discussions and for the best part of the S(ketch) GRAPH acronym (I just supplied the humps and hollows). Val Tyas for typing much of the manuscript, exploring the limits of RSI in the process, and coping cheerfully with my continual changes. The late Lynn Burton for initial work on the manuscript and diagrams. She was still fiddling with the diagrams only weeks before she succumbed to cancer after a long and spirited fight. I am especially indebted to her for her friendship and inspiration – she would chuckle at that.

I also benefited from an anonymous reviewer who went far beyond the call of duty in providing meticulous, invaluable comment – It's clear that (s)he is a good teacher. Of course, any remaining errors are my responsibility.

The team at Butterworth-Heinemann did a wonderful job in dealing with a complicated manuscript – sense of humour essential!

Last but not least I must mention the hundreds of students who have kept me in line over the years. I have tried to write the book that would help most of them. I hope they, and their successors, will be pleased with it.

Bill Cox, June 2001

To the Student

Whatever your previous background in mathematics, it is likely that when you begin your engineering studies at university you will need to consolidate your mathematical skills before moving on to new material. The first ten chapters of this book are designed to help with this 'transition' by providing you with individual pathways to quickly **review** your current skills and understanding, then **revise** and **reinforce** where necessary.

Chapters 1–10 have a three-part structure by which you:

- **Review** your present knowledge and skills, with a review test on key topics
- **Revise** as you need to
- **Reinforce** the essential skills that you will need for your particular programme, so that they are there when you need them.

The three sections are linked, so that you can choose your own pathway through the material and focus on your specific needs. **In the review test the single arrows forward you to the corresponding revision section and the review question solution, while the double arrows fast forward you to the reinforcement exercises.**

Some suggestions for working through these chapters may help:

- Use the lists of prerequisites and objectives to get an overview of the chapter – what have you seen before, what needs a reminder, what is completely new?
- Use the review test to establish your current understanding of the various topics.
- Where you are confident you may still be able to learn something from the revision section or polish your skills by working through the reinforcement questions.
- Where you are unsure and need a reminder, go to the relevant revise section for some hints before trying the review question, then consolidate your skills with the reinforcement exercises.
- Where the topic is perhaps new to you, start with the revision section, using the review questions as worked examples and the reinforcement exercises as further practice.

The remaining chapters (11–17), covering material appropriate to typical first-year courses in engineering mathematics, are designed to support you in developing independent learning skills for more advanced study. The order of material and the structure of these chapters are at first supportive with many examples, then gradually progress to a more concise and mature format. The focus is on the key material, and the text contains leading problems that encourage you to develop ideas for yourself. The structure of the book is therefore designed not only to ease the transition to university, but also to develop your independent learning skills and prepare you for the style of more advanced textbooks.

Each chapter has a number of 'Applications' exercises that provide illustrations of typical engineering applications, bring together the different topics of the chapter, or prepare the way for later material. Some are simple, while others provide significant and challenging projects.

The book is the core of a larger educational resource of web-based material enabling you to broaden and deepen your studies. The book web-site (www.bh.com/companions/0750650982) provides advice on learning mathematics, solutions to all of the reinforcement and applications exercises, develops some topics more thoroughly, and provides relevant examples and illustrations from different engineering disciplines.

1

Number and Arithmetic

In this chapter we review the key features of elementary numbers and arithmetic. The topics covered are those found to be most useful later on.

Prerequisites

It will be helpful if you know something about:

- simple types of numbers such as integers, fractions, negative numbers, decimals
- the concepts of 'greater than' and 'less than'
- elementary arithmetic: addition, subtraction, multiplication and division
- powers and indices notation, $2^3 = 2 \times 2 \times 2$, for example
- how to convert a simple fraction to a decimal and vice versa

Objectives

In this chapter you will find:

- different types of numbers and their properties (particularly zero)
- the use of inequality signs
- highest common factors and lowest common denominators
- manipulation of numbers (BODMAS)
- handling fractions
- factorial ($n!$) and combinatorial $\left({}^nC_r \text{ or } \binom{n}{r} \right)$ notation
- powers and indices
- decimal notation
- estimation of numerical expressions

Motivation

You may need the material of this chapter for:

- numerical manipulation and calculation in engineering applications
- checking and using scientific formulae
- illustrating and checking results used later in mathematics
- statistical calculations
- numerical estimation and 'back of an envelope' calculations

Understanding ENGINEERING Mathematics

A note about calculators

Calculators obviously have their place, particularly in applied mathematics, numerical methods and statistics. However, they are very rarely needed in this chapter, and the skills it aims to develop are better learnt without them.

1.1 Review

1.1.1 Types of numbers ▶5 27▶▶

A. For each number choose one or more descriptions from the following: (a) integer, (b) negative, (c) rational number (fraction), (d) real, (e) irrational, (f) decimal, (g) prime.
(i) is done as an example

(i) -1 (a, b, c, d) (ii) $\dfrac{1}{2}$ (iii) 0

(iv) 7 (v) $\dfrac{23}{5}$ (vi) $-\dfrac{3}{4}$

(vii) 0.73 (viii) 11 (ix) 8

(x) $\sqrt{2}$ (xi) -0.49 (xii) π

B. Which of the following descriptions apply to the expressions in (i)–(x) below?

(a) infinite (b) does not exist (c) negative
(d) zero (e) finite (f) non-zero

(i) 0×1 (d, e) (ii) $0 + 1$ (iii) $\dfrac{1}{0}$

(iv) $2 - 0$ (v) 0^2 (vi) $0 - 1$

(vii) $\dfrac{0}{0}$ (viii) $3 \times 0 + \dfrac{3}{0}$ (ix) $\dfrac{0^3}{0}$

(x) $\dfrac{2}{2}$

1.1.2 Use of inequality signs ▶7 27▶▶

Express symbolically:

(i) x is a positive, non-zero, number ($x > 0$)
(ii) x lies strictly between 1 and 2
(iii) x lies strictly between -1 and 3
(iv) x is equal to or greater than -2 and is less than 2
(v)   The absolute value of x is less than 2.

1.1.3 Highest common factor and lowest common multiple

A. Express in terms of prime factors

(i) 15 ($= 3 \times 5$) (ii) 21 (iii) 60
(iv) 121 (v) 405 (vi) 1024
(vii) 221

B. Find the highest common factor (HCF) of each of the following sets of numbers

(i) 24, 30 (6) (ii) 27, 99 (iii) 28, 98
(iv) 12, 54, 78 (v) 3, 6, 15, 27

C. Find the lowest common multiple (LCM) of each of the following sets of numbers

(i) 3, 7 (21) (ii) 3, 9 (iii) 12, 18
(iv) 3, 5, 9 (v) 2, 4, 6

1.1.4 Manipulation of numbers

Evaluate

(i) $2 + 3 - 7 \; (= -2)$ (ii) $4 \times 3 \div 2$ (iii) $3 + 2 \times 5$
(iv) $(3 + 2) \times 5$ (v) $3 + (2 \times 5)$ (vi) $18 \div 2 \times 3$
(vii) $18 \div (2 \times 3)$ (viii) $-2 - (4 - 5)$ (ix) $(4 \div (-2)) \times 3 - 4$
(x) $(3 + 7) \div 5 + (7 - 3) \times (2 - 4)$

1.1.5 Handling fractions

A. Simplify

(i) $\dfrac{4}{6} \left(= \dfrac{2}{3} \right)$ (ii) $\dfrac{18}{9}$ (iii) $\dfrac{7}{3} \times \dfrac{4}{7}$

(iv) $\dfrac{7}{5} \times \dfrac{3}{14}$ (v) $\dfrac{3}{4} \div \dfrac{4}{5}$ (vi) $\dfrac{1}{2} + \dfrac{1}{3}$

(vii) $\dfrac{1}{2} - \dfrac{1}{3}$ (viii) $\dfrac{4}{15} - \dfrac{7}{3}$ (ix) $1 + \dfrac{1}{2} + \dfrac{1}{3}$

(x) $\dfrac{2}{3} - \dfrac{3}{4} + \dfrac{1}{8}$

B. If the numbers a and b are in the ratio $a : b = 3 : 2$ and $a = 6$, what is b?

1.1.6 Factorial and combinatorial notation – permutations and combinations ▶16 29▶▶

A. Evaluate

(i) $3! \,(= 6)$ (ii) $6!$ (iii) $\dfrac{24!}{23!}$ (iv) $\dfrac{12!}{9!\,3!}$

B. (i) Evaluate (a) $^3C_2 \,(= 3)$ (b) $^6C_4$ (c) $^6P_3$
(ii) In how many ways can two distinct letters be chosen from ABCD?
(iii) How many permutations of the letters ABCDE are there?

1.1.7 Powers and indices ▶18 29▶▶

A. Reduce to simplest power form.

(i) $2^3 2^4 \,(= 2^7)$ (ii) $3^4/3^3$ (iii) $(5^2)^3$

(iv) $(3 \times 4)^4/(9 \times 2^3)$ (v) $16^2/4^4$ (vi) $(-6)^2(-\tfrac{3}{2})^3$

(vii) $(-ab^2)^3/a^2 b$ (viii) $2^2(\tfrac{1}{2})^{-3}$

B. Express in terms of simple surds such as $\sqrt{2}$, $\sqrt{3}$, etc.

(i) $\sqrt{50} \,(= 5\sqrt{2})$ (ii) $\sqrt{72} - \sqrt{8}$ (iii) $(\sqrt{27})^3$

(iv) $\left(\dfrac{\sqrt{2}\sqrt{3}}{4}\right)^2$ (v) $\dfrac{\sqrt{3}\sqrt{7}}{\sqrt{84}}$ (vi) $\dfrac{\sqrt{3}+2\sqrt{2}}{\sqrt{3}-\sqrt{2}}$

(vii) $\left(\dfrac{3^{1/3} 9^{1/3}}{27}\right)^2$

1.1.8 Decimal notation ▶22 30▶▶

A. Express in decimal form

(a) $\tfrac{1}{2}$ (b) $-\tfrac{3}{2}$ (c) $\tfrac{1}{3}$ (d) $\tfrac{1}{7}$

B. Express as fractions

(a) 0.3 (b) 0.67 (c) $0.\dot{6}$ (d) 3.142

C. Write the following numbers in scientific notation, stating the mantissa and exponent.

(i) 11.00132 (ii) 1.56 (iii) 203.45 (iv) 0.0000321

D. Write the numbers in **C** to three significant figures.

1.1.9 Estimation

Estimate the approximate value of each of

(i) $\dfrac{4.5 \times 10^5 \times 2.0012}{8.892 \times 10^4}$

(ii) $\dfrac{\sqrt{254 \times 10^4 + 28764.5}}{2.01 \times 10 - 254 \times 10^{-6}}$

1.2 Revision

1.2.1 Types of numbers

Numbers can be classified into different types:

- natural numbers
- zero
- directed numbers
- integers
- rational numbers (fractions)
- irrational numbers
- real numbers
- complex numbers (➤ Chapter 12).

The counting numbers

$$1, 2, 3, 4, \ldots$$

are called **natural numbers**.

Zero, 0, is really in a class of its own – we always have to be careful with it. It is an integer and also, of course, a real number. Essentially, zero enables us to define negative numbers. Thus, the negative of 3 is the number denoted $n = -3$ satisfying

$$3 + n = 0$$

This enables us to 'count in opposite directions' using **directed** or **negative numbers**

$$-1, -2, -3, -4, \ldots$$

The full set of numbers

$$\{\ldots -4, -3, -2, -1, 0, 1, 2, 3, 4, \ldots\}$$

is called the set of **integers**.
  Numbers that can be written in the form:

$$\dfrac{\text{integer}}{\text{non-zero integer}} \left(\text{e.g.} \; \dfrac{3}{4}, \dfrac{-1}{2} \right)$$

$$\left(\text{including integers, such as } 6 = \dfrac{6}{1} \right)$$

are called **rational numbers** or **fractions**. All measurements of a physical nature (length, time, voltage, etc.) can only be expressed in terms of such numbers. Numbers which are not rational, and cannot be expressed as ratios of integers, are called **irrational numbers**. Examples are $\sqrt{2}$ and π. We will prove that $\sqrt{2}$ is irrational in Chapter 14.

The set of all numbers: integers, rational and irrationals is called the set of **real numbers**. It can be shown that together these numbers can be used to 'label' every point on a continuous infinite line – the **real line**. So called 'complex numbers' are really equivalent to pairs of real numbers. They are studied in Chapter 12, and an introduction is provided in the Applications section of Chapter 2.

Note that zero, 0, is an exceptional number in that one cannot **divide** by it. It is not that 1/0 is 'infinity', but simply that **it does not exist at all**. **Infinity**, denoted ∞, is not really a number. It is a concept that indicates that no matter what positive (negative) number you choose, you can always find another positive (negative) number greater (less) than it. Crudely, ∞ denotes a 'number' that is as large as we wish.

Solution to review question 1.1.1

A. All numbers here are real, so d applies to them all.

(i) -1 is a negative natural number, i.e. an integer; (a, b, c, d)

(ii) $\frac{1}{2}$ is a ratio of integers and is therefore rational; (c, d)

(iii) 0 is an integer – the only one that is its own negative; (a, c, d)

(iv) 7 is a natural number and an integer. It is in fact also a prime number – that is, only divisible by itself or 1 (Section 1.2.3). It is also an **odd** number (cannot be exactly divided by 2); (a, c, d, g)

(v) $\frac{23}{5}$ is a rational number – actually an improper fraction (Section 1.2.5); (c, d)

(vi) $-\frac{3}{4}$ is a rational number – a proper fraction (Section 1.2.5); (b, c, d)

(vii) 0.73 is actually a decimal representation of a rational number

$$0.73 = \frac{73}{100}$$

sometimes called a decimal fraction, but usually simply a decimal (Section 1.2.8); (c, d, f)

(viii) 11 is a natural number and an integer – like 7 it is also prime, and is also odd as any prime greater than 2 must be; (a, c, d, g)

(ix) 8 is another natural number and integer – but it is not prime, since it can be written as $2 \times 2 \times 2 = 2^3$ (Section 1.2.7). It is also an even number; (a, c, d)

(x) The square root of 2, $\sqrt{2}$, is not a rational number. This can be shown by assuming that

$$\sqrt{2} = \frac{m}{n}$$

where m and n are two integers and deriving a contradiction. $\sqrt{2}$ is irrational and is a real number. Such numbers, square roots of prime numbers, are called **surds** (Section 1.2.7); (d, e)

(xi) -0.49 is a decimal representation (Section 1.2.8) of the negative rational number

$$-\frac{49}{100}$$

(b, c, d, f)

(xii) π, the ratio of the circumference of a circle to its diameter, is not a rational number – it is an **irrational number**. That is, it cannot be written as a fraction. 22/7, for example, is just an approximation to π; (d, e)

B. (i) $0 \times 1 = 0$, i.e. zero – which of course is also finite (d, e).

(ii) $0 + 1 = 1$, finite, non-zero (e, f).

(iii) $\frac{1}{0}$ does not exist – it is not infinite, negative, zero, finite or non-zero – it just does not exist (b).

(iv) $2 - 0 = 2$, finite and non-zero (e, f).

(v) $0^2 = 0 \times 0 = 0$, zero and finite (d, e).

(vi) $0 - 1 = -1$, negative, finite, non-zero (c, e, f).

(vii) $\frac{0}{0}$ does not exist (you can't 'cancel' the zeros!). It is not infinite, negative, zero, finite or non-zero – it just does not exist (b).

(viii) Because of the $\frac{3}{0}$ the expression $3 \times 0 + \frac{3}{0}$ does not exist (b).

(ix) $\dfrac{0^3}{0}$ again, does not exist (b).

(x) $\frac{2}{2} = 1$ – no problem here, finite and non-zero (e, f).

Note that none of the numbers in **B** is referred to as 'infinite'.

1.2.2 Use of inequality signs

◀2   27▶

The real numbers are **ordered**. That is, we can always say whether one number a is less than, equal to, or greater than another given number b. To denote this we use the 'comparator' symbols or **inequalities**, $<$ and $\leq$, $>$ and $\geq$. $a > b$ means a is greater than b; $a < b$ means a is less than b. Thus $6 > 5$, $4 < 5$. $a \geq b$ means a is greater than or equal to b, and similarly $a \leq b$ means a is less than or equal to b. Be very careful to distinguish between, for example $a > b$ and $a \geq b$. Sometimes it is also useful to use the 'not equal to' symbol, $\neq$.

Care is needed when changing signs and forming reciprocals with inequalities. For example, if $a > b > 0$, then $-a < -b$ and $\dfrac{1}{a} < \dfrac{1}{b}$. However, if $a > 0 > b$ then $-a < -b$ is still true, but $\dfrac{1}{a} > \dfrac{1}{b}$. Try a few numerical examples to check these statements. Most of us find inequalities difficult to handle and they require a lot of practice. However, in this book we will need only the basic properties of inequalities. We will say more about inequalities in Section 3.2.6.

Often we wish to refer to the **positive** or **absolute** value of a number x (for example in a rectified sine wave). We denote this by the **modulus of x**, $|x|$. For example

$$|-4| = 4$$

By definition $|x|$ is never negative, so $|x| \geq 0$. Also, note that $|x| < a$ means $-a < x < a$. For example:

$$|x| < 3$$

means

$$-3 < x < 3$$

Solution to review question 1.1.2

(i) 'x is a positive non-zero number' is expressed by $x > 0$
(ii) 'x lies strictly between 1 and 2' is expressed by $1 < x < 2$
(iii) 'x lies strictly between -1 and 3' is expressed by $-1 < x < 3$
(iv) 'x is equal to or greater than -2 and is less than 2' is expressed by $-2 \leq x < 2$
(v) If the absolute value of x is less than 2 then this means that if x is positive then $0 \leq x < 2$, but if x is negative then we must have $-2 < x \leq 0$. So, combining these we must have $-2 < x < 2$. This can also be expressed in terms of the modulus as $|x| < 2$.

1.2.3 Highest common factor and lowest common multiple

A **prime number** is a positive integer which cannot be expressed as a product of two or more smaller distinct positive integers. That is, a prime number cannot be divided exactly by any integer other than 1 or itself. From the definition, 1 is not a prime number. 6, for example, is not a prime, since it can be written as 2×3. The numbers 2 and 3 are called its (prime) **factors**. Another way of defining a prime number is to say that it is has no integer factors other than 1 and itself.

There are an infinite number of prime numbers:

$$2, 3, 5, 7, 11, 13, \ldots$$

but no formula for the nth prime has been discovered. Prime numbers are very important in the theory of codes and cryptography. They are also the 'building blocks' of numbers, since any given integer can be written uniquely as a product of primes:

$$12 = 2 \times 2 \times 3 = 2^2 3$$

This is called **factorising** the integer into its prime factors. It is an important operation, for example, in combining fractions.

The **highest common factor** (HCF) of a set of integers is the largest integer which is a factor of all numbers of the set. For small numbers we can find the HCF 'by inspection' – splitting the numbers into prime factors and constructing products of these primes that divide each number of the set, choosing the largest such product.

The **lowest common multiple** (LCM) of a set of integers is the smallest integer which is a multiple of all integers in the set. It can again be found by prime factorisation of the numbers. In this book you will only need to use the LCM in combining fractions and only for small, manageable numbers, so the LCM will usually be obvious 'by inspection'. In such cases one can normally guess the answer by looking at the prime factors of the numbers, and then check that each number divides the guess exactly.

Solution to review question 1.1.3

A. (i) $15 = 3 \times 5$
(ii) $21 = 3 \times 7$
(iii) $60 = 3 \times 20 = 3 \times 4 \times 5 = 2 \times 2 \times 3 \times 5 = 2^2 \times 3 \times 5$
(iv) $121 = 11 \times 11$
(v) $405 = 5 \times 81 = 5 \times 9 \times 9 = 5 \times 3 \times 3 \times 3 \times 3 = 3^4 \times 5$
(vi) $1024 = 4 \times 256 = 4 \times 16 \times 16$
$= 4 \times 4 \times 4 \times 4 \times 4 = 2 \times 2 \times 2 \times 2 \times 2 \times 2 \times 2$
$\times 2 \times 2 \times 2$
$= 2^{10}$ (or, anticipating the rules of indices $= 4^5 = (2^2)^5 = 2^{10}$)
(vii) $221 = 13 \times 17$

Notice that there may be more than one way of factorising, but that the final result is always the same. You may also have noticed that it gets increasingly difficult to factorise, compared to multiplying – thus in (vii), it is so much easier to multiply 13×17 than to discover these factors from 221. This fact is actually the key idea behind many powerful coding systems – the **trap-door principle** – in some cases it is much easier doing a mathematical operation than undoing it!

B. (i) 24, 30. In this case it is clear that the largest integer that exactly divides these two is 6 and so the HCF of 24 and 30 is 6.
(ii) 27, 99. Again the fairly obvious answer here is 9.
(iii) 28, 98. Perhaps not so obvious, so split each into prime factors:

$28 = 4 \times 7 = 2 \times 2 \times 7$

$98 = 2 \times 49 = 2 \times 7 \times 7$

from which we see that the HCF is $2 \times 7 = 14$.
(iv) 12, 54, 78. Splitting into prime factors will again give the answer here, but notice a short cut: 2 clearly divides them all, leaving 6, 27, 39. 3 divides all of these leaving 2, 9, 13. These clearly have no factors in common (except 1) and so we are done, and the HCF is $2 \times 3 = 6$.

> (v) 3, 6, 15, 27. Here 3 is the only number that divides them all and is the HCF in this case.
>
> C. (i) Since 3, 7 are both primes, the LCM is simply their product, 21.
> (ii) $9 = 3 \times 3$, so 3 and 9 both divide 9 and there is no smaller number that does so. The LCM is thus 9.
> (iii) Since $3 \times 12 = 36$ and $2 \times 18 = 36$, we see directly that the LCM is 36.
> (iv) We have to deal with 3, 5, and $9 = 3^2$. The LCM must contain at least two factors of 3 and one of 5. So the LCM is $5 \times 3^2 = 45$.
> (v) 2, 4, 6. 2 divides 4, so we must have a factor of 4 in the LCM. Also, 4 and 6 both divide 12, but no smaller number and so the LCM is 12.

1.2.4 Manipulation of numbers

Much of arithmetic is based on just a few operations: addition, subtraction, multiplication and division, satisfying a small number of rules. The extension of these rules to include **symbols** as well as numbers leads us on to **algebra** (Chapter 2).

Addition, denoted $+$, produces the **sum** of two numbers:

$$6 + 3 = 9 = 3 + 6 \quad \text{(addition is 'commutative')}$$

Subtraction, denoted $-$, produces the **difference** of two numbers:

$$6 - 3 = 3 = -(3 - 6) \quad \text{(minus sign changes signs in brackets)}$$

Multiplication, denoted by $a \times b$ or simply as ab in algebra, produces the **product** of two numbers:

$$6 \times 3 = (6)(3) = 18 = 3 \times 6 \quad \text{(multiplication is commutative)}$$

$a \cdot b$ is sometimes used to denote the product but can be confused with decimal notation in arithmetic.

Division, denoted by $a \div b$ or a/b or, better, $\dfrac{a}{b}$, produces the **quotient** of two numbers:

$$6 \div 3 = 6/3 = \tfrac{6}{3} = 2 \quad \text{(of course, } 6 \div 3 \neq 3 \div 6!\text{)}$$

Note that $\div$ and / are very rarely used in written calculations, where we use the form $\tfrac{6}{3}$ unless we need to call into play $\div$ or / because we have a large number of divisions. Also notice how we have simplified the quotient to 2. We always simplify such fractions to lowest form whenever we can (➤ 12).

The way the above and other arithmetic operations are combined is according to a set of conventional precedences – **the rules of arithmetic**. Thus we always perform multiplication before addition, so:

$$2 \times 3 + 5 = 6 + 5 = 11$$

Brackets can be used if we want to override such rules. For example:

$$2 \times (3 + 5) = 2 \times 8 = 16$$

In general, an arithmetic expression, containing numbers, ( ), x, $\div$, $+$, $-$, must be evaluated according to the following priorities:

BODMAS

Brackets ( )   first

Of (as in 'fraction of' — rarely used these days)
Division $\div$ second
Multiplication $\times$

Addition $+$
Subtraction $-$ third

If an expression contains only multiplication and division we work from left to right. If it contains only addition and subtraction we again work from left to right. If an expression contains powers or **indices** (Section 1.2.7) then these are evaluated after any brackets.

Products and quotients of negative numbers can be obtained using the following rules:

$$(+1)(+1) = +1 \quad (+1)(-1) = -1$$
$$(-1)(+1) = -1 \quad (-1)(-1) = +1$$
$$\frac{1}{(-1)} = -1$$

For example $(-2)(-3) = (-1)(-1)6 = 6$

Note that if you evaluate expressions on your calculator, it may not follow the BODMAS order, simply because of the way your calculator operates. However, BODMAS is the universal convention in Western mathematics and applies equally well to algebra, as we will see in Chapter 2.

Solution to review question 1.1.4

Following the BODMAS rule

(i) $2 + 3 - 7 = 5 - 7 = -2$
(ii) $4 \times 3 \div 2 = 12 \div 2 = 6$
(iii) $3 + 2 \times 5 = 3 + 10 = 13$
(iv) $(3 + 2) \times 5 = 5 \times 5 = 25$
(v) $3 + (2 \times 5) = 3 + 10 = 13$. In this case the brackets are actually unnecessary, since the BODMAS rules tell us to evaluate the multiplication first.
(vi) $18 \div 2 \times 3 = 9 \times 3 = 27$ following the convention of working from left to right.
(vii) $18 \div (2 \times 3) = 18 \div 6 = 3$ because the brackets override the left to right rule.

> (viii) $-2 - (4 - 5) = -2 - (-1)$
> $\qquad\qquad\qquad\quad = -2 + 1$
> $\qquad\qquad\qquad\quad = -1$
>
> (ix) $(4 \div (-2)) \times 3 - 4 = \left(\dfrac{4}{-2}\right) \times 3 - 4$
> $\qquad\qquad\qquad\qquad\quad = (-2) \times 3 - 4$
> $\qquad\qquad\qquad\qquad\quad = -6 - 4$
> $\qquad\qquad\qquad\qquad\quad = -10$
>
> (x) $(3 + 7) \div 5 + (7 - 3) \times (2 - 4)$
> $\qquad\qquad\qquad = 10 \div 5 + (4) \times (-2)$
> $\qquad\qquad\qquad = 2 - 8$
> $\qquad\qquad\qquad = -6$
>
> Notice the care taken in these examples, spelling out each step. This may seem to be a bit laboured, but I would encourage you to take similar care, particularly when we come to algebra. Slips with brackets and signs crop up frequently in most people's calculations (mine included!). Whereas this may only lose you one or two marks in an exam, in real life, an error in sign can convert a stable control system into an unstable one, or a healthy bank balance into an overdraft.

1.2.5 Handling fractions

◄ 3   28 ►

A **fraction** or **rational number** is any quantity of the form

$$\frac{m}{n} \quad n \neq 0$$

where m, n are integers but n is not equal to 0. It is of course essential that $n \neq 0$, because as noted in Section 1.2.1 **division by zero is not defined**.

$\qquad$ m is called the **numerator**

$\qquad$ n is the **denominator**

If $m \geq n$ the fraction is said to be **improper**, and if $m < n$ it is **proper**.

A number expressed in the form $2\frac{1}{2}$ (meaning $2 + \frac{1}{2}$) is called a **mixed fraction**. In mathematical expressions it is best to avoid this form altogether and write it as a **vulgar fraction**, $\frac{5}{2}$, instead, otherwise it might be mistaken for '$2 \times \frac{1}{2} = 1$', and it is also more difficult to do calculations such as multiplication and division using mixed fractions.

The numerator and denominator of a fraction may have common factors. These may be cancelled to reduce the fraction to its simplest or 'lowest' form:

$$\frac{6}{12} = \frac{2 \times 3}{3 \times 4} = \frac{2}{4} = \frac{1 \times 2}{2 \times 2} = \frac{1}{2}$$

Each of these forms are **equivalent fractions**, but clearly the last one is the simplest. However, sometimes one of the other forms may be convenient for particular purposes, such as adding fractions. A very common fraction where we tend **not** to cancel down in this way is the **percentage**. Thus we usually express 32/100 as '32 percent' rather than as its equivalent, '8 out of 25'!

Fractions are multiplied 'top by top and bottom by bottom' as you might expect:

$$\frac{m}{n} \times \frac{p}{q} = \frac{mp}{nq} \quad (n, q \neq 0) \quad \text{e.g.} \quad \frac{3}{2} \times \frac{5}{11} = \frac{15}{22}$$

with p and q also any integers. There may, of course, be common factors to cancel down, as for example in

$$\frac{3}{2} \times \frac{6}{7} = \frac{9}{7}$$

The inverse or **reciprocal** of a fraction is obtained by turning it upside down:

$$1 \bigg/ \left(\frac{m}{n}\right) = \frac{n}{m} \quad \text{e.g.} \quad 1 \bigg/ \left(\frac{3}{2}\right) = \frac{2}{3}$$

where both m and n must be non-zero. So dividing by a vulgar fraction is done by inverting it and multiplying:

$$\left(\frac{p}{q}\right) \div \left(\frac{m}{n}\right) = \left(\frac{p}{q}\right) \bigg/ \left(\frac{m}{n}\right) = \frac{p}{q} \times \frac{n}{m} = \frac{np}{mq}$$

e.g.
$$\frac{7}{2} \div \frac{14}{6} = \frac{7}{2} \times \frac{6}{14} = \frac{3}{2}$$

Multiplication and division of fractions are therefore quite simple. Addition and subtraction are less so.

Two fractions with the same denominator are easily added or subtracted:

$$\frac{m}{n} \pm \frac{p}{n} = \frac{m \pm p}{n} \quad \text{e.g.} \quad \frac{3}{2} - \frac{1}{2} = \frac{3-1}{2} = \frac{2}{2} = 1$$

So to add and subtract fractions in general we rewrite them all with the same **common denominator**, which is the **lowest common multiple** of all the denominators. For example

$$\frac{3}{4} - \frac{4}{3} = \frac{3 \times 3}{12} - \frac{4 \times 4}{12} = \frac{9 - 16}{12} = -\frac{7}{12}$$

In the example, 12 is the LCM of 3 and 4.

An electrical example – resistances in parallel

Three resistances R_1, R_2, R_3 connected in parallel are equivalent to a single resistance R given by

$$\frac{1}{R} = \frac{1}{R_1} + \frac{1}{R_2} + \frac{1}{R_3}$$

So, for example if $R_1 = 2\Omega$, $R_2 = \frac{1}{2}\Omega$, $R_3 = \frac{2}{3}\Omega$ then

$$\frac{1}{R} = \frac{1}{2} + 2 + \frac{2}{3} \quad \text{(units of inverse ohms)}$$

or, with 6 the LCM of 2 and 3

$$= \frac{3}{6} + \frac{12}{6} + \frac{4}{6} = \frac{19}{6} \; (\Omega^{-1})$$

and so the equivalent resistance is

$$R = \frac{6}{19} \; \Omega$$

Finally, on fractions, recall the ideas of **ratio** and **proportion**. These are met early in our mathematical education, yet often continue to confuse us later in life. Specifically, it is not uncommon to see someone make errors such as:

'$\dfrac{a}{b} = \dfrac{1}{3}$ means $a = 1$ and $b = 3$'

so it is worth having a quick review of this topic.

The notation $a : b$ is used to indicate that the numbers a and b are in a certain ratio or proportionality to each other.

$$a : b = 1 : 3$$

simply means that

$$\frac{a}{b} = \frac{1}{3}$$

and this most certainly does not mean $a = 1$ and $b = 3$. For example

$$3 : 9 = 2 : 6 = 7 : 21 = 1 : 3$$

All $a : b = 1 : 3$ means is that

$$a = \frac{b}{3}$$

i.e. a is a third of b. If we are given a (or b) then we can find b (or a). The review question illustrates this.

In general, if we can write $a = kb$ where k is some given constant then we say 'a is proportional to b' and write this as $a \propto b$. a and b are then in the ratio $a : b = 1 : k$. On the other hand if we can write $a = k/b$ then we say 'a is inversely proportional to b' and write $a \propto 1/b$.

> **Solution to review question 1.1.5**
>
> **A.** (i) $\dfrac{4}{6} = \dfrac{2 \times 2}{2 \times 3} = \dfrac{2}{3}$
>
> (ii) $\dfrac{18}{9} = \dfrac{9 \times 2}{9} = 2$

(iii) $\dfrac{\not{7}}{3} \times \dfrac{4}{\not{7}} = \dfrac{4}{3}$

(iv) $\dfrac{7}{5} \times \dfrac{3}{14} = \dfrac{1}{5} \times \dfrac{3}{2}$

(cancelling 7 from top and bottom, as in (iii))

$$= \dfrac{1 \times 3}{5 \times 2} = \dfrac{3}{10}$$

(v) $\dfrac{3}{4} \div \dfrac{4}{5} = \dfrac{3}{4} \times \dfrac{5}{4} = \dfrac{15}{16}$

(vi) $\dfrac{1}{2} + \dfrac{1}{3} = \dfrac{3}{2 \times 3} + \dfrac{2}{2 \times 3}$

on multiplying top and bottom appropriately to get the common denominator in both fractions,

$$= \dfrac{3}{6} + \dfrac{2}{6} = \dfrac{3+2}{6} = \dfrac{5}{6}$$

(vii) $\dfrac{1}{2} - \dfrac{1}{3} = \dfrac{3}{6} - \dfrac{2}{6}$

$$= \dfrac{3-2}{6} = \dfrac{1}{6}$$

(viii) $\dfrac{4}{15} - \dfrac{7}{3} = \dfrac{4}{15} - \dfrac{5 \times 7}{15}$

$$= \dfrac{4 - 35}{15}$$

$$= -\dfrac{31}{15}$$

(ix) $1 + \dfrac{1}{2} + \dfrac{1}{3} = \dfrac{6}{6} + \dfrac{3}{6} + \dfrac{2}{6}$

$$= \dfrac{11}{6}$$

Here we found the LCM of 2 and 3 (6) and put everything over this, including the 1.

(x) $\dfrac{2}{3} - \dfrac{3}{4} + \dfrac{1}{8}$

We want the LCM of 3, 4, 8. This is 24, so

$$\dfrac{2}{3} - \dfrac{3}{4} + \dfrac{1}{8} = \dfrac{2 \times 8}{24} - \dfrac{3 \times 6}{24} + \dfrac{3}{24}$$

$$= \dfrac{16 - 18 + 3}{24}$$

$$= \dfrac{1}{24}$$

> **B.** If $a : b = 3 : 2$ then $\dfrac{a}{b} = \dfrac{3}{2}$ so $b = \dfrac{2a}{3}$. So, if $a = 6$ then $b = \dfrac{2 \times 6}{3} = 4$.

1.2.6 Factorial and combinatorial notation – permutations and combinations ◀4 29▶

The factorial notation is a shorthand for a commonly-occurring expression involving positive integers. It provides some nice practice in manipulation of numbers and fractions, and gently introduces algebraic ideas. If n is some positive integer ≥ 1 then we write

$$n! = n(n-1)(n-2)\ldots 2 \times 1$$

read as 'n **factorial**'. For example

$$5! = 5 \times 4 \times 3 \times 2 \times 1 = 120$$

Notice that the factorial expression yields large values very quickly, that is $n!$ increases rapidly with n. In calculations involving factorials it is often useful to remember such results as

$$10! = 10 \times 9 \times 8 \times 7!$$

i.e. we can pick out a lower factorial if this is convenient, and this often helps with cancellations in expressions containing factorials.

Note that $1! = 1$. Also, while the above definition does not define $0!$, the **convention** is adopted that

$$0! = 1$$

The factorial notation is useful in the binomial theorem (▶ 71) and in statistics. It can be used to count the number **permutations** of n objects, i.e. the number of ways of arranging n objects in a given order:

> First object can be chosen in n ways
> Second object can be chosen in $(n-1)$ ways
> Third object can be chosen in $(n-2)$ ways
> $\vdots$
> Last object can only be chosen in 1 way.

So the total number of permutations of n objects is

$$n \times (n-1) \times (n-2)\ldots 2 \times 1 = n!$$

Note that $n! = n \times (n-1)!$

For 3 objects A, B, C, for example, there are $3! = 6$ permutations, which are:

ABC, ACB, BAC, BCA, CAB, CBA.

Each of these is the same **combination** of the objects A, B, C – that is a selection of three objects in which order is not important.

Now suppose we select just r objects from the n. Each such selection is a different combination of r objects from n. An obvious question is how many different permutations of r objects chosen from n can be formed in this way? This number is denoted by ^nP_r. It may be evaluated by repeating the previous counting procedure, but only until we have chosen r objects:

The first may be chosen in n ways
The second may be chosen in $(n - 1)$ ways
The third may be chosen in $(n - 2)$ ways
$\vdots$
The rth may be chosen in $(n - (r - 1))$ ways

So the total number of permutations will be

$$^nP_r = n \times (n - 1) \times (n - 2) \times \ldots \times (n - r + 1)$$
$$= \frac{n(n - 1)(n - 2) \ldots (n - r + 1)(n - r)(n - r - 1) \ldots 2 \times 1}{(n - r)(n - r - 1) \ldots 2 \times 1}$$
$$= \frac{n!}{(n - r)!}$$

For example the number of ways that we can permute 3 objects chosen from 5 distinct objects is

$$^5P_3 = 5 \times 4 \times 3 = \frac{5 \times 4 \times 3 \times 2 \times 1}{2 \times 1} = \frac{5!}{(5 - 3)!} = 60$$

Since the order does not matter in a **combination**, ^nP_r will include $r!$ permutations of the same combinations of r different objects. So the **number of combinations of r objects chosen from n** is

$$\frac{1}{r!}\,^nP_r = \frac{n!}{(n - r)!r!}$$

This is usually denoted by ^nC_r (called the '$n - C - r$' notation) or $\binom{n}{r}$ – 'choose r objects from n':

$$^nC_r = \binom{n}{r} = \frac{n!}{(n - r)!r!} \quad \text{e.g.} \quad ^5C_3 = \binom{5}{3} = \frac{5!}{(5 - 3)!3!} = 10$$

which is very useful in binomial expansions (see Section 2.2.13) and other areas, simply as a notation, regardless of its 'counting' significance.

> **Solution to review question 1.1.6**
>
> **A.** (i) $3! = 3(3-1)(3-2) = 3 \times 2 \times 1 = 6$
> (ii) $6! = 6 \times 5 \times 4 \times 3 \times 2 \times 1 = 720$
> (iii) $\dfrac{24!}{23!} = \dfrac{24 \times 23!}{23!} = 24$
> (iv) $\dfrac{12!}{9!3!} = \dfrac{12 \times 11 \times 10 \times 9!}{9!3!} = \dfrac{12 \times 11 \times 10}{6} = 220$
>
> **B.** (i) (a) $^3C_2 = \dfrac{3!}{(3-2)!2!} = \dfrac{3!}{1!2!} = 3$
> (b) $^6C_4 = \dfrac{6!}{(6-4)!4!} = \dfrac{6 \times 5 \times 4!}{2!4!} = 15$
> (c) $^6P_3 = \dfrac{6!}{3!} = 6 \times 5 \times 4 = 120$
>
> (ii) Two letters can be chosen from $ABCD$ in
> $$^4C_2 = \dfrac{4!}{2!2!} = 6 \text{ ways}$$
>
> (iii) There are $5! = 120$ permutations of five different letters.

1.2.7 Powers and indices

◀ 4   29 ▶

Powers, or indices, provide, in the first instance, a shorthand notation for multiplying a number by itself a given number of times:

$$2 \times 2 = 2^2$$
$$2 \times 2 \times 2 = 2^3$$
$$2 \times 2 \times 2 \times 2 = 2^4$$

etc.

For a given number a we have

$$a^n = \underbrace{a \times a \times a \times \ldots \times a}_{n \text{ times}}$$

a is called the **base**, n the **power** or **index**. a^1 is simply a. By convention we take $a^0 = 1$ ($a \neq 0$). We introduce a^{-1} to denote the **reciprocal** $\dfrac{1}{a}$, since then $1 = a \times \dfrac{1}{a} = a^1 \times \dfrac{1}{a} = a^1 \times a^{-1} = a^{1-1} = a^0$ follows. In general, $a^{-n} = \dfrac{1}{a^n}$. From these definitions we can derive the **rules of indices**:

$$a^m \times a^n = a^{m+n}$$

18

$$\frac{a^m}{a^n} = a^{m-n}$$

$$(a^m)^n = a^{mn}$$

$$(ab)^n = a^n b^n$$

Note that for any index n, $1^n = 1$.

Examples

$$2^3 \times 2^4 = 2^{3+4} = 2^7$$

$$3^5 \times 3^0 = 3^5$$

$$(5^2)^3 = 5^6$$

$$(2^2)^0 = 2^0 = 1$$

$$4^3/2^2 = (2^2)^3/2^2 = 2^6/2^2 = 2^4$$

A **square root** of a positive number a, is any number that, when squared, yields the number a. We use $\sqrt{a}$ to denote the positive value of the square root (although the notation has to be stretched when we get to complex numbers). For example

$$2 = \sqrt{4} \quad \text{since } 2^2 = 4$$

Since $-2 = -\sqrt{4}$ also satisfies $(-2)^2 = 4$, $-\sqrt{4}$ is also a square root of 4. So the square roots of 4 are $\pm\sqrt{4} = \pm 2$.

We can similarly have cube roots of a number a, which yield a when they are cubed. If a is positive then $\sqrt[3]{a}$ denotes the positive value of the cube root. For example

$$2 = \sqrt[3]{8} \quad \text{because } 2^3 = 8$$

In the case of taking an odd root of a negative number the convention is to let $\sqrt{}$ denote the negative root value, as in $\sqrt[3]{-8} = -2$, for example.

The corresponding nth root of a number a is denoted in general by

$$\sqrt[n]{a} \quad \text{(also called a } \textbf{radical}\text{)}$$

If n is even then a must be positive to yield a real root ($\sqrt{-1}$ is an **imaginary number**, forming the basis of complex numbers, see Chapter 12). In this case, because $(-1)^2 = 1$, there will be at least two values for the root differing only by sign. If n is odd then the nth root $\sqrt[n]{a}$ exists for both positive and negative values of a, as in $\sqrt[3]{-8} = -2$ above.

If a is a prime number such as 2, then $\sqrt{a}$ is an irrational number, i.e. it can't be expressed in rational form as a ratio of integers (6 ◄). This is not just a mathematical nicety. $\sqrt{2}$ for example, is the diagonal of the unit square, and yet because it is irrational, it can never be written down exactly as a rational number or fraction ($\sqrt{2} = 1.4142$ is, for example, only an approximation to $\sqrt{2}$ to four decimal places).

In terms of indices, roots are represented by fractional indices, for example:

$$\sqrt{a} = a^{\frac{1}{2}}$$

and in general

$$\sqrt[n]{a} = a^{\frac{1}{n}}$$

This fits in with the rules of indices, since

$$\left(a^{\frac{1}{n}}\right)^n = a^{\frac{1}{n} \times n} = a^1 = a$$

Fractional powers satisfy the same rules of indices as integer powers – but there are some new features:

- multiplicity of roots: $2^2 = (-2)^2 = 4$
- non-existence of certain roots of negative numbers: $\sqrt{-1}$ is not a real number
- irrational values for roots of primes and their multiples: $\sqrt{2}$ cannot be expressed as a fraction

Quantities such as $\sqrt{2}, \sqrt{3}, \ldots$ containing square roots of primes, are called **surds**. The term originates from the Greek word for mute, referring to a number that cannot 'speak' its value – because its decimal part never ends (see Section 1.2.8). In mathematical manipulation surds are always best left as they are – retaining the root sign. Any decimal form for them will simply be an approximation as noted for $\sqrt{2}$ above. Usually we try to manipulate surds so that the result is the simplest form, and none remain in denominators (although we would normally write, for example, $\sin 45° = \frac{1}{\sqrt{2}}$). To do this we can use the rules of indices, and also a process known as **rationalisation**, in which surds in denominators are moved to the numerator. The ideas are illustrated in the solution to the review question.

Solution to review question 1.1.7

A. (i) $2^3 2^4 = 2^{3+4} = 2^7$ (leave it as a power, like this)

(ii) $\dfrac{3^4}{3^3} = 3^{4-3} = 3^1 = 3$

(iii) $(5^2)^3 = 5^{2 \times 3} = 5^6$

(iv) $\dfrac{(3 \times 4)^4}{(9 \times 2^3)} = \dfrac{3^4 4^4}{9 \times 2^3}$ (note both 3 and 4 are raised to the power 4)

Note that $9 = 3^2$, $4 = 2^2$, so we can write,

$$= \frac{3^4 (2^2)^4}{3^2 2^3} = \frac{3^4 2^8}{3^2 2^3}$$

$$= 3^{4-2} 2^{8-3}$$

$$= 3^2 2^5$$

(v) $\dfrac{16^2}{4^4} = \dfrac{(4^2)^2}{4^4} = \dfrac{4^4}{4^4} = 1$

(vi) $(-6)^2 \left(-\dfrac{3}{2}\right)^3 = (-1)^2 6^2 (-1)^3 \dfrac{3^3}{2^3}$

$= -\dfrac{6^2 3^3}{2^3}$

on using $(-1)^2 = 1$, $(-1)^3 = -1$

$= -\dfrac{(2 \times 3)^2 3^3}{2^3} = -\dfrac{2^2 3^2 3^3}{2^3}$

$= -\dfrac{3^5}{2}$

(vii) If it helps, just think of a and b as given numbers:

$$\dfrac{(-ab^2)^3}{a^2 b} = \dfrac{(-1)^3 a^3 b^6}{a^2 b}$$

$$= -ab^5$$

(viii) $2^2 \left(\dfrac{1}{2}\right)^{-3} = 2^2 (2^{-1})^{-3}$

$= 2^2 2^3 = 2^5$

The steps to watch out for in such problems are the handling of the minus signs and brackets, and dealing with the negative powers and reciprocals.

Don't forget that a^n, $n \geq 0$, is not defined for $a = 0$.

B. We can get a long way simply by using $\sqrt{ab} = \sqrt{a}\sqrt{b}$

(i) $\sqrt{50} = \sqrt{25 \times 2} = \sqrt{5^2}\sqrt{2} = 5\sqrt{2}$

(ii) $\sqrt{72} - \sqrt{8} = \sqrt{36 \times 2} - \sqrt{4 \times 2} = 6\sqrt{2} - 2\sqrt{2} = 4\sqrt{2}$

(iii) $(\sqrt{27})^3 = (3\sqrt{3})^3 = 3^3(\sqrt{3})^3 = 3^3 3\sqrt{3} = 3^4 \sqrt{3} = 81\sqrt{3}$

(iv) $\left(\dfrac{\sqrt{2}\sqrt{3}}{4}\right)^2 = \left(\dfrac{\sqrt{3}}{2\sqrt{2}}\right)^2 = \dfrac{3}{4 \times 2} = \dfrac{3}{8}$

(v) $\dfrac{\sqrt{3}\sqrt{7}}{\sqrt{84}} = \dfrac{\sqrt{21}}{\sqrt{4 \times 21}} = \dfrac{\sqrt{21}}{2\sqrt{21}} = \dfrac{1}{2}$

(vi) To simplify $\dfrac{\sqrt{3} + 2\sqrt{2}}{\sqrt{3} - \sqrt{2}}$ we **rationalise** it by removing all surds from the denominator. To do this we use the algebraic identity:

$$(a - b)(a + b) \equiv a^2 - b^2$$

(see Section 2.2.1) and the removal of surds by squaring. For a $\sqrt{3} - \sqrt{2}$ on the bottom we multiply top and bottom by $\sqrt{3} + \sqrt{2}$, using:

$$(\sqrt{3} - \sqrt{2})(\sqrt{3} + \sqrt{2}) = (\sqrt{3})^2 - (\sqrt{2})^2 = 3 - 2 = 1$$

Thus:

$$\frac{(\sqrt{3}+2\sqrt{2})}{(\sqrt{3}-\sqrt{2})} \times \left(\frac{\sqrt{3}+\sqrt{2}}{\sqrt{3}+\sqrt{2}}\right)$$

$$= \frac{(\sqrt{3}+2\sqrt{2})(\sqrt{3}+\sqrt{2})}{(\sqrt{3}-\sqrt{2})(\sqrt{3}+\sqrt{2})}$$

$$= \frac{(\sqrt{3})^2 + 3\sqrt{2}\sqrt{3} + 2(\sqrt{2})^2}{(\sqrt{3})^2 - (\sqrt{2})^2}$$

$$= \frac{3 + 3\sqrt{2}\sqrt{3} + 4}{3-2}$$

$$= 7 + 3\sqrt{6}$$

A similar ploy is used in the rationalisation or division of complex numbers (➤ 355)

(viii) $\left(\dfrac{3^{1/3}9^{1/3}}{27}\right)^2 = \left(\dfrac{3^{1/3}3^{2/3}}{27}\right)^2 = \left(\dfrac{3}{27}\right)^2 = \left(\dfrac{1}{9}\right)^2 = \dfrac{1}{81}$

This topic, powers and indices, often gives beginners a lot of trouble. If you are still the slightest bit unsure, go to the reinforcement exercises for more practice – there is no other way. This is, literally, power training!

1.2.8 Decimal notation ◀4   30▶

You probably know that $\frac{1}{2}$ may be represented by the decimal 0.5, $\frac{1}{4}$ by 0.25 and so on. In fact any real number, a, $0 \leq a < 1$, has a **decimal representation**, written

$$a = 0.d_1 d_2 d_3 \ldots$$

where each d_i is one of the digits $0, 1, 2, \ldots, 9$, and the sequence may not terminate (see below). The term decimal actually refers to the **base** 10 and represents the fact that:

$$a = d_1 \times 10^{-1} + d_2 \times 10^{-2} + d_3 \times 10^{-3} \ldots$$

Note the importance of 'place value' here – the value of each of the digits depends on its place in the decimal.

Any real number can be represented by an integer part and such a decimal part. If, from some point on the decimal consists of a repeating string of one or more digits, then the decimal is said to be a **repeating** or **recurring** decimal. All rational numbers can be represented by a finite decimal representation or a recurring one. Irrational numbers cannot be represented in this way as a terminating or recurring decimal – thus the decimal representation of $\sqrt{2}$ is non-terminating:

$$\sqrt{2} = 1.4142135623\ldots$$

that is, the decimal part goes on forever.

All quantities measured in scientific or engineering experiments will have a finite decimal – every human observation of any kind is subject to a limited accuracy and so to a limited number of decimal places. Similarly any mechanical or electronic device can only yield a terminating decimal representation with a finite number of decimal places. In particular any number that you output on your calculator must represent a finite or recurring decimal – a rational number. So, for example no calculator or computer could ever yield the **exact** value of $\sqrt{2}$ or π. In practice even the most finicky engineer has limited need for decimal places – it can be shown that to measure the circumference of a circle girdling the known universe with an error no greater than the radius of a hydrogen atom requires the value of π to only 39 decimal places. π is actually known to many millions of decimal places. Nevertheless, irrational numbers such as $\sqrt{2}$, $\sqrt{3}$ actually occur frequently in engineering calculations, so we have to learn to handle them. $1/\sqrt{2}$ occurs for example in the rms value of an alternating current.

A useful way of expressing numerical value is by specifying a certain number of **significant digits**. To discuss these we need to be clear about zeros in numbers and what they represent. Some zeros are needed in a number simply as place holders – i.e. to tell us whether we are dealing with units, tens, hundreds, or tenths, hundredths, etc. For example in

$$1500, \ 0.00230, \ 2.1030$$

the bold zeros are essential to hold place value – the only way to avoid them is to write the number in scientific notation (see below). The underlined zeros in these numbers are not strictly necessary and should only be included if they are significant – i.e. they represent a level of accuracy. For example if the number 1.24 is only accurate to the three 'significant figures' given then it could lie between 1.235 and 1.245. But if we write 1.240 then we are saying that there are four significant figures of accuracy and the number must lie between 1.2395 and 1.2405. The two end zeros in 1500 may or may not represent an accuracy to four figures – we have no way of knowing without further information. Therefore unless you are given further information, such zeros are assumed to be not significant. Similarly, the two first zeros in 0.002320 are assumed to be not significant – they are just place holders.

To count the number of **significant figures** in a number, start from the first non-zero digit on the left and count all digits (zero or not) to the right, counting final zeros if they are to the right of the decimal point, but not otherwise. Final zeros to the left of the decimal point are assumed not significant unless more information is given.

Examples

3.214 (4 sf), 2.041 (4 sf), 12.03500 (7 sf), 420 (2 sf), 0.003 (1 sf), 0.0801 (3 sf), 2.030 (4 sf), 500.00 (5 sf)

Sometimes numbers are approximated by terminating the digits after a given number of digits and replacing them with zeros. If this is done with no regard to the size of the removed digits, then we say the number has been 'chopped' or 'truncated'. For example 324829.1 chopped to 3 significant figures is 324000. Another, more accurate, method of approximation is 'rounding', in which we take account of the size of the removed digits.

When we 'round' a number we change the last non-zero digit not removed according to the size of the digits dropped. Specifically:

- If the digit to be removed is >5 then the immediately preceding digit is increased by 1
- If the digit to be removed is <5 the immediately preceding digit is left unchanged
- If the digit to be removed is equal to 5 then you may round up or down – one 'fair' way to do this is to round up if the previous digit is odd and down otherwise, for example.

Although 'chopping' may seem to give bigger errors because, for example, 324829.1 is closer to 325000 than 324000, it is usually the preferred method in computer arithmetic because it is much quicker than the more accurate 'rounding'.

Examples

213.457 chopped/rounded to 4 sf is 213.4/213.5, 56.0011 chopped/rounded to 4 sf is 56.00/56.00

We often need to convert between fractions and decimal representations. We can go from fraction to decimal by ordinary division. Conversely, we can convert a terminating decimal to the corresponding rational number by multiplying top and bottom by an appropriate factor as in, for example

$$0.625 = \frac{625}{1000} = \frac{25}{40} = \frac{5}{8}$$

Any decimal number can be written as a decimal number between 1 and 10 (the **mantissa**) multiplied by an appropriate power (the **exponent**) of 10. For example:

$$74.932 = 7.4932 \times 10$$

$$\text{mantissa} = 7.4932$$

$$\text{exponent} = 1$$

The purpose of such representation, called **scientific notation**, is to reduce very large and very small numbers to manageable form. For example

$$573000000000000000 = 5.73 \times 10^{17}$$

$$0.0000000000000000000137 = 1.37 \times 10^{-20}$$

In engineering there is a variation on scientific notation that uses only multiples of 3 as exponents, i.e. as powers of 10. This is so that we can use the standard prefixes kilo, mega, micro, nano, etc.

Solution to review question 1.1.8

A. (i) (a) $\frac{1}{2} = 0.5$ (b) $-\frac{3}{2} = -1.5$ (c) $\frac{1}{3} = 0.\dot{3}$

where the dot above the final 3 denotes that this repeats forever: 0.3333.... All the results in (a), (b), (c) are exact.

(d) By long division we find:

$$\tfrac{1}{7} = 0.142857142857\ldots = 0.\dot{1}4285\dot{7}$$

and the string of digits 142857 recur indefinitely as denoted by the over dots at the ends of the sequence. To six decimal places we can write

$$\tfrac{1}{7} \simeq 0.142857$$

B. (ii) (a) $0.3 = \dfrac{3}{10}$ (b) $0.67 = \dfrac{67}{100}$

(c) $0.\dot{6} = 0.6666\ldots = \tfrac{2}{3}$

To see this, let $x = 0.6666\ldots$ then $10x = 6.6666\ldots$ and subtracting gives

$$9x = 6$$

so $\quad x = \tfrac{6}{9} = \tfrac{2}{3}$

(d) $3.142 = \dfrac{3142}{1000}$

C. (i) $11.00132 = 1.100132 \times 10$
mantissa $= 1.100132$
exponent $= 1$

(ii) $1.56 = 1.56 \times 10^0$
mantissa $= 1.56$
exponent $= 0$

(iii) $203.45 = 2.0345 \times 10^2$
mantissa $= 2.0345$
exponent $= 2$

(iv) $0.0000321 = 3.21 \times 10^{-5}$
mantissa $= 3.21$
exponent $= -5$

D. To three significant figures we have

(i) $11.00132 = 11.0$

(ii) 1.56

(iii) $203.45 = 203$

(iv) $0.000321 = 0.000321$

1.2.9 Estimation

◄ 5   31 ►

With the availability of calculators we are now used to having enormous number crunching capability at our fingertips. But there are occasions when we don't have our hands on a calculator, or we need to get a rough order of magnitude check on a messy calculation. In such situations the engineer's most powerful tool has always been an ability to mentally estimate quantities and perform quick 'back of the envelope' (we still have them, despite email!) calculations. The trick is to approximate the numbers you are dealing with so that the calculations become simple, yet some sort of rough accuracy is retained. It is a matter of judgement and practice. Absolute values of numbers are less important than their relative values – for example 1021 is significant in

$$3 \times 1021 + 40 \times 234$$

but is relatively insignificant in

$$\frac{1021}{10} - 103372415$$

So, inspect all the numbers occurring in an expression and approximate them each to an appropriate order of magnitude, rounding as necessary, then perform the (hopefully) simplified calculation with the results.

Solution to review question 1.1.9

(i) $\dfrac{4.5 \times 10^5 \times 2.0012}{8.892 \times 10^4} \simeq \dfrac{4.5 \times 10^5 \times 2}{9 \times 10^4}$

$\simeq \dfrac{10^5}{10^4} \simeq 10$

So if your calculator gave you 101.2753 ... then you know you have slipped up on a decimal point.

(ii) $\dfrac{\sqrt{254 \times 10^4 + 28764.5}}{2.01 \times 10 - 2.54 \times 10^{-6}} \simeq \dfrac{\sqrt{254 \times 10^4 + 3 \times 10^4}}{2 \times 10}$

neglecting 2.54×10^{-6} in comparison with 2.01×10

$= \dfrac{\sqrt{257 \times 10^4}}{2 \times 10} \simeq \dfrac{\sqrt{256 \times 10^4}}{2 \times 10}$

$= \dfrac{\sqrt{16^2 \times 10^4}}{2 \times 10}$

on replacing 257 by 256 for easy square rooting:

$= \dfrac{16 \times 10^2}{2 \times 10} = 80$

The answer to two decimal places is in fact 79.74.

1.3 Reinforcement

1.3.1 Types of numbers

A. Say what you can about the type and nature of the following numbers:

(i) 2 (ii) -3 (iii) 11

(iv) 21 (v) -0 (vi) $\dfrac{2}{3}$

(vii) $\dfrac{5}{2}$ (viii) $1\dfrac{2}{5}$ (ix) $-\dfrac{3}{7}$

(x) $\dfrac{18}{9}$ (xi) 0.0 (xii) 0.2

(xiii) -0.31 (xiv) 6.3 (xv) $\sqrt{3}$

(xvi) 3π (xvii) e (xviii) e^2

(xix) $-\sqrt{2}$ (xx) -1.371

(e is the base of natural logs – see Chapter 4)

B. Say all you can about the following expressions

(i) 0×3 (ii) $\dfrac{2}{0}$ (iii) $0 - 2$

(iv) $\dfrac{0}{-1}$ (v) $\dfrac{0+2}{0}$ (vi) 0^4

(vii) $0 \times \dfrac{1}{0}$ (viii) $-1 \times 0 + \dfrac{0}{2}$ (ix) $\dfrac{3 \times 0}{0}$

(x) 4^0 (xi) $0!$ (xii) $\dfrac{4!}{0!}$

1.3.2 Use of inequality signs

A. Using inequality signs, order all of the numbers in Q1.3.1A.

B. Suppose a, b and c are three non-zero positive numbers satisfying

$$a < b \leq c$$

What can you say about:

(i) a^2, b^2, c^2 (ii) $\dfrac{1}{a}, \dfrac{1}{b}, \dfrac{1}{c}$ (iii) $a+b, 2c$

(iv) $-a, -b, -c$ (v) $\sqrt{a}, \sqrt{b}, \sqrt{c}$?

1.3.3 Highest common factor and lowest common multiple

A. Express in terms of prime factors

   (i) 2   (ii) −6   (iii) 21

   (iv) 24   (v) −72   (vi) 81

   (vii) $\dfrac{27}{14}$ (viii) 143   (ix) 391

   (x) 205

B. Determine the highest common factor of each of the following sets of numbers

   (i) 11, 88   (ii) 28, 40   (iii) 25, 1001

   (iv) 20, 45, 90   (v) 14, 63, 95   (vi) 24, 72, 96

   (vii) 36, 42, 54

C. Find the lowest common multiple of each of the following sets of numbers

   (i) 2, 4   (ii) 5, 8   (iii) 12, 15

   (iv) 6, 9, 27   (v) 12, 42, 60, 70   (vi) 66, 144

1.3.4 Manipulation of numbers

A. Evaluate

   (i) $3 - 6 \times 7$ (ii) $3(4 - 1) - 2$ (iii) $(4 - 1) \div (6 - 3)$

   (iv) $6 - (3 - 6) \times 4$ (v) $24 \div (6 \div 2)$ (vi) $(24 \div 6) \div 2$

   (vii) $(2 \times (3 - 1)) \div (7 - 2(3 - 1))$

B. Evaluate

   (i) $10 - (2 - 3 \times 4^2)$ (ii) $100 - 3(7 - 10)^3$ (iii) $-((-(-((-1)))))$

   (iv) $-3(2 - (3 + 1)(-4 + 2) + 4 \times 3)$ (v) $1 - 4(-2)$

C. Evaluate $4 + 5 \times 2^3$ in its conventional meaning. Without these conventions how many pairs of brackets would be needed to make the meaning of the expression clear?

   Now insert one pair of brackets in as many different non-trivial ways as possible and evaluate the resulting expressions (retaining the other usual conventions). Can any other results be obtained by the insertion of a second pair of brackets?

1.3.5 Handling fractions

A. Find in simplest form as a fraction

(i) $\dfrac{10}{12}$ (ii) $\dfrac{36}{9}$ (iii) $\dfrac{7}{3} \times \dfrac{2}{5}$

(iv) $\dfrac{14}{15} \div \dfrac{5}{7}$ (v) $\dfrac{1}{2} + \dfrac{1}{4}$ (vi) $\dfrac{1}{3} - \dfrac{1}{7}$

(vii) $\dfrac{21}{5} - \dfrac{7}{10}$ (viii) $1 - \dfrac{1}{2} + \dfrac{1}{3}$ (ix) $\dfrac{3}{5} - \dfrac{1}{2} + \dfrac{1}{6}$

(x) $\left(\dfrac{3}{7} - \dfrac{2}{9}\right) \div \left(\dfrac{1}{5} - \dfrac{1}{2}\right)$ (xi) $\dfrac{12}{1/2}$

B. (a) If $a : b = 7 : 3$ determine a for the following values of b: (i) 4, (ii) 3, (iii) -7, (iv) 15.
 (b) Repeat for b with the same values for a.

C. If $a : b = 5 : 2$ evaluate as fractions

(i) $\dfrac{a}{b}$ (ii) $\dfrac{a}{a+b}$ (iii) $\dfrac{b}{a+b}$ (iv) $\dfrac{a-b}{a+b}$

(v) $\dfrac{a}{b} + \dfrac{b}{a}$ (vi) $\dfrac{a}{a+b} - \dfrac{b}{a-b}$ (vii) $\dfrac{a+b}{a-b} + \dfrac{a-b}{a+b}$

D. If a is proportional to b and $a = 6$ when $b = 4$, what are the values of the following

(i) a when $b = 3$ (ii) $\dfrac{a}{b} + \dfrac{b}{a}$ (iii) $\dfrac{a-b}{a+b}$ (iv) b when $a = 21$

1.3.6 Factorial and combinatorial notation – permutations and combinations ◀◀ 4   16 ◀

A. Evaluate

(i) $5!$ (ii) $10!$ (iii) $\dfrac{301!}{300!}$

(iv) $\dfrac{18!\,6!}{16!}$ (v) $\dfrac{14!}{(7!)^2 13}$ (vi) $\dfrac{10!}{4!\,6!}$

(vii) $10! + 11!$ (viii) $\dfrac{9!}{6!\,3!} - \dfrac{8!}{5!\,4!}$

B. Evaluate

(i) $^9C_2$ (ii) $^9C_7$ (iii) $^{11}C_4$

(iv) $^{10}C_4$ (v) $^{100}C_{100}$ (vi) $^7P_3$

(vii) $^6P_3\,^6C_3$

C. How many combinations of 4 different letters can be chosen from *ABCDEFG*?

1.3.7 Powers and indices ◀◀ 4   18 ◀

A. Evaluate in terms of powers of primes

(i) $2^2 2^3$ (ii) $3^4/3^2$ (iii) $6^3 \times 3^2/4$
(iv) $6^2 2^{-2} 3^2$ (v) $2^2 \times 4 \times 2^5$ (vi) $5^6/10^4$
(vii) $3^4 2^2 3^{-1}$ (viii) $49 \times 7/21^2$

B. Simplify (write as simplest products of powers of primes)

(i) $3^4 3^6 3^2$ (ii) $2^3 4^2/2^5$ (iii) $\dfrac{6^2 2^3 3^4 9}{2^2 3^3}$

(iv) $(4 \times 6)^6/(3^2 \times 4^2)$ (v) $27^5/9^5$ (vi) $(-4)^3/(-12)^4$

(vii) $\dfrac{3^2 4^6}{3^{-3} 2^{-1}}$ (viii) $\dfrac{8^{1/3} 27}{3^2 2^3}$

C. Show that the following are all the same number

$$\dfrac{2\sqrt{7}}{3\sqrt{5}}, \dfrac{\sqrt{28}}{\sqrt{45}}, \dfrac{14}{3\sqrt{35}}, \dfrac{2\sqrt{35}}{15}, \dfrac{2}{3}\sqrt{\dfrac{7}{5}}, \dfrac{2\sqrt{7}}{\sqrt{45}}, \dfrac{\sqrt{28}}{3\sqrt{5}}, \sqrt{\dfrac{28}{45}}$$

D. Express in terms of simplest surds

(i) $\sqrt{18}$ (ii) $\sqrt{20}$ (iii) $\sqrt{32}$ (iv) $\sqrt{52}$
(v) $\sqrt{512}$ (vi) $\sqrt{396}$ (vii) $\sqrt{108}$ (viii) $\sqrt{63}$

E. Rationalize

(i) $\dfrac{1}{\sqrt{7}}$ (ii) $-\dfrac{1}{\sqrt{3}}$ (iii) $\dfrac{1}{\sqrt{2}-1}$ (iv) $\dfrac{1}{4-\sqrt{10}}$

(v) $\dfrac{\sqrt{5}+1}{\sqrt{5}-1}$ (vi) $\dfrac{\sqrt{2}-2\sqrt{3}}{\sqrt{2}+\sqrt{3}}$ (vii) $\sqrt{\dfrac{1}{2}}+\sqrt{\dfrac{1}{4}}+\sqrt{\dfrac{1}{8}}$

(viii) $\sqrt{512}+\sqrt{128}+\sqrt{32}$

1.3.8 Decimal notation

◀◀4   22◀

A. Express in decimal form, to four decimal places in each case

(i) $-\dfrac{1}{2}$ (ii) $\dfrac{7}{2}$ (iii) $\dfrac{2}{3}$ (iv) $-\dfrac{2}{9}$
(v) $\dfrac{0}{6}$ (vi) $\dfrac{1}{8}$

B. Express as fractions

(i) 0.25   (ii) 0.125   (iii) 72.45   (iv) -0.312
(v) 0.17

C. Write the following numbers in scientific notation stating the mantissa and exponent

(i) 21.3241 (ii) 429.003 (iii) −0.000321 (iv) 0.00301
(v) 1,000,100 (vi) 300491.2

D. Write the numbers in **C** to (a) 3, (b) 6 significant figures.

1.3.9 Estimation

A. Assuming at $\pi \simeq 3.142$ and $e \simeq 2.718$ give approximate values for

(i) π^2 (ii) e^3 (iii) π^{-2} (iv) e^{-3}

B. Estimate the values of

(i) $\dfrac{0.0003 \times 3.1 \times 10^6}{9050}$

(ii) $\dfrac{2.01 + 403}{2.1 \times 10^{-3} - 29.9}$

(iii) $\dfrac{6 \times 10^5 + 1001e^3}{3.109 - 3.009}$

(iv) $\dfrac{3\pi^2 e^3}{63}$

1.4 Applications

1. An electrical circuit comprised of resistors only is illustrated in Figure 1.1

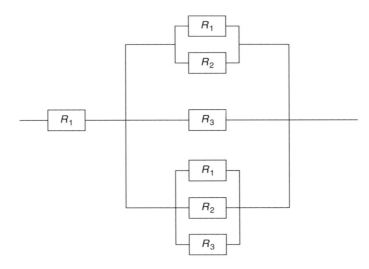

Figure 1.1

The equivalent resistance, R, of two resistors R_1 and R_2 in **series** is the **sum** of their resistances:

$$R = R_1 + R_2$$

The **reciprocal** of the equivalent resistance of two resistors in **parallel** is equal to the **sum of their reciprocals**:

$$\frac{1}{R} = \frac{1}{R_1} + \frac{1}{R_2}$$

(see Section 1.2.5)

(a) If, in the circuit of Figure 1.1, $R_1 = 1\Omega$, $R_2 = 2\Omega$, $R_3 = 3\Omega$ determine the overall equivalent resistance of the circuit, working entirely with fractions.
(b) Repeat the calculation using three decimal places accuracy and compare your result with (a).
(c) Suppose now that R_2 is not fixed, but can vary. Obtain the equivalent resistance in terms of R_2 using exact fractions.
(d) Using the result of (c), plot a graph of the equivalent resistance R for integer values of R_2 from 1 to 10.
(e) Discuss the calculations and results of (c), (d) in the light of the need to plot a useful graph.
(f) If we wish to increase slightly the equivalent resistance above that when $R_2 = 2\Omega$ what should we do with R_2, increase or decrease it?

2. Using the expressions for ^nC_r try a selection of values of n and r to investigate whether the following relations might be true.

(i) $^nC_r + {^nC_{n-r}} = {^rC_n}$ (ii) $^{n-1}C_r + {^{n-1}C_{n-r}} = {^nC_r}$
(iii) $^nC_r = {^nC_{n-r}}$

Prove any result that you suspect is true.

Answers to reinforcement exercises

1.3.1 Types of numbers

A.   (i)   2 is a positive prime and even integer.

(ii) -3 is a negative integer.

(iii)   11 – integer, positive, prime, odd.

(iv)   21 – integer, positive, odd, composite (3×7).

(v) -0 – zero, both positive and negative.

(vi) $\frac{2}{3}$ – proper fraction, rational number, positive.

(vii) $\frac{5}{2}$ – improper fraction, positive, rational.

(viii) $1\frac{2}{5}$ is a positive mixed fraction expressible as the improper fraction $\frac{7}{2}$.

(ix) $-\frac{3}{7}$ – negative proper fraction.

(x) $\frac{18}{9}$ – improper fraction which can be cancelled down to lowest terms as an integer 2.

(xi) 0.0 – decimal representation, to one decimal place, of zero.

(xii) 0.2 is a decimal fraction expressible as the proper fraction $\frac{2}{10} = \frac{1}{5}$.

(xiii) -0.31 is a negative decimal fraction expressible as the negative proper fraction $-\frac{31}{100}$.

(xiv) 6.3 is a positive decimal number, expressible as the mixed fraction $6\frac{3}{10}$ or the improper fraction $\frac{63}{10}$.

(xv) $\sqrt{3}$ is a positive irrational number.

(xvi) 3π – irrational.

(xvii) e – irrational.

(xviii) e^2 – irrational.

(xix) $-\sqrt{2}$ is a negative, irrational number.

(xx) -1.371 is a negative decimal fraction expressible as the negative improper fraction $-\frac{1371}{1000}$.

B. (i) zero (ii) not defined (iii) negative integer
(iv) zero (v) not defined (vi) zero
(vii) not defined (viii) zero (ix) not defined
(x) $4^0 = 1$ (xi) $0! = 1$ (xii) 24

1.3.2 Use of inequality signs

A. $21 > 11 > 3\pi > e^2 > 6.3 > e > \frac{5}{2} > 2$

$= \frac{18}{9} > \sqrt{3} > 1\frac{2}{5} > \frac{2}{3} > 0.2 > 0.0$

$= -0 > -0.31 > -\frac{3}{7} > -\sqrt{2} > -1.371 > -3$

B. (i) $a^2 < b^2 \le c^2$ (ii) $\frac{1}{a} > \frac{1}{b} \ge \frac{1}{c}$ (iii) $a + b < 2c$

(iv) $-c \le -b < -a$ (v) $\sqrt{a} < \sqrt{b} \le \sqrt{c}$

1.3.3 Highest common factor and lowest common multiple

A. (i) 2 is already prime (ii) $-6 = -1 \times 2 \times 3$ (iii) $21 = 3 \times 7$

33

(iv) $24 = 2^3 \times 3$ (v) $-72 = -1 \times 2^3 \times 3^2$ (vi) $81 = 3^4$
(vii) $2^{-1} \times 3^3 \times 7^{-1}$ (viii) 11×13 (ix) 17×23
(x) 5×41

B. (i) 11   (ii) 4   (iii) 1
(iv) 5   (v) 1   (vi) 24
(vii) 6

C. (i) 4   (ii) 40   (iii) 60
(iv) 54   (v) 420   (vi) 1584

1.3.4 Manipulation of numbers

A. (i) -39 (ii) 7   (iii) 1
(iv) 18   (v) 8   (vi) 2
(vii) $\frac{4}{3}$

B. (i) 56   (ii) 181   (iii) 1
(iv) -66 (v) 9

C. 44; 2 pairs of brackets, $4 + (5 \times (2^3))$
With one pair we can get 72, 1004, 2744
With two pairs we could also get 18^3

1.3.5 Handling fractions

A. (i) $\frac{5}{6}$ (ii) 4   (iii) $\frac{14}{15}$
(iv) $\frac{98}{75}$ (v) $\frac{3}{4}$ (vi) $\frac{4}{21}$
(vii) $\frac{7}{2}$ (viii) $\frac{5}{6}$ (ix) $\frac{4}{15}$
(x) $-\frac{130}{189}$ (xi) 24

B. (a) (i) $\frac{28}{3}$ (ii) 7   (iii) $-\frac{49}{7}$ (iv) 35
(b) (i) $\frac{12}{7}$ (ii) $\frac{9}{7}$ (iii) -3 (iv) $\frac{45}{7}$

C. (i) $\frac{5}{2}$ (ii) $\frac{5}{7}$ (iii) $\frac{2}{7}$ (iv) $\frac{3}{7}$ (v) $\frac{29}{10}$ (vi) $\frac{1}{21}$ (vii) $\frac{58}{21}$

D. (i) $\dfrac{9}{2}$ (ii) $\dfrac{13}{6}$ (iii) $\dfrac{2}{5}$ (iv) 14

1.3.6 Factorial and combinatorial notation

A. (i) 120   (ii) 3628800   (iii) 301
(iv) 220320   (v) 264   (vi) 210
(vii) 43545600   (viii) 70

B. (i) 36   (ii) 36   (iii) 330
(iv) 210   (v) 1   (vi) 210
(vii) 2400

C. 35

1.3.7 Powers and indices

A. (i) 2^5 (ii) 3^2 (iii) 2×3^5 (iv) 3^4 (v) 2^9 (vi) $5^2 2^{-4}$
(vii) $2^2 3^3$ (viii) 7×3^{-2}

B. (i) 3^{12} (ii) 2^2 (iii) $2^3 3^5$
(iv) $2^{14} 3^4$ (v) 3^5 (vi) $-2^{-2} 3^{-4}$
(vii) $2^{13} 3^5$ (viii) $2^{-2} 3 = \tfrac{3}{4}$

D. (i) $3\sqrt{2}$ (ii) $2\sqrt{5}$ (iii) $2\sqrt{2}$
(iv) $2\sqrt{13}$ (v) $16\sqrt{2}$ (vi) $6\sqrt{11}$
(vii) $6\sqrt{3}$ (viii) $3\sqrt{7}$

E. (i) $\dfrac{\sqrt{7}}{7}$ (ii) $-\dfrac{\sqrt{3}}{3}$ (iii) $\sqrt{2}+1$
(iv) $\dfrac{4+\sqrt{10}}{6}$ (v) $\dfrac{3+\sqrt{5}}{2}$ (vi) $3\sqrt{6}-8$
(vii) $\dfrac{1}{4}(2+3\sqrt{2})$ (viii) $28\sqrt{2}$

1.3.8 Decimal notation

A. (i) -0.5000 (ii) 3.5000 (iii) 0.6667
(iv) -0.2222 (v) 0.0000 (vi) 0.1250

B. (i) $\dfrac{1}{4}$ (ii) $\dfrac{1}{8}$ (iii) $\dfrac{1449}{20}$

(iv) $-\dfrac{39}{125}$ (v) $\dfrac{17}{100}$

C.

	Given	Scientific Notation	Mantissa	Exponent
(i)	21.3241	2.13241×10	2.13241	1
(ii)	429.003	4.29003×10^2	4.29003	2
(iii)	−0.000321	-3.21×10^{-4}	−3.21	−4
(iv)	0.00301	3.01×10^{-3}	3.01	−3
(v)	1,000,100	1.0001×10^6	1.0001	6
(vi)	300491.2	3.004912×10^5	3.004912	5

D. (a) (i) 21.3   (ii) 429   (iii) −0.000321   (iv) 0.00301   (v) 1,000,000
(vi) 300,000

(b) (i) 21.3241   (ii) 429.003   (iii) −0.000321   (iv) 0.003010   (v) 1,000,100
(vi) 300491

1.3.9 Estimation

A. (i) 10   (ii) 20   (iii) $\dfrac{1}{10}$ (iv) $\dfrac{1}{20}$

B. (i) 0.1   (ii) −13   (iii) 62×10^5 (iv) 10

2

Algebra

In this chapter we review the basic principles of algebra in some detail and introduce some more advanced topics. There is a lot of work in this chapter, reflecting the importance of algebra, but don't feel you have to do it all at once. Some topics, such as the binomial theorem, you may like to leave until you need them.

Prerequisites
It will be helpful if you know something about:

- the rules of arithmetic (10 ◄)
- the properties of zero (6 ◄)
- algebraic manipulation of symbols
- powers and indices (18 ◄)
- factorial notation (16 ◄)

Objectives
In this chapter you will find:

- multiplication of linear expressions
- addition and multiplication of simple polynomials
- factorisation of polynomials by inspection
- simultaneous linear equations
- definition and use of identities
- factors and roots of a polynomial and use of the factor theorem
- rational functions and their properties
- addition and multiplication of rational functions
- division of polynomial expressions and use of the remainder theorem
- partial fractions
- properties of quadratic expressions and equations
- powers and indices for algebraic expressions
- the use of the binomial expansions of $(a+b)^n$ and $(1+x)^n$ for positive integer n

Motivation
You may need the material in this chapter for:

- manipulation of mathematical expressions of scientific and engineering relationships

- further mathematical topics and techniques such as integration, coordinate geometry, Laplace transforms
- solving systems of linear equations occurring in electrical circuits, chemical reactions, etc.
- solving differential equations
- numerical methods and statistics

2.1 Review

2.1.1 Multiplication of linear expressions

➤ 40 73 ➤➤

Expand the brackets in the following expressions

(i) $-(x-3)$ (ii) $2x(x-1)$ (iii) $(a-3)(2a-4)$
(iv) $(t-3)(t+3)$ (v) $(u-2)^2$

2.1.2 Polynomials

➤ 43 74 ➤➤

Expand and simplify into polynomial form

(i) $x^3 - 2x + 1 + 3(x^4 + 2x^3 - 4x^2 - x - 1)$
(ii) $(x-2)(x+3)$
(iii) $(x-1)^2(x+1)$
(iv) $(2x-1)(x^2+x+1)$

2.1.3 Factorisation of polynomials by inspection

➤ 45 75 ➤➤

A. Factorise (i) $3x^2 + 6x$ (ii) $u^2 - 16$

B. Factorise (i) $x^2 - x - 2$ (ii) $x^3 - 2x^2 - x + 2$
(iii) $3x^2 + 5x - 2$

2.1.4 Simultaneous equations

➤ 48 75 ➤➤

Solve the simultaneous equations

$$2x - y = 1$$
$$x + 2y = 2$$

2.1.5 Equalities and identities

➤ 50 76 ➤➤

Determine A and B if

$$A(x-3) + B(x+2) \equiv 4$$

2.1.6 Roots and factors of a polynomial

A. Referring to 2.1.2(iii), what are the (i) **factors** and (ii) **roots** of the polynomial $x^3 - x^2 - x + 1$? What are the **solutions** of the equation $x^3 - x^2 - x + 1 = 0$?

B. Determine the factors of $x^3 + 2x^2 - 5x - 6$.

2.1.7 Rational functions

Which of the following are rational functions?

(i) $\dfrac{x-1}{x^2+1}$ (ii) $\sqrt{x} + \dfrac{1}{\sqrt{x}}$ (iii) $\dfrac{x-1}{x^2+x+1}$

(iv) $\sqrt{\dfrac{x-1}{x+1}}$

2.1.8 Algebra of rational functions

Put the following over a common denominator

(i) $\dfrac{2}{x+1} - \dfrac{3}{x-2}$

(ii) $\dfrac{1}{x-1} + \dfrac{1}{x+1} - \dfrac{1}{x+2}$

2.1.9 Division and the remainder theorem

A. Divide $2x^3 + x^2 - 6x + 9$ by $x + 2$.

B. Use the remainder theorem to find the remainder when $4x^3 - 2x^2 + 3x - 1$ is divided by $x - 2$.

2.1.10 Partial fractions

Split into partial fractions

$$\dfrac{x+1}{(x-1)(x+3)}$$

2.1.11 Properties of quadratic expressions and equations

A. Solve the quadratic equations

(i) $x^2 - 3x = 0$

(ii) $x^2 - 5x + 6 = 0$

(iii) $2x^2 + 3x - 2 = 0$

B. Complete the square for the quadratic $x^2 + x + 1$ and hence determine its minimum value.

C. What is the (i) sum (ii) product of the roots of the quadratic $x^2 + 2x + 3$?

39

2.1.12 Powers and indices for algebraic expressions

Simplify the following

(i) $\dfrac{a^2 c^4}{a^{-3} b^2 c}$

(ii) $\dfrac{(3x)^3 (2y)^{-1}}{(x^2 y)^{-1}}$

(iii) $\dfrac{(x-1)^{\frac{1}{4}} (16x+16)^{\frac{1}{4}}}{(x+1)(x^2-1)^{-\frac{3}{4}}}$

2.1.13 The binomial theorem

Expand $(2-x)^5$ by the binomial theorem.

2.2 Revision

2.2.1 Multiplication of linear expressions

Algebra is the branch of mathematics that generalises arithmetic by using **symbols** as well as numbers. Symbols, such as x, y, a, b, may represent any numbers, but there are certain conventions used which are useful to know: letters from the beginning of the alphabet: $a, b, c, \ldots$ usually represent **constants** – quantities having a fixed value; letters from the middle of the alphabet $\ldots l, m, n, \ldots$ usually represent **whole numbers** or **integers**; letters from the end of the alphabet $\ldots x, y, z$ usually represent **variables**, which may take a range of values. As mathematics becomes more advanced we need more symbols and the Greek alphabet is pressed into service – you have already seen an example in π, the Greek p.

An **algebraic expression** is any quantity built up from such a finite number of symbols using only the arithmetic operations of **addition, subtraction, multiplication** or **division**. This includes integer powers which are simply successive multiplication, and roots of variables, such as $\sqrt{x}$.

Examples

$$x+y, \quad \dfrac{a+2b}{a-b}, \quad x^2+2x+3, \quad \sqrt{t}+\dfrac{1}{\sqrt{t-1}}$$

are all algebraic expressions, in the symbols indicated. Be careful to distinguish between an **algebraic expression**, such as $x^2 - 1$, and an **algebraic equation** such as $x^2 - 1 = 0$. An expression tells you nothing about the variables involved, it stands alone, whereas an equation can fix the values of the variables.

Most of the expressions we will deal with in fact contain just one variable – called **functions of a single variable** – which is traditionally, but not exclusively, denoted by x. The simplest example is the **linear expression in** x such as $2x + 3$ (called linear because its graph is a straight line (➤ 212), or in general

$$ax + b$$

where a and b are given constants. If the use of symbols a, b to represent constants worries you, just think of them as numbers such as 2, 3, etc.

40

Another important example is a **quadratic expression in** x such as $2x^2 + 7x - 4$, of general form

$$ax^2 + bx + c$$

where a, b, c are again 'constants', independent of x.

In order to use algebra for more advanced topics such as factorisation or partial fractions, we need to be able to perform multiplication of simple algebraic expressions quickly and accurately. The key to this is dealing with multiplication of bracket expressions such as $(a + b)(c + d)$. All you need for this is to know that the expression $a(c + d)$ is 'expanded' by multiplying c and d by a:

$$a(c + d) = ac + ad$$

This is called the **distributive rule**: we say 'multiplication distributes over addition'. Real numbers also satisfy the **associative rule** $a(bc) = (ab)c$ and the **commutative rule** $ab = ba$. Another useful property of real numbers is that if $ab = 0$ then one or both of a, b must be zero. Note that these statements are **rules** of algebra, which will not necessarily hold for algebras not based on the real numbers – for example the commutative law does not hold in matrix algebra, the subject of Chapter 13. However, given these rules we can use them to handle more complicated expressions. For example, we can now expand out any number of bracket expressions.

Examples

$$(a + b)(c + d) = (a + b)c + (a + b)d$$
$$= ac + bc + ad + bd$$
$$(3x + 2)(x - 1) = (3x + 2)x + (3x + 2)(-1)$$
$$= 3x^2 + 2x - 3x - 2$$
$$= 3x^2 - x - 2$$

Note the collection of the 'like' terms, $2x - 3x = -x$. You should always tidy up calculations in this way.

Always be careful with the treatment of signs and brackets, particularly when they are mixed. If in doubt leave minus signs bracketed:

$$(-1)(x - 1) = (-1)x - (-1)1$$
$$= -x + 1$$

Never omit brackets out of laziness.

Later you will need to **reverse** some of the above operations. For example, you could be given $ac + bc + ad + bd$ and asked to 'factorise' it – that is, convert it to the form $(a + b)(c + d)$. This is much easier to do if you are highly proficient at multiplying such expressions in the first place – to do something 'standing on your head' it helps if you can do it the right way up first!! There is only one way to achieve this proficiency – practice, so try the reinforcement exercises until you are confident.

Understanding Engineering Mathematics

Two very important special cases of products of linear expressions are

$$(a - b)(a + b) = a^2 - b^2$$
$$(a + b)^2 = a^2 + 2ab + b^2$$

You should know these backwards – literally, given either side you should immediately be able to write down the other side.

Examples

Fill in the blanks:

$(x - 3)(x + 3) =$ (**expanding**)

$(x + 2)^2 =$ (**expanding**)

$= x^2 - 4x + 4$ (**factorising**)

$= x^2 - 16$ (**factorising**)

Solution to review question 2.1.1

(i) You just need to be careful with signs here. Take your time until you are fully confident:

$$-(x - 3) = (-1)(x - 3) = (-1)x - (-1)3$$
$$= -x + 3$$

(ii) $2x(x - 1) = 2x \times x - 2x = 2x^2 - 2x$

(iii) Here you have to remember to collect terms together at the end. We have

$$(a - 3)(2a - 4) = a(2a - 4) - 3(2a - 4)$$
$$= 2a^2 - 4a - 6a + 12$$
$$= 2a^2 - 10a + 12$$

(iv) $(t - 3)(t + 3) = t^2 + 3t - 3t - 3^2$
$$= t^2 - 9$$

This is just the difference of two squares result. Practice this until you can miss out the intermediate step.

(v) $(u - 2)^2 = (u + (-2))^2 = u^2 + 2(-2)u + (-2)^2$
$$= u^2 - 4u + 4$$

Again, this square of a linear term is so important that you should practice it until you are able to jump the intermediate steps and able to go from 'right to left' just as proficiently.

2.2.2 Polynomials

A **monomial** is an algebraic expression consisting of a single term, such as $3x$, while a **binomial** consists of a sum of two terms, such as $x + 3y$. A **polynomial** is an algebraic expression consisting of a sum of terms each of which is a product of a constant and one or more variables raised to a non-negative integer power. An example with a single variable, x, is

$$x^3 - 2x^2 + x + 4$$

and the general form of a polynomial in x is written:

$$p_n(x) = a_n x^n + a_{n-1} x^{n-1} + a_{n-2} x^{n-2} + \cdots + a_1 x + a_0$$

where the a_i, $i = 0, 1, \ldots, n$ are given numbers called the **coefficients** of the polynomial. We use $p_n(x)$ to denote a polynomial in x of degree n. The notation a_i is often used when we have a list of quantities to describe – i is called the **subscript**. If n is the highest power that occurs, as in the above, and if $a_n \neq 0$, then we say that the polynomial is of **nth degree**. An important property of a polynomial in x is that it **exists** (i.e. has a definite value) for every possible value of x.

A polynomial of degree zero is simply a **constant**:

$$p_0(x) = a_0$$

A polynomial of degree one:

$$p_1(x) = a_1 x + a_0$$

is a **linear polynomial**, or **linear function**.

A polynomial of degree two:

$$p_2(x) = a_2 x^2 + a_1 x + a_0$$

is a **quadratic polynomial**, or **quadratic function**.

Similarly for cubic (degree 3), quartic (degree 4), quintic (degree 5), $\ldots$, etc.

An equation of the form

$$p_n(x) = a_n x^n + a_{n-1} x^{n-1} + \cdots + a_1 x + a_0 = 0$$

is a **polynomial equation in x of degree n**.

Examples

(i) $p(x) = 2x + 1$ is a **linear polynomial** or **function**, while $2x + 1 = 0$ is a **linear equation**, which gives $x = -\frac{1}{2}$.

(ii) $p(x) = x^2 - 3x + 2$ is a **quadratic function**. $x^2 - 3x + 2 = 0$ is a **quadratic equation**, which can be solved by factorising (see Section 2.2.3):

$$x^2 - 3x + 2 = (x - 1)(x - 2) = 0$$

This can only be true if either $x - 1 = 0$ or $x - 2 = 0$, yielding two possible values of x:

$$x = 1, 2$$

Such solutions of a polynomial equation are called the **roots** or **zeros** of the corresponding polynomial (➤ 52).

Some polynomials do not have roots that are real numbers – for example there are no real numbers that satisfy

$$x^2 + 1 = 0$$

However, if we allow the possibility of **complex roots** (complex numbers are considered in Chapter 12), then there is a famous theorem of algebra (the **fundamental theorem of algebra**) which states that a polynomial of nth degree has exactly n roots – which may be real, equal or complex.

Polynomials may be added and multiplied to produce other polynomials. In doing so remember to gather like terms, and take care with signs.

Examples

$$(2x + 1) + (x^3 - 2x^2 - 3x - 7) = x^3 - 2x^2 + (2x - 3x) + (1 - 7)$$
$$= x^3 - 2x^2 - x - 6$$

$$(2x^2 + x - 1)(x^2 - 2) = 2x^2 x^2 + 2x^2(-2) + xx^2 + x(-2) - 1x^2 - 1(-2)$$
$$= 2x^4 - 4x^2 + x^3 - 2x - x^2 + 2$$
$$= 2x^4 + x^3 - 5x^2 - 2x + 2$$

Solution to review question 2.1.2

(i) We must gather together all terms of the same degree:

$$x^3 - 2x + 1 + 3(x^4 + 2x^3 - 4x^2 - x - 1)$$
$$= x^3 - 2x + 1 + 3x^4 + 6x^3 - 12x^2 - 3x - 3$$
$$= 3x^4 + 7x^3 - 12x^2 - 5x - 2$$

(ii) Expanding the brackets:

$$(x - 2)(x + 3) = x(x + 3) - 2(x + 3)$$
$$= x^2 + 3x - 2x - 6$$
$$= x^2 + x - 6$$

(iii) This example illustrates how good facility with elementary results can save you work. Thus, one way to expand in this case is:

$$(x - 1)^2(x + 1) = (x^2 - 2x + 1)(x + 1)$$
$$= x^3 - 2x^2 + x + x^2 - 2x + 1$$
$$= x^3 - x^2 - x + 1$$

This is fine, but it is much quicker if you notice that pairing off one of the $(x-1)$ factors with the $(x+1)$ gives $x^2 - 1$ and:

$$(x-1)^2(x+1) = (x-1)(x^2-1)$$
$$= x^3 - x^2 - x + 1$$

is obtained directly.

(iv) No tricks here, just a careful plod:

$$(2x-1)(x^2+x+1) = 2x^3 + 2x^2 + 2x - x^2 - x - 1$$
$$= 2x^3 + x^2 + x - 1$$

Note that in each of the above examples, we can always check our results by substituting 'obvious' values of x in the result. $x = 0$ is often a good start, whereas $x = \pm 1$ can be used in (iii), the point being that your calculated results must vanish for these values. $x = \frac{1}{2}$ will do a similar (but messy) job in (iv).

2.2.3 Factorisation of polynomials by inspection ◀38 75▶

As mentioned in Section 2.2.1 it is sometimes necessary to reverse the multiplication of polynomials or brackets to split an expression into **factors** of a particular type. This process is called **factorising**. It is useful in solving algebraic equations, or determining their properties. Usually we try to factorise into 'linear' factors of the form $x - a$.

Example

You can confirm that $x^2 + 2x + 1 = (x+1)^2$. This tells us that the quadratic equation $x^2 + 2x + 1 = 0$ has two equal roots $x = -1$. It also makes clear that $x^2 + 2x + 1$ is always positive whatever the value of x.

In general, factorising can be difficult. It pays to take it slowly and build up confidence with simple expressions.

Example

To factorise $x^2 + 2x$ note that x is a common factor of each of the terms in the polynomial. We can therefore 'take it out' and write

$$x^2 + 2x = x(x+2)$$

The approach in general is to inspect each term of an expression and check whether there are factors common to each. All we are really doing is reversing the **distributive rule** stated in Section 2.2.1.

Example

$$8x^2 - 2x^4 = 4(2x^2) - (2x^2)x^2$$
$$= 2x^2(4 - x^2)$$
$$= 2x^2(2-x)(2+x)$$

by using the difference of two squares (see Section 2.2.1).

When factorising more complicated polynomials it pays to remember that this is not always possible. For example, we cannot factorise $x^2 + 1$ in terms of real factors, and it must therefore be left as it is. In any event, factorisation is rarely easy, and by far the most potent weapon in factorising is a high degree of skill in multiplying out brackets. The result for the difference of two squares, and the square of a linear factor (Section 2.2.1) are particularly useful here. In the latter case try to learn to recognise that, provided the coefficient of x^2 is 1, the constant term in $(x + a)^2 = x^2 + 2ax + a^2$ (i.e. independent of x) is the square of half of the coefficient of the linear term in x.

Expressions can sometimes be factorised by looking for like terms and combining them, or for terms with common factors.

Example

$$2ac + ad - 6bc - 3bd$$

'By inspection' we notice that a may be taken out of the first two terms to give $a(2c + d)$. We then look at the last two terms to see if they conceal a $2c + d$ – sure enough they may be written $-6bc - 3bd = -3b(2c + d)$.

So we have

$$2ac + ad - 6bc - 3bd = a(2c + d) - 3b(2c + d)$$

and taking out the common factor $2c + d$ finally gives the factorised form

$$(a - 3b)(2c + d)$$

Note that this required some inspired guesswork and trial and error.

Solution to review question 2.1.3

A. (i) To factorise $3x^2 + 6x$ examine each term and find factors common to them. In this case both terms contain an x and a 3, from which

$$3x^2 + 6x = (3x)x + (3x)(2)$$
$$= 3x(x + 2)$$

This done you could, for practise, check the result by expanding out again, to go from right to left.

(ii) Know your difference of two squares backwards!

$$u^2 - 16 = u^2 - 4^2 = (u - 4)(u + 4)$$

B. (i) In factorising something like $x^2 - x - 2$ most of us use a sort of inspired trial and error. And the better you are at multiplying pairs of brackets, the easier the trial will be. We look at the -2. This can only come from multiplying something like $(x \pm 1)$ and $(x \mp 2)$ together. Whichever of these we choose must combine to give us the $-x$. All you have to do therefore is multiply such pairs of brackets and check which gives the right result. If your expansion of brackets is good, then you'll be able to do this mentally:

$$(x - 1)(x + 2) = x^2 + x - 2 \quad \times$$
$$(x + 1)(x - 2) = x^2 - x - 2 \quad \checkmark$$

So the factors are

$$(x + 1)(x - 2)$$

If you want a more formal approach, note that

$$(x + a)(x + b) = x^2 + (a + b)x + ab$$

So, to factorise $x^2 - x - 2$ we need to find a and b such that $ab = -2$ and $a + b = -1$, which is then done 'by inspection' (i.e. trial and error!) to give the result obtained above.

(ii) There is a routine device for factorising polynomials such as $x^3 - 2x^2 - x + 2$ or higher degree, based on the factor theorem (see Section 2.2.6). This only works if the roots of the polynomial are easy to spot. This example shows what can be done by looking at the terms and picking factors out directly. In this case a couple of elementary factorisations reveals that

$$x^3 - 2x^2 - x + 2 = x^2(x - 2) - (x - 2)$$

We continue by taking out the $(x - 2)$ factor now common to the two terms

$$= (x^2 - 1)(x - 2)$$

and finish by factorising the $x^2 - 1$

$$= (x - 1)(x + 1)(x - 2)$$

This example nicely illustrates how one often has to piece together elementary results and ideas to solve problems.

(iii) For $3x^2 + 5x - 2$ we have to deal with the 3, the coefficient of x. Because of this we guess that the factors are going to be something like $(3x + a)(x + b)$, with $ab = -2$, and $a + 3b = 5$. Trial and error (which can be systematised, but is hardly worth the effort) soon reveals that $b = 2$ and $a = -1$ works and we can then quickly check that $(3x - 1)(x + 2) = 3x^2 + 5x - 2$.

Note that in these solutions we don't want to give the impression that factorisation is an haphazard, hit and miss process – there are systematic routines, including symbolic computer algebra packages – we simply want to indicate what can be achieved with a good facility in simple algebra.

2.2.4 Simultaneous equations

We will say more about solving equations in Section 13.5, but here we need to cover some simple examples that we will use in subsequent sections. A simple linear equation in one variable, of the form

$$ax + b = 0 \quad (a \neq 0)$$

where a, b are given constants, is easy to solve:

$$ax = -b \text{ and so } x = -\frac{b}{a}$$

Often we have to deal with equations involving two variables, such as

$$ax + by = e$$
$$cx + dy = f$$

where a, b, c, d, e, f are all given constants. This is referred to as a **system of simultaneous linear equations in the two variables x and y**. Such a system can be solved by 'eliminating' one of the variables, say y, and hence determining the other.

Example

$$x - y = 1$$
$$x + 2y = -1$$

In this case it is actually easier to eliminate x first, because we notice that by subtracting the equations (i.e. doing the same subtraction on each side of the equation) we obtain

$$(x - y) - (x + 2y) = 1 - (-1)$$
$$x - y - x - 2y = 1 + 1 = 2$$

So

$$-3y = 2$$

and therefore

$$y = -\tfrac{2}{3}$$

We can now obtain x from the first equation:

$$x = 1 + y = 1 - \tfrac{2}{3} = \tfrac{1}{3}$$

So the solution is

$$x = \tfrac{1}{3}, \quad y = -\tfrac{2}{3}$$

which you should check in the original equations. Note that the answers are left as fractions rather than converting to decimals. Any decimal form of the solutions is likely to be an approximation and therefore incur errors, which may have serious consequences in an actual engineering application.

Complications can arise with such systems of equations, depending on the coefficients a, b, c, d, e, f. For example consider the system

$$x + y = 1$$
$$2x + 2y = 3$$

This system has no solutions for x and y because whereas the left-hand sides are related by a factor of two, the right-hand sides are not – we say that the equations are **incompatible**. To see the nonsense they lead to, subtract twice the first from the second to give

$$`0 = 2x + 2y - 2(x + y) = 3 - 2 = 1$$
$$\text{Therefore } 0 = 1\text{'}$$

An uncomfortable conclusion!
Now consider the pair

$$x + y = 1$$
$$2x + 2y = 2$$

These are not really two different equations – the second is just twice the first. In this case we can find any number of solutions to the system – just fix a value for, say, y and then determine the required value of x.

The following is an example of another important type of system

$$x + y = 0$$
$$2x + y = 0$$

Because of the zeros on the right-hand side such a system is said to be **homogeneous**. We will have a lot to say about them in Chapter 13, but for now convince yourself that the only possible solution of this system is $x = y = 0$, a so-called 'trivial' solution. In general, show that for the system

$$ax + by = 0$$
$$cx + dy = 0$$

to have a 'non-trivial' solution (i.e. x, y not both zero) for non-zero coefficients a, b, c, d we must have $ad = bc$.

The sorts of complications we have pointed to are really the concern of more advanced mathematics. For the moment all we need to be able to do is solve a given system by elimination, when no complications arise. Graphically, equations of the form $ax + by = c$ represent straight lines in a plane and systems of simultaneous equations of this type represent sets of such lines that may or may not intersect. We will say more about this in Chapter 7.

Solution to review question 2.1.4

$$2x - y = 1 \qquad \text{(i)}$$
$$x + 2y = 2 \qquad \text{(ii)}$$

No simple cancellation jumps out at us here. If we want to eliminate y, say, then the easiest way is to multiply (i) by 2 to get

$$4x - 2y = 2$$

and add this result to (ii) to obtain

$$4x - 2y + x + 2y = 5x = 2 + 2 = 4$$

So $x = 4/5$.
Now substitute back in (i) to find y

$$y = 2x - 1 = \tfrac{8}{5} - 1 = \tfrac{3}{5}$$

The solution is therefore

$$x = \tfrac{4}{5}, \quad y = \tfrac{3}{5}$$

2.2.5 Equalities and identities

◄ 38   76 ►

You might not have noticed but so far we have used the equals sign, $=$, in two distinct contexts. Look at the two 'equations':

$$3x + 6 = 12 \qquad \text{(i)}$$
$$(x - 1)(x + 1) = x^2 - 1 \qquad \text{(ii)}$$

The first is the usual sort of equation. It only holds for a particular value of x – i.e. $x = 6$. The equality in (i) enables us to determine x.

The second 'equation' actually tells us nothing about x – it is true for **any** value of x. We call this an **identity**. The expressions on either side of the equals sign are merely alternative forms of each other. To distinguish such **identities** from ordinary equalities we use the symbol $\equiv$ (a 'stronger' form of equals!) and write

$$(x - 1)(x + 1) \equiv x^2 - 1$$

$\equiv$ is read as 'is equivalent to', or 'is identical to'.

The powerful thing about an identity in x is that it must be true for **all values of x**. We can sometimes use this to gain useful information (a particularly important application occurs in partial fractions – see Section 2.2.10).

Example

Suppose we are given

$$x^2 - 3x + 2 \equiv Ax^2 + Bx + C$$

where the A, B, C are 'unknown'. In fact, the only way the left- and right-hand sides can be **identical** is if the coefficients of corresponding powers are the same, i.e.

$$A = 1, \quad B = -3, \quad C = 2$$

Another way of looking at this is to say that since the identity must hold true for all values of x, we can determine the A, B, C by substituting 'useful' values of x. For example $x = 0$ gives $C = 2$ immediately. Choosing any two other values for x will give two equations for A and B. In this case we may notice that the roots of the quadratic are 1 and 2 and so taking these values of x give

$$x = 1, \quad A + B + 2 = 0$$
$$x = 2, \quad 4A + 2B + 2 = 0$$

Solving these simultaneous equations gives

$$A = 1, \quad B = -3, \text{ as before.}$$

Solution to review question 2.1.5

To determine A and B in

$$A(x - 3) + B(x + 2) \equiv 4$$

use the fact that this must be true for **all** values of x. In particular it must hold if $x = 3$, giving

$$A(0) + B(5) = 4$$

or

$$B = \tfrac{4}{5}$$

Similarly, putting $x = -2$ gives

$$A(-5) = 4$$

or

$$A = -\tfrac{4}{5}$$

An alternative (longer) approach is to rewrite the identity as

$$(A + B)x + (-3A + 2B) = 4$$

and equate coefficients on each side to give

$$A + B = 0$$
$$-3A + 2B = 4$$

Now solve these to give the previous results.

2.2.6 Roots and factors of a polynomial

Shortly, we will be looking at more powerful methods for factorising polynomials. First, I want to clarify something that often confuses those new to algebra. This is the distinction between factors, roots and solutions of polynomials or polynomial equations. Remember the difference between an algebraic **expression** and an algebraic **equation** referred to in Section 2.2.1? In particular, a **polynomial** is an expression such as

$$x^2 + x - 2$$

whereas the corresponding **polynomial equation** would be

$$x^2 + x - 2 = 0$$

Now if we proceed to solve this equation by factorising:

$$x^2 + x - 2 \equiv (x - 1)(x + 2)$$

the expressions $x - 1$, $x + 2$ are factors of the **polynomial expression** alone, independently of the equation. The last identity is simply an alternative way of writing $x^2 + x - 2$ in terms of these factors. This is of course helpful in **solving** the corresponding **polynomial equation**:

$$x^2 + x - 2 \equiv (x - 1)(x + 2) = 0$$

Remembering that for real numbers $ab = 0$ means one or both of a, b must be zero (41 ◀), this factorised form tells us that the solutions to the equation can be obtained by putting the factors $x - 1$, $x + 2$ equal to zero. This leads to the **solutions** of the equation

$$x = 1, \quad x = -2$$

These values of x, which make the polynomial $x^2 + x - 2$ zero, are called the **roots** or **zeros** of the polynomial. So, if $x - 1$ is a **factor** of the quadratic expression $x^2 + x - 2$ then $x = 1$ is a **solution** of the quadratic equation $x^2 + x - 2 = 0$.

In general, if $x - a$ is a factor of a polynomial $p_n(x)$ then $x = a$ will be a solution of the polynomial equation $p_n(x) = 0$. The solution $x = a$ is often referred to as a **root** of the polynomial. This idea is the basis of the **factor theorem**:

If $p(x)$ is a polynomial and if $x = a$ is a root of the polynomial, i.e. $p(a) = 0$, then $x - a$ is a factor of $p(x)$. On the other hand, if $x - a$ is a factor of $p(x)$ then clearly $p(a) = 0$.

This result is fairly obvious from the fact that if $p(x)$ has a factor $x - a$ then it can be written as $p(x) = (x - a)q(x)$ where $q(x)$ is another polynomial. Then clearly $p(a) = 0$.

We can use the factor theorem in factorising more complicated polynomials. While we know from the example $x^2 + 1$ that not all polynomials can be factorised into linear factors $x - a$, it **can** be shown that **any polynomial with real coefficients can always be factorised into linear and/or quadratic factors with real coefficients**. It may then be possible to find the linear factors by trial and error using the factor theorem, if the roots are simple numbers, easy to spot.

Another useful result is that if the coefficient of the term of highest degree is unity the roots of a polynomial may be factors of the constant term, which gives us some clues about what the factors might be.

Example

$$p(x) \equiv x^4 + 2x^3 - x - 2$$

The roots must be factors of -2 and so the only integer roots are likely to be ± 1 or ± 2. We find by trial that

$$p(1) = 0 \text{ giving a factor } x - 1$$
$$p(-2) = 0 \text{ giving a factor } x + 2$$

However, -1 and 2 are not roots, so we have probably exhausted all possible real factors (although there may be repeated roots, see below). To find the quadratic factor remaining we write

$$x^4 + 2x^3 - x - 2 \equiv (x - 1)(x + 2)(x^2 + ax + b)$$
$$\equiv (x^2 + x - 2)(x^2 + ax + b)$$

and find, on multiplying out the right-hand side and equating coefficients in the resulting identity

$$-2b = -2, \quad \text{so } b = 1 \quad \text{(constant term)}$$
$$b - 2a = -1, \quad \text{so } a = 1 \quad \text{(coefficient of } x\text{)}$$

So the factorised form is

$$p(x) = (x - 1)(x + 2)(x^2 + x + 1)$$

The quadratic $x^2 + x + 1$ does not factor further into real factors – it is said to be **irreducible**.

It may not always be clear whether or not there are repeated factors in a polynomial. This does not really matter much since any repeated factors will become apparent from the remaining quadratic factor for simple cases. There is a result that if $x - a$ is a repeated factor of a polynomial $p(x)$ then it is also a factor of the derivative $p'(x)$, but this is rarely worth using.

Solution to review question 2.1.6

A. From Review Question 2.1.2(iii) we have

$$x^3 - x^2 - x + 1 = (x - 1)^2 (x + 1)$$

(i) The factors are therefore $(x - 1)$ (twice) and $x + 1$.
(ii) The roots are the values which make the polynomial vanish, which from the factors are $x = 1$ (twice) and $x = -1$.
Note that a factor is an expression containing x while a root is a particular value of x.
The solutions of the equation

$$x^3 - x^2 - x + 1 = 0$$

are, of course, simply the roots of the polynomial on the left-hand side, by definition, and so the solutions are $x = 1$ (twice) and $x = -1$.

B. No obvious factors present themselves in $p(x) = x^3 + 2x^2 - 5x - 6$. The best approach here is to use the factor theorem. We know that any linear factor of the form $x - a$ must be such that a is a factor of -6. Also, by the factor theorem, if $x - a$ is a factor of the polynomial $p(x)$, then $p(a) = 0$. All this suggests that we look for roots of the polynomial amongst the factors of -6, by substituting these in the polynomial. Trying $x = 1$:

$$p(1) = 1^3 + 2 \times 1^2 - 5 \times 1 - 6 = -8 \neq 0$$

So $(x - 1)$ is not a factor.
But $x = -1$ gives

$$(-1)^3 + 2(-1)^2 - 5(-1) - 6 = 0$$

So $(x - (-1)) = (x + 1)$ is a factor.
Similarly, you can check that $x = -2$, $x = 3$ are not roots, but $x = 2$ and $x = -3$ are, giving the final factorisation as

$$(x + 1)(x - 2)(x + 3)$$

NB. By far the most common mistake in this topic is to think that if $x = a$ is a root then $x + a$ is the corresponding factor. This error is particularly easily made if a is negative. Remember, the correct factor is $x - a$.

2.2.7 Rational functions

In writing down a polynomial we need only addition (or subtraction) and multiplication. The next step is to consider algebraic **fractions**, involving the division of polynomials. The simplest example is

$$\frac{1}{x}$$

This already gives us new problems. Whereas a polynomial in x exists for every value of x – that is, plug in a value of x and you will get out a value for the polynomial – the above fraction **does not exist for $x = 0$**. That is, there is no such number as $\frac{1}{0}$. As noted in Section 1.2.1 it simply does not exist – we say $\frac{1}{x}$ **is not defined at $x = 0$**. By this we mean that we are not allowed to put $x = 0$ in this expression. For this reason we should really write

$$\frac{1}{x} \quad x \neq 0$$

but the '$x \neq 0$' is often omitted, so long as it is clearly understood.

You may think that the fraction

$$\frac{x}{x}$$

does exist, and that it is equal to 1, on cancellation of the x. But again this is only true if $x \neq 0$. The expression is not defined for $x = 0$, that is, there is no number equal to $\frac{0}{0}$. In general then, **algebraic fractions are not defined for those values of x which make their denominator vanish**.

Examples

$\dfrac{1}{x-1}$ is not defined at $x = 1$

$\dfrac{x-2}{x-2}$ is not defined at $x = 2$

$\dfrac{x+1}{x^2 - 2x - 3}$ is not defined at $x = -1$ and $x = 3$ (why?)

Just as an arithmetic fraction such as $\frac{2}{3}$ is called a rational number, a general algebraic function of the form:

$$\frac{\text{Polynomial}}{\text{Polynomial}}$$

is called a **rational function**. The polynomial on the top is the **numerator** that on the bottom the **denominator** in analogy with numerical fractions.

Examples

$$\frac{x+1}{x+2}, \quad \frac{x^2 + 2x - 3}{x^2 + 4x + 2}$$

are rational functions, whereas

$$\frac{\sqrt{x}+1}{x+2}, \quad \frac{x^2 + 3x - 4}{x + \sqrt{x}}$$

are not because neither of $\sqrt{x}+1$ or $x + \sqrt{x}$ are polynomials.

Solution to review question 2.1.7

(i) $\dfrac{x-1}{x^2 + 1}$ and (iii) $\dfrac{x-1}{x^2 + x + 1}$ are both of the form

$$\frac{\text{polynomial}}{\text{polynomial}}$$

and are therefore rational functions.

(ii) $\sqrt{x} + \dfrac{1}{\sqrt{x}}$ and (iv) $\sqrt{\dfrac{x-1}{x+1}}$ are not rational functions because of the square roots.

2.2.8 Algebra of rational functions

Rational functions often cause a lot of problems for beginners in mathematics. There are some common errors that occur frequently, such as

$$`\frac{1}{a} + \frac{1}{b} = \frac{1}{a+b}`$$

To clear these up we will approach the algebra of rational functions gradually and systematically. First, to avoid such errors as that above, it may be helpful for you to get into the habit of checking such results numerically. For example, putting $a = b = 1$ in the above gives the nonsense

$$`\frac{1}{1} + \frac{1}{1} = 1 + 1 = 2 = \frac{1}{1+1} = \frac{1}{2}`$$

Algebraic fractions behave much like numerical fractions. For example, multiplication is straightforward:

$$\frac{a}{b} \times \frac{c}{d} = \frac{ac}{bd}$$

where a, b, c, d may be any polynomials.

Dividing by a rational function requires a little care, but essentially depends on the fact that

$$\frac{1}{1/a} = a \quad (a \neq 0)$$

– the reciprocal of the reciprocal undoes the reciprocal – as can be seen by multiplying top and bottom on the left-hand side by a. In general, division is carried out by inverting the dividing fraction and multiplying, thus: $\frac{a}{b} \Big/ \frac{c}{d} = \frac{a}{b} \times \frac{d}{c} = \frac{ad}{bc}$. Again, this can be seen by multiplying top and bottom of the left-hand expression by $\frac{d}{c}$.

Example

Multiply and divide the functions $x + 1$, $\frac{1}{x}$, stating when the corresponding operations are permissible.

We can have the product $(x + 1) \times \frac{1}{x} = \frac{x+1}{x} = 1 + \frac{1}{x}$ for $x \neq 0$.

For division we can have:

$$\frac{x+1}{1/x} = (x+1) \times \frac{1}{1/x} = (x+1) \times x$$

$$= x(x+1) \quad \text{for } x \neq 0$$

or $\quad \dfrac{1/x}{x+1} = \dfrac{1}{x} \times \dfrac{1}{x+1} = \dfrac{1}{x(x+1)} \quad$ for $x \neq 0$ or -1

Just as we can simplify the fraction $\frac{2}{4}$ to its 'lowest' form $\frac{1}{2}$, so we may be able to 'cancel down' algebraic fractions – but with slightly more care. For example on cancelling the x on top and bottom you may be happy with

$$\frac{x(x+1)}{x} = x+1$$

but remember that the functions on each side are **not** the same for all values of x. Whereas the right-hand side exists for all values of x, the left-hand side is not defined for $x = 0$. Provided you are careful about such things, algebraic cancellations present little difficulty.

Examples

$$\frac{x^2 - x}{x - 1} = \frac{x(x-1)}{x-1} = x \quad (x \neq 1)$$

$$\frac{3x - 6}{x^2 - x - 2} = \frac{3(x-2)}{(x+1)(x-2)} = \frac{3}{x+1} \quad (x \neq -1, 2)$$

In algebraic fractions it is perhaps not so much what you **can do** that needs emphasising, but what you **can't do**. For example, presented with something like $\frac{x+1}{x+2}$, you cannot 'cancel' the x's as follows:

$$`\frac{x+1}{x+2} = \frac{\cancel{x}+1}{\cancel{x}+2} = \frac{1+1}{1+2} = \frac{2}{3}`$$

The fact is that $\frac{x+1}{x+2}$ cannot be 'simplified' further – leave it as it is. Tread warily with algebraic fractions – check each step, with numerical examples if need be.

We now consider addition and subtraction of rational functions. First, get used to adding fractions with the same denominator:

$$\frac{a}{c} + \frac{b}{c} = \frac{a+b}{c}$$

This is the only way you can add fractions directly – when the denominators are the same (13 ◄).

Example

$$\frac{3}{x-1} - \frac{7}{x-1} = \frac{3-7}{x-1} = \frac{-4}{x-1}$$

To add fractions with different denominators we have to make each denominator the same by multiplying top and bottom by an appropriate factor. We use the result for constructing equivalent fractions:

$$\frac{1}{a} = \frac{1}{a} \times \frac{b}{b} = \frac{b}{ab} \quad (a, b \neq 0) \quad \text{e.g.} \quad \frac{1}{3} = \frac{5}{15}$$

So, for example,

$$\frac{1}{a} + \frac{1}{b} = \frac{b}{ab} + \frac{a}{ab} = \frac{b+a}{ab} = \frac{a+b}{ab}$$

This is called **putting the fractions over a common denominator**. It is what we did with numerical fractions in Section 1.2.5. Note that the restrictions $a \neq 0$ and $b \neq 0$ are necessary for the left-hand end to exist, and are all that is needed to convert to the final result.

Examples

(i) $1 + \dfrac{1}{x+1} = \dfrac{x+1}{x+1} + \dfrac{1}{x+1} = \dfrac{x+1+1}{x+1} = \dfrac{x+2}{x+1}$

(ii) $\dfrac{1}{x} + \dfrac{1}{x-1} \equiv \dfrac{1}{x} \times \dfrac{(x-1)}{(x-1)} + \dfrac{x}{x} \times \dfrac{1}{(x-1)}$

$= \dfrac{x-1}{x(x-1)} + \dfrac{x}{x(x-1)}$

$= \dfrac{x-1+x}{x(x-1)}$

$= \dfrac{2x-1}{x(x-1)}$

(iii) $\dfrac{2}{x+1} - \dfrac{3}{x+2} = \dfrac{2}{x+1} \times \dfrac{(x+2)}{(x+2)} - \dfrac{3}{(x+2)} \times \dfrac{(x+1)}{(x+1)}$

$= \dfrac{2(x+2)}{(x+1)(x+2)} - \dfrac{3(x+1)}{(x+1)(x+2)}$

$= \dfrac{2x+4-3x-3}{(x+1)(x+2)}$

$= \dfrac{-x+1}{(x+1)(x+2)} = \dfrac{1-x}{(x+1)(x+2)}$

In the above examples it has been pretty clear what the 'common denominator' should be. Consider the example

$$\frac{3}{x^2-1} + \frac{2x-1}{(x+1)^2}$$

In this case, you might be tempted to put both fractions over $(x^2-1)(x+1)^2$. But of course this ignores the fact that both $x^2-1 = (x-1)(x+1)$ and $(x+1)^2$ have a factor $x+1$ in common. The lowest common denominator in this case is thus $(x^2-1)(x+1) = (x-1)(x+1)^2$ and we write:

$$\frac{3}{x^2-1} + \frac{2x-1}{(x+1)^2} = \frac{3}{(x-1)(x+1)} + \frac{2x-1}{(x+1)^2}$$

$$= \frac{3(x+1)}{(x-1)(x+1)^2} + \frac{(2x-1)(x-1)}{(x-1)(x+1)^2}$$

$$= \frac{3(x+1) + (2x-1)(x-1)}{(x-1)(x+1)^2}$$

$$= \frac{3x + 3 + 2x^2 - 3x + 1}{(x-1)(x+1)^2}$$

$$= \frac{2x^2 + 4}{(x-1)(x+1)^2}$$

Here we have really used the LCM of the denominators.

Solution to review question 2.18

(i) We first of all express **each** fraction over a common denominator and **then** combine the numerators. The common denominator in this case is simply the product of the denominators and we write

$$\frac{2}{x+1} - \frac{3}{x-2} \equiv \frac{2(x-2)}{(x+1)(x-2)} - \frac{3(x+1)}{(x+1)(x-2)}$$

where the factors shown are introduced on top and bottom – essentially we are multiplying by 1 in each case.

$$\equiv \frac{2(x-2) - 3(x+1)}{(x+1)(x-2)}$$

$$\equiv \frac{-x - 7}{(x+1)(x-2)}$$

This approach extends naturally to three or more fractions as shown in the next solution.

(ii) The common denominator in this case is $(x-1)(x+1)(x+2)$ and we have

$$\frac{1}{x-1} + \frac{1}{x+1} - \frac{1}{x+2} \equiv \frac{(x+1)(x+2)}{(x-1)(x+1)(x+2)}$$

$$+ \frac{(x-1)(x+2)}{(x-1)(x+1)(x+2)} - \frac{(x-1)(x+1)}{(x-1)(x+1)(x+2)}$$

$$= \frac{(x+1)(x+2) + (x-1)(x+2) - (x-1)(x+1)}{(x-1)(x+1)(x+2)}$$

$$= \frac{x^2 + 3x + 2 + x^2 + x - 2 - (x^2 - 1)}{(x^2 - 1)(x+2)}$$

$$\equiv \frac{x^2 + 4x + 1}{(x^2 - 1)(x+2)}$$

If you really know your difference of two squares then you might spot a short cut here:

$$\frac{1}{x-1} + \frac{1}{x+1} - \frac{1}{x+2} = \frac{2x}{x^2-1} - \frac{1}{x+2}$$

$$= \frac{2x(x+2) - (x^2-1)}{(x^2-1)(x+2)}, \text{ etc.}$$

2.2.9 Division and the remainder theorem ◄ 39 77 ►

Consider the sum

$$x + 1 + \frac{3}{x-1}$$

Putting this over a common denominator gives

$$x + 1 + \frac{3}{x-1} = \frac{(x+1)(x-1) + 3}{x-1}$$

$$= \frac{x^2 - 1 + 3}{x-1}$$

$$= \frac{x^2 + 2}{x-1}$$

So adding polynomials to fractions presents few problems (it is analogous to, say, $2 + \frac{1}{2} = \frac{5}{2}$). But what if we want to go the other way – that is, divide $x - 1$ into $x^2 + 2$ to reproduce the original sum? This is called **long division**. There is a routine algorithmic (i.e. step by step) procedure for such division, but usually one can get away with a simplified procedure which relies on repeatedly pulling out from the numerator terms that contain the denominator. Our numerical example illustrates this:

$$\frac{5}{2} = \frac{4+1}{2} = \frac{2 \times 2 + 1}{2} = \frac{2 \times 2}{2} + \frac{1}{2}$$

$$= 2 + \frac{1}{2}$$

This approach requires only that we are good at spotting factors. So for example:

$$\frac{x^2 + 2}{x-1} = \frac{x^2 - 1 + 3}{x-1} = \frac{(x-1)(x+1) + 3}{x-1}$$

$$= \frac{(x-1)(x+1)}{x-1} + \frac{3}{x-1}$$

$$= x + 1 + \frac{3}{x-1}$$

This does require some algebraic intuition and clever grouping of terms, but can be very quick.

In this example division by $x - 1$ resulted in a remainder of $3/(x - 1)$. In general, when a polynomial $p(x)$ is divided by a linear factor $x - a$ the remainder can be found directly by the **remainder theorem**:

The remainder when a polynomial $p(x)$ is divided by $x - a$ is given by

$$\frac{p(x)}{x - a} = q(x) + \frac{r}{x - a}$$

where $r = p(a)$ and $q(x)$ is a polynomial of degree less than that of $p(x)$.

We can see this by re-writing the above result as

$$p(x) = (x - a)q(x) + r$$

and putting $x = a$ to give

$$p(a) = r$$

For example, the remainder when $x^2 + 2$ is divided by $x - 1$ is $1^2 + 2 = 3$, as found above.

Solution to review question 2.1.9

A. We proceed by repeatedly adding and subtracting terms in the numerator to give us factors containing the denominator, like so:

$$\frac{2x^3 + x^2 - 6x + 9}{x + 2} \equiv \frac{2x^3 + 4x^2 - 4x^2 + x^2 - 6x + 9}{x + 2}$$

$$\equiv \frac{2x^3(x + 2) - 3x^2 - 6x + 9}{x + 2}$$

$$\equiv 2x^3 - \frac{3x^2 + 6x - 9}{x + 2}$$

$$\equiv 2x^3 - \frac{3x(x - 2) - 9}{x + 2}$$

$$\equiv 2x^3 - 3x + \frac{9}{x + 2}$$

as above. This method gives you lots of practice in finding factors!

B. From the remainder theorem, the remainder when $p(x) = 4x^3 - 2x^2 + 3x - 1$ is divided by $x - 2$ is the value of $p(x)$ when $x = 2$, i.e.

$$r = 4.2^3 - 2.2^2 + 3.2 - 1 = 29$$

That is, we can write

$$\frac{4x^3 - 2x^2 + 3x - 1}{x - 2} \equiv \text{polynomial of degree 2} + \frac{29}{x - 2}$$

2.2.10 Partial fractions

Often we have to reverse the process of taking the common denominator, i.e. split an expression of the form

$$\frac{px + q}{(x - a)(x - b)} \quad (p, q \text{ numbers})$$

into its **partial fractions**

$$\frac{A}{x - a} + \frac{B}{x - b}$$

The expression is easier to differentiate and integrate in this form, for example. The method used is best illustrated by an example. We effectively have to reverse the process of taking the common denominator. But this is an example of a mathematical process that is much easier to do than it is to undo!

Consider

$$\frac{x + 1}{(x - 1)(x - 2)} \equiv \frac{A}{x - 1} + \frac{B}{x - 2}$$

where the object is to determine A and B on the right-hand side. There are a number of ways of doing this, but they all implicitly depend on combining the right-hand side over a common denominator and insisting that the numerators be identical on both sides:

$$\frac{A}{x - 1} + \frac{B}{x - 2} \equiv \frac{A(x - 2) + B(x - 1)}{(x - 1)(x - 2)}$$

$$\equiv \frac{x + 1}{(x - 1)(x - 2)}$$

This implies

$$A(x - 2) + B(x - 1) \equiv x + 1 \tag{1}$$

There are now two ways to go, both relying on the fact that the result is an **identity** (50 ◄) and must therefore be true for all values of x. In particular, it is true for

$$x = 1, \text{ which gives } A = -2$$

and

$$x = 2, \text{ which gives } B = 3$$

So

$$\frac{x + 1}{(x - 1)(x - 2)} \equiv \frac{3}{x - 2} - \frac{2}{x - 1}$$

NB. Checking such results by recombining the right-hand side to confirm that you do indeed get the left-hand side will help you to develop your algebraic skills.

Another means of finding A and B is to rewrite the identity (1) as

$$(A + B)x - (2A + B) \equiv x + 1$$

and use the fact that this must be an identity to get relations between A and B:

$$A + B = 1$$
$$-(2A + B) = 1$$

the solution of which reproduces $A = -2$, $B = 3$.

There is a short cut to all this, called the **cover-up rule**. Many teachers frown on such 'tricks' because they can be used blindly without real understanding. I certainly don't advocate that. Rather, they should be evidence of your mastery of the material, enabling you to save time. In this case the trick is really a mental implementation of the first method described above. We write

$$\frac{x+1}{(x-1)(x-2)} \equiv \frac{}{x-1} \quad \frac{}{x-2}$$

and fill in the numerators as follows. The numerator of the $\dfrac{1}{x-1}$ term is the value of the left-hand side obtained by 'covering up', or ignoring, the $x - 1$ factor and putting $x - 1 = 0$, i.e. $x = 1$ in the remaining expression. This gives

$$\left.\frac{x+1}{\boxed{(x-1)}(x-2)}\right|_{x=1} = -2$$

as found for A.

Similarly, for B we get

$$B = \left.\frac{x+1}{(x-1)\boxed{(x-2)}}\right|_{x=2} = 3$$

as before. With practice this may be done mentally and the result

$$\frac{x+1}{(x-1)(x-2)} \equiv \frac{-2}{x-1} + \frac{3}{x-2}$$

written down immediately.

More complicated rational functions may be similarly decomposed in terms of partial fractions, but the cover up rule may not apply, and we may have to resort to a combination of methods. The general forms for such partial fractions are:

$$\frac{p(x)}{(x-a)(x-b)(x-c)} \equiv \frac{A}{x-a} + \frac{B}{x-b} + \frac{C}{x-c}$$

$$\frac{p(x)}{(x-a)(x^2+b)} \equiv \frac{A}{x-a} + \frac{Bx+C}{x^2+b}$$

$$\frac{p(x)}{(x-a)(x-b)^3} \equiv \frac{A}{x-a} + \frac{B}{x-b} + \frac{C}{(x-b)^2} + \frac{D}{(x-b)^3}$$

with obvious generalisations. $p(x)$ is any polynomial of degree at least one less than the denominator. Note that the number of constants $A, B, \ldots$ to determine on the right-hand side is always the same as the total degree of the original denominator. In the above the cover-up rule may be used for A, B, C in the first result, for A in the second and for A and D in the third. There are many examples of the more complicated types in the reinforcement exercises, but the simplest case of linear factors should cover most of your needs.

Solution to review question 2.1.10

Write
$$\frac{x+1}{(x-1)(x+3)} \equiv \frac{A}{x-1} + \frac{B}{x+3}$$
$$\equiv \frac{A(x+3) + B(x-1)}{(x-1)(x+3)}$$

so
$$A(x+3) + B(x-1) \equiv x+1$$

You can now determine A and B from this identity by substituting $x = 1$ to give $A = \frac{1}{2}$ and $x = -3$ to give $B = \frac{1}{2}$. Or, equate coefficients and solve the equations

$$A + B = 1$$
$$3A - B = 1$$

Check these results by using the 'cover-up rule'. We finally get

$$\frac{x+1}{(x-1)(x+3)} \equiv \frac{1}{2(x-1)} + \frac{1}{2(x+3)}$$

2.2.11 Properties of quadratic expressions and equations ◀39 78▶

So far, we have dealt mainly with algebraic expressions and rearranging them – now we turn to solving simple algebraic equations. We will only be considering solution by **exact** rather than **numerical** means here.

The first point to emphasise is that an equation is a precise statement involving an equality, and any solution must reflect that precision, or else it is only an approximate solution. Consider for example the simple equation

$$3x - 2 = 0$$

The precise solution is

$$x = \frac{2}{3}$$

If you now use your calculator to give you the approximation $x = 0.666666667$, then this is **not** a solution of the equation. It is simply an approximation to it and you should write your answer as $x \simeq 0.666666667$. You may feel that this is nit-picking, with no real practical importance, since we are always using approximations in engineering and science. However, small errors can often assume serious significance in engineering systems, and in any event it only requires care to use $=$ and $\simeq$ appropriately. In algebra we consider only exact solutions, always use $=$ and avoid decimals wherever possible.

The simplest equation is of course

$$x = \alpha$$

where α is a given number. Here the solution is trivial and immediate. Writing the equation in the form

$$x - \alpha = 0$$

is only a small disguise, and it is often convenient to write equations in this way, with an algebraic expression equated to zero. The most general form of such a **linear equation** in x is

$$ax + b = 0$$

where a, b are given constants and $a \neq 0$. The solution is of course $x = -\dfrac{b}{a}$, as noted in Section 2.2.4.

The solution of more complicated polynomial equations, such as quadratics, usually amounts to trying to pick out such linear factors which can be solved separately, as above. This is most easily seen with quadratic equations in the case when they can be factorised.

The most general **quadratic equation** in x is

$$ax^2 + bx + c = 0$$

where $a (\neq 0)$, b, c are given numbers. If the left-hand side factorises then this gives us two linear equations to solve.

Example

Solve the equation $3x^2 + 5x - 2 = 0$

Factorising, we have

$$3x^2 + 5x - 2 \equiv (3x - 1)(x + 2) = 0$$

So the left-hand side can only be zero if

$$3x - 1 = 0 \text{ whence } x = \tfrac{1}{3}$$
$$\text{or} \quad x + 2 = 0 \text{ whence } x = -2$$

Notice the use of a key property of real numbers – if $ab = 0$ then one or both of a, b is zero – this property does not hold, for example, for matrices (Chapter 13).

So the two solutions are $x = -2, \tfrac{1}{3}$.

A quadratic equation must of course always have two solutions, although these may be equal, or complex (Chapter 12).

Example

Solve $x^2 + 4x + 4 = 0$

$$x^2 + 4x + 4 \equiv (x + 2)^2 = 0$$

giving two identical solutions $x = -2$.

The problem with a quadratic equation occurs when it won't factorise. But in all cases we can use the following formula for the solution of the quadratic

$$ax^2 + bx + c = 0$$

$$x = -\frac{b \pm \sqrt{b^2 - 4ac}}{2a}$$

We will derive this formula below – for now we look at its properties and how to apply it.

First notice that the division by a is permissible – if a were zero then we wouldn't have a quadratic. Next notice that the nature of the solutions depends on what is under the square root, i.e. on the object

$$\Delta = b^2 - 4ac$$

which is called the **discriminant** of the quadratic. There are three cases:

$$\Delta > 0, \text{ then } \sqrt{b^2 - 4ac} \text{ is a real number}$$

and we have two different real roots.

$$\Delta = 0, \text{ then } \sqrt{b^2 - 4ac} = 0$$

and we have two equal real roots.

$$\Delta < 0, \text{ then } \sqrt{b^2 - 4ac} \text{ is a complex number}$$

and we have two different complex (conjugate) roots (see Chapter 12).

In this chapter we are concerned only with the first two cases.

Example

In the equation $3x^2 + 5x - 2 = 0$ we have $a = 3$, $b = 5$, $c = -2$ (always remember to include signs). So the solutions are

$$x = \frac{-5 \pm \sqrt{5^2 - 4 \times 3 \times (-2)}}{2 \times 3}$$

$$= \frac{-5 \pm \sqrt{25 + 24}}{6}$$

$$= \frac{-5 \pm 7}{6} = \frac{2}{6} \text{ or } -\frac{12}{6}$$

i.e. $x = \frac{1}{3}$ or -2 as before.

To see where the formula for the solution of a quadratic **equation** comes from we **complete the square** of the quadratic **expression**. This procedure has many applications in elementary and further mathematics and gives good practice in algebraic manipulation. Basically, it is a technique for expressing the quadratic as a sum or difference of two squares, and it relies on the key result (42 ◄)

$$(x + a)^2 = x^2 + 2ax + a^2 \qquad (i)$$

We will do a numerical example in parallel with the symbolic form as a concrete illustration. We start with the general quadratic and factor out the a:

$$ax^2 + bx + c \equiv a\left[x^2 + \frac{b}{a}x + \frac{c}{a}\right] \qquad 3x^2 + 5x - 2 \equiv 3\left[x^2 + \frac{5}{3}x - \frac{2}{3}\right]$$

Now make $x^2 + \frac{b}{a}x$ into a complete square by adding and subtracting (i.e. adding zero) the square of half the coefficient of $x(b/2a)$:

$$\left(\frac{b}{2a}\right)^2 - \left(\frac{b}{2a}\right)^2 \qquad \left(\frac{5}{6}\right)^2 - \left(\frac{5}{6}\right)^2$$

so that the quadratic becomes

$$\equiv a\left[x^2 + \frac{b}{a}x + \left(\frac{b}{2a}\right)^2 - \left(\frac{b}{2a}\right)^2 + \frac{c}{a}\right] \qquad 3\left[x^2 + \frac{5}{3}x + \left(\frac{5}{6}\right)^2 - \left(\frac{5}{6}\right)^2 - \frac{2}{3}\right]$$

Now note that

$$x^2 + \frac{b}{a}x + \left(\frac{b}{2a}\right)^2 = \left(x + \frac{b}{2a}\right)^2 \qquad x^2 + \frac{5}{3}x + \left(\frac{5}{6}\right)^2 = \left(x + \frac{5}{6}\right)^2$$

giving for the original quadratic

$$\equiv a\left[\left(x + \frac{b}{2a}\right)^2 + \frac{c}{a} - \left(\frac{b}{2a}\right)^2\right] \qquad 3\left[\left(x + \frac{5}{6}\right)^2 - \frac{25}{36} - \frac{24}{36}\right]$$

$$\equiv a\left[\left(x + \frac{b}{2a}\right)^2 + \frac{4ac - b^2}{(2a)^2}\right] \qquad 3\left[\left(x + \frac{5}{6}\right)^2 - \frac{49}{36}\right]$$

$$\equiv a\left[\left(x + \frac{b}{2a}\right)^2 - \frac{\Delta}{(2a)^2}\right]$$

where Δ is the discriminant referred to above. $\Delta/(2a)^2$ will be either a positive, negative or zero number, so we can always write the final result in the form

$$a\left[\left(x + \frac{b}{2a}\right)^2 \pm p^2\right] \qquad 3\left[\left(x + \frac{5}{6}\right)^2 - \left(\frac{7}{6}\right)^2\right]$$

where p is some real number.

Note how the a (3 in the example) is retained throughout the calculation, right to the end. It is a common mistake to drop such factors.

Completing the square in this way gives us the formula for the solution of the quadratic:

$$ax^2 + bx + c = a\left[\left(x + \frac{b}{2a}\right)^2 - \frac{b^2 - 4ac}{(2a)^2}\right] = 0$$

Since $a \neq 0$ this gives

$$\left(x + \frac{b}{2a}\right)^2 = \frac{b^2 - 4ac}{(2a)^2}$$

Hence

$$x + \frac{b}{2a} = \pm \frac{\sqrt{b^2 - 4ac}}{2a}$$

so

$$x = -\frac{b}{2a} \pm \frac{\sqrt{b^2 - 4ac}}{2a}$$

$$= \frac{-b \pm \sqrt{b^2 - 4ac}}{2a}$$

Also, once the square is completed for a quadratic it becomes clear what its maximum or minimum values are as x varies. This is because x only occurs under a square, which is always positive. Looking at the general form:

$$a\left[\left(x + \frac{b}{2a}\right)^2 - \frac{b^2 - 4ac}{(2a)^2}\right]$$

we have two cases:

a positive (a > 0)

As x varies, the quadratic will go through a **minimum** value when $x + \frac{b}{2a} = 0$, because this yields the smallest value within the square brackets.

a negative (a < 0)

As x varies, the quadratic goes through a **maximum** value when $x + \frac{b}{2a} = 0$.

Example

For the minimum value of $3x^2 + 5x - 2$ we have

$$3x^2 + 5x - 2 \equiv 3\left[\left(x + \tfrac{5}{6}\right)^2 - \left(\tfrac{7}{6}\right)^2\right]$$

This has a minimum value (3 is positive) when $x = -\frac{5}{6}$. The minimum value is $3\left(-\left(\tfrac{7}{6}\right)^2\right)$
$= -\frac{49}{12}$.

There is a useful relationship between the roots of a quadratic equation and its coefficients. Thus, suppose α, β are the roots of the quadratic $x^2 + ax + b$. Then $x - \alpha$, $x - \beta$

are factors and we have

$$(x - \alpha)(x - \beta) \equiv x^2 + ax + b$$

Expanding the left-hand side gives

$$x^2 - (\alpha + \beta)x + \alpha\beta \equiv x^2 + ax + b$$

So

$b = $ product of roots $= \alpha\beta$

$a = -$sum of roots $= -(\alpha + \beta)$

Solution to review question 2.1.11

A. (i) You may have jumped straight in to the formula here. But, in fact, all we have to do is take out an x to give

$$x^2 - 3x = x(x - 3) = 0$$

The only possible solutions are then

$x = 0$ and $x = 3$ (don't forget the zero solution!)

i.e.

$x = 0$ or 3

(ii) A routine factorisation gives

$$x^2 - 5x + 6 = (x - 3)(x - 2) = 0$$

so

$x = 2$ or 3

(iii) This sort of problem, with a coefficient of x^2 that is not one sometimes gives problems when it comes to using factorisation. We have to try factors of both the 2 and the -2. Some trial and error leads to

$$2x^2 + 3x - 2 \equiv (2x - 1)(x + 2) = 0$$

So we get solutions $x = \frac{1}{2}, -2$

If you do get stuck on the factorisation then use the formula – **but** be careful in handling the coefficient of x^2. The solution of

$$ax^2 + bx + c = 0$$

is

$$x = \frac{-b \pm \sqrt{b^2 - 4ac}}{2a}$$

so in this case we get

$$x = \frac{-3 \pm \sqrt{3^2 - 4 \times 2 \times (-2)}}{2 \times 2}$$

$$= \frac{-3 \pm \sqrt{25}}{4} = \frac{-3 \pm 5}{4}$$

$$= \tfrac{1}{2} \text{ or } -2, \text{ as above}$$

B. To complete the square we proceed as follows (see text above):

- if necessary factor out the coefficient of x^2
- take half the coefficient of x
- square it
- add and subtract the result to the quadratic expression
- use the result $(x + a)^2 = x^2 + 2ax + a^2$ with the added bit

In other words 'add and subtract (half of the coefficient of $x)^{2'}$'. This does not change the value of the expression, because we are simply adding zero, but it sets up the expression for us to use the result $(x + a)^2 = x^2 + 2ax + a^2$. Thus we have

$$x^2 + x + 1 \equiv x^2 + x + 1 + \left(\tfrac{1}{2}\right)^2 - \left(\tfrac{1}{2}\right)^2$$

$$\equiv x^2 + x + \left(\tfrac{1}{2}\right)^2 + \tfrac{3}{4}$$

$$= \left(x + \tfrac{1}{2}\right)^2 + \tfrac{3}{4}$$

This form of the expression makes it clear that its minimum value must be $\tfrac{3}{4}$, which must occur when $x + \tfrac{1}{2} = 0$, i.e. $x = -\tfrac{1}{2}$.

C. If α, β are the roots of the quadratic $x^2 + 2x + 3$ then their sum is the negative of the coefficient of x, while their product is the constant term (including sign). So, we have

$$\alpha + \beta = -2$$

$$\alpha\beta = 3$$

2.2.12 Powers and indices for algebraic expressions ◀40 78▶

We introduced powers and indices in Section 1.2.7, mainly for numbers. We will extend them here to algebraic symbols. The rules, reproduced here, are exactly the same for algebraic functions:

$$a^m a^n = a^{m+n}$$

$$\frac{a^m}{a^n} = a^{m-n}$$

$$(a^m)^n = a^{mn}$$

$$(ab)^n = a^n b^n$$
$$a^0 = 1$$

where a, b are algebraic functions, m, n need not be integers, **but** care is needed when they are not. For example $\sqrt{x^2 + 1}$ exists for all values of x, since $x^2 + 1$ is positive, but $\sqrt{x^2 - 1}$ only exists (as a real number) if $x^2 \geq 1$, i.e. $x \geq 1$ or $x \leq -1$.

As always we need to be careful when brackets and signs are involved, for example $(-2x)^3 = (-1)^3 2^3 x^3 = -8x^3$.

Solution to review question 2.1.12

(i) $\dfrac{a^2 c^4}{a^{-3} b^2 c} = a^2 c^4 (a^{-3})^{-1} b^{-2} c^{-1} = a^2 a^3 b^{-2} c^4 c^{-1}$

$\qquad = a^5 b^{-2} c^3$

(ii) $\dfrac{(3x)^3 (2y)^{-1}}{(x^2 y)^{-1}} = 3^3 x^3 2^{-1} y^{-1} x^2 y$

$\qquad = \dfrac{27}{2} x^5$

(iii) Here, treat the algebraic expressions under the indices as single objects.

$$\frac{(x-1)^{\frac{1}{4}}(16x+16)^{\frac{1}{4}}}{(x+1)(x^2-1)^{-\frac{3}{4}}} = \frac{(x-1)^{\frac{1}{4}}(16(x+1))^{\frac{1}{4}}(x^2-1)^{\frac{3}{4}}}{(x+1)}$$

$$= \frac{16^{\frac{1}{4}}(x-1)^{\frac{1}{4}}(x+1)^{\frac{1}{4}}(x^2-1)^{\frac{3}{4}}}{x+1}$$

$$= \frac{16^{\frac{1}{4}}(x^2-1)^{\frac{1}{4}}(x^2-1)^{\frac{3}{4}}}{x+1}$$

$$= \frac{16^{\frac{1}{4}}(x^2-1)}{x+1}$$

$$= 2(x-1)$$

Note that in these formal manipulations using the rules of indices we have to remember that for real quantities throughout we must have $x > 1$.

2.2.13 Binomial theorem   ◀ 40   79 ▶

There are many occasions in applied mathematics when we need to expand expressions such as $(a + b)^6$. Very simple examples can be done long hand:

$$(a+b)^2 = a^2 + 2ab + b^2$$
$$(a+b)^3 = (a^2 + 2ab + b^2)(a+b)$$
$$\qquad = a^3 + 3a^2 b + 3ab^2 + b^3$$

but this soon becomes tedious. Fortunately there is a well established routine method for expanding such expressions. Since $a + b$ is a **binomial** this method is called the **binomial theorem**.

The binomial theorem deals with expanding expressions such as $(a+b)^n$ or $(1+x)^n$. Here we will restrict n to be a positive integer, but it works equally well for other values (leading to infinite series). This is perhaps a place where 'in at the deep end' is the best policy and so here it is, the binomial theorem:

$$(a+b)^n = a^n + na^{n-1}b + \frac{n(n-1)}{2!}a^{n-2}b^2 + \cdots$$
$$+ \frac{n(n-1)\ldots(n-r+1)}{r!}a^{n-r}b^r + \cdots + b^n$$

or, using the combinatorial notation ${}^nC_r$ (Section 1.2.6 ➤)

$$(a+b)^n = a^n + {}^nC_1 a^{n-1}b + {}^nC_2 a^{n-2}b^2 + \cdots$$
$$+ {}^nC_r a^{n-r}b^r + \cdots + b^n$$

You have two problems (at least!) here – remembering the result and understanding where it comes from. Memorising it is helped by noting the patterns:

- terms are of the form $a^{n-r}b^r$, i.e. n quantities multiplied together
- the coefficient of $a^{n-r}b^r$ is ${}^nC_r = \dfrac{n!}{(n-r)!r!} = \dfrac{n(n-1)\ldots(n-r+1)}{r!}$
- there is symmetry – for example terms of the form $a^{n-r}b^r$ and $a^r b^{n-r}$ are symmetrically placed, as are the coefficients, since ${}^nC_r = {}^nC_{n-r}$
- the coefficient of $a^{n-r}b^r$ has $r!$ in the denominator and $n(n-1)\ldots(n-r+1)$ in the numerator. The latter is perhaps best remembered by starting at the second term with n and reducing n in successive terms.

Example

$$(a+b)^6 = a^6 + 6a^5 b + \frac{6.5}{2!}a^4 b^2 + \frac{6.5.4}{3!}a^3 b^3$$
$$+ \frac{6.5.4.3}{4!}a^2 b^4 + \frac{6.5.4.3.2}{5!}ab^5 + \frac{6!}{6!}b^6$$
$$= a^6 + 6a^5 b + 15a^4 b^2 + 20a^3 b^3 + 15a^2 b^4 + 6ab^5 + b^6$$

Your second problem, understanding where the binomial expansion comes from, is best approached by noticing that in

$$(a+b)^n = \underbrace{(a+b) \times (a+b) \times \cdots \times (a+b)}_{n \text{ factors}}$$

we can form terms of the form $a^{n-r}b^r$ by choosing, say, b in r ways from n different brackets – there are ${}^nC_r$ ways to do this, giving ${}^nC_r$ such terms in all, contributing ${}^nC_r a^{n-r}b^r$ to the expansion.

The expansion for $(1+x)^n$ is now easily obtained from the above by putting $a = 1$, $b = x$ to get

$$(1+x)^n = 1 + nx + \frac{n(n-1)}{2!}x^2 + \frac{n(n-1)(n-2)}{3!}x^3 + \cdots + x^n$$

$$= 1 + {}^nC_1 x + {}^nC_2 x^2 + {}^nC_3 x^3 + \cdots + x^n$$

You may have previously seen the binomial theorem treated by Pascal's triangle. While this is fine for expanding small powers such as $(1 + x)^4$, it is cumbersome for increasingly higher powers and less useful for more advanced theory and use of the binomial theorem, as for example in determining a required coefficient in the expansion. Note also that while here we have taken n to be a positive integer, it is possible to extend the theorem to the cases of n negative or a fraction. In such cases the above series expansion never terminates (➤ 106).

Solution to review question 2.1.13

The sign needs care here:

$$(2-x)^5 = 2^5 + 5 \times 2^4(-x) + \frac{5 \times 4}{2!} 2^3(-x)^2$$

$$+ \frac{5 \times 4 \times 3}{3!} 2^2(-x)^3 + \frac{5 \times 4 \times 3 \times 2}{4!} 2(-x)^4$$

$$+ \frac{5 \times 4 \times 3 \times 2 \times 1}{5!}(-x)^5$$

$$= 32 - 5 \times 16x + 10 \times 8x^2 - 10 \times 4x^3 + 10x^4 - x^5$$

$$= 32 - 80x + 80x^2 - 40x^3 + 10x^4 - x^5$$

You may have obtained the coefficients in a different way – say by Pascal's triangle, or, you may have factored out the 2 and used the expansions of $(1 + x)^n$ instead of arriving directly at the penultimate line. In any event a steady hand is required to avoid losing signs or factors of 2. Finally notice a simple check for your result, by putting $x = 1$ in both sides. When $x = 1$, $(2 - x)^5 = 1^5 = 1$, while the expanded result gives

$$32 - 80 + 80 - 40 + 10 - 1 = 1$$

which checks.

2.3 Reinforcement

2.3.1 Multiplication of linear expressions

A. Identify which of the following are algebraic expressions in the variables involved:

(i) 14
(ii) $x - 3$
(iii) $2t + 1 = 0$
(iv) $2x + y$

(v) $x^3 - 2x^2 + x - 1$ (vi) $e^x + e^{-x}$
(vii) $x^2 - 3x + 2 = 0$ (viii) $2\sqrt{x} - x^2$
(ix) $\dfrac{2}{x+1}$ (x) $\sin x - 3\cos x$
(xi) $s^{\frac{1}{3}} - 2s^2 + s = 0$ (xii) $\dfrac{4x+1}{x^2 - 2x + 4}$
(xiii) $2\cos x = 1$ (xiv) $u^2 + 2uv - 3v$
(xv) $\dfrac{x-y}{x+y} = 0$ (xvi) $x \ln x + x^2 = 0$

B. Identify the algebraic equations in **A**.

C. For each of the pair of expressions, insert brackets in the one on the left to make it identically equal to the one on the right:

(i) $a + bc + d$ $a + bc + bd$ (vii) $x^2 - 3x + 4$ $x^2 - 3x - 12$
(ii) $a + bc + d$ $ac + bc + d$ (viii) $a + bc + bd$ $a + bc + b^2 d$
(iii) $a + bc + d$ $ac + bc + ad + bd$ (ix) $a + bc + bd$ $ad + bcd + bd$
(iv) $a - bc + d$ $a - bc - bd$ (x) $a - bc - d$ $ac - bc - ad + bd$
(v) $a - bc - d$ $ac - bc - d$ (xi) $a - bc - d$ $a - bc + bd$
(vi) $a - bc + d$ $ac - bc + d$ (xii) $x^2 - 3x + 4$ $x^3 - 3x + 4$

D. Remove the brackets in the following expressions:

(i) $2(x + 2)$ (ii) $3(x - 1) - (x - 4)$
(iii) $3t(t - 1)$ (iv) $(s - t)(s + 2t)$
(v) $a^2(a - 3)$ (vi) $(x^2 + 2x - 1)(x - 1)$
(vii) $-2u(u^2 + 3)$ (viii) $9(x^2 - 3) - 2(x + 4)$
(ix) $(a^2 - 1)(a + 2) - 3(a - 3)$ (x) $x(x - 1)(x + 2) - 3x^2$
(xi) $-(x - x^2)(x - 2)$ (xii) $-[(x^2 - 1)(x - 2) - (x - 3)(x + 2)]$
(xiii) $(1 - t)(1 - s)(1 - u)$ (xiv) $(a - 2b)^2 - (a + 2b)^2$
(xv) $(x - y)^2 + (x + y)^2$

E. Factorise each of your answers to Question **D** as far as possible.

2.3.2 Polynomials   ◀◀ 38   43 ◀

A. Which of the following are polynomials? For those that are give the degree and list the coefficients.

(i) $t^2 - t + 4$ (ii) 0 (iii) $\dfrac{u+1}{u-1}$
(iv) $7t^3 - 2t + 1$ (v) $4x^4 - 2x^3 + 3x - \dfrac{1}{x}$ (vi) $27x^4 - 3x^2 + 1$
(vii) $\dfrac{x^3 + 2x}{x}$ (viii) $x + \sqrt{x}$ (ix) $3x^2 + t^3$
(x) $x^2 y + \sqrt{y}$

B. Expand the following expressions, collecting like terms:

(i) $(x-1)(x+2)$
(ii) $(x-1)(x+2)(x+4)$
(iii) $(x-1)(x+1)^2$
(iv) $(x-2)(x-3)(x+1)(x+2)$
(v) $(u-1)^2(u+1)^2$
(vi) $(x-1)^3(x+2)$
(vii) $(t+1)(t-2)(t+2)$
(viii) $(u-2)(u+3)(u-3)$
(ix) $(s-2)^4$
(x) $(x-1)(x+2)(x-3)(x+4)$
(xi) $(x+2)^2(x-3)^2$
(xii) $(2t+1)(3t-4)$
(xiii) $(3s-1)(s+2)(4s+3)$
(xiv) $(3x+2)^2(2x-1)(x+2)$
(xv) $(3x+1)(3x-1)(x+3)$

Check each expansion with suitable numerical values.

2.3.3 Factorisation of polynomials by inspection

A. Factorise the following, retaining only real coefficients:

(i) $x^2 + x$
(ii) $3x^3 - 2x^2$
(iii) $-7x^2 + 42x^4$
(iv) $t^4 - 3t^3 + t^2$
(v) $u^2 - 9$
(vi) $t^2 - 121$
(vii) $s^{24} - 16s^{22}$
(viii) $4x^{12} - 64x^8$

B. Factorise the following polynomial expressions (hint: look back at Q2.3.2B):

(i) $t^2 + 5t + 6$
(ii) $t^3 + t^2 - 4t - 4$
(iii) $y^4 - 2y^3 - 7y^2 + 8y + 12$
(iv) $9x^3 + 27x^2 - x - 3$

2.3.4 Simultaneous equations

A. Solve the following systems of linear equations, verifying your solution by back substitution in each case.

(i) $x - y = 1$
$x + 2y = 0$
(ii) $A + B = 0$
$3A - B = 1$
(iii) $s + 3t = 1$
$s - 2t = 1$
(iv) $3x + 2y = 2$
$-2x + 3y = 1$
(v) $u + 4v = 1$
$u - v = 2$
(vi) $7x_1 - 2x_2 = 1$
$3x_1 - 2x_2 = 0$

B. Comment on the following systems of equations:

(i) $x + y = 1$
$3x + 3y = 3$
(ii) $2x - y = 3$
$4x - 2y = 1$
(iii) $x + y = 0$
$x - y = 0$
(iv) $2A + B = 1$
$4A + 2B = -1$
(v) $u + v = -1$
$3u + 3v = -3$
(vi) $x + y = 0$
$x^2 - y^2 = 1$

75

2.3.5 Equalities and identities

A. Determine the real values of A, B, C, D in the following identities:

(i) $(s-1)(s+2) \equiv As^2 + Bs + C$
(ii) $(x-1)^3 = Ax^3 + 2Bx^2 - 3Cx + D$
(iii) $(x-A)(x+B) \equiv x^2 - 4$
(iv) $(x+2)(x-3)(2x-1) \equiv A(x-1)^2 + Bx + Cx^3 - Dx^2$
(v) $A(x-1) + B(x+2) \equiv x - 3$
(vi) $(x+A)^2 + B^2 \equiv x^2 - 2x + 5$

B. Given that

$$A(x-a) + B(x-b) \equiv ax + b$$

determine expressions for A, B in terms of a, b by two different methods.

2.3.6 Roots and factors of a polynomial

Use the factor theorem to factorise the following polynomials:

(i) $u^3 - u^2 - 4u + 4$
(ii) $x^3 + 4x^2 + x - 6$
(iii) $t^4 - t^3 - 7t^2 + t + 6$
(iv) $x^3 - x^2 - x + 1$
(v) $x^3 + 3x^2 - 10x - 24$

2.3.7 Rational functions

A. Identify the rational functions and state (a) the denominator, (b) the numerator.

(i) $\dfrac{x+1}{x-1}$
(ii) $\dfrac{x+\sqrt{2}}{x^2-1}$
(iii) $\dfrac{\sqrt{x}+3}{x^3-2x+1}$
(iv) $\dfrac{x^3+2x-1}{x^4-2x+3}$
(v) $\dfrac{2x^4-1}{x^4+1}$

B. For what values of x does $\dfrac{1}{x-1} - \dfrac{1}{x-1}$ exist?

2.3.8 Algebra of rational functions

A. Put the following over a common denominator and check your results with $x = 0$ and one other appropriate value.

(i) $\dfrac{2}{x+2} - \dfrac{3}{x-1}$
(ii) $\dfrac{1}{x+3} - \dfrac{2}{x-4}$
(iii) $\dfrac{2}{x+3} - \dfrac{3}{x-2}$
(iv) $\dfrac{2}{x-2} + \dfrac{3}{x-3}$
(v) $\dfrac{1}{x-1} + \dfrac{1}{x-2} + \dfrac{1}{x-3}$
(vi) $\dfrac{3}{x+2} - \dfrac{3}{x+3} - \dfrac{4}{x-1}$

(vii) $\dfrac{x}{x^2-1} + \dfrac{1}{x-1}$ (viii) $x - \dfrac{1}{x+1} + \dfrac{2}{x+2}$

(ix) $\dfrac{3}{x^2+1} + \dfrac{2}{x^2+2}$ (x) $2x - 1 + \dfrac{2}{x-1} - \dfrac{3}{x+2}$

B. Put over a common denominator:

(i) $\dfrac{2}{x-4} + \dfrac{3}{x-1}$ (ii) $\dfrac{4}{x-4} - \dfrac{2}{x-4}$

(iii) $\dfrac{x-1}{(x-2)^2} + \dfrac{3}{x-1}$ (iv) $\dfrac{2x-1}{x^2+1} - \dfrac{4}{x-2}$

(v) $1 + \dfrac{3}{x-2}$

2.3.9 Division and the remainder theorem   ◀◀ 39   60 ◀

A. Perform the following divisions, whenever permissible:

(i) $\dfrac{x^2}{x^2}$ (ii) $\dfrac{x^2 - 2x + 2}{(x-1)^2}$

(iii) $\dfrac{x^4 - 2x^3 + 4x - 1}{x - 3}$ (iv) $\dfrac{x^3 + 1}{x + 1}$

(v) $\dfrac{2x^2 - 3x + 4}{x^2 - 1}$ (vi) $\dfrac{x^4 - y^4}{x^2 + y^2}$

B. Find the remainders when the following polynomials are divided by:

(a) $x - 1$ (b) $x + 2$ (c) x

(i) $3x^3 + 2x - 1$ (ii) $x^5 - 2x^2 + 2x - 1$

(iii) $x^4 - x^2$ (iv) $2x^7 - 3x^5 + 4x^3 - 2x^2 + 1$

2.3.10 Partial fractions   ◀◀ 39   62 ◀

A. For Q2.3.8A, B check your results by reversing the operation and resolving your answer into partial fractions. Usually, the answers should of course be what you started with in those questions. There are however a couple of cases where this is not so – explain these cases.

B. Split into partial fractions:

(i) $\dfrac{x-1}{(x+2)(x-3)}$ (ii) $\dfrac{4}{x^2-1}$ (iii) $\dfrac{x+1}{x(x-3)}$

(iv) $\dfrac{2x+1}{(x-1)^2(x+2)}$ (v) $\dfrac{3}{(x^2+1)(x+1)}$ (vi) $\dfrac{5x-4}{(x^2+4)(x-2)^2}$

(vii) $\dfrac{x+1}{(x-1)(x+3)(x-4)}$

77

2.3.11 Properties of quadratic expressions and equations

A. Factorise the quadratics:

(i) $x^2 + x - 2$
(ii) $x^2 + 6x + 9$
(iii) $x^2 - 81$
(iv) $2x^2 + 5x - 3$
(v) $2x^2 - 8x$

B. Solve the quadratic equations obtained by equating the expressions in **A** to zero.

C. Complete the square:

(i) $x^2 + 2x + 2$
(ii) $x^2 - 6x + 13$
(iii) $4x^2 + 4x - 8$
(iv) $4x^2 - 4x$

D. By completing the square determine the maximum or minimum values (as appropriate) of the following quadratics, and the values of x at which they occur.

(i) $x^2 + 2x + 4$
(ii) $16 - 4x - 4x^2$

E. Solve the following equations by both factorisation and formula.

(i) $x^2 + 3x + 2 = 0$
(ii) $2x^2 - 5x + 2 = 0$
(iii) $3x^2 - 11x + 6 = 0$
(iv) $x^2 + 10x + 16 = 0$
(v) $x^2 + 2x - 8 = 0$
(vi) $9x^2 + 6x - 3 = 0$

F. Evaluate the following exactly with as little labour as possible and without a calculator.

$$\frac{8(23.7)^2 - 10(23.7)(45.4) + 3(45.4)^2}{4(23.7) - 3(45.4)}$$

G. For Q2.3.11E confirm your results by calculating the (a) sum and (b) product of the roots.

2.3.12 Powers and indices for algebraic expressions

A. Express in the form a^n

(i) $a^2 a^4$
(ii) $a^3 a^2 a$
(iii) $aa^2 a^{-1}$
(iv) $a^7 a^3 / a^2$
(v) $(a^3)^2 a^{-2} a^3$
(vi) $a^{-21} a^2 (a^3)^6$
(vii) a^5 / a^{-3}
(viii) $(a^2)^3 / (a^3)^2$

B. Express in the form a^{2n}, stating the value of n.

(i) $a^9 a^{17}$
(ii) $(a^{40})^{1/4}$
(iii) $a^3((a^3)^6 a^7 a^5)^{1/2}$
(iv) $a^{27} a^{-3} / a^2$

C. Reduce to simplest form

(i) $\dfrac{a^2b^3c^2}{abc}$

(ii) $\dfrac{(a^2)^3 c^{12}}{ab^2}$

(iii) $\dfrac{ba^{12}b^7c^4}{(a^2b^4c^6)^{1/2}}$

(iv) $\dfrac{(a^3)^4 c^{12}}{b^{-3}c^{10}a^{-1}b^2}$

D. Simplify the following expressions

(i) $b^2 a^2 b^3 a b^3$

(ii) $\dfrac{t^3 xy}{x^2 yt}$

(iii) $\dfrac{(x^2)^4 y^{7/2}}{(x^6)^{1/3}\sqrt{y}}$

2.3.13 The binomial theorem ◂◂ 40 71 ◂

A. Write down the coefficients of x^4 in the following expansions

(i) $(1+x)^7$

(ii) $(1+3x)^6$

(iii) $(1-2x)^5$

(iv) $(3-2x)^8$

(v) $(3x+2y)^6$

B. Expand by the binomial theorem

(i) $(1-2x)^7$

(ii) $(a+b)^6$

(iii) $(2x+3y)^5$

(iv) $(2-3x)^6$

(v) $(3s-2t)^5$

C. Evaluate, without using a calculator:

(i) $(1+\sqrt{2})^3 + (1-\sqrt{2})^3$

(ii) $(2+\sqrt{3})^4 + (2-\sqrt{3})^4$

D. Use the binomial theorem to evaluate to three decimal places:

(i) $(1.01)^{10}$

(ii) $(0.998)^8$

Hint: write $1.01 = 1 + 0.01$ and $0.998 = 1 - 0.002$.

2.4 Applications

1. The time behaviour of a source free resistor (R)-inductor (L)-capacitor (C) series circuit can be described by a particular **differential equation**:

$$\dfrac{d^2V}{dt^2} + \dfrac{R}{L}\dfrac{dV}{dt} + \dfrac{1}{LC}V = 0$$

where V is the capacitor voltage. Such equations will be solved in Chapter 15, but for now note that in doing so it is necessary to solve the associated quadratic **auxiliary equation**

$$\lambda^2 + \frac{R}{L}\lambda + \frac{1}{LC} = 0$$

for the quantity λ, Greek 'lambda'. Write down the solutions to this quadratic for general values of R, L, C. Solve the quadratic for the following values of these parameters:

(i) $R = 10\Omega$, $C = 1/9F$, and $L = 1H$ (ii) $R = 6\Omega$, $C = 1/9F$, and $L = 1H$
(iii) $R = 2\Omega$, $C = 1/9F$, and $L = 1H$

In the case of (iii) you will obtain **complex roots** – such numbers are studied in Chapter 12, but the next application gives you a preview.

2. A **complex number** is one of the form $z = a + jb$ where a and b are ordinary, real numbers and j is a symbol for the 'imaginary number' $\sqrt{-1}$. All you need to know is that $j^2 = -1$ and that otherwise the normal rules of algebra apply. Investigate the 'algebra' of complex numbers, that is the rules for adding, subtracting, multiplying and dividing complex numbers. You will see this in Chapter 12, and as a taster you can try to derive the results:

(i) $(a + jb)(c + jd) = ac - bd + j(bc + ad)$ (ii) $\dfrac{a + jb}{c + jd} = \dfrac{ac + bd}{c^2 + d^2} + j\dfrac{bc - ad}{c^2 + d^2}$

In (ii) multiply top and bottom of the left-hand side by $c - jd$.

3. Emperor Yang Sun (ninth century CE) had to choose between two equally qualified clerks for a promotion, He set them the following problem to decide between them.

    A man walking in the woods heard thieves arguing over the division of rolls of cloth which they had stolen. They said that if each took six rolls there would be five left over; but if each took seven rolls, they would be eight short. How many thieves were there, and how many rolls of cloth?

    The candidates performed the calculation in the ancient Chinese way, by using bamboo rods on the tiled floor. This technique led to the early development of adding and subtracting rows and columns of arrays to solve linear equations. Would you get the job?

4. An open box is made from a 12 cm square card by cutting squares of equal size from the four corners and folding the edges up. Calculate the size of the squares to be cut to yield the box with the largest volume.

5. Vapour deposition methods are often used for thin film layers in the fabrication of semiconductor devices. Such methods involve exposing a semiconductor wafer to a mixture of gases in a heated vacuum chamber. The chemical composition of the surface layer, and the deposition time depend on the relative concentrations of the vapours in the chamber. If three gases A, B, C are to be combined to form a number of overlayers of new molecules ABC then the time dependence of the deposition of the new molecules

is described by the **differential equation**

$$\frac{dx}{dt} = k(x-a)(x-b)(x-c)$$

where x is the number of ABC molecules and a, b, c are the initial concentrations of the A, B, C gases respectively. k is a constant depending on the rate of the chemical reaction. Such equations are typical of a chemical reaction. In such a tri-molecular first order reaction the $x - a$ factor, for example, models the fact that the rate of production of the new molecule is proportional (14 ◀) to the current amount of each component substance available. As x approaches the smallest of the a, b, c values the rate of the reaction will slow down, until it stops completely when there is no more of that constituent left. The differential equation will be solved in Chapter 15 (➤ 477) and a crucial step in the solution is the splitting of

$$\frac{1}{(x-a)(x-b)(x-c)}$$

into partial fractions. Do this for the case of $a = 1$, $b = 2$, $c = 3$.
Obtain an expression for the partial fractions in the case when a, b, c have general values.

6. In the toss of a fair coin the probability of a head is $\frac{1}{2}$ and the same for a tail. Suppose we toss such a coin three times. We can get 3H, 2H1T, 1H2T, and 3T. Write down the probability of each possibility. Show that the sum of these probabilities can be written

$$\left(\tfrac{1}{2} + \tfrac{1}{2}\right)^3$$

i.e. as a binomial expansion. In general, if we have n independent 'trials', each resulting in a 'success' with probability p and a 'failure' with probability $q = 1 - p$ then the probability of s 'successes' is given by the binomial term

$$p(s) = {}^nC_s p^s (1-p)^{n-s}$$

Use the binomial theorem to show that the sum of all these probabilities, for all possibilities, is 1. $p(s)$ defines the **binomial distribution**, one of the most important in statistics.

Answers to reinforcement exercises

2.3.1 Multiplication of linear expressions

A. (i), (ii), (iv), (v), (viii), (ix), (xii), (xiv) are algebraic expressions, the rest are either not algebraic, such as $\sin x - 3\cos x$, or are equations such as $2t + 1 = 0$ rather than expressions.

B. (iii), (vii), (xi), (xv) are algebraic equations.

C. (i) $a + b(c + d)$ (ii) $(a+b)c + d$ (iii) $(a+b)(c+d)$
    (iv) $a - b(c + d)$ (v) $(a-b)c - d$ (vi) $(a-b)c + d$
    (vii) $x^2 - 3(x+4)$ (viii) $a + b(c + bd)$ (ix) $(a + bc + b)d$
    (x) $(a-b)(c-d)$ (xi) $a - b(c - d)$ (xii) $(x^2 - 3)x + 4$

D. (i) $2x+4$ (ii) $2x+1$ (iii) $3t^2-3t$
(iv) $s^2+st-2t^2$ (v) a^3-3a^2 (vi) x^3+x^2-3x+1
(vii) $-2u^3-6u$ (viii) $9x^2-2x-35$ (ix) a^3+2a^2-4a+7
(x) x^3-2x^2-2x (xi) x^3-3x^2+2x (xii) $-x^3+3x^2-8$
(xiii) $1-s-t-u+st+su+tu-sut$ (xiv) $-8ab$
(xv) $2x^2+2y^2$

E. In cases (i), (iii), (iv), (v), (vi), (vii), (xi), (xiii) you should retrieve the original question. (ii), (viii), (ix) do not factorise further with simple integer roots. (x) becomes $x(x^2-2x-2)$, (xii) can be written $-(x^3-3x^2+8)$ but does not easily factorise further. (xiv) is already factorised! (xv) can be written $2(x^2+y^2)$.

2.3.2 Polynomials

A.

	Polynomial?	Degree	Coefficients
(i)	Yes	2	$1, -1, 4$
(ii)	Yes	0	0
(iii)	No		
(iv)	Yes	3	$7, 0, -2, 1$
(v)	No		
(vi)	Yes	4	$27, 0, -3, 0, 1$
(vii)	No		
(viii)	No		
(ix)	Yes (in x and t)	3	Coefficients all zero except of $x^2(3)$ and $t^3(1)$
(x)	No		

B. (i) x^2+x-2 (ii) x^3+5x^2+2x-8
(iii) x^3+x^2-x-1 (iv) $x^4-2x^3-7x^2+8x+12$
(v) u^4-2u^2+1 (vi) $x^4-x^3-3x^2+5x-2$
(vii) t^3+t^2-4t-4 (viii) $u^3-2u^2-9u+18$
(ix) $s^4-8s^3+24s^2-32s+16$ (x) $x^4+2x^3-13x^2-14x+24$
(xi) $x^4-2x^3-11x^2+12x+36$ (xii) $6t^2-5t-4$
(xiii) $12s^3+29s^2+7s-6$ (xiv) $9x^3+27x^2-x-3$

2.3.3 Factorisation of polynomials by inspection

A. (i) $x(x+1)$ (ii) $x^2(3x-2)$ (iii) $7x^2(\sqrt{6}x-1)(\sqrt{6}x+1)$
(iv) $t^2(t^2-3t+1)$ (v) $(u-3)(u+3)$ (vi) $(t-11)(t+11)$
(vii) $s^{22}(s-4)(s+4)$ (viii) $4x^8(x-2)(x+2)(x^2+4)$

B. (i) $(t+2)(t+3)$ (ii) $(t+1)(t-2)(t+2)$ (See Q2.3.2B(vii))
(iii) $(y-2)(y-3)(y+1)(y+2)$ (See Q2.3.2B(iv))
(iv) $(3x-1)(3x+1)(x+3)$ (See Q2.3.2B(xv))

2.3.4 Simultaneous equations

B. (i) Infinite number of solutions $x = 1 - s$, $y = s$
(ii) No solutions
(iii) Trivial solution $x = y = 0$
(iv) No solutions
(v) Infinity of solutions $u = -(1+s)$, $u = s$
(vi) No solutions

2.3.5 Equalities and identities

A. (i) $A = 1$, $B = 1$, $C = -2$
(ii) $A = 1$, $B = -\frac{3}{2}$, $C = -1$, $D = -1$
(iii) $A = 4$ and $B = 4$ **or** $A = -4$ and $B = 4$
(iv) $A = 6$, $B = 1$, $C = 2$, $D = 9$
(v) $A = \frac{5}{3}$, $B = -\frac{2}{3}$
(vi) $A = -1$ and $B = \pm 2$

B. $A = \dfrac{(a+1)b}{b-a}$ $B = \dfrac{a^2 + b}{a - b}$

2.3.6 Roots and factors of a polynomial

(i) $(u-1)(u+2)(u-2)$ (ii) $(x-1)(x+2)(x+3)$
(iii) $(t-1)(t+1)(t+2)(t-3)$ (iv) $(x-1)^2(x+1)$
(v) $(x+2)(x-3)(x+4)$

2.3.7 Rational functions

A.

	Rational function?	(a) denominator	(b) numerator
(i)	Yes	$x - 1$	$x + 1$
(ii)	Yes	$x^2 - 1$	$x + \sqrt{2}$
(iii)	No		
(iv)	Yes	$x^4 - 2x + 3$	$x^3 + 2x - 1$
(v)	Yes	$x^4 + 1$	$2x^4 - 1$

B. All values except $x = 1$.

2.3.8 Algebra of rational functions

A. (i) $\dfrac{-x-8}{(x+2)(x-1)}$ (ii) $\dfrac{-x-10}{(x+3)(x-4)}$ (iii) $\dfrac{-x-13}{(x+3)(x-2)}$

(iv) $\dfrac{5x-12}{(x-2)(x-3)}$ (v) $\dfrac{3x^2-12x+11}{(x-1)(x-2)(x-3)}$ (vi) $\dfrac{-4x^2-17x-27}{(x+2)(x+3)(x-1)}$

(vii) $\dfrac{2x+1}{x^2-1}$ (viii) $\dfrac{x^3+3x^2+3x}{(x+1)(x+2)}$ (ix) $\dfrac{5x^2+8}{(x^2+1)(x^2+2)}$

(x) $\dfrac{2x^3+x^2-6x+9}{(x-1)(x+2)}$

B. (i) $\dfrac{5x-14}{(x-1)(x+2)}$ (ii) $\dfrac{2}{x-4}$ (iii) $\dfrac{4x^2-14x+13}{(x-1)(x-2)^2}$

(iv) $\dfrac{-2x^2-5x-2}{(x^2+1)(x-2)}$ (v) $\dfrac{x+1}{x-2}$

2.3.9 Division and the remainder theorem

A. (i) $1, \quad x \neq 0$ (ii) $1 + \dfrac{1}{(x-1)^2} \quad x \neq 1$

(iii) $x^3 + x^2 + 3x + 13 + \dfrac{38}{x-3} \quad x \neq 3$

(iv) $x^2 - x + 1 \quad x \neq -1$ (v) $2 - \dfrac{3(x-2)}{x^2-1} \quad x \neq \pm 1$

(vi) $x^2 - y^2$

B. (i) (a) 4   (b) −29   (c) −1
(ii) (a) 0   (b) −45   (c) −1
(iii) (a) 0   (b) 28   (c) 0
(iv) (a) 0   (b) −199 (c) 1

2.3.10 Partial fractions

A. For answers see Q2.3.8A, B.

B. (i) $\dfrac{3}{5(x+2)} + \dfrac{2}{5(x-3)}$ (ii) $\dfrac{2}{x-1} - \dfrac{2}{x+1}$

(iii) $\dfrac{4}{3(x-3)} - \dfrac{1}{3x}$ (iv) $\dfrac{1}{(x-1)^2} - \dfrac{1}{x-1} - \dfrac{3}{x+2}$

(v) $\dfrac{3}{2(x+1)} - \dfrac{3}{2}\dfrac{(x-1)}{x^2+1}$

(vi) $\dfrac{1}{4(x-2)} + \dfrac{3}{4(x-2)^2} + \dfrac{1}{4}\dfrac{(x+5)}{x^2+4}$

(vii) $-\dfrac{1}{6(x-1)} - \dfrac{1}{14(x+3)} + \dfrac{5}{28(x-4)}$

2.3.11 Properties of quadratic expressions and equations

A. (i) $(x-1)(x+2)$ (ii) $(x+3)^2$ (iii) $(x-9)(x+9)$
   (iv) $(x+3)(2x-1)$ (v) $2x(x-4)$

B. (i) $1, -2$ (ii) -3 $(2\times)$ (iii) ± 9
   (iv) $-3, \frac{1}{2}$ (v) $0, 4$

C. (i) $(x+1)^2 + 1$ (ii) $(x-3)^2 + 2^2$ (iii) $4\left(x+\frac{1}{2}\right)^2 - 3^2$
   (iv) $4\left(x-\frac{1}{2}\right)^2 - 1$

D. (i) Minimum of 3 when $x = -1$
   (ii) Maximum of 17 when $x = -\frac{1}{2}$

E. (i) $-1, -2$ (ii) $2, \frac{1}{2}$ (iii) $\frac{2}{3}, 3$
   (iv) $-2, -8$ (v) $2, -4$ (vi) $-1, \frac{1}{3}$

F. 2

G. (i) (a) -3 (b) 2 (ii) (a) $\frac{5}{2}$ (b) 1 (iii) (a) $\frac{11}{3}$ (b) 2
   (iv) (a) -10 (b) 16 (v) (a) -2 (b) -8 (vi) (a) $\frac{2}{3}$ (b) $-\frac{1}{3}$

2.3.12 Powers and indices for algebraic expressions

A. (i) a^6 (ii) a^6 (iii) a^2 (iv) a^8 (v) a^7 (vi) a^{-1} (vii) a^8 (viii) 1
B. (i) 13 (ii) 5 (iii) 9 (iv) 11
C. (i) $ab^2 c$ (ii) $a^5 b^{-2} c^{12}$ (iii) $a^{11} b^6 c$ (iv) $a^{13} bc^2$
D. (i) $a^3 b^8$ (ii) $\dfrac{t^2}{x} = t^2 x^{-1}$ (iii) $x^6 y^3$

2.3.13 The binomial theorem

A. (i) 35 (ii) 1215 (iii) 80
   (iv) 90720 (v) 4860

B. (i) $1 - 14x + 84x^2 - 280x^3 + 560x^4 - 672x^5 + 448x^6 - 128x^7$
   (ii) $a^6 + 6a^5 b + 15a^4 b^2 + 20a^3 b^3 + 15a^2 b^4 + 6ab^5 + b^6$
   (iii) $32x^5 + 240x^4 y + 720x^3 y^2 + 1080x^2 y^3 + 810xy^4 + 243y^5$

(iv) $64 - 576x + 2160x^2 - 4320x^3 + 19440x^4 - 2916x^5 + 729x^6$

(v) $243s^5 - 810s^4t + 1080s^3t^2 - 720s^2t^3 + 240st^4 - 32t^5$

C. (i) 14   (ii) 226
D. (i) 1.105   (ii) 0.984

3

Functions and Series

This chapter gathers together a range of topics relating to functions and their general properties. Some elementary work on series is also included here, because they most often arise in the context of expressing or approximating functions.

Prerequisites

It will be helpful if you know something about:

- function notation
- plotting graphs
- use of formulae
- inequalities (7 ◄)
- sequences and series
- binomial theorem for the case of positive integer power (71 ◄)

Objectives

In this chapter you will find:

- function notation and its use
- plotting the graph of a function
- changing the subject in a formula
- odd and even functions
- the composition of two or more functions
- the use of inequalities
- definition and evaluation of the inverse of a function
- definition and use of sigma notation
- summation of a finite geometric series
- summation of an infinite geometric series
- the binomial theorem for negative and fractional exponents

Motivation

You may need the topics of this chapter for:

- using formulae in your engineering subjects
- sketching graphs of simple functions (Chapter 10)

- use of inequalities in such topics as linear programming
- manipulation of functions in other maths topics such as calculus
- summation and use of series in approximations and statistics
- iteration in numerical methods
- infinite series expansions for functions
- approximations using the binomial expansion

A note about rigour

An engineer will probably be more concerned about using mathematics, rather than proof and rigour. However, this chapter contains some topics that do really need careful treatment even to use them. I have tried to make such things as palatable as possible!

3.1 Review

3.1.1 Definition of a function

If $f(x) = \dfrac{x+1}{x^2+2}$ evaluate (i) $f(0)$ (ii) $f(-1)$

3.1.2 Plotting the graph of a function

Choosing perpendicular x-, y-axes and suitable scales plot the graph of the function $y = x^2 - 2x - 3$ for integer values of x in $-2 \leq x \leq 4$. Describe the shape of the function and discuss the points where it crosses the axes, and the minimum point.

3.1.3 Formulae

The focal length f of a mirror is given by

$$\frac{1}{f} = \frac{1}{u} + \frac{1}{v}$$

where u is the lens-object distance and v the lens-image distance. Express v as a function of u and f.

3.1.4 Odd and even functions

Classify the following functions as odd, even or neither.

(i) $3x^3 - x$ (ii) $\dfrac{x^2}{1+x^2}$ (iii) $\dfrac{2x}{x^2-1}$ (iv) $\dfrac{x^2}{x+1}$

3.1.5 Composition of functions

If $f(x) = x + 1$ and $g(x) = \dfrac{1}{x-1}$ express in their simplest form:

(i) $f(g(x))$ (ii) $g(f(x))$

3.1.6 Inequalities

Find the range of values of x for which

$$1 < 4x - 3 < 4$$

3.1.7 Inverse of a function

Find the inverse function of the function

$$f(x) = \frac{x}{x-1} \quad x \neq 1$$

3.1.8 Series and sigma notation

Write out the following series in full

$$\sum_{r=1}^{6} \frac{r}{r+1}$$

3.1.9 Finite series

Sum the finite geometric series

$$\frac{1}{2} + \frac{1}{2^2} + \frac{1}{2^3} + \frac{1}{2^4} + \frac{1}{2^5}$$

3.1.10 Infinite series

Sum the infinite geometric series

$$\frac{1}{2} + \frac{1}{2^2} + \frac{1}{2^3} + \cdots$$

3.1.11 Infinite binomial series

Find the first four terms of the binomial expansion of $(1 + 3x)^{-2}$.

3.2 Revision

3.2.1 Definition of a function

A **function** is a relation which expresses how the value of one quantity, the **dependent variable**, depends on the value of another, the **independent variable**. Formally, this is usually expressed

$$y = f(x)$$

dependent ↑   ↑ independent
variable   variable

where x is some quantity in a particular set of numbers which, when substituted into the function f, produces the quantity y, which may be in a completely different set of numbers.

Example

Suppose $x = n$ is a non-zero integer, then the **reciprocal function** is defined by:

$$y = f(x) = f(n) = \frac{1}{n} \quad n \neq 0$$

and y can be a rational number of magnitude less than one.

The important point about a function is that it must have a single **unique** value, y, for **every** value, x, for which the function is defined.

x is also called the **argument** of the function $f(x)$. The set X of all values for which the function f is defined is called its **domain**. The set of all corresponding values of $y = f(x)$, Y, is called the **range** of $f(x)$. We sometimes express a function as a **mapping** between the sets X, Y denoted $f : X \to Y$. The value of a particular function for a particular value of x, say $x = a$, is called the **image** of a under f, denoted $f(a)$.

Examples

(i) $y = f(x) = x^2 + 1$ is defined for all values of x, and for all such values y is greater than or equal to 1. The image of -1 for this function is

$$f(-1) = (-1)^2 + 1 = 2$$

(ii) $y = g(x) = \dfrac{3}{x-1}$ is defined for all x except $x = 1$, and y can vary over all real non-zero numbers. The image of $x = 0$ for example is $g(0) = \dfrac{3}{0-1} = -3$.

We have already met the common algebraic functions in Chapter 2:

    Linear: $f(x) = ax + b$ (defined for all x)

    Quadratic: $f(x) = ax^2 + bx + c$ (all x)

    Cubic: $f(x) = ax^3 + bx^2 + cx + d$ (all x)

Polynomial: $f(x) = a_n x^n + a_{n-1} x^{n-1} + \cdots + a_1 x + a_0$ (all x)

Reciprocal: $f(x) = \dfrac{1}{x}$ $x \neq 0$

Rational function – for example:

$$f(x) = \frac{x+1}{(x-1)(x+2)} \quad x \neq 1, -2$$

Here the coefficients, $a, b, c, d, a_n, \ldots$ etc. are constants, or more strictly independent of x. Such algebraic functions occur frequently in mechanics and electrical circuits, for example. Rational functions are particularly important in control theory, where the roots of the denominators (the so-called **poles**) have physical meaning for the stability of a system.

The other standard functions – exponential and trigonometric – will be studied later. They all reflect some important characteristic physical behaviour – for example the exponential function represents the 'law of natural growth' whereby a quantity increases at a rate proportional to the amount present. The trigonometric functions $\sin x$ and $\cosine x$ describe wavelike behaviour such as occurs in an oscillating electrical circuit.

The form $y = f(x)$ defines an **explicit form** of a function of x. It is sometimes convenient to define a function **implicitly**. For example the reciprocal $y = 1/x$ may be defined implicitly as $xy = 1$.

Another form of representation of a function is by means of a parameter in terms of which both x and y are expressed – this is called a **parametric representation**. For example the position (x, y) of a projectile in a plane at time t may be expressed by $x = 3 + 2t$, $y = 4 + 2t - t^2$.

Solution to review question 3.1.1

If $f(x) = \dfrac{x+1}{x^2 + 2}$ then

(i) $f(0) = \dfrac{0+1}{0^2 + 2} = \dfrac{1}{2}$

(ii) $f(-1) = \dfrac{-1+1}{(-1)^2 + 1} = 0$

3.2.2 Plotting the graph of a function   ◄ 88   108 ►

We can learn a lot about a function $y = f(x)$ from its **graph** plotted against x-, y-axes, in a plane **Cartesian coordinate system**. We choose an appropriate sequence of values of x and calculate the corresponding values of y and plot the resulting **x, y coordinates** against the axes. We may then be able to draw a continuous curve through the points. Figure 3.1 illustrates this for the straight line graph of the linear function $y = 2x + 1$. For this we only need two points, say (0, 1) and (1, 3), to define the resulting straight line graph.

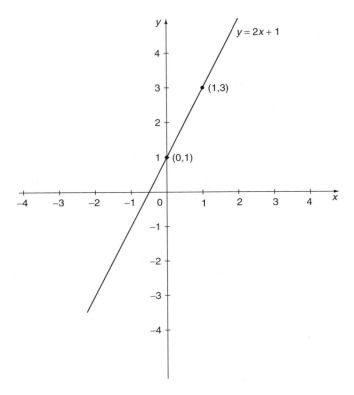

Figure 3.1 The graph of $y = 2x + 1$.

Like the linear function all the common functions have characteristic graphs. Some are given in the reinforcement exercises and others you will meet as they arise in the book. One important point to note is that here we are talking about **plotting** rather than **sketching** a graph (see Chapter 10).

Solution to review question 3.1.2

Evaluating the function $y = x^2 - 2x - 3$ at the given values of x yields the coordinates:

x	-2	-1	0	1	2	3	4
y	5	0	-3	-4	-3	0	5

The graph drawn through these points is shown in Figure 3.2. Remember to label your axes!
This is a quadratic function. It crosses the x-axis where $x^2 - 2x - 3 = 0$. We can solve this quadratic equation by factorisation:

$$x^2 - 2x - 3 = (x - 3)(x + 1) = 0$$

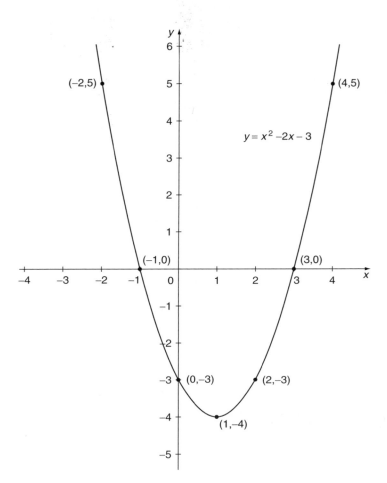

Figure 3.2 The graph of $y = x^2 - 2x - 3$.

to give $x = 3$ and $x = -1$ as the crossing points ('gateways' through the x-axis), as we observe from the graph. Also, we can complete the square to get:

$$x^2 - 2x - 3 = (x - 1)^2 - 4$$

which tells us that the minimum value is -4 and occurs at $x = 1$. Again, we see this from the graph. (Note that any quadratic curve – or **parabola** – is symmetric about its minimum or maximum point.)

3.2.3 Formulae

Perhaps the most common way you have been exposed to functions before is in the use of **formulae** to express the value of some variable in terms of other given quantities. For example for an ideal gas pressure P can be expressed in terms of the volume V and temperature T by the **ideal gas law**:

$$P = \frac{nRT}{V}$$

where nR is a constant dependent on the gas. If T is fixed then we can regard this formula as expressing P as a function f, say, of V:

$$P = f(V)$$

If V is fixed then P can be regarded as a (different) function, g, of T:

$$P = g(T)$$

If V and T can vary then we can regard P as a **function, F, of the two variables T and V** (we study such functions in Chapter 16):

$$P = F(T, V)$$

So the term **function** is really a precise formulation of the idea of a formula.

In the above examples it may, for example, be that P and T are known and it is required to find V. We do this by rearranging or **transposing** the formula to give V as a function of P and T – in this case:

$$V = \frac{nRT}{P}$$

We say this is **making V the subject of the formula**.

Solution to review question 3.1.3

Given that

$$\frac{1}{f} = \frac{1}{u} + \frac{1}{v}$$

then if we know u and f we can obtain v as follows:

$$\frac{1}{v} = \frac{1}{f} - \frac{1}{u} = \frac{u - f}{uf}$$

So

$$v = \frac{uf}{u - f}$$

making v the subject of the formula.

3.2.4 Odd and even functions

An **even function** is one which is unchanged when the sign of its argument changes,

$$f(-x) = f(x)$$

Examples include:

$$f(x) = 5x^2, \quad 3x^4 + x^2$$

i.e. polynomials with only even powers of x.

An **odd function** changes sign with its argument

$$f(-x) = -f(x)$$

Examples are $f(x) = 3x$, $2x^3 - x$, i.e. polynomials with only odd powers of x.

The trig function $\cos x$, studied in Chapter 6 is even:

$$\cos(-x) = \cos x$$

On the other hand $\sin x$ is odd:

$$\sin(-x) = -\sin x$$

The graph of an even function is symmetric about the y-axis, while the graph of an odd function is unchanged under a rotation of $180°$ – see Figure 3.3.

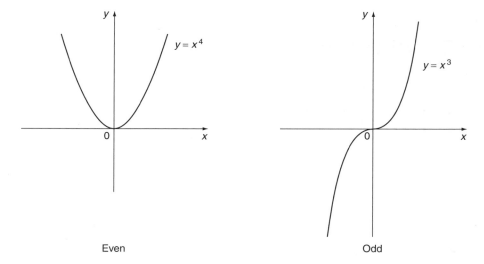

Figure 3.3 Even and odd functions.

Any function $f(x)$ for which $f(-x)$ exists can be expressed as the sum of an even and an odd function:

$$f(x) = \underbrace{\frac{f(x) + f(-x)}{2}}_{\text{even}} + \underbrace{\frac{f(x) - f(-x)}{2}}_{\text{odd}}$$

note the 'something for nothing' trick of adding zero in the form

$$0 = \frac{f(-x)}{2} - \frac{f(-x)}{2}$$

We did a similar thing in completing the square (66 ◀).
The **modulus function** $f(x) = |x|$ is defined by (8 ◀)

$$|x| = x \text{ if } x \geq 0$$
$$= -x \text{ if } x < 0$$

Its graph is shown in Figure 3.4 and it is clearly even.

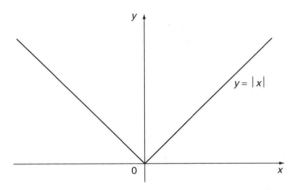

Figure 3.4 The modulus function $y = |x|$.

Solution to review question 3.1.4

(i) For $f(x) = 3x^3 - x$ we have
$$f(-x) = 3(-x)^3 - (-x) = -3x^3 + x$$
$$= -(3x^3 - x) = -f(x)$$
So $3x^3 - x$ is an odd function.

(ii) For $f(x) = \dfrac{x^2}{1 + x^2}$
$$f(-x) = \frac{(-x)^2}{1 + (-x)^2} = \frac{x^2}{1 + x^2} = f(x)$$
so this is even.

(iii) If $f(x) = \dfrac{2x}{x^2 - 1}$ then
$$f(-x) = \frac{2(-x)}{(-x)^2 - 1} = -\frac{2x}{x^2 - 1} = -f(x)$$
so $f(x)$ is odd.

(iv) If $f(x) = \dfrac{x^2}{x + 1}$ we have
$$f(-x) = \frac{(-x)^2}{-x + 1} = \frac{x^2}{1 - x}$$
This is not equal to $f(x)$ or $-f(x)$ and so this function is neither odd nor even.

3.2.5 Composition of functions

Often we need to combine functions. For example in differentiating a function such as $\frac{1}{x^2+1}$ it is useful to regard it as a **function** (the reciprocal) **of a function** (x^2+1) (Chapter 8). More specifically, if $f(x)$ and $g(x)$ are functions with appropriate domains and ranges we can define their **composition** $f(g(x))$ which is the result of evaluating f at the values given by $g(x)$ for each value of x. Note that $f(g(x))$ and $g(f(x))$ are different in general.

Example

$$f(x) = x^2 + 1, \quad g(x) = \frac{1}{x}.$$

$$f(g(x)) = \left(\frac{1}{x}\right)^2 + 1 = \frac{1}{x^2} + 1$$

while

$$g(f(x)) = \frac{1}{f(x)} = \frac{1}{x^2 + 1}$$

> **Solution to review question 3.1.5**
>
> (i) $f(g(x)) = g(x) + 1 = \dfrac{1}{x-1} + 1 = \dfrac{x}{x-1}$.
>
> (ii) $g(f(x)) = \dfrac{1}{f(x) - 1} = \dfrac{1}{x+1-1} = \dfrac{1}{x}$.
>
> Remember to tidy up your result at the end.

3.2.6 Inequalities

We introduced inequalities in Section 1.2.2. Their basic properties are the following:

- If $a < b$ and $c < d$ then $a + c < b + d$

That is, if a is less than b and c is less than d then the sum of a and c will be less than the sum of b and d. This is a fairly obvious property of numbers in general, whether positive or negative.

- If $a < b$ and $c > d$ then $a - c < b - d$

Perhaps the easiest way to see this is to plot the numbers a, b, c, d satisfying the above conditions, on the real line. Then, $a - c$ is the 'distance' between a and c and $b - d$ is the 'distance' between b and d, and it is then pictorially obvious that the former is less than the latter. Or, since $a < b$ we can write $a = b - p$ where p is a positive number and we can similarly write $c = d + q$ where q is another positive number. Then $a - c = b - p - (d + q) = b - d - (p + q)$ from which, since $p + q$ is positive, it follows that $a - c < b - d$, as required.

- If $a < b$ and $b < c$ then $a < c$

This property is another fairly obvious one for real numbers – if Jack is shorter than Jill and Jill is shorter than John then Jack is shorter than John!

- If $a < b$ then $a + c < b + c$

Again fairly obvious – Jack doesn't become any taller than Jill if they both stand on the same chair.

- If $a < b$ and $c > 0$ then $ac < bc$

This just says that if you multiply two numbers a, b by a **positive** number c then the order relation is unchanged – if you look at Jack and Jill through the same telescope, Jack is still shorter than Jill.

- If $a < b$ and $c < 0$ then $ac > bc$

This is usually where our troubles start with inequalities, when we bring in negative multipliers. This property says that changing the sign throughout reverses the inequality. It is pretty obvious if you just think of an example: $4 > 3$, but $-4 < -3$ and $-8 < -6$, and plotting a, b, $-a$ and $-b$ on the real line may make it clear generally. Jack and Jill can help by hanging from the ceiling – Jack is now 'higher than Jill'! More rigorously, note that we can write $b = a + p$ where p is a positive number and so if c is a negative number then $bc = ac + pc$, so $ac = bc - pc$. But if p is positive and c is negative, $-pc$ is positive, from which it follows that $ac > bc$.

- If $a < b$ and $ab > 0$ then $\dfrac{1}{a} > \dfrac{1}{b}$

Our troubles are mounting now – reciprocals and inequalities cause problems for most of us. Again it helps to think of numbers: $2 < 3$, and $\frac{1}{2} > \frac{1}{3}$ **but** while $-2 < 3$, $\frac{1}{-2} = -\frac{1}{2} < \frac{1}{3}$. So the difference in sign affects the inequality when we take reciprocals. The condition $ab > 0$ is included to make sure that a and b have the same sign, both positive, or both negative. Only then does the inequality reverse if we take the reciprocals. The contortions that Jack and Jill have to get up to to convince you of this don't bear thinking about. Perhaps it's best if you simply satisfy yourself with a few examples!

Most of the above results may be extended, with care, to $\leq$ and $\geq$.

Inequalities can be used to specify **intervals** of numbers, for which a special notation is sometimes used. Specifically, $[a, b]$ denotes the **closed** interval $a \leq x \leq b$ and (a, b) denotes the **open** interval $a < x < b$.

Sometimes we have a problem in which we need to find values of x such that a given function $f(x)$ satisfies an inequality such as

$$f(x) < 0$$

We call this **solving the inequality**. In such problems we may have to combine algebra, knowledge of the function, and properties of inequalities. It is a common error to try to solve such an inequality by considering the **equality** $f(x) = 0$ instead. Whilst doing this may be useful, remember that it will only give the **boundaries** of the inequality region, and extra work is needed to define the entire region.

An inequality in which $f(x)$ is a linear (quadratic) function is called a **linear (quadratic) inequality**, for example:

$$ax + b > 0 \quad \text{linear inequality}$$
$$ax^2 + bx + c > 0 \quad \text{quadratic inequality}$$

Example

If
$$1 < 3x + 2 < 4$$

then subtracting 2 from each term gives

$$-1 - 2 = -3 < 3x < 4 - 2 = 2$$

So
$$-3 < 3x < 2$$

and so, dividing by the positive quantity 3, we obtain

$$-1 < x < \tfrac{2}{3}$$

Example

Find the range of values of x for which $2x^2 - 3x - 2 > 0$.

It helps to factorise if we can, or find the roots of the quadratic:

$$(x - 2)(2x + 1) > 0$$

This will be true if either:

$$x - 2 > 0 \text{ and } 2x + 1 > 0, \quad \text{i.e. } x > 2$$

or if

$$x - 2 < 0 \text{ and } 2x + 1 < 0, \quad \text{i.e. } x < -\tfrac{1}{2}$$

So $2x^2 - 3x - 2 > 0$ for $x < -\tfrac{1}{2}$ and $x > 2$. You can also see this from the graph of the function (▶RE (Reinforcement exercise) 3.3.2B), and graphical considerations often help in dealing with inequalities generally. However, it is important that you gain practice in the manipulation of inequalities, and the associated algebra.

Care is needed for inequalities involving the modulus. For example, suppose

$$|x - 3| < 2$$

Then this means that

$$-2 < (x - 3) < 2$$
$$\text{or} \quad -2 + 3 = 1 < x < 2 + 3 = 5$$
$$\text{so} \quad 1 < x < 5$$

> **Solution to review question 3.1.6**
>
> If $1 < 4x - 3 < 4$ then adding 3 throughout gives
>
> $$1 + 3 < 4x < 4 + 3$$
>
> or $\quad 4 < 4x < 7$
>
> so $\quad 1 < x < \frac{7}{4}$

3.2.7 Inverse of a function

Suppose we are given some function $y = f(x)$. Then given a value of x we can evaluate, in principle, the corresponding value of y. But often we are given the reverse problem – we know y and want to find the value of x. What we have in fact is an equation for x which we have to solve. We call this **inverting the function** f and we write

$$x = f^{-1}(y)$$

The **inverse** of a function f is the function f^{-1} such that

$$f^{-1}(f(x)) = x$$

i.e.

$$f^{-1} \text{ 'undoes' } f$$

It is important to note that the inverse of a function only exists if the function is 'one-to-one' that is, each value of x gives one value of y and each value of y gives one value of x. Sometimes this means that we have to be careful with the domains of our functions. For example, the function $y = x^2$ is only 'one-to-one' if, say, we restrict x to the domain $x > 0$. Then the inverse function is $x = \sqrt{y}$ or $f^{-1}(x) = \sqrt{x}$.

Example

$$y = f(x) = \frac{1}{x - 2} \quad (x \neq 2)$$

If $y = 4$, find x.
We have

$$4 = \frac{1}{x - 2}$$

so

$$4(x - 2) = 1 = 4x - 8$$

and therefore

$$4x = 9$$

and

$$x = \frac{9}{4}$$

In fact we can do this generally for any value of y by finding the inverse function of f:

$$y = f(x) = \frac{1}{x-2}$$

Multiplying through by $x - 2$ gives

$$y(x-2) = 1 = yx - 2y$$

and so
$$yx = 1 + 2y$$

and solving for x we obtain

$$x = \frac{1+2y}{y}$$

But if $y = f(x)$ then by definition $x = f^{-1}(y)$ so

$$x = f^{-1}(y) = \frac{1+2y}{y}$$

We may also write this as $f^{-1}(x) = \frac{1+2x}{x}$, with $x \neq 0$ of course.

We can now check the numerical example given above

$$f^{-1}(4) = \frac{1 + 2 \times 4}{4} = \frac{9}{4}$$

as we found previously.

The inverse of a function is often as important as the function itself. For example the inverse of the **exponential function** is the **logarithm function** (Chapter 4). Graphically, the graph of the inverse function can be obtained by reflecting the graph of the function in the straight line $y = x$.

Note that the -1 in $f^{-1}(x)$ does not mean forming the reciprocal – i.e. $f^{-1}(x) \neq 1/f(x)$.

> **Solution to review question 3.1.7**
>
> If $y = f(x) = \dfrac{x}{x-1}$, then the inverse of $f(x)$ is given by $x = f^{-1}(y)$.
> So, writing
>
> $$y = \frac{x}{x-1}$$
>
> we have $y(x-1) = x$ or $(y-1)x = y$.
> So
>
> $$x = \frac{y}{y-1} = f^{-1}(y)$$
>
> Hence
>
> $$f^{-1}(x) = \frac{x}{x-1} \qquad x \neq 1$$

> Here the inverse function is exactly the same as the original function. There is nothing strange about this, but it does emphasise the fact that the inverse of a function is **not** the same thing as the reciprocal of a function!

3.2.8 Series and sigma notation

We will say more about series (particularly infinite series) in Chapter 14. Here we want to introduce the basic ideas by looking at two types of infinite series, the **geometric** and the **binomial**. In fact, these two are extremely important, in both the principles and practice of series, and they occur frequently in applications in science and engineering.

A **series** is a sum of a **sequence** (i.e. a list) of terms which may be finite or infinite in number, for example

$$2 + 4 + 8 + 16 + 32 \equiv 2 + 2^2 + 2^3 + 2^4 + 2^5$$

is a **finite** series. Dealing with finite series is essentially algebra.

Infinite series are usually implicitly defined by indicating that the series continues in a particular way, for example:

$$1 + \frac{1}{2} + \frac{1}{4} + \frac{1}{8} + \frac{1}{16} + \frac{1}{32} + \cdots + \underset{\uparrow}{\frac{1}{2^{n-1}}} + \cdots$$

nth term

A useful shorthand for writing such series is the **sigma notation**. The Greek capital letter sigma $\sum$ denotes summation. Thus, if a_r ('a subscript r') denotes some mathematical object, such as a number or algebraic expression, then the expression

$$\sum_{r=m}^{n} a_r = a_m + a_{m+1} + \cdots + a_{n-1} + a_n$$

is 'the sum of all a_r as r goes from m to n'.

The letter r is called the **index of the summation**, while m, n are called the **lower** and **upper limits** of summation respectively. Note that the letter r is in fact a **dummy index** – it can be replaced by any other:

$$\sum_{r=m}^{n} a_r = \sum_{i=m}^{n} a_i$$

for example, in this respect r is similar to the integration variable in definite integration (Chapter 9):

$$\int_a^b f(x)\,dx = \int_a^b f(t)\,dt$$

Example

The general nth degree polynomial (43 ◀) may be written in sigma notation as:

$$a_n x^n + a_{n-1} x^{n-1} + \cdots + a_1 x + a_0 = \sum_{r=0}^{n} a_r x^r$$

Solution to review question 3.1.8

The series in full is:

$$\sum_{r=1}^{6} \frac{r}{r+1} = \frac{1}{1+1} + \frac{2}{2+1} + \frac{3}{3+1} + \frac{4}{4+1} + \frac{5}{5+1} + \frac{6}{6+1}$$

$$= \frac{1}{2} + \frac{2}{3} + \frac{3}{4} + \frac{4}{5} + \frac{5}{6} + \frac{6}{7}$$

Note that such series can be represented in many different ways in terms of the sigma notation. For example the above series could be written just as well as

$$\sum_{r=0}^{5} \frac{r+1}{r+2}$$

3.2.9 Finite series

◀ 89   109 ▶

The sigma notation can be used to write series, as for example in Section 3.2.8:

$$\sum_{r=1}^{7} \frac{1}{2^{r-1}} = \frac{1}{2^0} + \frac{1}{2^1} + \frac{1}{2^2} + \frac{1}{2^3} + \frac{1}{2^4} + \frac{1}{2^5} + \frac{1}{2^6}$$

$$= 1 + \frac{1}{2} + \frac{1}{4} + \frac{1}{8} + \frac{1}{16} + \frac{1}{32}$$

The most well known and useful elementary series is the sum of a **geometric progression** (GP):

$$a, ar, ar^2, ar^3, \ldots, ar^{n-1}, \ldots$$

where a is the **first term** and r the **common ratio**. Notice that the nth term is ar^{n-1}. Summing a finite geometric progression uses a nice argument. Let

$$S_n = a + ar + ar^2 + \cdots + ar^{n-1}$$

be the sum to n terms of such a GP. Now multiply the series through by r to get

$$r S_n = ar + ar^2 + ar^3 + \cdots + ar^n$$

Then on subtracting the two series we obtain

$$S_n - r S_n = (1-r) S_n = ar - ar^n$$

103

because all terms but the first and last cancel out.
So
$$S_n = \frac{a(1-r^n)}{1-r}$$

Example

For the series

$$\sum_{r=1}^{5} \frac{1}{3^{r-1}} = 1 + \frac{1}{3} + \frac{1}{9} + \frac{1}{27} + \frac{1}{81}$$

$$a = 1, \quad r = \frac{1}{3}, \quad S_5 = \frac{1\left(1-\left(\frac{1}{3}\right)^5\right)}{1-\frac{1}{3}} = \frac{121}{81}$$

You can practice your skills in handling fractions by checking this directly.

Another interesting series is the **arithmetic series**

$$S = a + (a+d) + (a+2d) \ldots$$

where each successive term is formed by adding a 'common difference' d, so that the nth term is $a + (n-1)d$. In this case we have

$$S_n = a + (a+d) + (a+2d) + \cdots + (a+(n-1)d)$$

and a neat way to sum this is to reverse the series and add corresponding terms of the two results:

$$S_n = a + (a+d) + (a+2d) + \cdots + (a+(n-1)d)$$
$$= (a+(n-1)d) + (a+(n-2)d) + \cdots + a$$

So

$$2S_n = (2a+(n-1)d) + (2a+(n-1)d) + \cdots + (2a+(n-1)d)$$
$$= n(2a+(n-1)d)$$

and therefore

$$S_n = \tfrac{1}{2}n(2a+(n-1)d)$$

for the arithmetic series.

Solution to review question 3.1.9

The sum to n terms of the finite geometric series with common ratio r, first term a is

$$S_n = \frac{a(1-r^n)}{1-r}$$

So, for the series

$$\frac{1}{2} + \frac{1}{2^2} + \frac{1}{2^3} + \frac{1}{2^4} + \frac{1}{2^5}$$

we have $a = \frac{1}{2}$ and $n = 5$, and therefore the sum is

$$S_5 = \frac{\frac{1}{2}\left(1 - \left(\frac{1}{2}\right)^5\right)}{1 - \frac{1}{2}} = 1 - \left(\frac{1}{2}\right)^5 = 1 - \frac{1}{32} = \frac{31}{32}$$

3.2.10 Infinite series

◀ 89   110 ▶

Merely the mention of the term 'infinite' sets alarm bells ringing with many of us. An infinite series is one whose terms continue forever, or indefinitely. We tend to indicate this by such notation as

$$S_\infty = 1 + \frac{1}{3} + \frac{1}{9} + \frac{1}{27} + \cdots$$

where the dots (technically called an **ellipsis** – a set of three dots indicating an omission of something understood) infer that the terms continue forever following the same pattern, which it is hoped you have spotted. In this case we see that the **nth term** is $\left(\frac{1}{3}\right)^{n-1}$.

Of course, if we are adding up an infinite number of terms we might expect that the sum would yield an infinite amount (i.e. a number as large as we please). This would certainly be the case for $1 + 1 + 1 + \cdots$ for example. Maybe if the terms get smaller and smaller this would not be so? Maybe, for example, the series

$$1 + \tfrac{1}{2} + \tfrac{1}{3} + \tfrac{1}{4} + \cdots$$

'converges' to a finite total? After all, the terms seem to fade away to zero 'eventually'. In fact the answer is no – it in fact adds up to an infinite amount, as we shall see in Chapter 14, by one of the prettiest proofs of elementary mathematics. In fact the question of the 'convergence' of such infinite series is a very difficult topic belonging to an area of mathematics called **analysis**. We will leave such questions until Chapter 14. For the moment we will take the pragmatic view that if we can find an expression for the sum of a series by fair means then that series converges and we can use the sum with safety.

In particular, we can **sum a GP to infinity** provided that the common ratio, r, satisfies $|r| < 1$. We have for the infinite GP:

$$S_\infty = a + ar + ar^2 + ar^3 + \cdots = \sum_{n=1}^{\infty} ar^{n-1}$$

The sum of the first n terms of this is, from Section 3.2.9

$$S_n = \frac{a(1 - r^n)}{1 - r} = \frac{a}{1 - r} - \frac{ar^n}{1 - r}$$

The only place n occurs here is in the r^n. Now, if $|r| < 1$, then r^n 'tends to' zero as n gets larger and larger, i.e. r^n tends to zero as n tends to infinity, or in symbols, $r^n \to 0$ as $n \to \infty$ if $|r| < 1$. We will look further into such 'limits' in Chapter 14. So as n tends to infinity the second term in expression for the sum S_n goes to zero and the sum to infinity is

$$S_\infty = \frac{a}{1-r}$$

Note that if $r \geq 1$ then the series will not converge to a finite quantity because r^n increases indefinitely in value as n increases if $|r| > 1$, while $a/(1-r)$ does not exist at all if $r = 1$.

Example

Consider the series we started with:

$$S_\infty = 1 + \frac{1}{3} + \frac{1}{9} + \frac{1}{27} + \cdots = 1 + \frac{1}{3} + \left(\frac{1}{3}\right)^2 + \left(\frac{1}{3}\right)^3 + \cdots$$

Here the first term is 1 and the common ratio is $\frac{1}{3}$:

$$a = 1, \quad r = \tfrac{1}{3}$$

so

$$S_\infty = \frac{1}{1 - \tfrac{1}{3}} = \frac{3}{2}$$

Solution to review question 3.1.10

Using the result that $S_\infty = a/(1-r)$, for the sum of an infinite GP with first term $a = \tfrac{1}{2}$ and common ratio $r = \tfrac{1}{2}$ we have

$$S_\infty = \frac{\tfrac{1}{2}}{1 - \tfrac{1}{2}} = 1$$

3.2.11 Infinite binomial series

Another important type of infinite series is that obtained by applying the binomial theorem with negative or fractional power. Referring back to Section 2.2.13 we can write for the binomial expansion for **n an integer**

$$(1+x)^n = 1 + nx + \cdots + \frac{n(n-1)}{2!}x^2 + \frac{n(n-1)(n-2)}{3!}x^3$$
$$+ \cdots + \frac{n(n-1)\ldots(n-r+1)}{r!}x^r + \cdots$$

Now we have certainly not proved that this holds if n is **not an integer**. For example if n is a negative integer, or a fraction, then it is by no means obvious that the result will still hold. However, it can be shown that it does hold, but there is one dramatic difference if n is not a positive integer. In this case the series will never terminate because only positive integers are subtracted from n and $n - r + 1$ will never become zero because r is an integer whereas n is not. So the expansion becomes an infinite series. In such cases it is called an **infinite power series** in x. Such series can be used to express functions in a convenient form for calculation, and we return to them in Chapter 14. For now, just accept the formal generalisation to non-integer values of n. It can be shown that the resulting infinite series **converges** if $|x| < 1$, **diverges** if $|x| > 1$ and if $|x| = 1$ it may diverge or converge.

Example

$$\frac{1}{1-x} = (1-x)^{-1} = 1 + x + x^2 + x^3 + \cdots \quad \text{provided } |x| < 1$$

In this case it is obvious that the series will diverge for $|x| \geq 1$.

This series is a standard result of such importance that you should memorise it. This is not too difficult – it is in fact the sum of an infinite GP with common ratio x. It is useful in making approximations such as

$$\frac{1}{1-x} \simeq 1 + x \text{ for } x \text{ very small}$$

Solution to review question 3.1.11

Applying the binomial expansion

$$(1+x)^n = 1 + nx + \cdots + \frac{n(n-1)}{2!}x^2 + \frac{n(n-1)(n-2)}{3!}x^3$$
$$+ \cdots + \frac{n(n-1)\ldots(n-r+1)}{r!}x^r + \cdots$$

we have

$$(1+3x)^{-2} = 1 + (-2)(3x) + \frac{(-2)(-3)}{2!}(3x)^2$$
$$+ \frac{(-2)(-3)(-4)}{3!}(3x)^3 + \cdots$$

Note the careful treatment of the signs and the $3x$ – it is usually a good policy to keep such things in brackets until the final stages of the calculation. Tidying up the results we find for the first four terms

$$(1+3x)^{-2} = 1 - 6x + 27x^2 - 108x^3 + \cdots$$

3.3 Reinforcement

3.3.1 Definition of a function

◀◀ 88   90 ◀

A. Find the values of the following functions at the points (a) -1, (b) 0, (c) 3 (d) $-\frac{1}{3}$.

(i) $f(x) = 2x$ (ii) $f(x) = 3x^2 - 1$

(iii) $f(x) = 2x^2 - x + 1$ (iv) $f(x) = \dfrac{1}{x+2}$ $x \neq -2$

(v) $f(x) = \dfrac{3x-1}{2x+1}$ $x \neq -\dfrac{1}{2}$

B. If $f(x) = \dfrac{x^2 - 1}{x + 2}$ obtain expressions for

(i) $f(u)$ (ii) $f(t+1)$ (iii) $f\left(\dfrac{s}{t}\right)$

(iv) $f(a+h)$ (v) $f(x + \delta x)$

3.3.2 Plotting the graph of a function   ◀◀88 91◀

A. Plot the graphs of the following functions over the ranges indicated:

(i) $2x^3$ $-3 \leq x \leq 3$ (ii) $\dfrac{3}{x}$ $-6 \leq x \leq 6$ (iii) 2^x $-4 \leq x \leq 4$

B. Plot the graph of the function $f(x) = 2x^2 - 3x - 2$ and confirm the statement made in Section 3.3.6 about the range of values of x for which this function is positive.

3.3.3 Formulae   ◀◀88 93◀

The following are examples of standard formulae occurring in engineering, with standard notation. In each case define the variables, explain what the formula tells us, and describe the type of function that the left-hand side is of the bracketed variables. Make the indicated variable the subject of the formula.

(i) $V = IR$ (R) (ii) $P = I^2 R$ (R)
(iii) $E = \tfrac{1}{2}mv^2$ (v) (iv) $s = ut + \tfrac{1}{2}at^2$ (t)
(v) $E = mc^2$ (c)

3.3.4 Odd and even functions   ◀◀88 94◀

State whether the following functions are even, odd or neither.

(i) $2x$ (ii) $3x^2 - 1$
(iii) $2x^3 - x$ (iv) $x^2 + 2x + 1$
(v) $\cos x$ (vi) $x^4 + 2x^2 + 1$
(vii) $\sin x$ (viii) e^x

3.3.5 Composition of functions   ◀◀89 97◀

If $g(x) = x^2 + 1$, $f(x) = \dfrac{x}{x-1}$ determine the compositions $f(g(x))$ and $g(f(x))$.

3.3.6 Inequalities

Find the ranges of values of x for which the following are satisfied:

(i) $2x - 3 > 2$

(ii) $\dfrac{2x}{x - 3} < 1$

(iii) $x^2 - x + 1 \geq 3$

(iv) $x^2 + 2x + 2 < 5$

(v) $|2x - 1| \leq 2$

(vi) $\dfrac{2}{|x - 4|} < 4$

(vii) $\dfrac{1}{x - 1} < -3$

3.3.7 Inverse of a function

Find the inverse functions for each of the following functions, specifying the values for which they exist.

(i) $2x + 1$

(ii) $\dfrac{x - 1}{x + 2}$, $x \neq -2$

(iii) $x^2 + 1$, $x \geq 0$

3.3.8 Series and sigma notation

A. Write the following in sigma notation:

(i) $1 + 8 + 27 + 64 + 125$

(ii) $3 + 6 + 9 + 12 + 15 + \cdots + 99$

(iii) $\dfrac{1}{2} + \dfrac{1}{3} + \dfrac{1}{4} + \dfrac{1}{5} + \cdots + \dfrac{1}{50}$

(iv) $1 - \dfrac{1}{3} + \dfrac{1}{9} - \dfrac{1}{27} + \cdots$

B. Write down the first four terms in the series:

(i) $\displaystyle\sum_{r=1}^{\infty} \dfrac{1}{r}$

(ii) $\displaystyle\sum_{r=0}^{20} r!$

(iii) $\displaystyle\sum_{r=1}^{\infty} \dfrac{1}{r(r+1)}$

3.3.9 Finite series

Sum the geometric series:

(i) $1 + \dfrac{1}{2} + \dfrac{1}{4} + \dfrac{1}{8} + \cdots + \dfrac{1}{32}$

(ii) $\displaystyle\sum_{n=1}^{6} (0.1)^n$

(iii) $\displaystyle\sum_{n=1}^{8} 2^n$

(iv) $\displaystyle\sum_{n=1}^{4} \left(\dfrac{1}{3}\right)^n$

(v) $1 + 0.1 + 0.01 + 0.001 + 0.0001$

3.3.10 Infinite series

A. Identify which are geometric sequences, and sum them to infinity.

(i) $1, 2, 3, 4, \ldots$ (ii) $-1, 3, 5, 9, \ldots$

(iii) $1, 1, 1, 1, \ldots$ (iv) $\frac{1}{2}, \frac{1}{3}, \frac{1}{4}, \frac{1}{5}, \ldots$

(v) $1, \frac{1}{3}, \frac{1}{9}, \frac{1}{27}, \ldots$ (vi) $0.1, 0.3, 0.5, 0.7, \ldots$

(vii) $2, -4, 8, -16, \ldots$ (viii) $0.2, 0.04, 0.008, 0.0016, \ldots$

B. Find the sum of the infinite geometric series with first terms and common ratios given respectively by

(i) $1, 2$ (ii) $2, \frac{1}{2}$

(iii) $-1, 1$ (iv) $1, \frac{1}{3}$

3.3.11 Infinite binomial series

Find the first four terms of the binomial expansion of the following:

(i) $(1+x)^{-1}$ (ii) $(1-3x)^{-1}$

(iii) $(1+4x)^{-2}$ (iv) $(1-x)^{\frac{1}{2}}$

3.4 Applications

1. For each of the following functions, obtain simplest forms of the expressions for

(a) $f(a+h) - f(a)$ (b) $\dfrac{f(a+h) - f(a)}{h}$

(i) $2x$ (ii) x^2 (iii) $4x^3 - 6x^2 + 3x - 1$

In each case give the result when you let h tend to zero. These sorts of calculations are required in **differentiation from first principles** (Chapter 8).

2. In statistics the **mean**, $\bar{x}$, and **variance**, σ^2, of a set of n numbers $x_1, x_2, \ldots, x_n$ are defined by

$$\bar{x} = \frac{1}{n}\sum_{i=1}^{n} x_i \qquad \sigma^2 = \frac{1}{n}\sum_{i=1}^{n}(x_i - \bar{x})^2$$

respectively. Show that σ^2 may be alternatively written as

$$\sigma^2 = \frac{1}{n}\left(\sum_{i=1}^{n} x_i^2 - n\bar{x}^2\right)$$

3. The following summation is used in the multiplication of **matrices** (see Chapter 13):

$$\sum_{k=1}^{n} a_{ik} b_{kj}$$

where a_{ik}, b_{kj}, are elements of a matrix. Write out the summations in full in the following cases.

(i) $n = 2$ (ii) $n = 3$, $i = 1$, $j = 2$
(iii) $n = 4$, $i = 3$, $j = 3$ (iv) $n = 4$, $i = 2$, $j = 4$

Repeat the exercise for

$$\sum_{l=1}^{n} a_{il} b_{lj}$$

with k replaced by l.

4. A spring vibrates 50 mm in the first oscillation and subsequently 85% of its previous value in each succeeding oscillation. How many complete oscillations will occur before the vibration is less than 10 mm and how far will the spring have travelled in this time? Assuming that the spring can perform an infinite sequence of oscillations, how far does it move before finally stopping? Repeat these calculations with 85% replaced by a general factor of $k\%$ and various initial and final conditions.

5. The emitter efficiency, γ, in an n-p diode is given by

$$\gamma = \frac{I_n}{I_n + I_p}$$

where I_p is the hole current and I_n is the electron current crossing the emitter-base junction. In practice, I_p is made much smaller than I_n. Show that in this case

$$\gamma \simeq 1 - \frac{I_p}{I_n}$$

6. In the next chapter we will meet the **exponential function**. The definition of this important function is perhaps a little more sophisticated than you are used to, either by a **limit** or by a **series**. The connection between these definitions relies on a use of a binomial expansion of a particular kind. Show that if n is an integer (➤ 126):

$$\left(1 + \frac{x}{n}\right)^n = 1 + x + \frac{1\left(1 - \frac{1}{n}\right)}{2!} x^2 + \frac{1\left(1 - \frac{1}{n}\right)\left(1 - \frac{2}{n}\right)}{3!} x^3 + \cdots$$

By letting n 'go to infinity' – i.e. letting n become as large as we like – obtain a series expression for the limit

$$\lim_{n \to \infty} \left(1 + \frac{x}{n}\right)^n$$

This is in fact the definition of the exponential function e^x.

Answers to reinforcement exercises

3.3.1 Definition of a function

A. (i)   (a) -2 (b) 0 (c) 6 (d) $-\frac{2}{3}$

(ii)   (a) 2 (b) -1 (c) 26 (d) $-\frac{2}{3}$

(iii)   (a) 2 (b) 1 (c) 16 (d) $\frac{14}{9}$

(iv)   (a) 1 (b) $\frac{1}{2}$ (c) $\frac{1}{5}$ (d) $\frac{3}{5}$

(v)   (a) 4 (b) -1 (c) $\frac{8}{7}$ (d) -6

B. (i) $\dfrac{u^2-1}{u+2}$ (ii) $\dfrac{t(t+2)}{t+3}$ (iii) $\dfrac{s^2-t^2}{t(s+2t)}$

(iv) $\dfrac{a^2+2ah+h^2-1}{a+h+2}$ (v) $\dfrac{x^2+2x\delta x+(\delta x)^2-1}{x+\delta x+2}$

3.3.2 Plotting the graph of a function

A. (i)

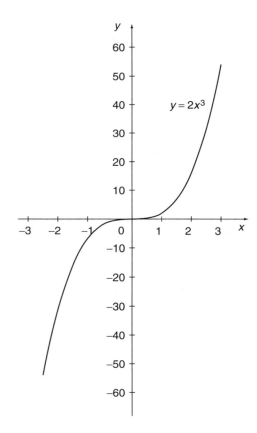

112

(ii)

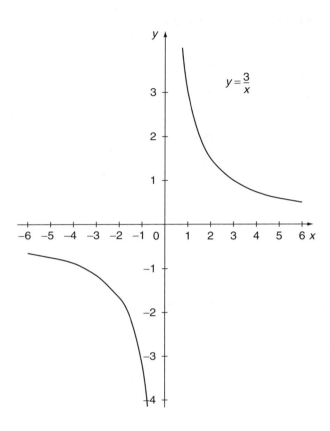

(iii)

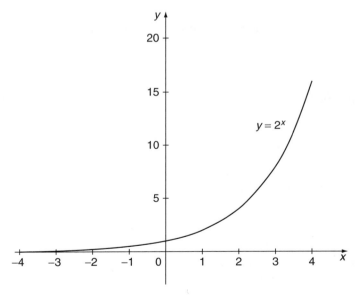

B.

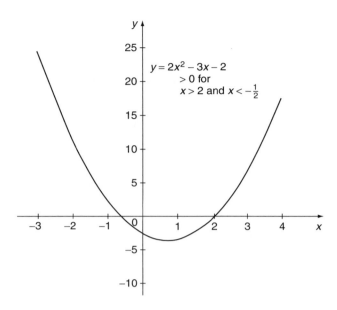

3.3.3 Formulae

(i) Ohm's law; V = voltage, I = current, R = resistance. V is a linear function of R – in fact it is proportional to R.

$$R = \frac{V}{I}$$

(ii) Electrical power P, resistance R, current I. P is proportional to R.

$$R = \frac{P}{I^2}$$

(iii) Kinetic Energy E, m = mass, v = velocity. E is a quadratic function of v.

$$v = \pm\sqrt{\frac{2E}{m}}$$

(iv) Distance – time function; s = distance, t = time, u = initial velocity, a = acceleration. s is a quadratic function of t.

$$t = \frac{-u \pm \sqrt{u^2 + 2as}}{a}$$

(v) Einstein's mass – energy formula; E = energy, m = mass, c = speed of light (positive!). E is proportional to c^2.

$$c = \sqrt{\frac{E}{m}}$$

3.3.4 Odd and even functions

(i) odd (ii) even (iii) odd
(iv) neither (v) even (vi) even
(vii) odd (viii) neither

3.3.5 Composition of functions

$$f(g(x)) = \frac{x^2+1}{(x^2+1)-1} = \frac{x^2+1}{x^2}$$

$$g(f(x)) = \left(\frac{x}{x-1}\right)^2 + 1 = \frac{2x^2-2x+1}{(x-1)^2}$$

3.3.6 Inequalities

(i) $x > \frac{5}{2}$ (ii) $-3 < x < 3$ (iii) $x \geq 2$, $x \leq -1$
(iv) $-3 < x < 1$ (v) $-\frac{1}{2} \leq x \leq \frac{3}{2}$ (vi) $x > \frac{9}{2}$, $x < \frac{7}{2}$
(vii) $\frac{2}{3} < x < 1$

3.3.7 Inverse of a function

(i) $\dfrac{x-1}{2}$ domain: all values; range: all values

(ii) $\dfrac{1+2x}{1-x}$ domain: all values $\neq 1$; range: all values $\neq -2$

(iii) $\sqrt{x-1}$ domain: any value ≥ 1; range: any value ≥ 0

3.3.8 Series and sigma notation

A. (i) $\displaystyle\sum_{n=1}^{5} n^3$ (ii) $\displaystyle\sum_{n=1}^{33} 3n$ (iii) $\displaystyle\sum_{n=2}^{50} \frac{1}{n}$

(iv) $\displaystyle\sum_{n=0}^{\infty} (-1)^n \frac{1}{3^n}$

Note that alternative forms are permissible – for example an equally acceptable answer to (iii) is $\displaystyle\sum_{n=1}^{49} \frac{1}{n+1}$.

B. (i) $1 + \dfrac{1}{2} + \dfrac{1}{3} + \dfrac{1}{4}$ (ii) $1+1+2+6$ (note $0! = 1$)

(iii) $\dfrac{1}{2} + \dfrac{1}{6} + \dfrac{1}{12} + \dfrac{1}{20}$

3.3.9 Finite series

(i) $a = 1$, $r = \dfrac{1}{2}$ $\qquad$ $S_6 = 2\left(1 - \dfrac{1}{2^6}\right) = \dfrac{63}{32}$

(ii) $a = 0.1$, $r = 0.1$ $\qquad$ $S_6 = \dfrac{1}{9}\left(1 - \dfrac{1}{10^6}\right) = 0.1111111$

(iii) $a = 2$, $r = 2$ $\qquad$ $S_8 = 2(2^8 - 1) = 510$

(iv) $a = \dfrac{1}{3}$, $r = \dfrac{1}{3}$ $\qquad$ $S_4 = \dfrac{1}{2}\left(1 - \dfrac{1}{3^4}\right) = \dfrac{40}{81}$

(v) $a = 1$, $r = 0.1$ $\qquad$ $S_5 = \dfrac{10}{9}\left(1 - \dfrac{1}{10^5}\right) = 1.1111$

3.3.10 Infinite series

A. $S_\infty = \dfrac{a}{1 - r}$

(i) No
(ii) No
(iii) Yes, diverges to ∞
(iv) No
(v) Yes $(a = 1, r = \tfrac{1}{3})$ $S_\infty = \tfrac{3}{2}$
(vi) No
(vii) Yes $(a = 2, r = -2)$ diverges
(viii) Yes $(a = 0.2, r = 0.2) S_\infty = \tfrac{1}{4}$

B. (i) ∞ $\qquad$ (ii) 4 $\qquad$ (iii) diverges to $-\infty$
(iv) $\tfrac{3}{2}$

3.3.11 Infinite binomial series

(i) $1 + (-1)x + \dfrac{(-1)(-2)(x)^2}{2!} + \dfrac{(-1)(-2)(-3)(x)^3}{3!}$
$= 1 - x + x^2 - x^3$

(ii) $1 + (-1)(-3x) + \dfrac{(-1)(-2)(-3x)^2}{2!} + \dfrac{(-1)(-2)(-3)(-3x)^3}{3!}$
$= 1 + 3x + 9x^2 + 27x^3$

(iii) $1 + (-2)(4x) + \dfrac{(-2)(-3)(4x)^2}{2!} + \dfrac{(-2)(-3)(-4)(4x)^3}{3!}$

$= 1 - 8x + 48x^2 - 256x^3$

(iv) $1 + \left(\dfrac{1}{2}\right)(-x) + \dfrac{\left(-\dfrac{1}{2}\right)\left(\dfrac{1}{2}\right)(-x)^2}{2!} + \dfrac{\left(\dfrac{1}{2}\right)\left(-\dfrac{1}{2}\right)\left(-\dfrac{3}{2}\right)(-x)^3}{3!}$

$= 1 - \dfrac{x}{2} - \dfrac{x^2}{8} - \dfrac{x^3}{16}$

4

Exponential and Logarithm Functions

The exponential function is one of the most important in engineering. It describes behaviour that is of a rapidly increasing or decreasing nature – for example bacterial growth or radioactive decay. Yet the exponential function is found to be one of the most troublesome for newcomers to engineering mathematics. This chapter therefore introduces the key ideas quite gently.

Prerequisites

It will be helpful if you know something about:

- powers and indices (70 ◄)
- plotting graphs (91 ◄)
- irrational numbers (6 ◄)
- binomial theorem (106 ◄)
- inverse of a function (100 ◄)

Objectives

In this chapter you will find

- functions of the form $y = a^n$ where a is given and n is an integer
- the general exponential function a^x
- the natural exponential function e^x
- manipulation of the exponential function
- logarithms to general base a
- manipulation of logarithms
- some applications of logarithms

Motivation

You may need the topics of this chapter for:

- the exponential form of complex numbers (Chapter 12)
- solving differential equations (Chapter 15)
- modelling engineering, scientific and other situations where some quantity grows or decays very rapidly – 'exponentially'
- converting power laws to linear form

4.1 Review

4.1.1 $y = a^n$, $n =$ an integer

(i) Plot the values of 2^n for $n = -4, -3, -2, -1, 0, 1, 2, 3, 4$ using rectangular Cartesian axes with n on the horizontal axis and 2^n on the vertical axis.

(ii) Repeat (i) with $\left(\frac{1}{2}\right)^n = 2^{-n}$ and compare the results.

(iii) Sketch the graph of (a) $y = 2^x$, (b) $y = 2^{-x}$, (c) $y = 3^x$, (d) $y = 3^{-x}$ on rectangular x-, y-axes.

4.1.2 The general exponential function a^x

A. Use the laws of indices to show that if $f(x) = a^x$, where a is a positive constant, then

$$f(x) \cdot f(y) = f(x+y)$$

What is $f(x-y)$?

B. Simplify the following ($a > 0$)

(i) $a^x a^{-x}$

(ii) $\dfrac{a^{3x} a^x}{a^{2x}}$

(iii) $\dfrac{(a^x)^3 a^{-2x}}{(a^4)^x}$

(iv) $\dfrac{a^{x^2} a^{-2x}}{a^{(x-1)^2}}$

(v) $\dfrac{2^x 16^{-3x}}{4^{2x} 8^{-x+1}}$

4.1.3 The natural exponential function e^x

A. Define the base of natural logarithms, e. Write down the value of e to 3 decimal places. Can you write down the exact value of e?

B. Given that $e^x = 2$, evaluate

(i) e^{2x}

(ii) e^{-x}

(iii) e^{3x}

(iv) $e^{4x} - 4e^{2x}$

C. Plot the graphs of $y = e^x$, e^{-x}, e^{2x}, e^{-3x} on the same axes.

4.1.4 Manipulation of the exponential function

A. Express each of the following as a single exponential

(i) $e^A e^B$

(ii) e^A / e^{-B}

(iii) $e^{2B}(e^{3B})^2$

(iv) $\dfrac{e^A e^{2B} e^{-C}}{e^{2A} e^B e^C}$

(v) $e^{-B} e^{-C} e^B$

(vi) $(e^A)^3 e^{-2A}$

B. Express in simplest form

(i) $\dfrac{e^{2A} - e^{2B}}{e^A + e^B}$

(ii) $(e^A - e^{-A})(e^A + e^{-A})$

(iii) $\dfrac{e^{2A} + 1}{e^{2A} + e^{-2A} + 2}$

(iv) $(e^A + e^{-A})^2 - (e^A - e^{-A})^2$

4.1.5 Logarithms to general base ▶ 130 137 ▶▶

Evaluate (ln denotes logs to base e)

(i) $\log_{10} 1$
(ii) $\log_2 2$
(iii) $\log_3 27$
(iv) $\log_2 \left(\tfrac{1}{4}\right)$
(v) $\log_a (a^4)$
(vi) $\log_a (a^x)$
(vii) $\log_3 1$
(viii) $\ln e$
(ix) $\ln \sqrt{e}$
(x) $\ln e^2$
(xi) $\ln 1$

4.1.6 Manipulation of logarithms ▶ 131 137 ▶▶

A. Express each of the following as a single logarithm ($\ln x$ is to base e, $\log_a x$ to base a)

(i) $\ln x + 2 \ln y$
(ii) $3 \ln x - 4 \ln y$
(iii) $2 \ln x - 3 \ln(2x) + 4 \ln x^3$
(iv) $3 \log_a x + 2 \log_a x^2$
(v) $a \log_a x + 3 \log_a (ax)$

B. If $\log_2 x = 6$, what is $\log_8 x$?

C. If $\ln y = 2 \ln x^{-1} + \ln(x - 1) + \ln(x + 1)$ obtain an expression for y explicitly in terms of x, stating any conditions required on x, y.

4.1.7 Some applications of logarithms ▶ 134 138 ▶▶

A. Solve the equation $2^{x+1} = 5$ giving x to three decimal places.

B. By making an appropriate transformation of the variables convert the equation

$$y = 3x^6$$

to one which has the forms of a straight line – i.e. a **linear form**. What is the gradient and the intercept of the line?

4.2 Revise

4.2.1 $y = a^n$, $n =$ an integer ◀ 119 136 ▶

The exponential function is essentially a power function in which the exponent is the variable. As such it obeys all the usual rules of indices. We can get some idea of the behaviour of the exponential function by looking at the behaviour of the power function for different values of the index, as shown in the review question.

Solution to review question 4.1.1

(i) If you have done RE 3.3.2A(iii) then you have already met this. The values of $y = 2^n$ are:

n	-4	-3	-2	-1	0	1	2	3	4
2^n	$\frac{1}{16}$	$\frac{1}{8}$	$\frac{1}{4}$	$\frac{1}{2}$	1	2	4	8	16

The corresponding graph (Figure 4.1) rises very steeply (note that the scales on the axes differ).

(ii) The graph for $y = 2^{-n}$ is also shown in Figure 4.1.

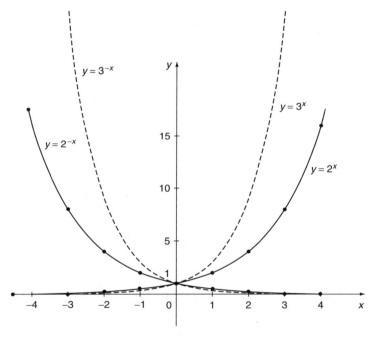

Figure 4.1 Graphs of exponential functions e^{2x}, e^{3x}, e^{-2x}, e^{-3x}.

(iii) Drawing a smooth curve through the points for $y = 2^n$ gives the curve for $y = 2^x$. Similarly for $y = 2^{-x}$. $y = 3^x (3^{-x})$ will be similar but will increase (decrease) more rapidly (see Figure 4.1).

4.2.2 The general exponential function a^x

In Section 2.2.12 we covered the laws of indices, including such things as $a^m \times a^n = a^{m+n}$. Although it was not the intention, it was perhaps easy to get the impression that the indices

121

m, n are 'constant' and to think in terms of a **power function** x^n, etc. where the base, x, is variable and n is given. However, it is just as possible to let the base be fixed and let the power be any variable. This gives a function of the form

$$f(x) = a^x \quad a \neq 0$$

where the **base** a is now regarded as constant, and the **exponent** x can vary. Such a function is called **exponential**. But of course x is still just an index, and obeys all the usual rules of indices (70 ◄). One important point to note at this stage is that since x can take **fractional** values it is essential that a be **positive** to deliver real values of the function for all values of x. x itself can of course be positive or negative. The results of Section 4.2.1 give us a feel for the shape of the graph of exponential functions. Basically such graphs can take three different forms:

- base $a > 1$, the graph increases steadily from left to right. We say a^x **increases monotonically** with x;
- $a < 1$ the graph decreases steadily from left to right $-a^x$ **decreases monotonically** with x;
- if $a = 1$ we get the straight line $y = 1$.

These results are illustrated in the graphs of Figure 4.2.

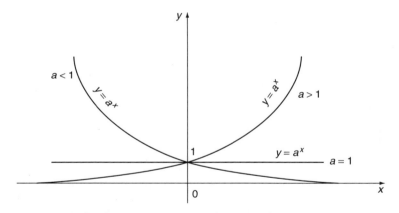

Figure 4.2 The exponential functions a^x, a^{-x}.

The exponential function satisfies all the usual laws of indices, which are worth repeating here in the new notation:

$$a^x a^y = a^{x+y}$$
$$\frac{a^x}{a^y} = a^{x-y}$$
$$(a^x)^y = a^{xy}$$
$$(ab)^x = a^x b^x$$
$$a^{-x} = \frac{1}{a^x}$$

We need to say something about the definition of a^x for different values of x, building on the work in Section 1.2.7. If $x = n$ is a positive integer, then for any real number a we know that we can define

$$a^x = a^n \equiv \underbrace{a \times a \times \cdots \times a}_{n \text{ factors}}$$

simply in terms of elementary algebraic operations.

Similarly, if $x = -n$ is a negative integer then we can define

$$a^x = a^{-n} = (a^{-1})^n = \underbrace{a^{-1} \times a^{-1} \times \cdots \times a^{-1}}_{n \text{ factors}}$$

And of course, by definition

$$a^0 = 1$$

In these definitions there is no need for any qualification on the real number a – it can be positive, negative, rational, irrational.

We can extend the definition of a^x to the case when x is a fraction or rational number by defining $a^{\frac{1}{q}}$, where q is an integer, as the positive real number such that when raised to the power q it yields a:

$$(a^{\frac{1}{q}})^q \equiv a$$

We say $a^{\frac{1}{q}}$ is the **qth root** of a, sometimes written $\sqrt[q]{a}$. However, complications arise if a is negative, so henceforth we insist that a is positive, then a^x will always be a real function.

For $x = p/q$ a rational number we can now define a^x by

$$a^x = a^{\frac{p}{q}} = (a^{\frac{1}{q}})^p$$

So, provided x is a rational number, and $a > 0$, the **exponential function**, a^x, is a well defined function of x. The extension to the case when x is an **irrational number** is not a trivial step, but we will assume that it can be done and therefore that a^x is in fact defined for all real values of x, rational or irrational, provided a is positive.

Solution to review question 4.1.2

A. If $f(x) = a^x$ then $f(y) = a^y$ and so

$$f(x)f(y) = a^x a^y = a^{x+y} = f(x + y)$$
$$f(x - y) = a^{x-y} = a^x/a^y = f(x)/f(y)$$

B. We simply apply the rules of indices but with the indices as functions of x.

(i) $a^x a^{-x} = a^{x-x} = a^0 = 1$

(ii) $\dfrac{a^{3x}a^x}{a^{2x}} = a^{3x+x-2x} = a^{2x}$

(iii) $\dfrac{(a^x)^3 a^{-2x}}{(a^4)^x} = \dfrac{a^{3x}a^{-2x}}{a^{4x}} = a^{-3x}$

(iv) $\dfrac{a^{x^2} a^{-2x}}{a^{(x-1)^2}} = a^{x^2 - 2x - (x-1)^2} = a^{-1}$ using a bit of elementary algebra

(v) $\dfrac{2^x 16^{-3x}}{4^{2x} 8^{-x+1}} = 2^x 2^{-12x} 2^{-4x} 2^{3x-3} = 2^{-12x-3}$

4.2.3 The natural exponential function e^x

The most commonly used exponential function – usually referred to as **the exponential function** – is e^x, where e is a number whose value to 16 decimal places (!) is $e \simeq 2.7182818284590452$. e is called the **base of natural logarithms**. It is an irrational number. Like π, it cannot be expressed as a fraction, or as a terminating or repeating decimal. Its decimal part goes on indefinitely. There are a number of equivalent definitions we could give for e^x, but it has to be said that none of them is easy to appreciate at the elementary level. Below we will study in detail perhaps the most natural derivation of an expression for e^x, by considering compound interest. For now, accept the following definitions as a general introduction to the properties of the exponential function.

We will see later that the particular exponential function e^x can be defined by its infinite series

$$e^x = 1 + x + \frac{x^2}{2!} + \frac{x^3}{3!} + \cdots$$

Using this form, if you know some elementary differentiation, then you can discover for yourself why the exponential function is so important:

Problem 1
Differentiate the series for e^x term by term – what do you get? You may assume

$$\frac{d}{dx}(x^n) = nx^{n-1}$$

You should find you get the same series – i.e.

$$\frac{de^x}{dx} = e^x$$

So the derivative of the exponential function is the function itself. This explains the special role played by e^x and its importance in, for example, differential equations (Chapter 15). There are many situations where rate of change is proportional to amount present – e.g. bacterial growth, radioactive decay (e^{-x} is the relevant function in the latter case).

One rigorous method of defining the exponential function e^x is by means of a **limit**. This is quite an advanced concept that we only address in Chapter 14 and there is a temptation to gloss over the idea here. However, the work is so important (and not without interest!) that we will give it a try – don't worry if it is too much for you at this stage, you should be able to handle the exponential function well enough without studying the next couple of pages. However, if you **can** find your way through this work it will be of great benefit to you, as well as providing practice in some techniques of algebra. We are going to approach the limit involved in the exponential function by considering a limiting process arising in the study of **compound interest**. You can try to work out some of the details yourself.

Problem 2

Suppose you borrow £C at an interest of $I\%$ compounded monthly, so that the interest added at the end of each month is at $\dfrac{I}{12}\%$, and you do not pay off any of the loan until after a year
How much debt do you have at the end of the year?

At the end of the first month you owe

$$£\left(C + \frac{I}{12}\frac{C}{100}\right) = £C\left(1 + \frac{I}{1200}\right)$$

At the end of the second month you owe

$$£C\left(1 + \frac{I}{1200}\right)\left(1 + \frac{I}{1200}\right) = £C\left(1 + \frac{I}{1200}\right)^2$$

and so on.
So after 12 months you owe

$$£C\left(1 + \frac{I}{1200}\right)^{12}$$

Now suppose the terms were changed and interest was compounded daily. Then at the end of a non-leap year you would owe

$$£C\left(1 + \frac{I}{36500}\right)^{365}$$

By the hour:

$$£C\left(1 + \frac{I}{876000}\right)^{8760}$$

and I will leave you to give the result if interest is calculated by the minute, or second (RE 4.3.3A).

In general, if the interest is compounded at a constant rate over each of n equal time intervals in the year, then the amount owing at the end of the year will be

$$£C\left(1 + \frac{I}{100n}\right)^n$$

This is all very well – but how do we actually **calculate** this? As n gets larger and larger it gets more and more difficult. In particular, what would happen if n became infinitely large – equivalently, interest is being added continuously. In this case the result can be expressed as a **limit**:

$$£ \lim_{n\to\infty} C\left(1 + \frac{I}{100n}\right)^n$$

Putting $x = \dfrac{I}{100}$ and dropping the constant factor C this shows that a limit of the form

$$\lim_{n\to\infty}\left(1 + \frac{x}{n}\right)^n$$

is very important. This limit is a function of x (because the n disappears on taking the limit), and is **in fact a definition of e^x**:

$$e^x = \lim_{n\to\infty}\left(1 + \frac{x}{n}\right)^n \tag{4.1}$$

This may seem an unusual way to define a function, but it does show how e^x is intimately related to an important process of accumulating compound interest and natural growth and decay in general. In fact it can be shown that e^x defined in this way as a limit does indeed satisfy all the laws of indices (see Section 4.2.4 below).

We therefore take the limit (Equation 4.1) as our definition of the **exponential function** e^x, which is also written $\exp(x)$. This definition can in fact be used to obtain the infinite series for e^x. You can try this yourself, using the binomial theorem:

Problem 3
Show that

$$\left(1+\frac{x}{n}\right)^n = 1 + x + \frac{1\left(1-\frac{1}{n}\right)}{2!}x^2 + \frac{1\left(1-\frac{1}{n}\right)\left(1-\frac{2}{n}\right)}{3!}x^3 + \cdots$$

By the binomial theorem (106, 111 ◄):

$$\left(1+\frac{x}{n}\right)^n = 1 + n\frac{x}{n} + \frac{n(n-1)}{2!}\left(\frac{x}{n}\right)^2 + \frac{n(n-1)\ldots(n-2)}{3!}\left(\frac{x}{n}\right)^3 + \cdots$$

$$= 1 + x + \frac{\left(1-\frac{1}{n}\right)}{2!}x^2 + \frac{1\cdot\left(1-\frac{1}{n}\right)\left(1-\frac{2}{n}\right)}{3!}x^3 + \cdots$$

Problem 4
Hence deduce that e^x can be represented by the infinite series

$$e^x = 1 + x + \frac{x^2}{2!} + \frac{x^3}{3!} + \cdots + \frac{x^r}{r!} + \cdots$$

All we need here is to note that such things as $1/n, 2/n, \ldots$ 'tend to' zero as n 'tends to' infinity, i.e. gets infinitely large. So:

$$e^x = \lim_{n \to \infty} \left(1 + \frac{x}{n}\right)^n$$

$$= \lim_{n \to \infty} \left(1 + x + \frac{1\left(1 - \frac{1}{n}\right)}{2!} x^2 + \frac{1\left(1 - \frac{1}{n}\right)\left(1 - \frac{2}{n}\right)}{3!} x^3 + \cdots \right.$$

$$\left. + \frac{1\left(1 - \frac{1}{n}\right) \cdots \left(1 - \frac{r-1}{n}\right)}{r!} x^r + \cdots \right)$$

$$= 1 + x + \frac{x^2}{2!} + \frac{x^3}{3!} + \cdots + \frac{x^r}{r!} + \cdots$$

On letting all the terms with $1/n, 2/n, \ldots$ tend to zero. Hence we obtain the series form for e^x from the limit definition:

$$e^x = 1 + x + \frac{x^2}{2!} + \frac{x^3}{3!} + \cdots + \frac{x^r}{r!} + \cdots$$

$$= \sum_{r=0}^{\infty} \frac{x^r}{r!}$$

In particular, we now have for $e = e^1$:

$$e = 1 + 1 + \frac{1}{2!} + \frac{1}{3!} + \cdots + \frac{1}{r!} + \cdots$$

We can calculate the value of e to any required accuracy by taking a sufficient number of terms of this series. For example, summing the first 11 terms of the above series (i.e. $r = 10$) gives e as 2.71828 to 5 decimal places. As noted earlier e is an **irrational number**. When your calculator gives you a 'value' for e, it is only an approximation to the available number of decimal places – to give the exact value of e it would need to have a display of infinite length.

The graph of an exponential function is very simple, see Figure 4.3. In view of the way we have derived e^x – by considering the growth of debt, it should now be no surprise to you that the function

$$y = e^x$$

is often described as representing the **law of natural growth**. For example, it might describe unrestrained bacterial growth. The function

$$y = e^{-x} = (e^{-1})^x$$

on the other hand defines the **law of natural decay**. For example it describes the decrease in mass of a radioactive element over time. Its graph is also shown in Figure 4.3.

Solution to review question 4.1.3

A. We may define e by the limit

$$e = \lim_{n \to \infty} \left(1 + \frac{1}{n}\right)^n$$

but it is more easily evaluated from the equivalent series

$$e = 1 + 1 + \frac{1}{2!} + \frac{1}{3!} + \frac{1}{4!} + \cdots + \frac{1}{r!}$$
$$+ \cdots (e^x \text{ with } x = 1)$$

To 3 decimal places $e = 2.718$. We cannot write down the exact numerical value of e, since it is an irrational number (6 ◄). Its decimal never terminates or recurs.

B. (i) We only have to remember that the exponential function behaves like any other power (18 ◄). We have

$$e^{2x} = (e^x)^2 = (2)^2 = 4$$

(ii) $e^{-x} = (e^x)^{-1} = (2)^{-1} = \frac{1}{2}$

(iii) $e^{3x} = (e^x)^3 = (2)^3 = 8$

(iv) We only need (i) now:

$$e^{4x} - 4e^{2x} = (e^{2x})^2 - 4e^{2x} = (4)^2 - 4 \times 4 = 0$$

C. Since $e > 1$, $y = e^x$ is an increasing function of x and e^{-x} is a decreasing function. Similarly, $y = e^{2x}$ is increasing, $y = e^{-3x}$ is decreasing. The graphs are illustrated in Figure 4.3.

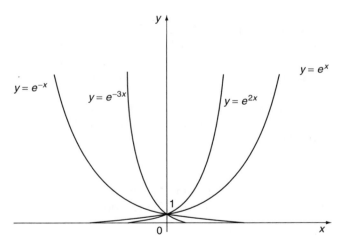

Figure 4.3 The exponential functions e^x, e^{-x}, e^{2x}, e^{-3x}.

4.2.4 Manipulation of the exponential function

As noted earlier, any exponential function, including e^x, satisfies the usual rules of indices (18 ◄):

$$e^x e^y = e^{x+y}$$

$$\frac{e^x}{e^y} = e^{x-y}$$

$$(e^x)^y = e^{xy}$$

$$e^{-x} = \frac{1}{e^x}$$

$$e^0 = 1$$

You need to be very proficient at using these rules. You will frequently need to manipulate functions of exponential functions, whatever area of engineering you enter.

Perhaps the most important thing to remember about the exponential function is that e^{A+B} is **not** equal to $e^A + e^B$. This is an error made by many beginners. The correct result is of course

$$e^{A+B} = e^A e^B$$

Solution to review question 4.1.4

A. (i) $e^A e^B = e^{A+B}$ (ii) $e^A/e^{-B} = e^A e^B = e^{A+B}$

(iii) $e^{2B}(e^{3B})^2 = e^{2B} e^{6B} = e^{8B}$

(iv) $\dfrac{e^A e^{2B} e^{-C}}{e^{2A} e^B e^C} = e^{-A} e^B e^{-2C} = e^{-A+B-2C}$

(v) $e^{-B} e^{-C} e^B = e^{-C}$ using $e^{-B} e^B = 1$

(vi) $(e^A)^3 e^{-2A} = e^{3A} e^{-2A} = e^A$

B. (i) $\dfrac{e^{2A} - e^{2B}}{e^A + e^B} = \dfrac{(e^A)^2 - (e^B)^2}{e^A + e^B}$

$$= \frac{(e^A - e^B)(e^A + e^B)}{e^A + e^B}$$

$$= e^A - e^B$$

(ii) $(e^A - e^{-A})(e^A + e^{-A}) = (e^A)^2 - (e^{-A})^2$

$$= e^{2A} - e^{-2A}$$

(iii) $\dfrac{e^{2A} + 1}{e^{2A} + e^{-2A} + 2} = \dfrac{e^{2A} + 1}{(e^A)^2 + 2 + (e^{-A})^2}$

$$= \frac{e^{2A} + 1}{(e^A + e^{-A})^2}$$

$$= \frac{e^A(e^A + e^{-A})}{(e^A + e^{-A})^2}$$

$$= \frac{e^A}{e^A + e^{-A}} = \frac{e^{2A}}{e^{2A} + 1}$$

(iv) $(e^A + e^{-A})^2 - (e^A - e^{-A})^2 = (e^A)^2 + 2e^A e^{-A}$
$\qquad + (e^{-A})^2 - ((e^A)^2 - 2e^A e^{-A} + (e^{-A})^2)$
$\qquad = 4e^A e^{-A} = 4$

4.2.5 Logarithms to general base

◀ 120 137 ▶

What about the **inverse function** (100 ◀) of the exponential function? That is, if

$$y = a^x$$

then what is x in terms of y?

By definition, we call the inverse of the exponential function the **logarithm to base** a and write

$$x = \log_a y$$

Note that a must be positive if we are to avoid complex numbers and it must not be equal to unity, since 1 raised to any power will again be 1.

In the special case of the exponential e^x, we call the inverse the **natural logarithm**, denoted ln. Thus if

$$y = e^x, \text{ then } x = \log_e y = \ln y$$

An equivalent definition of the **logarithm of a number** x **to base** a is as that power to which the base must be raised to give x. That is,

$$x = a^{\log_a x}$$

In particular

$$x = e^{\ln x}$$

Note that since a^x can never be negative, then $\log_a y$ is only defined for positive values of y. This is sometimes emphasised by writing $\log_a |y|$, although often we omit the modulus signs and simply take it for granted that all real quantities under a logarithm are to be assumed positive.

From the definition of $\log_a x$ it also follows that

$$x = \log_a a^x$$

and in particular

$$x = \ln e^x$$

The two results

$$x = e^{\ln x} \text{ and } x = \ln e^x$$

expressing the fact that the exponential and the log are inverse functions of each other are extremely important in advanced mathematics and are used repeatedly in, for example, the solution of differential equations (Chapter 15).

> **Solution to review question 4.1.5**
>
> It is perhaps easiest to think of $\log_a x$ as the power to which a must be raised to give x – then if we can express x in the form a^y, y will be the value we want.
>
> (i) $\log_{10} 1 = \log_{10} 10^0 = 0$
>
> In fact, log of 1 to any base (except 1!) is zero
>
> (ii) $\log_2 2 = \log_2 2^1 = 1$
>
> (iii) $\log_3 27 = \log_3 3^3 = 3$
>
> (iv) $\log_2 \left(\frac{1}{4}\right) = \log_2(2^{-2}) = -2$
>
> Note that we can have negative logs – we just can't take the log of a negative number and expect a real number.
>
> (v) $\log_a(a^4) = 4$
>
> (vi) $\log_a(a^x) = x$
>
> (vii) $\log_3 1 = \log_3 3^0 = 0$
>
> (viii) $\ln e = \log_e e = 1$
>
> Of course, e is no different to any other log base in this respect.
>
> (ix) $\ln \sqrt{e} = \ln e^{\frac{1}{2}} = \frac{1}{2}$
>
> (x) $\ln e^2 = 2$
>
> (xi) $\ln 1 = 0$

4.2.6 Manipulation of logarithms

Since the log is basically just an index or power, the properties of logs can be deduced from the laws of indices. We find, for any base a (positive and not equal 1):

- $\log 1 = 0$ because $a^0 = 1$
- $\log(xy) = \log x + \log y$

    Put $x = a^s$, $y = a^t$ so $s = \log x$ and $t = \log y$ then $xy = a^{s+t}$ so $s + t = \log x + \log y = \log(xy)$

- $\log(x/y) = \log x - \log y$

    Put $x = a^s$, $y = a^t$ so $s = \log x$ and $t = \log y$ then $x/y = a^{s-t}$ so $s - t = \log x - \log y = \log(x/y)$

- $\log x^\alpha = \alpha \log x$

  Put $x = a^s$, so $s = \log x$, then $x^\alpha = (a^s)^\alpha = a^{s\alpha} = a^{\alpha s}$ and so $\alpha s = \log x^\alpha = \alpha \log x$

The last result holds for any real number α, positive or negative, rational or irrational.

Sometimes we need to change the base of a logarithm. Thus, suppose we have $\log_a x$ and we wish to convert this to a form involving $\log_b x$, $b \neq a$, we have

so
$$\left. \begin{array}{l} y = \log_a x \\ x = a^y \end{array} \right\} \qquad (4.2)$$

and therefore

$$\log_b x = \log_b a^y = y \log_b a$$
$$= \log_a x \, \log_b a$$

So
$$\log_a x = \frac{\log_b x}{\log_b a}$$

In particular, if $x = b$ this gives

$$\log_a b = \frac{1}{\log_b a}$$

For completeness we will anticipate Chapter 8 here and mention that the derivative of the natural log function is simply the reciprocal:

$$\frac{d}{dx}(\ln x) = \frac{1}{x}$$

Equations of the form

$$a^x = b$$

occur frequently in engineering applications and may be solved by using logs. Thus, taking logs to base a we have

$$\log_a(a^x) = \log_a b$$
$$= x \log_a a = x$$

So
$$x = \log_a b$$

The graph of the logarithm function follows easily from that of the exponential function, its inverse, by reflecting the latter in the line $y = x$ (101 ◄). This is shown in Figure 4.4.

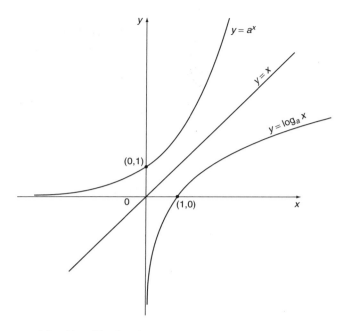

Figure 4.4 The exponential and logarithm functions.

Note that as observed above, $\log_a x$ does not exist for negative values of x.

Solution to review question 4.1.6

A. We simply apply the laws of logarithms given above:

(i) $\ln x + 2 \ln y = \ln x + \ln y^2$
$= \ln(xy^2)$

(ii) $3 \ln x - 4 \ln y = \ln x^3 - \ln y^4$
$= \ln(x^3/y^4)$

(iii) $2 \ln x - 3 \ln(2x) + 4 \ln x^3$
$= \ln x^2 - \ln(8x^3) + \ln x^{12}$
$= \ln \left(\dfrac{x^2 \cdot x^{12}}{8x^3} \right) = \ln \left(\dfrac{x^{11}}{8} \right)$

(iv) $3 \log_a x + 2 \log_a x^2 = \log_a x^3 + \log_a x^4 = \log_a(x^3 \times x^4)$
$= \log_a x^7$

(v) $a \log_a x + 3 \log_a(ax) = \log_a x^a + \log_a(ax)^3$
$= \log_a x^a + \log_a(a^3 x^3)$
$= \log_a(x^a a^3 x^3)$
$= \log_a(a^3 x^{3+a})$

B. If $\log_2 x = 6$ then from the change of base formula

$$\log_a x = \frac{\log_b x}{\log_b a}$$

we have

$$\log_8 x = \frac{\log_2 x}{\log_2 8} = \frac{\log_2 x}{3} = \frac{6}{3} = 2$$

Alternatively, if $\log_2 x = 6$ then $x = 2^6 = (2^3)^2 = 8^2$, so $\log_8 x = 2$, as above.

C. $\ln y = 2\ln x^{-1} + \ln(x-1) + \ln(x+1)$
$= \ln x^{-2} + \ln(x-1) + \ln(x+1)$
$= \ln(x^{-2}(x-1)(x+1))$
$= \ln\left(\frac{x^2-1}{x^2}\right)$

So $y = \dfrac{x^2-1}{x^2}$

4.2.7 Some applications of logarithms ◀ 120 138 ▶

When an unknown occurs in an index in an equation, as for example in

$$a^x = b$$

then we may be able to solve the equation by taking logs. If the base a happened to be e then, of course, the 'natural' thing to do would be to take natural logs, but we can use the same idea whatever the base. Thus, taking natural logs we obtain

$$\ln(a^x) = x \ln a = \ln b$$

so

$$x = \frac{\ln b}{\ln a}$$

Conversely, if an unknown occurs in a logarithm, then we can sometimes solve by taking an exponential. For example, the equation

$$\log_2(x+1) = 3 + \log_2 x$$

can be solved by first gathering the logs together to give

$$\log_2(x+1) - \log_2 x = \log_2\left[\frac{x+1}{x}\right] = 3$$

We can now remove the log by exponentiating with 2 – i.e. raising 2 to the power of each side to give (2 to the power of $\log_2 x$ is x)

$$\frac{x+1}{x} = 2^3 = 8$$

from which we find $x = \frac{1}{7}$.

Logs can also be used to simplify graphical representation of certain functions. Thus, given any function of the form

$$y = kx^\alpha$$

We can take logs to any base and obtain

$$\log y = \log(kx^\alpha)$$
$$= \log k + \alpha \log x$$

If we now put

$$X = \log x \quad Y = \log y$$

then we obtain the equation:

$$Y = \alpha X + \log k$$

If Y is plotted against X on rectangular Cartesian axes, as in Section 3.2.2, then this is a straight line. As we will see in Section 7.2.4 this line has gradient, or slope, α and an intercept on the y axis of $\log k$.

Solution to review question 4.1.7

A. If $2^{x+1} = 5$ then taking natural logs of both sides gives

$$\ln(2^{x+1}) = (x+1)\ln 2 = \ln 5$$

So

$$x + 1 = \frac{\ln 5}{\ln 2} = 2.322$$

to three decimal places, so $x = 1.322$

B. If $y = 3x^6$
then taking logs to any convenient base a we have

$$\log_a y = \log_a(3x^6)$$
$$= \log_a(x^6) + \log_a 3$$
$$= 6\log_a x + \log_a 3$$

Put

$$X = \log_a x \quad Y = \log_a y$$

to get the form of a straight line equation:

$$Y = 6X + \log_a 3$$

The gradient of this line is 6 and its intercept on the y-axis is $\log_a 3$ (➤ 212).

4.3 Reinforcement

4.3.1 $y = a^n$, $n =$ an integer

A. Plot the values of 3^n for $n = -2, -1, 0, 1, 2$, using Cartesian axes with n on the horizontal axis and 3^n on the vertical axis.
B. Plot the graphs of $y = 3^x$ and $y = 4^x$ on the same axes. Sketch the graph of $y = \pi^x$.

4.3.2 The general exponential function a^x

Simplify

(i) $\dfrac{8^x \times 2^{3x}}{4^{3x}}$

(ii) $\dfrac{6^{\frac{x}{2}} \times 12^{x+1} \times 27^{-\frac{x}{2}}}{32^{\frac{x}{2}}}$

(iii) $\dfrac{x^{-\frac{1}{3}} y^{-\frac{2}{3}}}{(x^2 y^4)^{-\frac{1}{6}}}$

(iv) $\dfrac{x^2(x^2+1)^{-\frac{1}{2}} - (x^2+1)^{\frac{1}{2}}}{x^2}$

(v) $\dfrac{e^x e^{-x^2}}{e^{x-1} e^{(x+1)^2}}$

(vi) $\dfrac{a^3 a^{-(x+1)^2}}{a^{-x^2} a^{-2x}}$

(vii) $\dfrac{a^{\cos 2x} a^{-\cos^2 x}}{a^{3-\sin^2 x}}$

4.3.3 The natural exponential function e^x

A. Referring to the 'interesting' Problem 2 in Section 4.2.3, determine the debt owing if interest is reckoned by the (i) minute (ii) second.
B. Use the series for e^x to evaluate to 4 decimal places (i) e, (ii) $e^{0.1}$, (iii) e^2.
C. Sketch the curves

(i) $y = e^x - 1$

(ii) $y = 1 - e^{-x}$

D. Solve the equation

$$e^{2x} - 2e^x + 1 = 0$$

4.3.4 Manipulation of the exponential function

Simplify

(i) $e^A (e^B)^2 e^{3C}$

(ii) $\dfrac{e^x e^{3y} (e^x)^2}{e^{-x} e^y}$

(iii) $\dfrac{(e^x)^3 e^{x^2} (e^y)^2}{e^{4y} e^{x+3}}$

(iv) $\dfrac{(e^B)^2 e^A e^{B-2}}{e^{2-B} e^{2A}}$

(v) $\dfrac{(e^A + e^{-A})^2}{e^{2A}}$

(vi) $\dfrac{(e^A + e^{3B})(e^{-A} + e^{-B})}{e^{2B} e^A}$

4.3.5 Logarithms to general base

A. Find x if

(i) $8 = \log_2 x$ (ii) $3 = \log_2 x$ (iii) $4 = \ln x$

(iv) $6 = \log_3 x$ (v) $4 = \log_3 x$ (vi) $2 = \ln x$

B. Evaluate

(i) $\ln e^3$ (ii) $\log_4(256)$ (iii) $\log_3 27$

(iv) $\log_9 81$ (v) $\log_4 2$ (vi) $\ln(e^2)^2$

(vii) $\ln e^7$ (viii) $\log_3(243)$

4.3.6 Manipulation of logarithms

A. Simplify as a single log

(i) $2\ln e^4 + 3\ln e^3$ (ii) $3\log_2 x + \log_2 x^2$

(iii) $\log_a x + \log_a(2y)$ (iv) $\ln(3x) - \frac{1}{2}\ln(9x^2)$

(v) $2\log_a x + 3\ln x$ (vi) $\log_a x - \log_{2a} x$ (vii) $\ln x + 2\log_a x^2$

B. Expand each as a linear combination of numbers and logs in simplest form

(i) $\ln(3x^2 y)$ (ii) $\log_2(8x^2 y^3)$ (iii) $\ln(e^A/e^B)$

(iv) $\log_a(a^x y^a)$ (v) $\log_{2a}(8a^3 x^2 y^4)$ (vi) $\ln(x^2 y^2 z^2)$

C. Simplify

(i) $\dfrac{a^{\log_a 6} x^2 \ln(e^{3x})}{2ax \log_2(4^x)}$ (ii) $\dfrac{a^{(x-1)^2} a^{2x} a^3}{(a^x)^2 a^{x^2} \ln(e^{2a})}$

D. Evaluate

(i) $\log_2 32$ (ii) $\log_{10} 100$ (iii) $\log_7 49$

(iv) $\log_5 625$ (v) $\log_a a^{\frac{1}{2}}$ (vi) $\ln e^{2001}$

(vii) $\log_{1/8} 64$ (viii) $\ln \dfrac{1}{e}$ (ix) $\log_8 2$

E. Simplify (log to any base)

(i) $\dfrac{\log 81}{\log 9}$ (ii) $\dfrac{\log 8}{\log 2}$ (iii) $\dfrac{\log 49}{\log 343}$

(iv) $5\log 2 - 3\log 32$ (v) $\frac{1}{2}\log 49$

F. Given that

$$\ln \frac{1}{y} = \frac{1}{2}\ln(x+1) - \frac{1}{2}\ln(x-1) + 3x + \ln x + C$$

where C is an arbitrary constant, obtain an explicit expression for y in terms of x.

4.3.7 Some applications of logarithms ◀◀ 120 134 ◀

A. Solve the following equations, giving your answers to 3 decimal places.

(i) $3^x = 16$ (ii) $4^{2x} = 9$ (iii) $4 \times 5^{-2x} = 3 \times 7^{x-2}$ (iv) $3^x = 4^{2x-1}$

B. Convert the following equations to straight line form

(i) $y = 4x^7$ (ii) $y = 3x^{-4}$ (iii) $y = \dfrac{5}{x^3}$ (iv) $y = 20e^{-2x}$ (v) $y = 2^{4x-1}$

4.4 Applications

You will see many applications of the exponential function in later chapters, particularly in complex numbers, differential equations and the Laplace transform.

1. The **hyperbolic functions** cosh, sinh, tanh are defined by

$$\cosh x = \frac{e^x + e^{-x}}{2} \quad \text{(hyperbolic cosine)}$$

$$\sinh x = \frac{e^x - e^{-x}}{2} \quad \text{(hyperbolic sine)}$$

$$\tanh x = \frac{\sinh x}{\cosh x} \quad \text{(hyperbolic tan)}$$

These are frequently occurring functions in engineering – for example the shape of a cable suspended at two ends can be described by the **catenary**, which is essentially the $\cosh x$ curve. The hyperbolic functions obey very similar identities to the trig functions. In particular, show that

(i) $\cosh^2 x - \sinh^2 x = 1$ $(\cosh^2 x = (\cosh x)^2$, etc.).
(ii) $\sinh(A + B) = \sinh A \cosh B + \sinh B \cosh A$.

Use trig identities to suggest other hyperbolic identities and use the above definitions to confirm them.

(iii) Evaluate $\cosh 0$, $\sinh 0$.
(iv) Given that $\dfrac{d}{dx}(e^x) = e^x$ and $\dfrac{d(e^{-x})}{dx} = -e^{-x}$ deduce the derivatives of $\sinh x$ and $\cosh x$.
(v) Plot the graphs of $\sinh x$ and $\cosh x$.

2. The current/voltage characteristic of a rectifying contact in a semiconductor device is given by

$$I = I_o \left[\exp\left(\frac{eV}{kT}\right) - 1 \right]$$

At room temperature kT/e is about 25 mV. Plot the curve of I/I_o for V between -60 and 60 mV in this case. Show that for values of V greater than 75 mV the increase in current I is approximately exponential, i.e

$$I \simeq I_o \exp\left(\frac{eV}{kT}\right)$$

In the general case obtain an expression for V in terms of I.

3. The speed, v, of the signal in a submarine cable depends on the radius R, in mm, of the cable's covering according to

$$v = \frac{16}{R^2} \ln\left(\frac{R}{4}\right)$$

Plot the graph of this function from $R = 4$ mm to $R = 12$ mm. Estimate the maximum speed, and the value of R for which this occurs.

4. According to Benford's law (an American physicist who noticed that in his university library the first pages of the now old fashioned tables of logarithms were more worn than the rest) the probability that a number drawn randomly from a sufficiently smooth range of numbers spread over several orders of magnitudes begins with the digit n is

$$P(n) = \log_{10}(n+1) - \log_{10} n$$

Plot this function for $n = 1$ to 9. Discuss the implications and conduct some experiments to see if the results agree with experience.

Answers to reinforcement exercises

4.3.1 $y = a^n$, $n =$ an integer

A. and B.

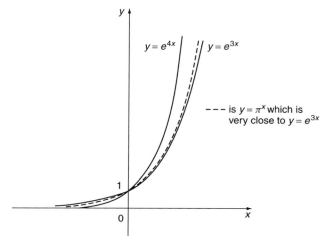

Note that $3 < \pi < 4$.

4.3.2 The general exponential function a^x

(i) 1 (ii) 12 (iii) 1
(iv) $-\dfrac{1}{x\sqrt{x^2+1}}$ (v) $e^{-2x(x+1)}$ (vi) a^2
(vii) a^{-3}

4.3.3 The natural exponential function e^x

A. (i) $£C\left(1+\dfrac{I}{52560000}\right)^{525600}$

(ii) $£C\left(1+\dfrac{I}{3153600000}\right)^{31536000}$

B. (i) 2.7183 (ii) 1.1052 (iii) 7.3891

C.

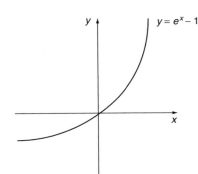
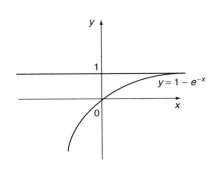

D. $x = 0$.

4.3.4 Manipulation of the exponential function

(i) $e^{A+2B+3C}$ (ii) $e^{2(2x+y)}$ (iii) $e^{x^2+2x-2y-3}$
(iv) $e^{-A+4B-4}$ (v) $(1+e^{-2A})^2$ (vi) $e^{-A-2B}+e^{-3B}+e^{B-2A}+e^{-A}$

4.3.5 Logarithms to general base

A. (i) 256 (ii) 8 (iii) e^4 (iv) 729
(v) 81 (vi) e^2

B. (i) 3 (ii) 4 (iii) 3 (iv) 2
(v) $\frac{1}{2}$ (vi) 4 (vii) 7 (viii) 5

4.3.6 Manipulation of logarithms

A. (i) $\ln e^{17} = 17$ (ii) $\log_2(x^5)$ (iii) $\log_a(2xy)$ (iv) 0
(v) $\ln(x^{(3+2/\ln a)})$ (vi) $\log_a\left(x^{1-\frac{1}{\log_a 2a}}\right)$ (vii) $\ln(x^{1+\frac{4}{\ln a}})$

B. (i) $\ln 3 + 2\ln x + \ln y$ (ii) $3 + 2\log_2 x + 3\log_2 y$
(iii) $A - B$ (iv) $x + a\log_a y$
(v) $3 + 2\log_{2a} x + 4\log_{2a} y$ (vi) $2\ln x + 2\ln y + 2\ln z$

C. (i) $\dfrac{9x}{4a}$ (ii) $\dfrac{a^{-2x+3}}{2}$

D. (i) 5 (ii) 2 (iii) 2
(iv) 4 (v) $\tfrac{1}{2}$ (vi) 2001
(vii) -2 (viii) -1 (ix) $\tfrac{1}{3}$

E. (i) 2 (ii) 3 (iii) $\tfrac{2}{3}$
(iv) $-10\log 2$ (v) $\log 7$

F. $y = \dfrac{1}{x}\sqrt{\dfrac{x-1}{x+1}}\, e^{-3x-C}$

4.3.7 Some applications of logarithms

A. (i) 2.528 (ii) 0.792 (iii) 0.809 (iv) 0.828

B. (i) $\ln y = \ln 4 + 7\ln x$ (ii) $\ln y = \ln 3 - 4\ln x$ (iii) $\ln y = \ln 5 - 3\ln x$
(iv) $\ln y = \ln 20 - 2x$ (v) $\ln y = (4\ln 2)x - \ln 2$

5

Geometry of Lines, Triangles and Circles

One can hardly get more practical than surveying and designing buildings and other structures and geometry underpins all these and much more, such as computer aided design. This chapter consolidates the key areas of basic geometry. It is relatively elementary in content, but not in concepts, some of which are found to give students considerable difficulty. Here these are covered concisely, offering plenty of practice.

> ## Prerequisites
>
> It will be helpful if you know something about:
>
> - ratio and proportion (14 ◄)
> - measurement of angles
> - elementary properties of triangles
> - Pythagoras' theorem
> - area and perimeter of a circle
> - solution of algebraic equations (Chapter 2 ◄)
> - elementary algebra (Chapter 2 ◄)

> ## Objectives
>
> In this chapter you will find:
>
> - division of a line in a given ratio
> - intersecting and parallel lines and angular measurement
> - triangles and their elementary properties
> - congruent triangles
> - similar triangles
> - the intercept theorem
> - the angle bisector theorem
> - Pythagoras' theorem
> - lines and angles in a circle
> - cyclic quadrilaterals

Understanding Engineering Mathematics

> **Motivation**
>
> You may need the material of this chapter for:
>
> - the study of structures – from molecular and crystalline to 'big' engineering
> - computer aided design
> - problems in statics
> - coordinate geometry (Chapter 7)
> - vector algebra and its applications (Chapter 11)
> - applications in differentiation and integration (Chapters 8, 9, 10)

5.1 Review

5.1.1 Division of a line in a given ratio

A line AB of length 30 cm is divided internally by a point P in the ratio 3 : 2. Find the lengths of AP and PB. Repeat for the case when P divides AB externally in the same ratio.

5.1.2 Intersecting and parallel lines and angular measurement

A. Give the angles in (a) degrees (b) radians corresponding to the following fractions of a full revolution

(i) $\dfrac{1}{2}$ (ii) $\dfrac{1}{3}$ (iii) $\dfrac{1}{4}$

(iv) $\dfrac{1}{12}$ (v) $\dfrac{3}{4}$ (vi) $\dfrac{5}{8}$

B. If l_1, l_2 are straight lines, name and evaluate the missing angles.

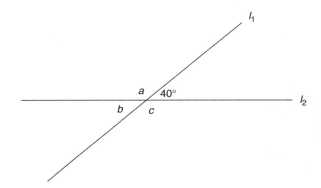

Figure 5.1

C. For (i) and (ii) determine the lettered angles a, b, c, d, e. Equal arrows denote parallel lines.

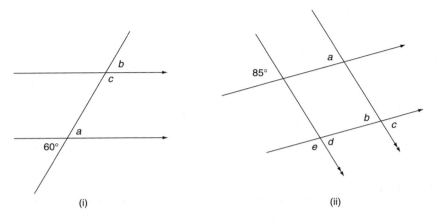

Figure 5.2

5.1.3 Triangles and their elementary properties

> 150 160 >>

Determine the lettered angles. Crossbars on lines denote that they have equal length.

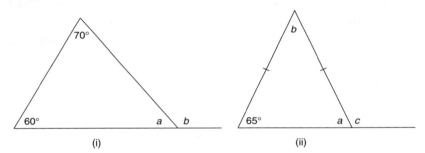

Figure 5.3

5.1.4 Congruent triangles

> 152 161 >>

For each of (i), (ii) state whether the pair of triangles is congruent. Equal angle arcs denote equal angles.

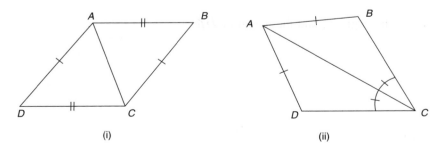

Figure 5.4

5.1.5 Similar triangles

The triangles ABC and DEF are similar.

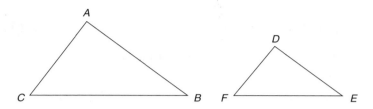

Figure 5.5

If $BC = 5.0$ cm, $AB = 4.0$ cm, $AC = 3.0$ cm and $FE = 3.0$ cm find:

(i) the ratio between the sides of the two triangles;
(ii) the lengths of the remaining sides, DF, DE.

What is the angle BAC?

5.1.6 The intercept theorem

Find the missing length, DB, in each of the figures below:

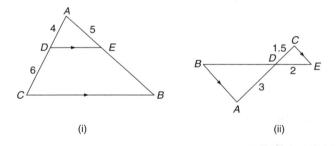

Figure 5.6

5.1.7 The angle bisector theorem

If AD bisects the angle A, find the length of DB.

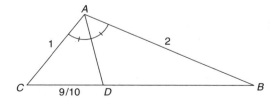

Figure 5.7

5.1.8 Pythagoras' theorem

The largest sides of a right angled triangle have lengths 5 and 6 units. What is the length of the third side?

5.1.9 Lines and angles in a circle

Determine the angles indicated (O is the centre of the circle)

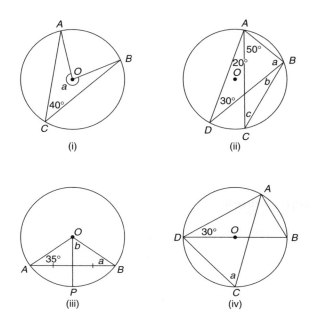

Figure 5.8

5.1.10 Cyclic quadrilaterals

Determine the angles indicated.

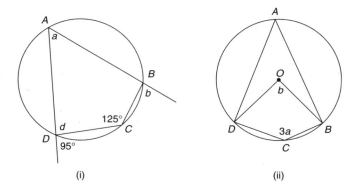

Figure 5.9

5.2 Revision

5.2.1 Division of a line in a given ratio ◀ 143 160 ▶

Points and lines are regarded as **undefined primitive concepts** (that is, we assume we all know what is meant by them) in geometry, in terms of which the **axioms** or rules of geometry may then be expressed. A **point** is a geometrical element which has a position, but no size or extent. A **line** is a straight one-dimensional geometrical figure of infinite length and no thickness. There is a unique straight line passing through two specified points A and B. A **line segment** is a finite portion of a line between two fixed points. Its length is the shortest distance between the points in a plane. Note that we are talking simply about geometry on a plane here, rather than, for example spherical geometry, which deals with geometrical properties on the surface of a sphere.

Much elementary geometry depends on the division of a line by a point. We say a point P on a line AB divides the line **internally in the ratio $p : q$** if P is between A and B and $AP : PB = p : q$, or $\dfrac{AP}{PB} = \dfrac{p}{q}$ (14 ◀).

If a point P is on AB produced (that is, extended in the direction AB), then P is said to divide AB **externally in the ratio $p : q$** if $AP : PB = p : q$ or $\dfrac{AP}{PB} = \dfrac{p}{q}$.

Solution to review question 5.1.1

For division internally we have, letting $AP = x$ cm, say:

```
     x              30 − x
•————————————•————————————•
A              P              B
```

Then $PB = (30 - x)$ cm, and since

$$AP : PB = 3 : 2$$

we have

$$\frac{x}{30-x} = \frac{3}{2} \text{ or } 2x = 3(30-x) = 90 - 3x$$

or

$$5x = 90$$

So $x = 18$ and hence

$$AP = 18 \text{ cm and } PB = 12 \text{cm}.$$

Alternatively, if we let the length of AP be $3y$ units and that of PB $2y$ units, then $3y + 2y = 5y = 30$, from which $y = 6$. Thus, AP is 18 cm and PB 12 cm as before.

For division **externally** let $AP = x$ then, from the figure below,

$$\frac{AP}{PB} = \frac{x}{x-30} = \frac{3}{2}$$

or

$$x = 90$$

So $AP = 90$ cm and $PB = 60$ cm.

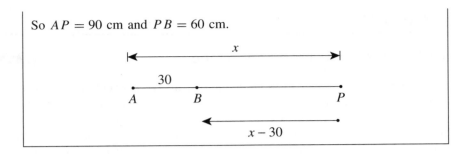

5.2.2 Intersecting and parallel lines and angular measurement

Given two intersecting lines, in a plane, the **angle** between them is the amount of rotation required to superimpose one on the other, measured in some conventional (usually anticlockwise as shown) direction. We denote the angle by $\angle AOB$ (see Figure 5.10).

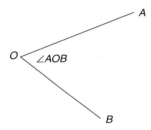

Figure 5.10 Definition of angle.

Sometimes, if it is helpful, we use an arc to denote an angle, but usually we try to avoid cluttering up a diagram. The angle is usually measured in **degrees**, with one degree, denoted $1°$, being 1/360 of a full rotation. Alternatively, we can measure angles in **radians**. One radian (1 rad or 1^c for 'circular measure') is essentially the angle subtended at the centre of a circle by an arc of length equal to the radius. If this rather obtruse definition does little for you, just note that the number of radians in a complete revolution will be the number of times that the radius divides the circumference, which is simply 2π. So there are 2π radians in a complete revolution and therefore

$$2\pi \text{ radians} = 360 \text{ degrees} = 360°$$

Since $180°$ is half a full rotation, it follows that angles which add together to make $180°$ form a straight line – such angles are called **supplementary**.

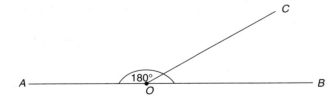

Figure 5.11 Supplementary angles AOC and COB.

In Figure 5.11 $\angle AOC$ and $\angle COB$ are supplementary. An angle of 90° is a rotation through a quarter circle and is called a **right angle**. In diagrams a right angle between two lines is always denoted by a small square at their intersection – if this is not present then you cannot **assume** the angle is 90°, even though it may look like it on the diagram. Angles which add up to 90° are called **complementary angles**. Two lines which intersect at right angles are said to be **perpendicular** to each other. Angles between 0° and 90° are called **acute**. Angles between 90° and 180° are called **obtuse**. Angles greater than 180° are called **reflex**.

When two lines intersect, **vertically opposite angles** are equal:

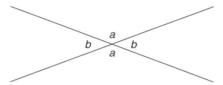

Figure 5.12 Intersecting lines.

Two lines are **parallel** if they do not intersect, no matter how far they are extended – we say they 'meet at infinity (wherever that is)'. We denote parallel lines by equal numbers of arrow-heads as in Figure 5.13. Equal angles are denoted by equal numbers of crossbars on the angle arcs.

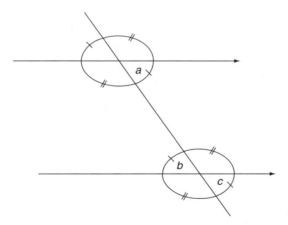

Figure 5.13 Angles on parallel lines.

As noted earlier, remember that we are talking about **plane geometry** here. Parallel lines drawn on the surface of a sphere for example do not satisfy the sorts of properties we will be discussing for parallel lines in a plane.

The line drawn crossing the parallel lines in Figure 5.13 is called a **transversal**. In Figure 5.13 pairs of angles such as a and b, on opposite sides of the transversal are called **alternate angles**, while pairs of angles a and c are called **corresponding angles**. For two parallel lines, as shown in Figure 5.13 alternate angles are equal, as are corresponding angles, i.e. $a = b = c$.

149

Understanding Engineering Mathematics

Solution to review question 5.1.2

A. (a) A full revolution is 360°, so a fraction (12 ◄) $\frac{p}{q}$ of a full revolution is $\frac{p}{q}360°$, i.e.:

(i) $\frac{1}{2} \times 360° = 180°$ (ii) $\frac{1}{3} \times 360° = 120°$ (iii) $\frac{1}{4} \times 360° = 90°$

(iv) 30°   (v) 270°   (vi) 225°

A full revolution is 2π radians, so a fraction $\frac{p}{q}$ of a full revolution is $\frac{p}{q} \times 2\pi$, i.e.

(i) $\frac{1}{2} \times 2\pi = \pi$ radians   (ii) $\frac{1}{3} \times 2\pi = \frac{2\pi}{3}$ radians

(iii) $\frac{1}{4} \times 2\pi = \frac{\pi}{2}$ radians   (iv) $\frac{\pi}{6}$ radians

(v) $\frac{3\pi}{2}$ radians   (vi) $\frac{5\pi}{4}$ radians

B. Referring to Figure 5.1 a and c are both **supplementary angles** to 40° and so $a = c = 140°$. b is the **vertically opposite** angle to 40° and is thus given by $b = 40°$.

C. (i) By vertically opposite angles for intersecting lines we have $a = 60°$ in Figure 5.2(i). Then by corresponding angles $b = a = 60°$. Finally, supplementary angles give $c = 180° - 60° = 120°$.

(ii) Corresponding angles in Figure 5.2(ii) give $a = 85°$, then $b = 85°$, whence opposite angles gives $c = b = 85°$. Corresponding angles then gives $d = c = 85°$. Supplementary angles finally gives $e = 180° - 85° = 95°$.

5.2.3 Triangles and their elementary properties   ◄ 144 160 ►

A **triangle** is any plane figure with three sides formed from line segments. There are three types of interest – a **scalene triangle** has three sides, and angles, of different lengths; an **isosceles triangle** has two sides of equal length standing on a base side with which the two equal sides make equal angles; an **equilateral triangle** has all three sides of equal length and three equal angles (all 60°). A triangle in which all angles are acute is called an **acute-angled triangle**. If one angle is 90° it is called a **right-angled triangle**. If one angle exceeds 90° we say the triangle is **obtuse**.

The most basic property of a plane triangle is that **the sum of its angles is 180°**. (This is **not** true for 'spherical triangles', drawn on the surface of a sphere – can you think of a 'triangle' on a sphere that has 270°?) This property of plane triangles is worth proving in terms of elementary facts of which we are already sure. We can do this for a general triangle ABC, acute-angle or obtuse, as follows. Draw a straight line, through A, parallel to the opposite side BC – see Figure 5.14. Since we already know that for two parallel lines, alternate angles are equal, then we can say $a = d$ and $c = e$.

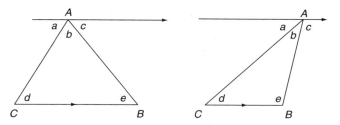

Figure 5.14 Sum of the angles in a triangle.

Also, from angles on a straight line we have

$$a + b + c = 180°$$

Hence

$$d + b + e = 180°$$

which proves the result.

Another important property of triangles is that **an exterior angle of a triangle is equal to the sum of the two opposite interior angles**.

Again, the proof involves additional construction and is not difficult, as shown in Figure 5.15.

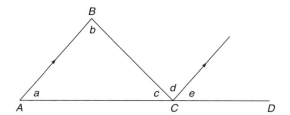

Figure 5.15 Exterior angle.

AC is 'produced' or extended to D. An additional line is also drawn through C parallel to AB. Then, by alternate angles we have $d = b$, and by corresponding angles we have $e = a$. So, $d + e$ = external angle at $C = a + b$, as required, i.e. $\angle BCD = \angle BAC + \angle CBA$.

Solution to review question 5.1.3

(i) Referring to Figure 5.3(i) we have $a = 180° - 70° - 60° = 50°$ by angles in a triangle. Then $b = 70° + 60° = 130°$ by the external angle result, or alternatively by angles on a straight line $b = 180° - a = 180° - 50° = 130°$.

(ii) Since the triangle in Figure 5.3(ii) is isosceles, $a = 65°$. So $b = 180° - 2 \times 65° = 50°$ and $c = b + 65° = 115°$ by the external angle result or by angles on a line $c = 180° - 65° = 115°$.

5.2.4 Congruent triangles

We often need to compare pairs of triangles. Two triangles are called **congruent** if they are identical. They may not **look** identical, because they may be rotated with respect to each other, so we need a test to determine when two triangles are identical. For this, the corresponding angles of the two triangles must be identical, and the corresponding sides must be of the same length. However, it is not necessary to check all these conditions to ensure that two triangles are congruent. The following provide alternative tests for congruent triangles:

- three sides of one triangle must be equal to three sides of the other
- two sides and the angle between them in one triangle must be equal to two sides and the enclosed angle in the other triangle
- two angles and one side in one triangle must be equal to two angles and the corresponding side in the second triangle
- for right-angled triangles, the right angle, hypotenuse and side in the first triangle must be equal to the right angle, hypotenuse and the corresponding side in the second triangle.

Solution to review question 5.1.4

(i) In Figure 5.4(i) AD is equal to BC, and DC is equal to AB. AC is a common side. So ACD and ABC are congruent triangles, since all corresponding sides are equal.

(ii) With AC as a common side and two identical sides AB, AD, the triangles ABC and ACD in Figure 5.4(ii) have two sides of identical length. They also have two equal angles $\angle DCA$ and $\angle ACB$. However, these angles are not those between the pair of identical sides in the two triangles (second test in the text above) so the triangles are not necessarily congruent.

5.2.5 Similar triangles

If one triangle is simply an enlargement or rotation of another triangle, then we say the two triangles are **similar**. So the three angles of one triangle are equal to the three angles of the other, similar triangle, and the corresponding sides of the two triangles are in the same ratio. However, we don't need to test both of these conditions, since it can be shown that they are equivalent, i.e. that **if two triangles contain the same angles then their corresponding sides are in the same ratio**.

Solution to review question 5.1.5

Since ABC and DEF in Figure 5.5 are similar, then the ratios of corresponding sides are equal.

(i) In this case the ratio is equal to $BC/FE = 5/3$.
(ii) Thus, $AB/DE = 4/DE = 5/3$ whence $DE = 12/5$ cm. Similarly(!) $AC/DF = 3/DF = 5/3$, so $DF = 9/5$ cm.

> Angle BAC looks like a right angle, but this doesn't mean that it is. In fact in this case it **is**, because, as you may have noticed, the sides satisfy Pythagoras: $3^2 + 4^2 = 5^2$(➤ 154).

5.2.6 The intercept theorem

◀ 145 162 ▶

The intercept theorem states that **a straight line drawn parallel to one side of a triangle divides the other two sides in the same ratio** (14 ◀) (see Figure 5.16) i.e. $AD/DC = AE/EB$.

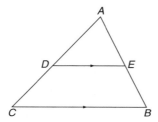

Figure 5.16 The intercept theorem.

Thus, for example, if we know AD, DC, AE say then we can deduce EB.

The proof of this theorem may be found in most standard books on geometry and relies on the fact that ABC and AED are similar triangles.

Solution to review question 5.1.6

(i) By the intercept theorem, since ED is parallel to CB in Figure 5.6(i), the line ED divides sides AC and AB in the same ratio. Hence

$$\frac{AE}{EC} = \frac{4}{6} = \frac{AD}{DB} = \frac{5}{DB}$$

So $DB = 15/2$ units.

(ii) This is the case where the parallel intercept is actually outside the triangle (Figure 5.6(ii)). This in fact makes no difference to the result and again, by the intercept theorem,

$$\frac{AD}{DC} = \frac{3}{1.5} = \frac{BD}{DE} = \frac{BD}{2}$$

So $DB = 4$ units.

5.2.7 The angle bisector theorem

◀ 145 163 ▶

The intercept theorem is about dividing the sides of triangles. Here we consider a theorem which divides an angle of a triangle: **the line bisecting an angle of a triangle divides the side opposite to that angle in the ratio** (14 ◀) **of the sides containing the angle** (see Figure 5.17).

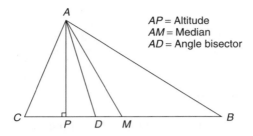

Figure 5.17 Altitude, median and angle bisector.

So the angle bisector theorem states that

$$\frac{AC}{AB} = \frac{CD}{DB}$$

The angle bisector should not be confused with two other lines dropped from a vertex of a triangle – the altitude and the median, also shown in Figure 5.17. The **altitude** is the line drawn from a vertex of a triangle, perpendicular to the opposite side. A **median** of a triangle is the line joining a vertex to the midpoint of the opposite side. You might like to explore the circumstances under which two or more of these lines are in fact the same thing.

Solution to review question 5.1.7

By the angle bisector theorem we have, in Figure 5.7.

$$\frac{AC}{AB} = \frac{CD}{DB}$$

So $DB = \dfrac{CD \times AB}{AC} = \dfrac{2 \times 9/10}{1} = \dfrac{9}{5}$

5.2.8 Pythagoras' theorem

◀ 146 163 ▶

For any right-angled triangle ABC Pythagoras' theorem tells us that

$$(BC)^2 + (AC)^2 = (AB)^2$$

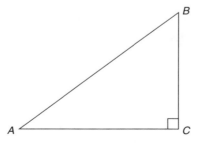

Figure 5.18 Pythagoras' theorem.

We can see that this works by checking it for a few simple triangles – the 3 : 4 : 5 being the classic example. But this is not a **proof** – it is merely **induction** from a few cases, not **deduction** from fundamental axioms. There are numerous proofs of Pythagoras' theorem – we will give one which has the sort of common sense appeal an engineer might enjoy.

Take four identical copies of any right-angled triangle, sides a, b, c and arrange them as shown in Figure 5.19, to form a square of side a. While you are at it, just check that all your **visual** arguments can be expressed solely in purely symbolic or geometrical terms – imagine, for example, that you could not see or draw the diagram, and that you had to describe and justify the whole thing to a friend.

Now from triangles AED and DHC, $DH = c$ and $DE = b$, so HE, the side of the small internal square is $b - c$. The same argument applies to all pairs of adjacent triangles.

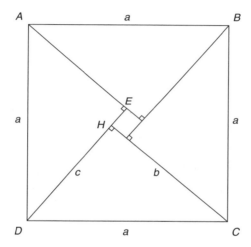

Figure 5.19 Proof of Pythagoras' theorem.

The total area of the big square is four times the area of the triangle ('half base times height' $= \frac{1}{2}bc$), plus the area of the small square. So

$$a^2 = 4 \times \tfrac{1}{2}bc + (b - c)^2$$

which with a bit of algebra (42 ◀) simplifies to

$$a^2 = b^2 + c^2$$

Solution to review question 5.1.8

Of the two given sides, the longest, 6, must be the hypotenuse. If the shortest side is x then we have

$$x^2 + 5^2 = 6^2$$

so

$$x = \sqrt{6^2 - 5^2} = \sqrt{11}$$

5.2.9 Lines and angles in a circle

Figure 5.20 reminds you of the main definitions relating to a circle.

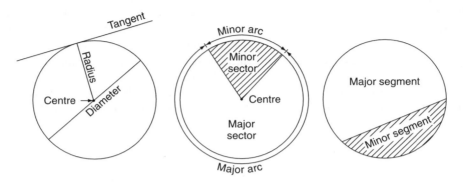

Figure 5.20 Definitions in circles.

There is not much that can be said about the circle in isolation, other than to note its perfect symmetry and the standard perimeter formula '$2\pi r$' and area formula 'πr^2' where r is the radius. It is when you start considering parts or arcs of the circle, or particular lines related to the circle that things get interesting. First note that, in geometry, we are not concerned so much with **measuring** aspects of the circle or its parts. We are more concerned about relations between them, independent of numerical values. An example would be the result that chords which are equidistant from the centre are of the same length. This result does not help in calculating the lengths of chords, it simply states a relationship between certain types of chords. Two obvious lines of importance are the radius and tangent. As shown in Figure 5.20 the tangent is perpendicular to the radius at its point of contact.

Consider an **arc** of a circle – a connected part of the circumference. If the arc has length less than half of the circumference then it is called a **minor arc**, otherwise it is a **major arc**. In Figure 5.21

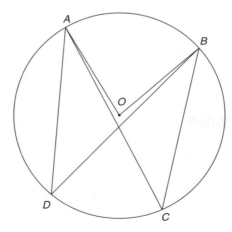

Figure 5.21 Angles subtended by arcs.

the angle **subtended by the minor arc** AB at the centre O is $\angle AOB$. We say the angle $\angle ACB$ **stands on the minor arc** AB, which is said to **subtend** an angle $\angle ACB$ at the circumference. Exactly the same statements can be made replacing C by D. Implicit in the definitions is the assumption that there is just **one** angle subtended at the circumference – i.e. that $\angle ACB = \angle ADB$. As we state below, this is in fact the case.

Now let's add a chord to the circle. There are two important results to note regarding chords.

- **Chords equidistant from the centre are of equal lengths**

To see this consider the triangles formed by connecting the ends of two such chords to the centre. These two triangles have two corresponding sides (the radii) equal, and the corresponding enclosed angles are equal. The triangles are therefore congruent, and so their third sides (i.e. the length of the chords) are also equal.

- **The perpendicular bisector of a chord of a circle passes through the centre of the circle**

The proof is similar to that of the first result using congruent triangles, but the result is sufficiently 'self evident' to gloss over here.

The next results concern the angles subtended by arcs at the centre and on the circumference. Whilst it is important that you understand and can use the results, the proofs are not important to us here.

- **The angle subtended by an arc at the centre of a circle is twice the angle subtended at the circumference (Figure 5.22)**

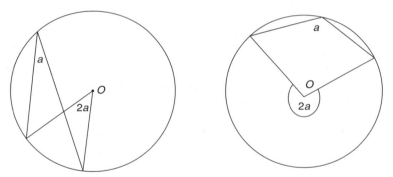

Figure 5.22 Angle subtended at the centre is twice that on the circumference.

- **All angles subtended at the circumference by the same arc are equal**
- **The angle in a semicircle is 90° – i.e. the angle subtended by a diameter is a right angle**

The last result follows from the fact that a semicircle subtends an angle of 180° at the centre of a circle. Any angle subtended at the circumference must, by the above results, be equal to half this, namely 90°.

Finally, we give some important results relating to the properties of tangents to circles. A **tangent** to a circle is a straight line which touches it at exactly one point – the **point of contact**. A straight line which cuts a circle at two distinct points is called a **secant** (a chord is thus a segment of a secant). An important property of the tangent is:

- **A tangent to a circle is perpendicular to the radius drawn through the point of contact.**

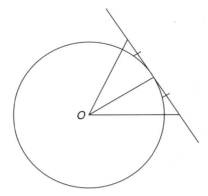

Figure 5.23 Tangent perpendicular to radius.

One can see this by constructing two radii extended to any two points either side and equidistant from the point of contact (Figure 5.23). By symmetry, the resulting triangles have corresponding sides equal and are therefore congruent (152 ◄) and must have corresponding angles equal. This implies that the angle of intersection between the tangent and radius at point of contact is $\frac{1}{2} \times 180° = 90°$.

Two other important results are

- **The two tangents drawn from an external point to a circle are equal in length.**
- **The angle between a tangent and a chord through the point of contact is equal to the angle subtended by the chord in the alternate segment (Figure 5.24).**

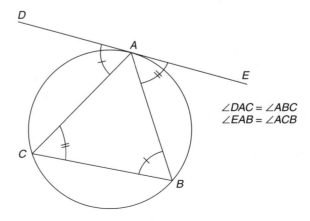

Figure 5.24 Angle in alternate segment.

Solution to review question 5.1.9

(i) Referring to Figure 5.8, $a = 360° - \angle AOB$, and $\angle AOB = 2\angle ACB = 2 \times 40° = 80°$. So $a = 280°$.

(ii) Equal angles subtended by equal arcs gives $\angle ACB = c = \angle ADB = 30°$. Angles in triangle ABD then gives $a = 180° - 30° - 70° = 80°$. Equal angles subtended by equal arcs now gives $\angle DAC = 20° = \angle DBC = b$. So $a = 80°$, $b = 20°$, $c = 30°$.

(iii) $OA = OB$, so triangle ABO is isosceles and therefore $a = 35°$ and $b = 180° - 35° - 90° = 55°$, since the chord bisector OP is perpendicular to the chord AB.

(iv) The angle $\angle DAB$ is subtended by a diameter and is therefore $90°$. So from triangle ABD, $\angle ABD = 180° - 30° - 90° = 60°$. But the angles subtended by the chord AD at B and C are the same and so $a = 60°$ also.

5.2.10 Cyclic quadrilaterals

The results so far essentially rely on fitting triangles into circles – we say the **circle circumscribes** the triangle. In general, when a circle circumscribes a polygon the vertices of the polygon lie on the circle. (A circle is **circumscribed** by a figure, or **inscribed** in a figure when all sides of the figure touch the circle.) The next step up is to consider circumscribed quadrilaterals – i.e. quadrilaterals whose vertices lie on a circle, and sides inside the circle. This is called a **cyclic quadrilateral**. See Figure 5.25 showing the cyclic quadrilateral $ABCD$.

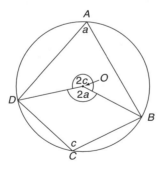

Figure 5.25 Cyclic quadrilateral.

The main result on cyclic quadrilaterals is that:

- **The opposite angles of a cyclic quadrilateral are supplementary – i.e. add up to 180°.**

The proof is interesting and not difficult. In Figure 5.25, with obvious notation we have

$$2a + 2c = 360° = 2(a + c)$$
$$= 2(\angle DAB + \angle DCB)$$

Hence

$$\angle DAB + \angle DCB = 180°$$

> **Solution to review question 5.1.10**
>
> (i) In Figure 5.9(i) we have $\angle BAD + \angle BCD = 180° = a + 125°$. So $a = 55°$ by supplementary angles in a cyclic quadrilateral. By angles on a line, $d = 180° - 95° = 85°$. Then by supplementary angles $\angle ABC = 180 - d = 95°$ and angles on a line gives $b = 180° - 95° = 85°$.
>
> (ii) In Figure 5.9(ii) opposite angles in a cyclic quadrilateral gives
>
> $$a + 3a = 4a = 180°$$
>
> So $a = 45°$. Then $b = 2a = 90°$.

5.3 Reinforcement

5.3.1 Division of a line in a given ratio   ◀◀ 143 147 ◀

A. The line AB is 4 cm long. P is a point which divides AB internally in the ratio $AP : PB = p : q$ for the values of p and q given below. In each case determine the lengths AP, PB. Repeat the exercise when the division is external, in the same ratios. Leave your answers as fractions in simplest form.

(i) 2 : 1   (ii) 3 : 2   (iii) 5 : 4   (iv) 3 : 7   (v) 1 : 5

B. With the same ratios as in Question **A**, it is given that $AP = 2$ cm. Determine PB and AB in each case.

5.3.2 Intersecting and parallel lines and angular measurement
◀◀ 143 148 ◀

Fill in all remaining angles:

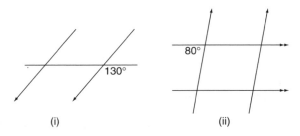

(i)   (ii)

5.3.3 Triangles and their elementary properties   ◀◀ 144 150 ◀

A. In each case the pairs of angles are angles in a triangle. Determine the other angle, and the corresponding external angle.

(i) 32°, 45°   (ii) 73°, 21°   (iii) 15°, 21°   (iv) 85°, 65°

B. Determine all the angles in triangles with the following angles and opposite external angles respectively

(i) 25°, 114°   (ii) 63°, 80°   (iii) 45°, 65°   (iv) 27°, 115°

C. (i) Find angles a, b, c, d:

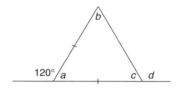

(ii) Find a, b, c, d, e, f:

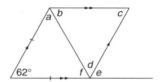

5.3.4 Congruent triangles

◂◂ 144 152 ◂

Identify, where possible, a pair of congruent triangles:

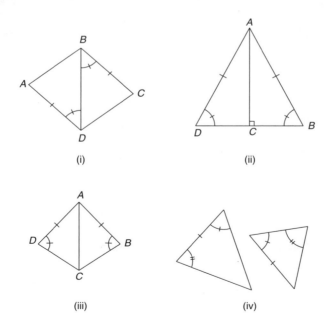

(i)   (ii)

(iii)   (iv)

161

5.3.5 Similar triangles

◀◀ 145 152 ◀

A. The triangles ABC and PQR are similar. Find the lengths of all sides of PQR for the values of x given by (i) 1 cm (ii) 2 cm (iii) 3.5 cm (iv) 4 m:

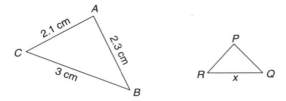

B. Triangles ABC and PQR are similar. Given that $AB/PQ = p/q$, determine all the sides of both triangles for p/q values (i) 3/2 (ii) 7/3 (iii) 3/8:

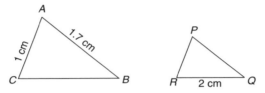

5.3.6 The intercept theorem

◀◀ 145 153 ◀

Find the missing lengths x, y, z, u in the figures below, to two decimal places:

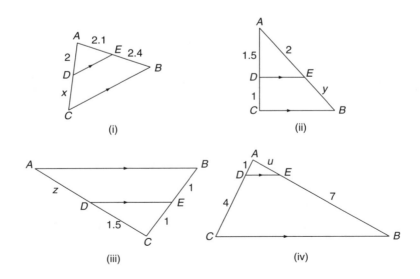

5.3.7 The angle bisector theorem

Find the missing lengths x, y, z, u, v:

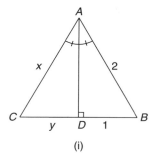
(i)

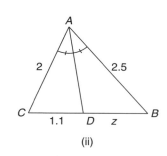
(ii)

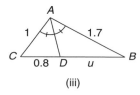
(iii)

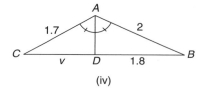
(iv)

5.3.8 Pythagoras' theorem

A. Regarding the pairs of lengths given in (i) to (v) as the (a) shortest (b) longest sides of a right angled triangle (in consistent units) determine the third side in each case. Leave your answers in surd form.

(i) 3, 5   (ii) 5, 12   (iii) 24, 25   (iv) 8, 17   (v) 3, 4

B. Determine the longest rod that may be placed in a rectangular box of edges 3, 4, 6 units.

5.3.9 Lines and angles in a circle

A. A, B, C are three points on a circle centre O, taken clockwise in that order. Given the following angles, determine the others named.

(i) $\angle ACB = 35°$, $\angle AOB$ (ii) $\angle ACB = 70°$, $\angle AOB$

(iii) $\angle AOB = 60°$, $\angle ACB$ (iv) $\angle AOB = 86°$, $\angle ACB$

B. Referring to Question **A** now assume the points A, B, C are such that $AC = BC$. For each case specified in Question **A**, determine all angles in the triangles AOC, AOB and BOC.

C. Determine the angles labelled with a letter:

Understanding Engineering Mathematics

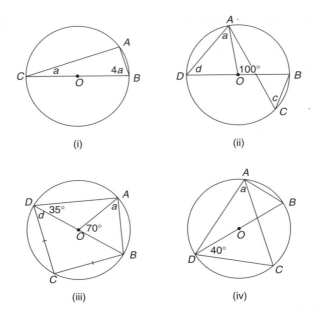

(i) (ii) (iii) (iv)

D. Determine the labelled angles:

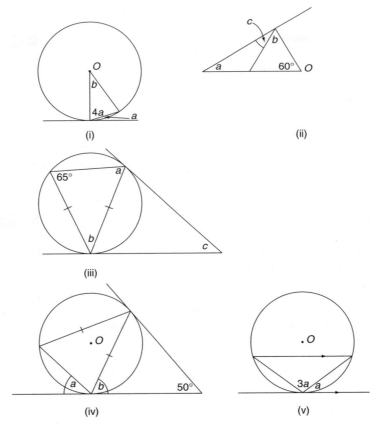

(i) (ii) (iii) (iv) (v)

E. An equilateral triangle of side 30 cm circumscribes a circle. Find the radius of the circle.

F. A circle of unit radius is inscribed in a right-angled isosceles triangle. Determine the lengths of the sides of the triangle.

5.3.10 Cyclic quadrilaterals

Determine angle a:

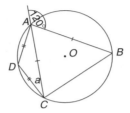

5.4 Applications

1. A major problem in surveying occurs when there is an obstacle, such as a river, that the survey line has to cross, but which we cannot easily walk round or put a tape across. There are three common methods to solve this, relying on elementary geometry – mainly similar triangles. These are illustrated in Figure 5.26(a), (b), (c).

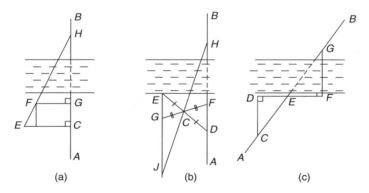

Figure 5.26

(a) A ranging pole is put at H on the far bank. CE, on the near bank, is set off perpendicular to AB, which is constructed perpendicular to the river. A pole is ranged in to a point F on EH and a perpendicular is dropped from F onto AB at G. Show that

$$GH = \frac{CG \times FG}{EC - FG}$$

(b) A line DE is set out on the near bank and bisected at C. FCG is now set out such that $FC = CG$. With a pole H on AB on the far bank, a pole can be set at J on the intersection of the lines EG and HC produced backwards. Show that $JG = FH$.

(c) The line AB crosses the river on the skew and poles placed at F and G on the near and far banks respectively. DF is set out along the near bank, so GF is perpendicular to GF. A perpendicular from D is constructed to meet AB at C. Show that

$$EG = \frac{CE \times EF}{ED}$$

Discuss the relative merits of each of the methods.

2. A railway track floor is cut into the side of a hill with slope 1 in k. Both sides of the cutting have slope 1 in m. If produced into the earth the slopes of the cutting intersect at a point G (see Figure 5.27). The flat horizontal bed of the cutting is at a depth h below the point where the centre line GL intersects the line of the slope of the hill. The width of the cutting floor is b.

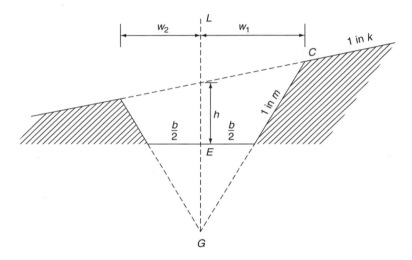

Figure 5.27

Show that the distance, w_1 from the up-slope edge of the cutting to the centre line GL is given by

$$w_1 = \left(\frac{b}{2} + mh\right)\left(\frac{k}{k-m}\right)$$

and that the distance w_2 between the centre line and the down-slope side is

$$w_2 = \left(\frac{b}{2} + mh\right)\left(\frac{k}{k+m}\right)$$

Show that the area of the cutting is

$$\frac{1}{2m}\left[\left(\frac{b}{2} + mh\right)(w_1 + w_2) - \frac{b^2}{2}\right]$$

Find the difference in level between the floor of the cutting and the top of the bank on each side.

Answers to reinforcement exercises

5.3.1 Division of a line in a given ratio

A. *Internal division* (AP, PB, all in cm)

(i) $\frac{8}{3}, \frac{4}{3}$ (ii) $\frac{12}{5}, \frac{8}{5}$ (iii) $\frac{20}{9}, \frac{16}{9}$

(iv) $\frac{6}{5}, \frac{14}{5}$ (v) $\frac{2}{3}, \frac{10}{3}$

External division (in cm)

(i) 4, 8   (ii) 8, 12   (iii) 16, 20

(iv) 3, 7   (v) 1, 5

B. *Internal division* (in cm)

(i) 3, 1   (ii) $\frac{10}{3}, \frac{4}{3}$ (iii) $\frac{18}{5}, \frac{8}{5}$

(iv) $\frac{20}{3}, \frac{14}{3}$ (v) 12, 10

External division (in cm)

(i) 1, 1   (ii) $\frac{2}{3}, \frac{4}{3}$ (iii) $\frac{2}{5}, \frac{8}{5}$

(iv) $\frac{8}{3}, \frac{14}{3}$ (v) 8, 10

5.3.2 Intersecting and parallel lines and angular measurement

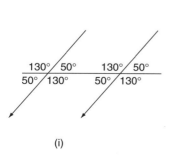

(i)

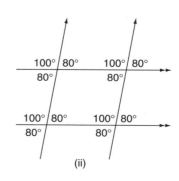

(ii)

5.3.3 Triangles and their elementary properties

A.
	Other angle	External angle
(i)	103°	77°
(ii)	86°	94°
(iii)	144°	36°
(iv)	30°	150°

B. (i) 25°, 66°, 89° (ii) 63°, 100°, 17°
   (iii) 45°, 115°, 20° (iv) 27°, 65°, 88°

C. (i) $a = b = c = 60°$, $d = 120°$ (ii) $a = b = d = f = 59°$, $c = e = 62°$

5.3.4 Congruent triangles

(i) ADB and CBD are congruent

(ii) ADC, ABC congruent

(iii) Not congruent

(iv) Not congruent

5.3.5 Similar triangles

A. (i) $PR = 0.7$ cm $PQ = 0.7\dot{6}$ cm
   (ii) $PR = 1.4$ cm $PQ = 1.5\dot{3}$ cm
   (iii) $PR = 2.45$ cm $PQ = 2.68\dot{3}$ cm
   (iv) $PR = 2.1$ m $PQ = 3.0\dot{6}$ m

B. (i) $PQ = 1.1\dot{3}$ cm $PR = 0.\dot{6}$ cm $BC = 3$ cm
   (ii) $PQ = 0.73$ cm (2dp) $PR = 0.43$ cm $BC = 4.\dot{6}$ cm
   (iii) $PQ = 4.5\dot{3}$ cm $PR = 2.\dot{6}$ cm $BC = 0.75$ cm

5.3.6 The intercept theorem

(i) $DC = 2.29$ (ii) $EB = 1.\dot{3}$ (iii) $AD = 1.5$
(iv) $AE = 1.75$

5.3.7 The angle bisector theorem

(i) $x = 2$, $y = 1$ (ii) $z = 1.375$
(iii) $u = 1.36$ (iv) $CD = 1.53$

5.3.8 Pythagoras' theorem

A. (a) (i) $\sqrt{34}$ (ii) 13   (iii) $\sqrt{1201}$
      (iv) $\sqrt{353}$ (v) 5

(b) (i) 4   (ii) $\sqrt{119}$ (iii) 7
(iv) 15   (v) $\sqrt{7}$

B. $\sqrt{61}$

5.3.9 Lines and angles in a circle

A. (i) 70°   (ii) 140°   (iii) 30°
(iv) 43°

B. (i) AOC and BOC are 17.5°, 145°, 175°; AOB is 55°, 70°, 55°
(ii) AOC and BOC are 35°, 110°, 35°; AOB is 20°, 140°, 20°
(iii) AOC and BOC are 15°, 150°, 15°; AOB is 60°, 60°, 60°
(iv) AOC and BOC are 21.5°, 137°, 21.5°; AOB is 47°, 86°, 47°

C. (i) $a = 18°$ (ii) $a = c = d = 50°$
(iii) $a = 55°, d = 45°$ (iv) $a = 50°$

D. (i) $a = 18°, b = 36°$ (ii) $a = 30°, b = 60°, c = 30°$
(iii) $a = 65°, b = 50°, c = 50°$ (iv) $a = 50°, b = 65°$
(v) $a = 36°$

E. $5\sqrt{3}$ cm

F. $\sqrt{2}(1 + \sqrt{2})$

5.3.10 Cyclic quadrilaterals

$a = 30°$

6

Trigonometry

Trigonometry (literally: 'the measurement of triangles' – 'trig' from now on) is a fairly straightforward topic conceptually, but there always seems a lot to remember. In fact you only have to remember a few key results – but you have to remember them very well. For example look at the function

$$\frac{\cos 2\theta}{\cos\theta + \sin\theta}$$

If you have to consult a formula book to remind yourself that $\cos 2\theta = \cos^2\theta - \sin^2\theta$ and the difference of squares identity $a^2 - b^2 = (a-b)(a+b)$ then it might not even occur to you to simplify this to the form

$$\frac{\cos^2\theta - \sin^2\theta}{\cos\theta + \sin\theta} = \frac{(\cos\theta - \sin\theta)(\cos\theta + \sin\theta)}{\cos\theta + \sin\theta}$$
$$= \cos\theta - \sin\theta$$

However, there is no need to remember the double angle formulae in detail because you can get them easily from the compound angle formulae for cosine and sine, which you **should** remember well. In this chapter we will cover such fundamental topics of trigonometry, and encourage you to learn just a few key formulae very well so that you can use these to derive other formulae as you need them.

Prerequisites

It will be helpful if you know something about:

- angular measurement using degrees and radians (148 ◄)
- ratio and proportion (14 ◄)
- properties of triangles (150 ◄)
- Pythagoras' theorem (154 ◄)
- plotting graphs (91 ◄)
- inverse of a function (100 ◄)
- surds (20 ◄)

Objectives

In this chapter you will find:

- radian measure and the circle
- definitions of the trig ratios

- the sine and cosine rules and solution of triangles
- graphs of the trig functions
- inverse trig functions
- the Pythagorean identities such as

$$\cos^2\theta + \sin^2\theta = 1$$

- compound angle formulae such as

$$\sin(A+B) = \sin A \cos B + \sin B \cos A$$

and their consequences such as the double angle formulae
- solution of simple trig equations
- The $a\cos\theta + b\sin\theta$ form

Motivation

You may need the material of this chapter for:

- solution of triangles in statics, surveying, etc.
- describing, analysing and combining oscillations and waves (➤ Chapter 17)
- evaluating trig and other integrals (➤ Chapter 9)
- describing angular motion
- describing alternating current circuits

6.1 Review

6.1.1 Radian measure and the circle   ➤ 173 194 ➤➤

A. Express as radians (i) 90° (ii) −30° (iii) 45° (iv) 270° (v) 60°.

B. Express the following radian measures in degrees

(i) $\dfrac{5\pi}{6}$ (ii) $\dfrac{3\pi}{2}$ (iii) $-\dfrac{7\pi}{4}$ (iv) 4π (v) $-\dfrac{2\pi}{3}$

C. An arc of a circle of radius 2 subtends an angle of 30° at the centre. Find

(i) the length of the arc
(ii) the area of the sector enclosed by the arc and the bounding radii.

6.1.2 Definition of the trig ratios   ➤ 174 194 ➤➤

Write down the **exact** values of the following (i.e. in surd form)

(i) $\cos 0$ (ii) $\cos 2\pi$ (iii) $\sin 90°$
(iv) $\sin \dfrac{\pi}{4}$ (v) $\cos \dfrac{\pi}{2}$ (vi) $\sin 45°$
(vii) $\tan 90°$ (viii) $\sin 0$ (ix) $\sin 60°$

(x) $\sin\dfrac{2\pi}{3}$ (xi) $\cos\dfrac{\pi}{3}$ (xii) $\tan 45°$

(xiii) $\cos 30°$ (xiv) $\sin 30°$ (xv) $\tan\dfrac{\pi}{3}$

(xvi) $\cos 45°$ (xvii) $\cos\dfrac{3\pi}{2}$ (xviii) $\tan(-60°)$

(xix) $\sin(-120°)$ (xx) $\cos\left(-\dfrac{\pi}{3}\right)$ (xxi) $\sin(585°)$

(xxii) $\cos(225°)$ (xxiii) $\tan(-135°)$ (xxiv) $\sec 30°$

(xxv) $\csc \pi/4$ (xxvi) $\cot 60°$ (xxvii) $\sec 120°$

(xxviii) $\cos\left(-\dfrac{\pi}{2}\right)$ (xxix) $\csc(-60°)$

6.1.3 Sine and cosine rules and solutions of triangles

For the triangle below find (i) a (ii) θ.

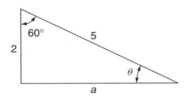

6.1.4 Graphs of the trigonometric functions

Sketch the graphs of (i) $3\sin\left(t - \dfrac{\pi}{2}\right)$ (ii) $4\cos\left(2t - \dfrac{\pi}{6}\right)$ (iii) $\tan\left(t + \dfrac{\pi}{2}\right)$

6.1.5 Inverse trigonometric functions

Evaluate the following inverse trigonometric ratios in the range $0 \le \theta \le 90°$:

(i) $\sin^{-1}\left(\dfrac{1}{2}\right)$ (ii) $\sin^{-1}\left(\dfrac{\sqrt{3}}{2}\right)$ (iii) $\cos^{-1}\left(\dfrac{\sqrt{3}}{2}\right)$

(iv) $\cos^{-1}\left(\dfrac{1}{2}\right)$ (v) $\tan^{-1}\left(\dfrac{1}{\sqrt{3}}\right)$ (vi) $\tan^{-1}(\sqrt{3})$

6.1.6 The Pythagorean identities – $\cos^2 + \sin^2 = 1$

Complete the following table in which each angle is in the first quadrant:

	$\sin\theta$	$\cos\theta$	$\tan\theta$
(i)	$\dfrac{1}{7}$		
(ii)		$\dfrac{1}{\sqrt{3}}$	
(iii)			$\dfrac{1}{\sqrt{2}}$

6.1.7 Compound angle formulae

A. Expand $\sin(A + B)$ in terms of sine and cosine of A and B.
B. From **A** derive similar expansions for

(i) $\sin(A - B)$ (ii) $\cos(A - B)$ (iii) $\tan(A + B)$

C. Given that $\sin 45° = \cos 45° = \dfrac{1}{\sqrt{2}}$, $\cos 60° = \dfrac{1}{2}$, $\sin 60° = \dfrac{\sqrt{3}}{2}$, evaluate

(i) $\cos 75°$ (ii) $\sin 105°$ (iii) $\tan(-75°)$

D. Express $\cos 2A$ in trigonometric ratios of A.
E. Given that $\cos 30° = \dfrac{\sqrt{3}}{2}$, $\sin 30° = \dfrac{1}{2}$, evaluate

(i) $\sin 15°$ (ii) $\tan 15°$

6.1.8 Trigonometric equations

Find the general solution of each of the equations:

(i) $\sin \theta + 2 \sin \theta \cos \theta = 0$
(ii) $\cos 3\theta = \cos \theta$

6.1.9 The $a \cos \theta + b \sin \theta$ form

Express $\cos \theta + \sin \theta$ in the form   (i) $r \sin(\theta + \alpha)$ (ii) $r \cos(\theta + \alpha)$

6.2 Revision

6.2.1 Radian measure and the circle

As noted in Section 5.2.2, a **radian** is the angle subtended at the centre of a circle by an arc with length equal to that of the radius. It follows that

$$\theta \text{ radians} = \frac{180}{\pi} \theta \text{ degrees}$$

The length of arc of a circle of radius r, subtending angle θ radians at the centre is by definition $s = r\theta$. See Figure 6.1. The area of the enclosed sector is $A = \frac{1}{2} r^2 \theta$. This follows because the total area of the circle is πr^2 and a sector with angle θ radians forms a fraction $\dfrac{\theta}{2\pi}$ of that area.

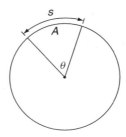

Figure 6.1 Radians, arc, sectors.

Solution to review question 6.1.1

A. (i) $90° = \dfrac{90}{180} \times \pi = \dfrac{\pi}{2}$ radians

(ii) $-30° = -\dfrac{30}{180} \times \pi = -\dfrac{\pi}{6}$ radians

(iii) $45° = \dfrac{45}{180} \times \pi = \dfrac{\pi}{4}$ radians

(iv) $270° = \dfrac{270}{180} \times \pi = \dfrac{3}{2}\pi$ radians

(v) $60° = \dfrac{60}{180} \times \pi = \dfrac{\pi}{3}$ radians

B. (i) $\dfrac{5\pi}{6}$ radians $= \dfrac{5\pi}{6} \times \dfrac{180}{\pi} = 150°$

(ii) $\dfrac{3\pi}{2}$ radians $= \dfrac{3\pi}{2} \times \dfrac{180}{\pi} = 270°$

(iii) $-\dfrac{7\pi}{4}$ radians $= -\dfrac{7\pi}{4} \times \dfrac{180}{\pi} = -315°$

(iv) 4π radians $= 4\pi \times \dfrac{180}{\pi} = 720°$

(v) $-\dfrac{2\pi}{3}$ radians $= -\dfrac{2\pi}{3} \times \dfrac{180}{\pi} = -120°$

C. (i) Length of arc $s = r\theta$, θ in radians.

$30° = 30° \times \dfrac{\pi}{180} = \dfrac{\pi}{6}$ radians

So $s = 2 \times \dfrac{\pi}{6} = \dfrac{\pi}{3}$

(ii) Area of sector $A = \dfrac{1}{2}r^2\theta = \dfrac{1}{2} \cdot 2^2 \cdot \dfrac{\pi}{6} = \dfrac{\pi}{3}$

6.2.2 Definition of the trig ratios

◀ 171 194 ▶

If ABC is the right-angled triangle, shown in Figure 6.2, then the **trigonometric ratios** are defined by:

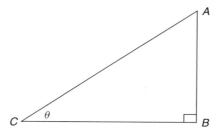

Figure 6.2

$$\frac{AB}{AC} = \sin\theta \text{ (sine of } \theta)\qquad \frac{AC}{AB} = \text{cosec}\,\theta \text{ (cosecant of } \theta) = \frac{1}{\sin\theta}$$

$$\frac{BC}{AC} = \cos\theta \text{ (cosine of } \theta)\qquad \frac{AC}{BC} = \sec\theta \text{ (secant of } \theta) = \frac{1}{\cos\theta}$$

$$\frac{AB}{BC} = \tan\theta \text{ (tangent of } \theta)\qquad \frac{BC}{AB} = \cot\theta \text{ (cotangent of } \theta) = \frac{1}{\tan\theta}$$

Note that $\tan\theta = \dfrac{\sin\theta}{\cos\theta}$.

Treating θ as an independent variable, these might also be regarded as **trigonometric functions** (90 ◄) of θ. For general angles, greater than 90° the sign of each ratio depends on the quadrant it is in – an example in the second quadrant is shown in Figure 6.3.

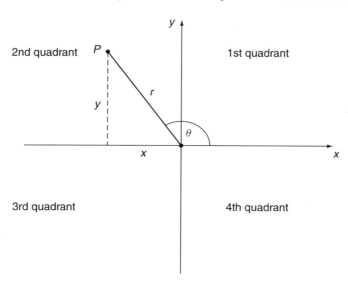

Figure 6.3 General angles.

$$\sin\theta = \frac{y}{r} = \frac{NP}{OP}\qquad \cos\theta = \frac{x}{r} = \frac{-ON}{OP}\qquad \tan\theta = \frac{y}{x} = \frac{NP}{-ON}$$

where $r = OP = \sqrt{x^2 + y^2} > 0$, ON and NP are the positive lengths indicated and x, y are the coordinates of P, including appropriate signs. θ is measured anticlockwise – the 'positive direction' – as shown. Note that for any angle, θ, $|\sin\theta|$ and $|\cos\theta|$ are both ≤ 1. The negative of θ, $-\theta$, means a rotation through angle θ in the clockwise direction from the positive x-axis. From Figure 6.3 it then follows that $\cos\theta$ is an **even** function of θ, $\cos(-\theta) = \cos\theta$ and $\sin\theta$ is an **odd** function of θ, $\sin(-\theta) = -\sin\theta$ (94 ◀). So, is $\tan\theta$ even or odd?

The signs of the trig ratios in the different quadrants can be remembered from 'All Silly Tom Cats' (or you might have learnt 'CAST', which is not so jolly), going anticlockwise round the diagram below

Sine only positive	All positive
Tan only positive	Cosine only positive

which tells us which ratios are positive in the respective quadrants.

It is useful to memorise the trig ratios for the commonly occurring angles 30°, 45°, 60°, which can be conveniently obtained from the triangles shown in Figure 6.4.

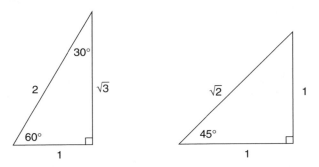

Figure 6.4 30–60 and 45 triangles.

From Figure 6.4 we see directly that, for example $\sin 60° = \dfrac{\sqrt{3}}{2}$, $\cos 60° = \dfrac{1}{2}$, $\tan 30° = \dfrac{1}{\sqrt{3}}$.

The term 'cosine' is not accidental – cosine of θ is the sine of $(90 - \theta)$. As noted in Section 5.2.2, if two angles α, β, sum to 90° they are said to be **complementary**. β is the **complement** of α and vice versa. From the triangle shown in Figure 6.5 it is readily

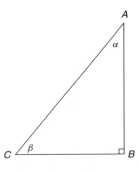

Figure 6.5 α and β are complementary angles.

seen that

$$\sin \alpha = \cos \beta = \cos(90° - \alpha)$$
$$\cos \alpha = \sin \beta = \sin(90° - \alpha)$$
$$\tan \alpha = \cot \beta = \cot(90° - \alpha)$$
$$\cot \alpha = \tan \beta = \tan(90° - \alpha)$$

i.e. the 'co-trig ratio' is the ratio of the complementary angle.

Solution to review question 6.1.2

You will find it very useful to commit as many of these as possible to memory. Many of these results can be obtained from Figure 6.4.

(i) $\cos 0 = 1$
(ii) $\cos 2\pi = 1$
(iii) $\sin 90° = 1$
(iv) $\sin \dfrac{\pi}{4} = \dfrac{1}{\sqrt{2}}$ (this is the **exact** value – see Figure 6.4)
(v) $\cos \dfrac{\pi}{2} = 0$
(vi) $\sin 45° = \sin \dfrac{\pi}{4} = \dfrac{1}{\sqrt{2}}$
(vii) $\tan 90°$ is, strictly, not defined but it is usual to take it as ∞
(viii) $\sin 0 = 0$
(ix) $\sin 60° = \dfrac{\sqrt{3}}{2}$
(x) $\sin \dfrac{2\pi}{3} = \sin \dfrac{\pi}{3} = \dfrac{\sqrt{3}}{2}$
(xi) $\cos \dfrac{\pi}{3} = \dfrac{1}{2}$
(xii) $\tan 45° = 1$
(xiii) $\cos 30° = \dfrac{\sqrt{3}}{2}$
(xiv) $\sin 30° = \dfrac{1}{2}$
(xv) $\tan \dfrac{\pi}{3} = \sqrt{3}$
(xvi) $\cos 45° = \dfrac{1}{\sqrt{2}}$
(xvii) $\cos \dfrac{3\pi}{2} = \cos \dfrac{\pi}{2} = 0$
(xviii) $\tan(-60°) = -\tan 60° = -\sqrt{3}$
(xix) $\sin(-120°) = -\sin 120° = -\dfrac{\sqrt{3}}{2}$

(xx) $\cos\left(-\frac{\pi}{3}\right) = \cos\left(\frac{\pi}{3}\right) = \frac{1}{2}$

(xxi) $\sin(585°) = \sin(225°) = -\sin(45°) = -\frac{1}{\sqrt{2}}$

(xxii) $\cos(225°) = -\cos(45°) = -\frac{1}{\sqrt{2}}$

(xxiii) $\tan(-135°) = -\tan(135°) = -(-\tan 45°) = -(-1) = 1$

(xxiv) $\sec 30° = 1/\cos 30° = 2/\sqrt{3}$

(xxv) $\cosec \pi/4 = 1/\sin(\pi/4) = \sqrt{2}$

(xxvi) $\cot 60° = 1/\tan 60° = 1/\sqrt{3}$

(xxvii) $\sec 120° = 1/\cos 120° = -1/\cos 60° = -2$

(xxviii) $\cot\left(-\frac{\pi}{2}\right) = -\cot(\pi/2) = 0$

(xxix) $\cosec(-60°) = -1/\sin 60° = -2/\sqrt{3}$

6.2.3 Sine and cosine rules and solutions of triangles

◀ 172 194 ▶

For general triangles two important rules apply. We use the standard notation – capital A, B, C for the angles with corresponding lower case letters labelling the opposite sides (Figure 6.6).

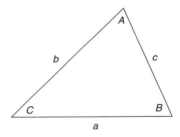

Figure 6.6 Standard labelling of angles and sides.

The **sine rule** states that

$$\frac{a}{\sin A} = \frac{b}{\sin B} = \frac{c}{\sin C}$$

i.e. the sides of a triangle are proportional to the sines of the opposite angles. We can see this by using the result for the area of a triangle $-\frac{1}{2}$ base × height. We can calculate this area in three different ways, giving:

$$\tfrac{1}{2}bc \sin A = \tfrac{1}{2}ac \sin B = \tfrac{1}{2}ab \sin C$$

The sine rule then follows by dividing by $\frac{1}{2}abc$ and taking the reciprocals of the results.

Incidentally, there is another useful formula (**Heron's formula**) for calculating the area of a triangle with sides a, b, c:

$$\text{area} = \sqrt{s(s-a)(s-b)(s-c)}$$

where $s = (a + b + c)/2$, the semi-perimeter of the triangle. This result is very useful in surveying, for example.

Note that the sine rule can only be used if we know at least one side and the angle opposite to that side. Failing such information, we may be able to use the **cosine rule**, which states that

$$a^2 = b^2 + c^2 - 2bc \cos A$$

The proof for this is instructive and illustrates yet again the power of Pythagoras – we only consider the case of acute angles, but the rule holds for all angles.

Consider the triangle shown in Figure 6.7, with altitude h (154 ◄).

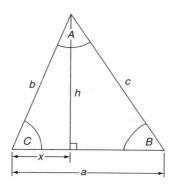

Figure 6.7 Proof of cosine rule.

We have, by Pythagoras' theorem (154 ◄):

$$h^2 = b^2 - x^2$$
$$= c^2 - (a - x)^2$$

which simplifies to (42 ◄)

$$c^2 = a^2 + b^2 - 2ax$$

But $x = b \cos C$, so

$$c^2 = a^2 + b^2 - 2ab \cos C$$

The result is clearly 'symmetric' under the rotation of labels of the sides and corresponding angles, so we get also

$$a^2 = b^2 + c^2 - 2bc \cos A \text{ and } b^2 = a^2 + c^2 - 2ac \cos B$$

It may help to think of it is as

$$(\text{side})^2 = \text{sum of squares of opposite sides}$$
$$- 2\times \text{ product of opposite sides} \times \cos \text{ of their included angle.}$$

These results may be used to solve triangles given appropriate information. A triangle can be 'solved' for all angles and sides if we know either:

- three sides
- two sides and an included angle
- two angles and one side
- two sides and a non-included angle – but in this case the answers can be ambiguous (see RE6.3.3(ii)).

Solution to review question 6.1.3

This is the case of two sides and an included acute angle.
(i) By the cosine rule, with $b = 2$, $c = 5$ and $A = 60°$ we have

$$a^2 = 2^2 + 5^2 - 2 \times 2 \times 5 \times \cos 60°$$
$$= 4 + 25 - 2 \times 10 \times \frac{1}{2}$$
$$= 19$$

So
$$a = \sqrt{19}$$

(ii) We can now find $\sin\theta$ from the sine rule

$$\frac{a}{\sin 60°} = \frac{\sqrt{19}}{\sin 60°} = \frac{2}{\sin\theta}$$

So
$$\sin\theta = \frac{2\sin 60°}{\sqrt{19}} = \frac{\sqrt{3}}{\sqrt{19}} = 0.3974 \text{ to 4 decimal places}$$

Hence $\theta \cong 23.4°$.

Note: The other angle is obtuse $(180° - 23.4°)$, but the sine rule applied to this angle would not tell us this – it is always safest to go for the angle that is obviously acute when using the sine rule to solve triangles.

6.2.4 Graphs of the trigonometric functions

◀ 172 195 ▶

We now focus on the trig ratios as **functions** (90 ◀) and look at their graphs (91 ◀). This requires careful thought about what happens to the trig functions as the independent variable, θ, changes. As the trig functions are functions of an angular variable their values will keep repeating – because, for example, $\theta + 360°$ is in the same angular position as θ, and therefore $\sin(\theta + 360°) = \sin\theta$, $\cos(\theta + 360°) = \cos\theta$, etc. We express this generally by saying that the trigonometric functions cos, sin, and tan are **periodic with period 360°** (**or 2π**) by which we mean that for such functions (NB: in future we will usually use radian measure for angles, rather than degrees – it is by far the safest policy when we are regarding the trig ratios as functions, particularly in calculus – as a rule of thumb, if a question is given in terms of degrees/radians, then the answer should be given in the same form)

$$f(\theta + 360°) \equiv f(\theta + 2\pi) = f(\theta)$$

Thus, in particular

$$\cos(\theta + 2\pi) = \cos\theta \text{ and } \tan(\theta + 2\pi) = \tan\theta$$

This is reflected in the oscillatory nature of the graphs of the functions – the form of the graph repeats itself at intervals of 2π for sin and cos, and at intervals of π for tan. This 'minimum repeating interval' is called the **period** of the function. In general, if

$$f(x + p) = f(x) \text{ for all } x$$

for some constant p then we say $f(x)$ is **periodic with period** t. Of course, it also follows that if $f(x)$ has period p then $f(x + np) = f(x)$ for any positive integer n (provided f is defined at $x + np$), but the term 'period' is usually reserved for the smallest value of the period over which the function repeats. What, then, is the **actual** period of $\tan\theta$?

The cosine and sine functions are particularly important in engineering and science, representing, as they do, general wave-like behaviour. Indeed if you think about the really fundamental types of physical change that we commonly experience, they are either 'exponential' growth or decay (127 ◀), or wave-like/oscillatory, which we can express in terms of trig functions. Of course, in reality most oscillatory behaviour is 'damped', and this is often expressed by multiplying the trig function by a decaying exponential.

The graphs of the trigonometric functions $\cos\theta$, $\sin\theta$ and $\tan\theta$ are shown below. cos and sin repeat every cycle of 2π and oscillate between -1 and $+1$. tan repeats every cycle of π, and is 'discontinuous' (▶ 414) at odd multiples of $\pi/2$, tending to either $-\infty$ or $+\infty$. Also, notice that the cosine graph is simply the sine graph shifted along the axis by $\pi/2$.

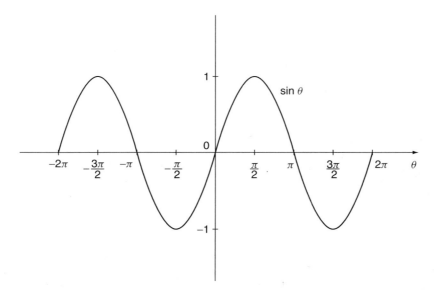

Figure 6.8 $\sin\theta$, $-2\pi \leq \theta \leq 2\pi$.

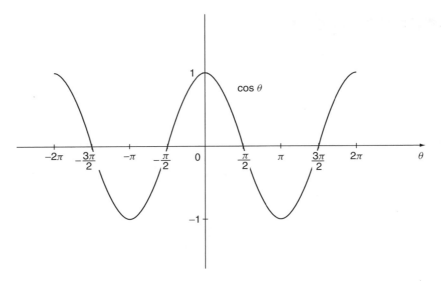

Figure 6.9 $\cos\theta \quad -2\pi \leq \theta \leq 2\pi$.

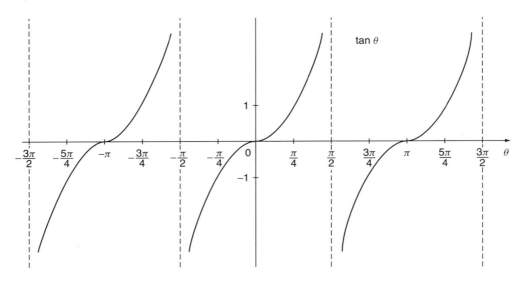

Figure 6.10 $\tan\theta \quad -\dfrac{3\pi}{2} \leq \theta \leq \dfrac{3\pi}{2}$.

Trig functions are very often used in describing time varying wave-like behaviour, where a typical functional form might be the **sinusoidal function** of t:

$$y = A\sin(\omega t + \alpha)$$

Such a wave-like graph is called a **sine wave** (whether it is cos or sin). A is called the **amplitude** of the wave, $2\pi/\omega$ its **period**, since it repeats after this cycle, $\omega/2$ its **frequency** and α its **phase**. This terminology is standard in the theory of waves in science and engineering. The solutions to the review question provide a number of examples.

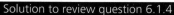

Solution to review question 6.1.4

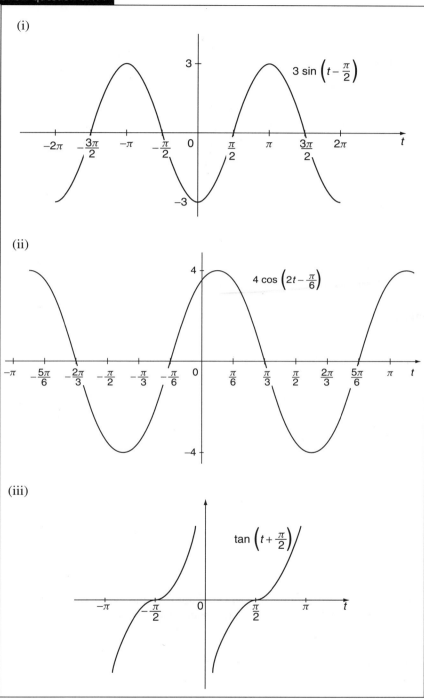

6.2.5 Inverse trigonometric functions

The inverse function (100 ◄) of $\sin x$ is denoted $\sin^{-1} x$ (Sometimes the notation 'arcsin x' is used). That is if $y = \sin x$ then $x = \sin^{-1} y$. So $\sin^{-1} x$ is **the angle whose sine is x**. Similarly

$$\cos^{-1} x = \text{angle whose cosine is } x$$
$$\tan^{-1} x = \text{angle whose tangent is } x$$
$$\sec^{-1} x = \text{angle whose sec is } x$$
$$\operatorname{cosec}^{-1} x = \text{angle whose cosec is } x$$
$$\cot^{-1} x = \text{angle whose cot is } x$$

Note that because, for example,

$$\sin 135° = \sin 45° = \frac{1}{\sqrt{2}}$$

the quantity $\sin^{-1}\left(\dfrac{1}{\sqrt{2}}\right)$ can take more than one value – we say it is **multi-valued**. In general, because of periodicity all the inverse trigonometric functions are multi-valued. Thus, for a given x, $\sin^{-1} x$ will yield an infinite range of values. In order to restrict to a unique value of $\sin^{-1} x$ for each value of x we take what is called the **principal value** of $\sin^{-1} x$, which is the value lying in the range

$$-\frac{\pi}{2} \leq \sin^{-1} x \leq \frac{\pi}{2}$$

So, for example, the principal value of $\sin^{-1}\left(\dfrac{1}{\sqrt{2}}\right)$ is 45°. This is shown in Figure 6.11.

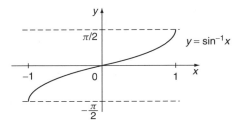

Figure 6.11 Inverse sine, $\sin^{-1} x$.

Similarly we can define principal values for the other inverse trigonometric functions – Figure 6.12 illustrates the principal value ranges for $\cos^{-1} x$ and $\tan^{-1} x$.

Since the inverse functions are multi-valued we need expressions for the general solution of equations such as $\sin y = x$, i.e. we need to obtain expressions for the most general angle given by $\sin^{-1} x$ and other inverse functions. We will simply state the results here:

$$\sin^{-1} x = n\pi + (-1)^n \text{PV}$$
$$\cos^{-1} x = 2n\pi \pm \text{PV}$$
$$\tan^{-1} x = n\pi + \text{PV}$$

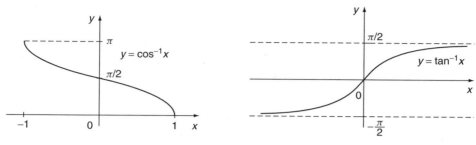

Figure 6.12 Inverse cosine, $\cos^{-1} x$ and inverse tan, $\tan^{-1} x$.

where PV is the principal value. Sometimes the principal value is distinguished by using a capital initial letter, for example we might denote the PV of $\sin^{-1} x$ by $\text{Sin}^{-1} x$.

> **Solution to review question 6.1.5**
>
> In the range $0° \leq \theta \leq 90°$ we have, referring to Figure 6.4:
>
> (i) $\sin^{-1}\left(\dfrac{1}{2}\right) = 30°$ (ii) $\sin^{-1}\left(\dfrac{\sqrt{3}}{2}\right) = 60°$
>
> (iii) $\cos^{-1}\left(\dfrac{\sqrt{3}}{2}\right) = 30°$ (iv) $\cos^{-1}\left(\dfrac{1}{2}\right) = 60°$
>
> (v) $\tan^{-1}\left(\dfrac{1}{\sqrt{3}}\right) = 30°$ (vi) $\tan^{-1}(\sqrt{3}) = 60°$

6.2.6 The Pythagorean identities – $\cos^2 + \sin^2 = 1$ ◀ 172 195 ▶

The standard trigonometric identities are normally found on a formulae sheet. However, the key identities should actually be **memorised** – and indeed should be second nature. In fact, you only have to remember one or two, from which the rest may then be derived. A sufficient minimal set to remember is in fact:

$$\cos^2 \theta + \sin^2 \theta = 1$$

and

$$\sin(A + B) = \sin A \cos B + \sin B \cos A$$

If you have these at your fingertips most of the others are easily recalled or derived from them. The first is the subject of this section, the second is treated in Section 6.2.7.

The **Pythagorean identities** are

$$\cos^2 \theta + \sin^2 \theta \equiv 1$$
$$1 + \tan^2 \theta \equiv \sec^2 \theta$$
$$\cot^2 \theta + 1 \equiv \operatorname{cosec}^2 \theta$$

These can be obtained, as their name suggests, from Pythagoras' theorem (154 ◀):

185

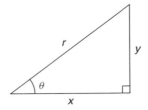

We have

$$x^2 + y^2 = r^2$$

Dividing by r^2 gives

$$\left(\frac{x}{r}\right)^2 + \left(\frac{y}{r}\right)^2 = 1$$

or $\cos^2 \theta + \sin^2 \theta = 1$

This identity should definitely become second nature to you – it is absolutely vital. By dividing through by $\cos^2 \theta$ we get $1 + \tan^2 \theta = \sec^2 \theta$ and dividing through by $\sin^2 \theta$ gives $\cot^2 \theta + 1 = \csc^2 \theta$, so there is no need to remember all three identities.

Solution to review question 6.1.6

(i) Using $\cos^2 \theta + \sin^2 \theta = 1$ we have

$$\cos^2 \theta = 1 - \sin^2 \theta = 1 - \frac{1}{49} = \frac{48}{49}$$

so

$$\cos \theta = \sqrt{\frac{48}{49}} \text{ in first quadrant}$$

$$= \frac{4\sqrt{3}}{7}$$

Then $\tan \theta = \dfrac{\sin \theta}{\cos \theta} = \dfrac{1/7}{4\sqrt{3}/7} = \dfrac{1}{4\sqrt{3}}$

(ii) $\sin \theta = \sqrt{1 - \cos^2 \theta} = \sqrt{1 - \dfrac{1}{3}} = \sqrt{\dfrac{2}{3}}$

$$\tan \theta = \frac{\sqrt{\dfrac{2}{3}}}{\dfrac{1}{\sqrt{3}}} = \sqrt{2}$$

(iii) $\sec^2 \theta = \dfrac{1}{\cos^2 \theta} = 1 + \tan^2 \theta$

$$= 1 + \left(\frac{1}{\sqrt{2}}\right)^2 = \frac{3}{2}$$

so
$$\cos^2\theta = \frac{2}{3}$$
$$\cos\theta = \sqrt{\frac{2}{3}}$$
$$\sin^2\theta = 1 - \cos^2\theta = 1 - \frac{2}{3} = \frac{1}{3}$$

and therefore
$$\sin\theta = \frac{1}{\sqrt{3}}$$

6.2.7 Compound angle formulae

◀ 173 196 ▶

The basic compound angle formulae are:

$$\sin(A + B) \equiv \sin A \cos B + \cos A \sin B$$
$$\sin(A - B) \equiv \sin A \cos B - \cos A \sin B$$
$$\cos(A + B) \equiv \cos A \cos B - \sin A \sin B$$
$$\cos(A - B) \equiv \cos A \cos B + \sin A \sin B$$
$$\tan(A + B) \equiv \frac{\tan A + \tan B}{1 - \tan A \tan B}$$
$$\tan(A - B) \equiv \frac{\tan A - \tan B}{1 + \tan A \tan B}$$

Note that there is no need to remember all these results separately. We can get all of them from the result for $\sin(A + B)$, and the elementary properties of the trig ratios (see solution to review question). Also, get used to 'knowing them backwards' that is going from right to left, recognising how the right-hand side simplifies to the left-hand side. Since the $\sin(A + B)$ result is so important we will take the trouble to prove it. The proof is instructive since it contains lots of ideas already covered.

We use one of those mystical constructions which makes geometry so pretty. Consider Figure 6.13.

PQ is drawn at Q perpendicular to the line making angle A with the base line OT, and meets the corresponding line defining the compound angle $A + B$ at P. PS and QT are dropped perpendicular to OT and RQ is then drawn parallel to OT. By alternate angles (149 ◀) $\angle OQR = A$. Since PQR is a right angled triangle $\angle QPR$ is the complement (149 ◀) of $\angle RQP$ and is therefore equal to $\angle OQR = \angle QOT = A$. Also, $RS = QT$.

$$\text{Now } \sin(A + B) = \frac{PS}{OP} = \frac{PR + RS}{OP} = \frac{PR + QT}{OP}$$
$$= \frac{PR}{OP} + \frac{QT}{OP} = \frac{QT}{OQ} \cdot \frac{OQ}{OP} + \frac{PR}{QP} \cdot \frac{QP}{OP}$$
$$= \sin A \cos B + \cos A \sin B$$

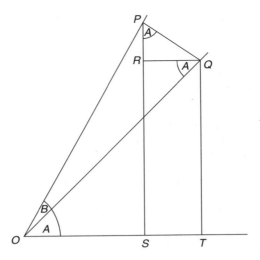

Figure 6.13 Proof of $\sin(A+B) = \sin A \cos B + \cos A \sin B$.

Putting $A = B$ in the compound angle identities immediately gives the **double angle formulae**:

$$\sin 2A \equiv \sin A \cos A$$
$$\cos 2A \equiv \cos^2 - \sin^2 A$$
$$\equiv 2\cos^2 A - 1$$
$$\equiv 1 - 2\sin^2 A$$
$$\tan A \equiv \frac{2\tan A}{1 - \tan^2 A}$$

The $\cos 2A$ results are often more useful in the form

$$\sin^2 A = \tfrac{1}{2}(1 - \cos 2A)$$
$$\cos^2 A = \tfrac{1}{2}(1 + \cos 2A)$$

From the double angle formulae we easily deduce the **half angle identities**. If $t = \tan \theta/2$, then:

$$\tan \theta \equiv \frac{2t}{1-t^2} \qquad \sin \theta \equiv \frac{2t}{1+t^2} \qquad \cos \theta \equiv \frac{1-t^2}{1+t^2}$$

For example:

$$\tan \theta = \tan(2 \times \theta/2) = \frac{2\tan \theta/2}{1 - \tan^2 \theta/2} = \frac{2t}{1-t^2}$$

from the double angle formula.

Solution to review question 6.1.7

A. $\sin(A+B) = \sin A \cos B + \sin B \cos A$ is so important it should be at your fingertips.

B. (i) $\sin(A - B) = \sin(A + (-B))$
$= \sin A \cos(-B) + \sin(-B) \cos A$
$= \sin A \cos B - \sin B \cos A$

on using

$\cos(-x) = \cos x$ and $\sin(-x) = -\sin x$

(ii) The result for $\cos(A - B)$ may now be obtained using complementary angles (149 ◄):

$\cos(A - B) = \sin(90 - (A - B))$
$= \sin(90 - A + B)$
$= \sin(90 - A) \cos B + \cos(90 - A) \sin B$
$= \cos A \cos B + \sin A \sin B$

Or, perhaps quicker, simply differentiate (➤ 233) the result for $\sin(A - B)$ with respect to B – a common trick for obtaining 'cos-results' from 'sin-results'.

(iii) The result for $\tan(A + B)$ can be obtained by division:

$$\tan(A + B) = \frac{\sin(A + B)}{\cos(A + B)} = \frac{\sin A \cos B + \sin B \cos A}{\cos A \cos B - \sin A \sin B}$$

$$= \frac{\tan A + \tan B}{1 - \tan A \tan B}$$

on dividing top and bottom by $\cos A \cos B$.

The key point here is that while **certain** results have to be second nature to you, such as the expansion of $\sin(A + B)$, it is just as important that you know how to derive others from them, rather than have to remember all the identities.

C. Here we make good use of the compound angle formulae

(i) $\cos 75° = \cos(45° + 30°)$
$= \cos 45° \cos 30° - \sin 45° \sin 30°$
$= \dfrac{1}{\sqrt{2}} \dfrac{\sqrt{3}}{2} - \dfrac{1}{\sqrt{2}} \dfrac{1}{2} = (\sqrt{3} - 1) \dfrac{\sqrt{2}}{4}$

(ii) $\sin 105° = \sin(60° + 45°)$
$= \sin 60° \cos 45° + \sin 45° \cos 60°$
$= \dfrac{\sqrt{3}}{2} \dfrac{1}{\sqrt{2}} + \dfrac{1}{\sqrt{2}} \dfrac{1}{2}$
$= \dfrac{\sqrt{3} + 1}{2\sqrt{2}} = \dfrac{(\sqrt{3} + 1)\sqrt{2}}{4}$

(iii) $\tan(-75°) = -\tan 75°$
$= -\tan(45° + 30°)$

$$= \frac{-(\tan 45° + \tan 30°)}{1 - \tan 45° \tan 30°}$$

$$= -\frac{\left(1 + \frac{1}{\sqrt{3}}\right)}{\left(1 - \frac{1}{\sqrt{3}}\right)} = \frac{1 + \sqrt{3}}{1 - \sqrt{3}} = -2 + \sqrt{3} \quad (21 \blacktriangleleft)$$

D. From $\cos(A + B) = \cos A \cos B - \sin A \sin B$, with $A = B$ we get

$$\cos 2A = \cos^2 A - \sin^2 A$$

Using $\sin^2 A + \cos^2 A = 1$ this can be expressed in two alternative forms:

$$\cos 2A = 2\cos^2 A - 1$$
$$= 1 - 2\sin^2 A$$

E. (i) Using $\cos 2\theta = 1 - 2\sin^2 \theta$

$$\cos 30° = 1 - 2\sin^2 15°$$

so $\sin^2 15° = \frac{1}{2}(1 - \cos 30°)$

$$= \frac{1}{2}\left(1 - \frac{\sqrt{3}}{2}\right)$$

$$= \frac{2 - \sqrt{3}}{4}$$

so $\sin 15° = \dfrac{\sqrt{2 - \sqrt{3}}}{2}$

(ii) $\tan 2A = \dfrac{2\tan A}{1 - \tan^2 A}$ with $A = 15°$ gives

$$\tan 30° = \frac{1}{\sqrt{3}} = \frac{2\tan 15°}{1 - \tan^2 15°} = \frac{2x}{1 - x^2}$$

where $x = \tan 15°$

So $1 - x^2 = 2\sqrt{3}x$

or $x^2 + 2\sqrt{3}x - 1 = 0$ giving

$$x = \frac{-2\sqrt{3} \pm \sqrt{4 \times 3 + 4}}{2}$$

$$= \frac{-2\sqrt{3} \pm 4}{2}$$

$$= 2 - \sqrt{3} \quad (\tan 15° \text{ is positive})$$

6.2.8 Trigonometric equations

A trigonometric equation is any equation containing ratios of an 'unknown' angle θ, to be determined from the equation. One attempts to solve such an equation by manipulating it (possibly using trig identities) until it can be solved by solutions of one or more equations of the form

$$\sin \theta = a$$
$$\cos \theta = b$$
$$\tan \theta = c$$

As noted in Section 6.2.5, in general the solution to such equations will not be unique, but will lead to multiple values of θ. The **general solution** is that which specifies **all possible solutions**. Such general solutions can always be expressed as simple extensions of **principal value solutions**, which are those for which θ is confined to a specific range in which the above basic equations have unique solutions. Thus, from Section 6.2.5 the principal values for the elementary trigonometric ratios are:

$$\sin \theta, \quad \theta \in \left[-\frac{\pi}{2}, \frac{\pi}{2}\right] \text{ which means } -\frac{\pi}{2} \leq \theta \leq \frac{\pi}{2}$$
$$\cos \theta, \quad \theta \in [0, \pi] \text{ or } 0 \leq \theta \leq \pi$$
$$\tan \theta, \quad \theta \in \left[-\frac{\pi}{2}, \frac{\pi}{2}\right] \text{ or } -\frac{\pi}{2} \leq \theta \leq \frac{\pi}{2}$$

In each case there is only **one** principal value solution to each of the equations given above. If we are asked for a solution in a specific range of θ then we have to examine such equations as $\sin \theta = a$ to determine which solutions fall in this range.

An important type of equation is

$$f(A) = f(B)$$

where f is an elementary trig function. The general solutions for such equations are

For $\cos A = \cos B$, we have $A = 2n\pi \pm B$

For $\sin A = \sin B$, we have $A = 2n\pi + B$ or $(2k+1)\pi - B$

For $\tan A = \tan B$, we have $A = n\pi + B$

where in each case, n is an arbitrary integer. These results can be confirmed by inspecting the graphs in Figures 6.8–10. The solution to the review question provides an example.

Solution to review question 6.1.8

(i) $\sin \theta + 2 \sin \theta \cos \theta = \sin \theta (1 + 2 \cos \theta) = 0$
implies either $\sin \theta = 0$ or $1 + 2 \cos \theta = 0$
$\sin \theta = 0$ has solutions $\theta = k\pi$ where k is an integer
The equation $1 + 2 \cos \theta = 0$ or $\cos \theta = -\frac{1}{2}$ has general solution

$$\theta = \frac{2\pi}{3} + 2m\pi \text{ or } -\frac{2\pi}{3} + 2n\pi$$

where m, n are integers.
So the general solution is
$$\theta = k\pi, \text{ or } \frac{2\pi}{3} + 2m\pi, \text{ or } -\frac{2\pi}{3} + 2n\pi \qquad k, m, n \text{ are integers}$$
NB – it is a common error to forget the $\sin\theta = 0$ part of the solution.

(ii) $\cos 3\theta = \cos\theta$ is an example of $\cos A = \cos B$ for which we know
$3\theta = 2k\pi \pm \theta$ where $k =$ an integer
So $4\theta = 2k\pi$ or $2\theta = 2k\pi$

We therefore obtain
$$\theta = \frac{2k\pi}{4} = \frac{k\pi}{2} \quad \text{or} \quad \frac{2k\pi}{2} = k\pi$$

and so the general solution is $\frac{k\pi}{2}$ or $k\pi$, with $k =$ an integer.

6.2.9 The $a\cos\theta + b\sin\theta$ form

◀ 173 197 ▶

The compound angle identities may be used, with a little help from Pythagoras (154 ◀), to convert an expression of the form $a\cos\theta + b\sin\theta$ to one of the more manageable forms $r\cos(\theta \pm \alpha)$ or $r\sin(\theta \pm \alpha)$. The latter forms, for example, tell us immediately by inspection the maximum and minimum values of such expressions, and where they occur. This section uses a lot of what we have done, in particular the 'minimal set' of results to remember – $\cos^2\theta + \sin^2\theta = 1$, and the expansions of $\sin(A + B)$ and $\cos(A + B)$. To illustrate conversion of this form, we consider the example of

$$a\cos\theta + b\sin\theta \equiv r\sin(\theta + \alpha)$$

Using the compound angle formula this implies

$$a\cos\theta + b\sin\theta \equiv r\cos\alpha\sin\theta + r\sin\alpha\cos\theta$$

This has got to be true for all possible values of θ, and to ensure this we equate the coefficients of $\cos\theta$, $\sin\theta$ on each side to get

$r\cos\alpha = a$

$r\sin\alpha = b$

where, by squaring and adding (185 ◀), $r = \sqrt{a^2 + b^2}$ and

$$\cos\alpha = \frac{a}{r}$$
$$\sin\alpha = \frac{b}{r}$$

from which α may be determined. Care is needed here if either a or b is negative – we have to look carefully at the quadrant that α lies in.

Such a conversion simplifies the solution of equations such as

$$a \cos\theta + b \sin\theta = c$$

by conversion to the form

$$\cos(\theta + \alpha) = \frac{c}{r} = \frac{c}{\sqrt{a^2 + b^2}}$$

for example.

Solution to review question 6.1.9

(i) We assume that we can write

$$\cos\theta + \sin\theta = r\sin(\theta + \alpha)$$
$$= r\sin\theta\cos\alpha + r\cos\theta\sin\alpha$$

then
$$r\cos\alpha = 1$$
$$r\sin\alpha = 1$$

so $r^2 = 1^2 + 1^2 = 2$

and $r = \sqrt{2}$

Then $\cos\alpha = \dfrac{1}{\sqrt{2}}$

$\sin\alpha = \dfrac{1}{\sqrt{2}}$

so $\tan\alpha = 1$, in the first quadrant, and hence

$$\alpha = 45°$$

and $\cos\theta + \sin\theta = \sqrt{2}\sin(\theta + 45°)$

(ii) In this case

$$\cos\theta + \sin\theta = r\cos(\theta + \alpha)$$
$$= r\cos\theta\cos\alpha - r\sin\theta\sin\alpha$$

so $r\cos\alpha = 1$

$r\sin\alpha = -1$

As before $r = \sqrt{2}$, so

$$\cos\alpha = \frac{1}{\sqrt{2}} \quad \sin\alpha = -\frac{1}{\sqrt{2}}$$

The appropriate solution in this case is

$$\alpha = -45°$$

so
$$\cos\theta + \sin\theta = \sqrt{2}\cos(\theta - 45°)$$

6.3 Reinforcement

6.3.1 Radian measure and the circle

A. Express as radians:

(i) 36° (ii) 101° (iii) 120° (iv) 250°
(v) 340° (vi) −45° (vii) −110° (viii) 15°
(ix) 27° (x) 273°

B. Express the following radian measures in degrees in the range $0° \leq \theta < 360°$.

(i) $\dfrac{2\pi}{3}$ (ii) 14π (iii) $-\dfrac{\pi}{2}$ (iv) $\dfrac{\pi}{3}$

(v) $\dfrac{\pi}{6}$ (vi) $\dfrac{5\pi}{2}$ (vii) $\dfrac{2\pi}{9}$ (viii) $\dfrac{5\pi}{4}$

(ix) $-\dfrac{2\pi}{5}$ (x) $\dfrac{\pi}{12}$

C. Determine the length of arc and the area of the sector subtended by the following angles in a circle of radius 4 cm.

(i) 15° (ii) 30° (iii) 45° (iv) 60°
(v) 90° (vi) 120° (vii) 160° (viii) 180°

6.3.2 Definitions of the trig ratios

A. Write down the exact values of sine, tan, sec, and complementary ratios for all 'special' angles $0, \pi/2, \pi/3, \pi/4, \pi/6$.

B. Classify as odd or even functions: cos, sin, tan, sec, cosec, cot.

C. Express as trig functions of x (n is an integer):

(i) $\sin(x + n\pi)$ (ii) $\cos(x + n\pi)$ (iii) $\tan(x + n\pi)$

(iv) $\sin\left(x + \dfrac{n\pi}{2}\right)$ (v) $\cos\left(x + \dfrac{n\pi}{2}\right)$ (vi) $\tan\left(x + \dfrac{n\pi}{2}\right)$

where n is an integer.

6.3.3 Sine and cosine rules and the solution of triangles

With the standard notation solve the following triangles.

(i) $A = 70°$, $C = 60°$, $b = 6$
(ii) $B = 40°$, $b = 8$, $c = 10$
(iii) $A = 40°$, $a = 5$, $c = 2$
(iv) $a = 5$, $b = 6$, $c = 7$

(v) $A = 40°$, $\quad\quad\quad\quad\quad b = 5,$ $\quad\quad\quad\quad\quad c = 6$
(vi) $A = 120°$, $\quad\quad\quad\quad b = 3,$ $\quad\quad\quad\quad\quad c = 5$

6.3.4 Graphs of trigonometric functions ◀◀ 172 180 ◀

Sketch the graphs of

(i) $2\sin\left(2t + \dfrac{\pi}{3}\right)$ $\quad\quad\quad\quad\quad\quad\quad$ (ii) $3\cos\left(3t - \dfrac{\pi}{2}\right)$

6.3.5 Inverse trigonometric functions ◀◀ 172 184 ◀

A. Find the principal value and general solutions for $\sin^{-1} x$ and $\cos^{-1} x$ for the following values:

(i) 0 $\quad\quad\quad\quad$ (ii) $\dfrac{1}{2}$ $\quad\quad\quad\quad$ (iii) $-\dfrac{1}{2}$

(iv) $\dfrac{\sqrt{3}}{2}$ $\quad\quad\quad$ (v) $\dfrac{1}{\sqrt{2}}$ $\quad\quad\quad$ (vi) $-\dfrac{1}{\sqrt{2}}$

B. Find the principal value and general solutions for $\tan^{-1} x$ for the following values of x:

(i) 0 $\quad\quad$ (ii) 1 $\quad\quad$ (iii) $\sqrt{3}$ $\quad\quad$ (iv) -1 $\quad\quad$ (v) $-\dfrac{1}{\sqrt{3}}$

6.3.6 The Pythagorean identities – $\cos^2 + \sin^2 = 1$ ◀◀ 172 185 ◀

A. (i) For $c = \cos\theta$, simplify

(a) $\sqrt{1 - c^2}$ $\quad\quad$ (b) $\dfrac{c}{\sqrt{1 - c^2}}$ $\quad\quad$ (c) $\dfrac{1 - c^2}{c^2}$

(ii) For $s = \sin\theta$, simplify

(a) $\sqrt{1 - s^2}$ $\quad\quad$ (b) $\dfrac{\sqrt{1 - s^2}}{s^2}$ $\quad\quad$ (c) $\dfrac{s}{1 - s^2}$

(iii) For $t = \tan\theta$, simplify

(a) $\sqrt{1 + t^2}$ $\quad\quad$ (b) $\dfrac{t}{(1 + t^2)}$ $\quad\quad$ (c) $\dfrac{1}{t\sqrt{1 + t^2}}$

B. Eliminate θ from the equations

(i) $x = a\cos\theta$, $y = b\sin\theta$ $\quad\quad\quad$ (ii) $x = a\sin\theta$, $y = b\tan\theta$

C. For the following values of $\sin\theta$, find the corresponding values of $\cos\theta$ and $\tan\theta$ without using your calculator, giving your answers in surd form, and assuming that θ is acute.

(i) $\dfrac{2}{5}$ $\quad\quad\quad\quad$ (ii) $\dfrac{1}{13}$ $\quad\quad\quad\quad$ (iii) $\dfrac{7}{25}$

D. If $r\cos\theta = 3$ and $r\sin\theta = 4$ determine the positive value of r, and the principal value of θ.

6.3.7 Compound angle formulae ◄◄ 173 187 ◄

A. Prove the following

(i) $\sin 3\theta = 3\sin\theta - 4\sin^3\theta$

(ii) $\cos 3\theta = 4\cos^3\theta - 3\cos\theta$

(iii) $\dfrac{\cos 2\theta}{\cos\theta + \sin\theta} = \cos\theta - \sin\theta$

(iv) $\cot\theta - \tan\theta = 2\cot 2\theta$

(v) $\cot 2\theta = \dfrac{\cot^2\theta - 1}{2\cot\theta}$

B. Without using a calculator or tables evaluate

(i) $\sin 15° \cos 15°$ (ii) $\sin 15°$ (iii) $\tan(\pi/12)$ (iv) $\cos(11\pi/12)$

(v) $\tan(7\pi/12)$ (vi) $\cos 75°$

C. Evaluate

(i) $\sin 22.5°$ (ii) $\cos 22.5°$ (iii) $\tan 22.5°$

given that $\cos 45° = 1/\sqrt{2}$.

D. Express the following products as sums or differences of sines and/or cosines of multiple angles

(i) $\sin 2x \cos 3x$ (ii) $\sin x \sin 4x$ (iii) $\cos 2x \sin x$ (iv) $\cos 4x \cos 5x$

E. Prove the following identities (Hint: put $P = (A+B)/2$, $Q = (A-B)/2$ in the left-hand sides, expand and simplify and re-express in terms of P and Q)

(i) $\sin P + \sin Q \equiv 2\sin\left(\dfrac{P+Q}{2}\right)\cos\left(\dfrac{P-Q}{2}\right)$

(ii) $\sin P - \sin Q \equiv 2\cos\left(\dfrac{P+Q}{2}\right)\sin\left(\dfrac{P-Q}{2}\right)$

(iii) $\cos P + \cos Q \equiv 2\cos\left(\dfrac{P+Q}{2}\right)\cos\left(\dfrac{P-Q}{2}\right)$

(iv) $\cos P - \cos Q \equiv -2\sin\left(\dfrac{P+Q}{2}\right)\sin\left(\dfrac{P-Q}{2}\right)$

6.3.8 Trigonometric equations ◄◄ 173 191 ◄

A. Give the general solution to each of the equations:

(i) $\cos\theta = 0$ (ii) $\cos\theta = -1$ (iii) $\cos\theta = -\dfrac{\sqrt{3}}{2}$

(iv) $\sin\theta = 0$ (v) $\sin\theta = -1$ (vi) $\sin\theta = \sqrt{3}$
(vii) $\tan\theta = 0$ (viii) $\tan\theta = -1$ (ix) $\tan\theta = \sqrt{3}$

B. Find the general solutions of the equations

(i) $\cos 2\theta = 1$ (ii) $\sin 2\theta = \sin\theta$
(iii) $\cos 2\theta + \sin\theta = 0$ (iv) $\cos 2\theta + \cos 3\theta = 0$
(v) $\sec^2\theta = 3\tan\theta - 1$

6.3.9 The $a\cos\theta + b\sin\theta$ form

A. Write the following in the form (a) $r\sin(\theta + \alpha)$, (b) $r\cos(\theta + \alpha)$

(i) $\sin\theta - \cos\theta$ (ii) $\sqrt{3}\cos\theta + \sin\theta$ (iii) $\sqrt{3}\sin\theta - \cos\theta$
(iv) $3\cos\theta + 4\sin\theta$

B. Determine the maximum and minimum values of each of the expressions in Question **A**, stating the values where they occur, in the range $0 \leq \theta \leq 2\pi$.

C. Find the solutions, in the range $0 \leq \theta \leq 2\pi$, of the equations obtained by equating the expressions in Question **A** to (a) 1 (b) -1.

6.4 Applications

1. In the theory of the elasticity of solids a crucial step in establishing the connection between the elastic constants and Poisson's ratio is to determine a relation between the normal strain in the 45° direction, ε, and the shear strain for a square section of the sort shown in Figure 6.14. By considering Figure 6.14b, where the diagonal BD is stretched by a factor $1 + \varepsilon$, and using the fact that for small angles γ (in radians) $\sin\gamma \cong \gamma$, show that

$$\varepsilon \cong \frac{\gamma}{2}$$

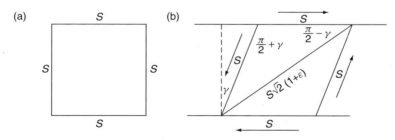

Figure 6.14

2. In alternating current theory we often need to add sinusoidal waves of the same frequency but different amplitude and phase, such as $A_1 \sin(\omega t + \alpha_1)$ and

$A_2 \sin(\omega t + \alpha_2)$. When we come to complex numbers we will see how this sort of thing can be done using objects called **phasors** (371 ◄), but really such methods are simply shorthand for the ideas covered in this chapter. In particular we can add two waves that are 90° out of phase using the results of Section 6.2.9, since $A_1 \sin(\omega t + 90°) + A_2 \sin(\omega t)$ is equivalent to $a \cos(\omega t) + b \sin(\omega t)$. For additions such as $A_1 \sin(\omega t + \alpha) + A_2 \sin(\omega t)$, where α is other than 90°, phasor methods are equivalent to constructing a parallelogram with sides A_1 and A_2 and included angle α and taking the combined amplitude as the length of the diagonal, r, and the combined phase to be the angle, θ, made by the diagonal with the side A_2 (see Figure 6.15).

$$A_1 \sin(\omega t + \alpha) + A_2 \sin(\omega t) = r \sin(\omega t + \theta)$$

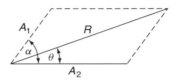

Figure 6.15

Use the above methods to find the sine waves representing:

(i) $4 \sin(\omega t) + 3 \cos(\omega t)$
(ii) $6 \sin(\omega t) + 4 \sin(\omega t + 45°)$

Answers to reinforcement exercises

6.3.1 Radian measure and the circle

A. (i) $\dfrac{\pi}{5}$ (ii) $\dfrac{101\pi}{180}$ (iii) $\dfrac{2\pi}{3}$ (iv) $\dfrac{25\pi}{18}$

(v) $\dfrac{17\pi}{9}$ (vi) $-\dfrac{\pi}{4}$ (vii) $-\dfrac{11\pi}{18}$ (viii) $-\dfrac{\pi}{12}$

(ix) $\dfrac{3\pi}{20}$ (x) $\dfrac{91\pi}{60}$

B. (i) 120° (ii) 0° (iii) 270° (iv) 60°

(v) 30° (vi) 90° (vii) 40° (viii) 225°

(ix) 308° (x) 15°

C. (i) $\dfrac{\pi}{3}, \dfrac{2\pi}{3}$ (ii) $\dfrac{2\pi}{3}, \dfrac{4\pi}{3}$ (iii) $\pi, 2\pi$ (iv) $\dfrac{4\pi}{3}, \dfrac{8\pi}{3}$

(v) $2\pi, 4\pi$ (vi) $\dfrac{8\pi}{3}, \dfrac{16\pi}{3}$ (vii) $\dfrac{32\pi}{9}, \dfrac{64\pi}{9}$ (viii) $4\pi, 8\pi$

Note: If you are adept with ratios you will have noticed that with the given radius the area will always have a magnitude double that of the arc length.

6.3.2 Definitions of the trig ratios

A.

θ	0	$\pi/2$	$\pi/3$	$\pi/4$	$\pi/6$
$\sin\theta$	0	1	$\dfrac{\sqrt{3}}{2}$	$\dfrac{1}{\sqrt{2}}$	$\dfrac{1}{2}$
$\tan\theta$	0	nd	$\sqrt{3}$	1	$\dfrac{1}{\sqrt{3}}$
$\sec\theta$	1	nd	2	$\sqrt{2}$	$\dfrac{2}{\sqrt{3}}$
$\cos\theta$	1	0	$\dfrac{1}{2}$	$\dfrac{1}{\sqrt{2}}$	$\dfrac{\sqrt{3}}{2}$
$\cot\theta$	nd	0	$\dfrac{1}{\sqrt{3}}$	1	$\sqrt{3}$
$\csc\theta$	nd	1	$\dfrac{2}{\sqrt{3}}$	$\sqrt{2}$	2

(nd = not defined)

B. cos and sec are even; sin, cosec, tan, cot all odd.

C. (i) $(-1)^n \sin x$ (ii) $(-1)^n \cos x$ (iii) $\tan x$

(iv) $(-1)^{(n-1)/2} \cos x$ if n odd, $(-1)^{n/2} \sin x$ if n even

(v) $(-1)^{n/2} \cos x$ if n even, $(-1)^{\frac{n+1}{2}} \sin x$ if n odd

(vi) $\tan x$ if n even, $-\cot x$ if n odd.

6.3.3 Sine and cosine rules and the solution of triangles

(i) $B = 50°$ $a = 7.36$, $c = 6.78$

(all answers to 2 decimal places)

(ii) Two solutions:

$C = 53.46$ $A = 86.54°$, $a = 12.42$

$C = 126.54°$, $A = 13.36°$, $a = 2.9$

(iii) $C = 14.9°$, $B = 125.1°$, $b = 6.36$

(the obtuse solution for C is inadmissible in this case)

(iv) $A = 44.42°$, $B = 57.11°$, $C = 78.47°$

(v) $a = 3.88$, $B = 55.98°$, $C = 84.02°$

(vi) $a = 7$, $B = 21.77°$, $C = 38.21°$

Understanding Engineering Mathematics

6.3.4 Graphs of trigonometric functions

(i)

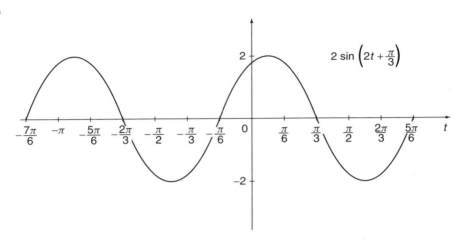

(ii)

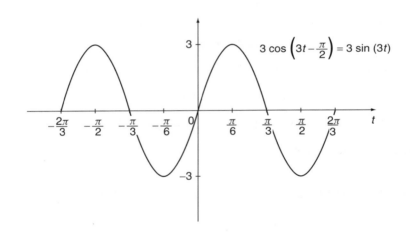

6.3.5 Inverse trigonometric functions

A.

	x	$\sin^{-1} x$ PV	$\sin^{-1} x$ GS	$\cos^{-1} x$ PV	$\cos^{-1} x$ GS
(i)	0	0	$n\pi$	$\dfrac{\pi}{2}$	$2n\pi \pm \dfrac{\pi}{2}$
(ii)	$\dfrac{1}{2}$	$\dfrac{\pi}{6}$	$n\pi + (-1)^n \dfrac{\pi}{6}$	$\dfrac{\pi}{3}$	$2n\pi \pm \dfrac{\pi}{3}$
(iii)	$-\dfrac{1}{2}$	$-\dfrac{\pi}{6}$	$n\pi + (-1)^{n+1} \dfrac{\pi}{6}$	$\dfrac{2\pi}{3}$	$2n\pi \pm \dfrac{2\pi}{3}$

(iv)	$\dfrac{\sqrt{3}}{2}$	$\dfrac{\pi}{3}$	$n\pi + (-1)^n \dfrac{\pi}{3}$	$\dfrac{\pi}{6}$	$2n\pi \pm \dfrac{\pi}{6}$
(v)	$\dfrac{1}{\sqrt{2}}$	$\dfrac{\pi}{4}$	$n\pi + (-1)^n \dfrac{\pi}{4}$	$\dfrac{\pi}{4}$	$2n\pi \pm \dfrac{\pi}{4}$
(vi)	$-\dfrac{1}{\sqrt{2}}$	$-\dfrac{\pi}{4}$	$n\pi + (-1)^{n+1} \dfrac{\pi}{4}$	$\dfrac{3\pi}{4}$	$2n\pi \pm \dfrac{3\pi}{4}$

where n is an integer

B.

$\tan^{-1} x$	0	1	$\sqrt{3}$	-1	$-\dfrac{1}{\sqrt{3}}$
PV	0	$\dfrac{\pi}{4}$	$\dfrac{\pi}{3}$	$-\dfrac{\pi}{4}$	$-\dfrac{\pi}{6}$
GS	$n\pi$	$n\pi + \dfrac{\pi}{4}$	$n\pi + \dfrac{\pi}{3}$	$n\pi - \dfrac{\pi}{4}$	$n\pi - \dfrac{\pi}{6}$

6.3.6 The Pythagorean identities – $\cos^2 + \sin^2 = 1$

A. You may obtain different forms of the answers – consider it a further exercise to check their equivalence to the following!

(i) (a) $\sin\theta$ (b) $\cot\theta$ (c) $\tan^2\theta$

(ii) (a) $\cos\theta$ (b) $\csc\theta \cot\theta$ (c) $\sec\theta \tan\theta$

(iii) (a) $\sec\theta$ (b) $\sin\theta \cos\theta$ (c) $\cos\theta \cot\theta$

B. (i) $\dfrac{x^2}{a^2} + \dfrac{y^2}{b^2} = 1$ (ii) $y = \dfrac{bx}{\sqrt{a^2 - x^2}}$

C. (i) $\dfrac{\sqrt{21}}{5}, \dfrac{2}{\sqrt{21}}$ (ii) $\dfrac{2\sqrt{42}}{13}, \dfrac{1}{2\sqrt{42}}$ (iii) $\dfrac{24}{25}, \dfrac{7}{24}$

D. 5, 53.13°

6.3.7 Compound angle formulae

B. (i) $\dfrac{1}{4}$ (ii) $\dfrac{\sqrt{2}(\sqrt{3} - 1)}{4}$ (iii) $2 - \sqrt{3}$

(iv) $-\dfrac{\sqrt{2}(\sqrt{3} + 1)}{4}$ (v) $-(2 + \sqrt{3})$ (vi) $\dfrac{\sqrt{2}(\sqrt{3} - 1)}{4}$

C. (i) $\dfrac{\sqrt{(2 - \sqrt{2})}}{2}$ (ii) $\dfrac{\sqrt{2 + \sqrt{2}}}{2}$ (iii) $\dfrac{\sqrt{6}(2 - \sqrt{2})}{6}$

D. (i) $\frac{1}{2}(\sin 5x - \sin x)$ (ii) $\frac{1}{2}(\cos 3x - \cos 5x)$ (iii) $\frac{1}{2}(\sin 3x - \sin x)$
(iv) $\frac{1}{2}(\cos 9x + \cos x)$

6.3.8 Trigonometric equations

A. (i) $2n\pi \pm \frac{\pi}{2}$ (ii) Undefined, or '∞' (iii) $2n\pi \pm \frac{5\pi}{6}$ (iv) $n\pi$
(v) $n\pi + (-1)^n \frac{\pi}{2}$ (vi) No solutions (vii) $n\pi$ (viii) $n\pi + \frac{3\pi}{4}$

B. (i) $n\pi$ (ii) $2n\pi$ or $\frac{(2n+1)}{3}\pi$ (iii) $\frac{2}{3}n\pi + \frac{\pi}{2}$
(iv) $\frac{(2n+1)}{5}\pi$ (v) $n\pi + \frac{\pi}{4}$ (vi) $n\pi + 1.11$

6.3.9 The $a\cos\theta + b\sin\theta$ form

A. (i) (a) $\sqrt{2}\sin\left(\theta - \frac{\pi}{4}\right)$ (b) $\sqrt{2}\cos\left(\theta - \frac{3\pi}{4}\right)$
(ii) (a) $2\sin\left(\theta + \frac{\pi}{3}\right)$ (b) $2\cos\left(\theta - \frac{\pi}{6}\right)$
(iii) (a) $2\sin\left(\theta - \frac{\pi}{6}\right)$ (b) $2\cos\left(\theta - \frac{5\pi}{6}\right)$
(iv) (a) $5\sin(\theta + 36.9°)$ (b) $5\cos(\theta - 53.1°)$

B. (i) Max of $\sqrt{2}$ at $\frac{3\pi}{4}$ Min of $-\sqrt{2}$ at $\frac{7\pi}{4}$
(ii) Max of 2 at $\frac{\pi}{6}$ Min of -2 at $\frac{7\pi}{6}$
(iii) Max of 2 at $\frac{2\pi}{3}$ Min of -2 at $\frac{5\pi}{3}$
(iv) Max of 5 at 2.21 rads   Min of -5 at 5.35 rads

C. (i) (a) $\frac{\pi}{2}, \frac{3\pi}{2}$ (b) $0, \pi$
(ii) (a) $\frac{\pi}{2}, \frac{5\pi}{3}$ (b) $\frac{5\pi}{6}, \frac{3\pi}{2}$
(iii) (a) $\frac{\pi}{3}, \pi$ (b) $\frac{4\pi}{3}, 2\pi$
(iv) (a) 2.3 rads, 5.84 rads (b) 2.7 rads, 5.44 rads

7

Coordinate Geometry

Coordinate Geometry turns 'diagrammatic' geometry of the sort discussed in Chapter 5 into algebraic and numerical equations. It is the language that we have to use when 'talking' to computers about geometry. Here we focus on the essential basics that are needed in science and engineering mathematics.

Prerequisites

It will be helpful if you know something about:

- plotting graphs and simple plane coordinate systems (91 ◄)
- Pythagoras' theorem (154 ◄)
- ratio and proportion (14 ◄)
- systems of linear equations (48 ◄)
- intercept theorem (153 ◄)
- simple trig ratios and identities (175 ◄)
- completing the square (66 ◄)

Objectives

In this chapter you will find:

- coordinate systems in a plane
- distance between two points in Cartesian coordinates
- midpoint and gradient of a line segment
- equation of a straight line
- parallel and perpendicular lines
- intersection of lines
- equation of a circle
- parametric representation of curves

Motivation

You may need the material in this chapter for:

- geometrical aspects of calculus (➤ Chapter 10)
- numerical methods
- study of laws converted to straight line form (135 ◄)
- linear regression in statistics
- vectors (➤ Chapter 11)

7.1 Review

7.1.1 Coordinate systems in a plane

A. Plot the points with Cartesian coordinates

(i) (0, 0)   (ii) (0, 1)   (iii) (−1, 3)

(iv) (−2, 4)   (v) (−2, −3)   (vi) (0, −1)

(vii) (3, 3)   (viii) (3, −2)

B. Plot the points with polar coordinates (r, θ)

(i) (0, 0)   (ii) (0, π)   (iii) (1, 0)   (iv) $\left(1, \dfrac{\pi}{2}\right)$

(v) $\left(3, -\dfrac{\pi}{2}\right)$ (vi) $\left(2, \dfrac{\pi}{6}\right)$ (vii) $\left(6, \dfrac{5\pi}{4}\right)$ (viii) $\left(5, \dfrac{2\pi}{3}\right)$

7.1.2 Distance between two points

Find the distance between the following pairs of points referred to rectangular Cartesian axes.

(i) (0, 0), (1, 1)   (ii) (1, 2), (1, 3)   (iii) (−2, 4), (1, −3)

(iv) (−1, −1), (−2, −3)

7.1.3 Midpoint and gradient of a line

Find the (a) mid-point and (b) the gradient of the line segments joining the pairs of points in Question 7.1.2.

7.1.4 Equation of a straight line

A. Determine the equations of the straight lines through the pairs of points in Question 7.1.2.

B. Find the gradient and the intercepts on the axes of the lines

(i) $y = 3x + 2$ (ii) $2x + 3y = 1$ (iii) $y = 4x$

(iv) $x + y + 1 = 0$

7.1.5 Parallel and perpendicular lines

A. Find the equation of the straight line parallel to the line $y = 1 - 3x$ and which passes through the point $(-1, 2)$

B. Find the equation of the straight line perpendicular to the line through the points $(-1, 2)$, $(0, 4)$ and passing through the first point.

7.1.6 Intersecting lines

Find all points where the following straight lines intersect.

(i) $x + y = 3$ (ii) $2x + 2y = -1$ (iii) $y = 3x - 1$

7.1.7 Equation of a circle

A. (a) Write down the equation of the circles with the centres and radii given

   (i) $(0, 0)$, 2   (ii) $(1, 2)$, 1   (iii) $(-1, 4)$, 3

(b) Determine the centre and radius of the circles given by the equations:

   (i) $x^2 + y^2 - 2x + 6y + 6 = 0$ (ii) $x^2 + y^2 + 4x + 4y - 1 = 0$

7.1.8 Parametric representation of curves

Eliminate the parameter t and obtain the Cartesian equation of the curve, in terms of x and y:

(i) $x = 3t + 1$, $y = t + 2$ (ii) $x = 6\cos 2t$, $y = 6\sin 2t$
(iii) $x = 2\cos t - 1$, $y = 2\sin t$ (iv) $x = \cos 2t$, $y = \cos t$

7.2 Revision

7.2.1 Coordinate systems in a plane

In Chapters 5 and 6 we never really had to fix our geometric objects in the plane in any way. A point such as the centre of a circle is special so far as the circle is concerned, but where it is located in the plane has no bearing on the properties of the circle. This characterises what is called **synthetic** or **axiomatic** geometry, in which the description of geometric objects such as lines and circles depends on their relation to each other.

However, by locating points in space by means of a **coordinate system**, assigning numbers, or **coordinates** to each point, we can represent geometrical ideas by **mathematical equations**. This is the subject of **coordinate geometry** – also called **analytical geometry**.

The simplest and most common coordinate system is the **Cartesian coordinate system** or **rectangular coordinate system** in a plane. An **origin** is chosen and two perpendicular lines through the origin are chosen as the **x- and y-axes** (91 ◄). The scale on each axis is taken to be the same in this chapter. Any point in the plane can then be specified by giving its x coordinate and y coordinate as an ordered pair (x, y). See, for examples, the solution to the review question.

Another two-dimensional coordinate system in common use is the **polar coordinate system**, especially convenient when dealing with engineering problems relating to circular motion for example. In polar coordinates an origin and an 'initial line' Ox radiating from this origin are chosen. The position of any point P is then referred to by specifying the distance, OP, from O to the point along a 'radius vector' whose angle, θ, with the initial line is also specified. Then the position of a point is specified by **polar coordinates** (r, θ) (Figure 7.1).

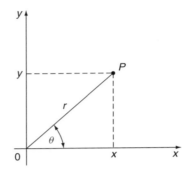

Figure 7.1 Cartesian (x, y) and polar (r, θ) coordinates in a plane.

In Figure 7.1 the polar coordinates are shown superimposed on a rectangular coordinate system where the positive x axis coincides with the initial line. With this arrangement the relation between the coordinates is seen to be (175 ◀):

$$x = r \cos \theta \quad y = r \sin \theta$$

Notice that in polar coordinates θ is measured anticlockwise and its value is restricted to $0 \leq \theta < 2\pi$ or sometimes $-\pi < \theta \leq \pi$ (175 ◀). Also, the origin is somewhat special ('singular' is the technical term) in polar coordinates – its r coordinate is $r = 0$, but its θ coordinate is not defined.

Example

To plot the curve with polar equation

$$r = 1 + 2\cos\theta$$

take some appropriate values of θ, and calculate $1 + 2\cos\theta$, as below.

θ	0	$\dfrac{\pi}{6}$	$\dfrac{\pi}{4}$	$\dfrac{\pi}{3}$	$\dfrac{\pi}{2}$	$\dfrac{2\pi}{3}$	$\dfrac{3\pi}{4}$	$\dfrac{5\pi}{6}$	π
$2\cos\theta$	2	1.732	1.414	1	0	-1	-1.414	-1.732	-2
$1 + 2\cos\theta$	3	2.732	2.414	2	1	0	-0.414	-0.732	-1

These values are plotted in Figure 7.2. Note that negative values of r are plotted in the opposite direction to the positive values, so for example $(-1, \pi)$ is plotted as $(1, 0)$ – in the opposite direction to π. Now for any angle θ, $\cos(-\theta) = \cos\theta$, therefore the same values of r will be obtained for negative values of θ, and we have only to reflect the curve in the initial line to obtain the complete plot.

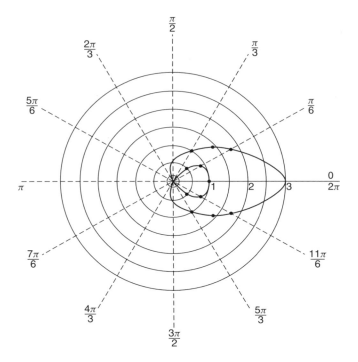

Figure 7.2 Polar plot of $r = 1 + 2\cos\theta$.

Solution to review question 7.1.1

A.

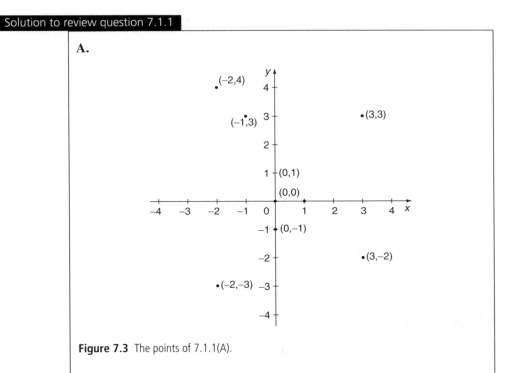

Figure 7.3 The points of 7.1.1(A).

B.

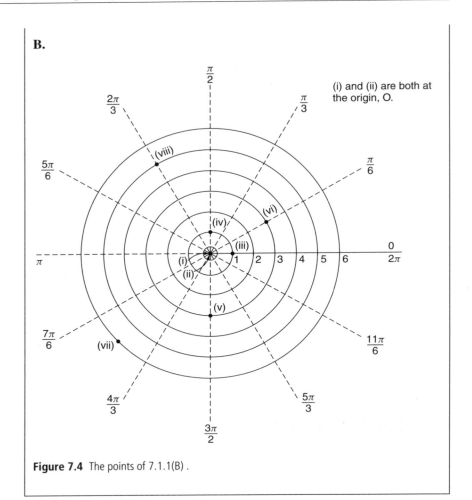

Figure 7.4 The points of 7.1.1(B).

7.2.2 Distance between two points

◀ 204 221 ▶

By Pythagoras' theorem (154 ◀) the distance between the two points (x_1, y_1), (x_2, y_2) is

$$d = \sqrt{(x_1 - x_2)^2 + (y_1 - y_2)^2}$$

as illustrated in Figure 7.5.
From the diagram:

$$d^2 = AB^2 = AC^2 + BC^2$$
$$= (x_2 - x_1)^2 + (y_2 - y_1)^2$$

Solution to review question 7.12

The distance between the points (x_1, y_1), (x_2, y_2), denoted $d((x_1, y_1), (x_2, y_2))$ is

$$d((x_1, y_1), (x_2, y_2)) = \sqrt{(x_2 - x_1)^2 + (y_2 - y_1)^2}$$

(i) $d((0, 0), (1, 1)) = \sqrt{(1 - 0)^2 + (1 - 0)^2} = \sqrt{2}$

(ii) $d((1, 2), (1, 3)) = \sqrt{(1 - 1)^2 + (3 - 2)^2} = \sqrt{1} = 1$

Note that in general

$$d((a, y_1), (a, y_2)) = |y_2 - y_1|$$

and

$$d((x_1, b), (x_2, b)) = |x_2 - x_1|$$

(iii) $d((-2, 4), (1, -3)) = \sqrt{(1 - (-2))^2 + (-3 - 4)^2} = \sqrt{58}$

(iv) $d((-1, -1), (-2, -3)) = \sqrt{(-2 + 1)^2 + (-3 + 1)^2} = \sqrt{5}$

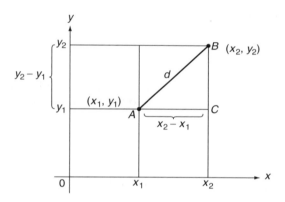

Figure 7.5 The distance between two points.

7.2.3 Midpoint and gradient of a line

Given two points (x_1, y_1), (x_2, y_2) the midpoint of the line segment between them is at the point

$$\left(\frac{x_1 + x_2}{2}, \frac{y_1 + y_2}{2} \right)$$

as may be seen from Figure 7.6.

To see this, note that by the intercept theorem (153 ◄) the x coordinate of the midpoint M of AB is the same as that of Q which divides BC in two. This is

$$x_1 + \frac{x_2 - x_1}{2} = \frac{x_1 + x_2}{2}$$

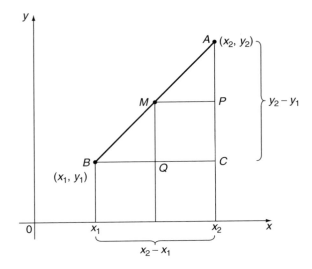

Figure 7.6 Midpoint of a line segment.

Similarly for the y coordinate – it is the same as that of P, which is

$$y_1 + \frac{y_2 - y_1}{2} = \frac{y_1 + y_2}{2}$$

More generally, again by using the intercept theorem or similar triangles (152 ◀), the point M that divides the line joining (x_1, y_1) to (x_2, y_2) in the ratio (14 ◀) $\mu : \lambda$ is

$$\left(\frac{\lambda x_1 + \mu x_2}{\lambda + \mu}, \frac{\lambda y_1 + \mu y_2}{\lambda + \mu} \right)$$

It may help in remembering this to note that the λ and μ are adjacent to the 'opposite' points:

```
           μ        M        λ
    •───────────────•───────────────•
  (x₁, y₁)                       (x₂, y₂)
```

The **gradient**, or **slope** of the line is given by

$$m = \frac{AC}{BC} = \frac{y_2 - y_1}{x_2 - x_1}$$

Note that this is the same as $\tan(\angle ABC)$. This holds for all values of x_1, x_2, y_1, y_2 except of course that $x_2 - x_1$ must be non zero, i.e. $x_1 \neq x_2$. The case where $x_1 = x_2$ actually corresponds to a vertical line, parallel to the y-axis. Crudely, you might think of it as having an infinite gradient.

If a line slopes upwards from right to left, i.e. y increases as x increases, then we have a positive gradient. If it slopes downwards to the right, or y decreases as x increases then the gradient is negative.

Solution to review question 7.1.3

(a) The midpoints of the line segment joining (x_1, y_1) and (x_2, y_2) has coordinates

$$\left(\frac{x_1 + x_2}{2}, \frac{y_1 + y_2}{2}\right)$$

(i) Applying this to the pair of points (0, 0), (1, 1), gives for the midpoint

$$\left(\frac{0+1}{2}, \frac{0+1}{2}\right) = \left(\frac{1}{2}, \frac{1}{2}\right)$$

(ii) For (1, 2), (1, 3) we get

$$\left(\frac{1+1}{2}, \frac{2+3}{2}\right) = \left(1, \frac{5}{2}\right)$$

(iii) For (−2, 4), (1, −3) we get

$$\left(\frac{-2+1}{2}, \frac{4-3}{2}\right) = \left(-\frac{1}{2}, \frac{1}{2}\right) \left(\begin{array}{c}\text{watch out} \\ \text{for the signs}\end{array}\right)$$

(iv) For (−1, −1), (−2, −3) we get

$$\left(\frac{-1-2}{2}, \frac{-1-3}{2}\right) = \left(-\frac{3}{2}, -2\right)$$

(b) The gradient of the line segment joining the points (x_1, y_1), (x_2, y_2) is

$$m = \frac{y_2 - y_1}{x_2 - x_1}$$

(i) For the points (0, 0), (1, 1) we get $m = \dfrac{1-0}{1-0} = 1$.

(ii) If we strictly apply the formula in the case of the points (1, 2), (1, 3) we get a 'gradient'

$$m = \frac{3-2}{1-1} = \frac{1}{0}$$

which of course is not defined. What is happening here is that the line segment is in fact **vertical** because the x-coordinates of the two points are the same. The gradient is 'infinite'. In such cases we have to use the formula with a bit of common sense.

(iii) For (−2, 4), (1 −3) the gradient is

$$m = \frac{-3-4}{1-(-2)} = -\frac{7}{3}$$

(iv) For $(-1, -1)$, $(-2, -3)$, the gradient is

$$m = \frac{-3+1}{-2+1} = 2$$

7.2.4 Equation of a straight line

Knowing that the **gradient** of the straight-line segment AB is

$$m = \frac{AC}{BC} = \frac{y_2 - y_1}{x_2 - x_1}$$

we can now derive an equation for the straight line through the two points A, B by equating its gradient to m. For a general point $P(x, y)$ on the line, its gradient is given by $(y - y_1)/(x - x_1)$, say, which also gives m:

$$\frac{y - y_1}{x - x_1} = \frac{y_2 - y_1}{x_2 - x_1} = m$$

Rearranging this (see RE7.3.4C) gives the equation of the straight line as

$$y - y_1 = \left(\frac{y_2 - y_1}{x_2 - x_1}\right)(x - x_1)$$

or

$$y - y_1 = m(x - x_1)$$

We can also write this in the form $y = mx + y_1 - mx_1$ and so the general form of the equation of a straight line can be written as:

$$y = mx + c$$

where m is the gradient of the line, and c is the **intercept on the y-axis**. This is illustrated in Figure 7.7.

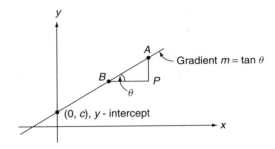

Figure 7.7 Gradient and intercept of a line $y = mx + c$.

The gradient is $m = AP/BP = \tan\theta$.

Two special cases require comment. For a horizontal line (i.e. parallel to the x-axis) the equation is $y = c$ (a constant), since the gradient is zero. For a vertical line (i.e. parallel to the y-axis) the equation is given by $x = c$.

Sometimes, if we want to tidy up fractional coefficients for example, we write the equation of a line as

$$ax + by + c = 0$$

Solution to review question 7.1.4

A. The equation of the straight line through the points (x_1, y_1), (x_2, y_2) is

$$y - y_1 = \left(\frac{y_2 - y_1}{x_2 - x_1}\right)(x - x_1) = m(x - x_1)$$

where m is the gradient. We can choose either of the given points for (x_1, y_1).

(i) With $(x_1, y_1) = (0, 0)$, $(x_2, y_2) = (1, 1)$ the gradient is $m = 1$ (210 ◄) so the required equation is

$$y - 0 = 1(x - 0)$$

or $\quad y = x$

(ii) This question illustrates the limitations of general formulae. Using the general result would give

$$y - 2 = \infty(x - 1)$$

which is not defined (6 ◄). In such circumstances we go back to first principles, and look again at the points, (1, 2), (1, 3). Here the x-coordinate always remains the same, so the line must be parallel to the y-axis. It is in fact the line $x = 1$.

(iii) For the points $(-2, 4)$, $(1, -3)$ the gradient is $-\frac{7}{3}$ and the equation becomes

$$y - 1 = -\frac{7}{3}(x - 3)$$

or, tidying this up

$$7x + 3y - 24 = 0$$

(iv) For the points $(-1, -1)$, $(-2, -3)$ the gradient is $m = 2$ and the line is

$$y - (-1) = 2(x - (-1))$$

or

$$2x - y + 1 = 0$$

B. Rewriting the lines in the form

$$y = mx + c$$

the gradient in m, the intercept on the y-axis is $c(x = 0, y = c)$ and the intercept on the x-axis is $x = -c/m$, $y = 0$.

(i) $m = 3$, $c = 2$, and x-intercept $= -2/3$
(ii) $y = \frac{1}{3}(1 - 2x)$
$m = -\frac{2}{3}$, $c = \frac{1}{3}$ and x-intercept $x = 1/2$
(iii) $m = 4$, x and y intercepts both the origin $(0, 0)$
(iv) $y = -x - 1$, $m = -1$, intercepts $(0, -1)$, $(-1, 0)$

7.2.5 Parallel and perpendicular lines

Consider two parallel lines, l_1, l_2 as shown in Figure 7.8.

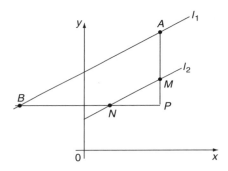

Figure 7.8 Parallel lines have the same gradient.

As we might expect, parallel lines have the same gradient (149 ◄). We can see this by a nice application of the intercept theorem(153 ◄). The gradient of l_1 is $\dfrac{AP}{BP}$ and that of l_2 is $\dfrac{MP}{NP}$. But, by the intercept theorem, since AB and MN are parallel, MN divides AP and BP in the same ratio, so $AP/BP = MP/NP$, i.e. l_1, l_2 have the same gradient.

Lines that are perpendicular clearly do not have the same gradient, but they are related. If lines l_1, l_2 are perpendicular, and have gradients m_1, m_2 respectively then

$$m_1 m_2 = -1$$

We can see this from Figure 7.9.
The lines l_1, l_2 are perpendicular.
First, note that since $\alpha + \theta = \dfrac{\pi}{2} = \alpha + \beta$ then $\beta = \theta$. If the gradient of l_1 is m, then

$$m = \frac{BD}{AD} = \tan\theta$$

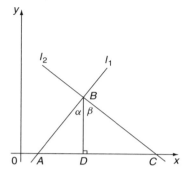

Figure 7.9 Gradients of perpendicular lines multiply to -1.

On the other hand the gradient of l_2 is

$$-\frac{BD}{CD} = -\frac{1}{\tan\beta} = -\frac{1}{\tan\theta} = -\frac{1}{m}$$

Solution to review question 7.1.5

A. Any line parallel to $y = 1 - 3x$ has the same gradient, $m = -3$ and so can be written as

$$y = -3x + c$$

We now have to choose the intercept c so that the line passes through $(-1, 2)$, which we do by substituting these values in the equation:

$$2 = -3(-1) + c$$

So
$$c = -1 \text{ and the equation required is}$$
$$y = -3x - 1$$

B. If the line l through $(-1, 2)$, $(0, 4)$ has gradient m then any line perpendicular to it has gradient $-1/m$. The gradient of l is

$$m = \frac{4-2}{0-(-1)} = 2$$

so any line perpendicular to it has gradient $-\frac{1}{2}$ and has an equation of the form

$$y = -\tfrac{1}{2}x + c$$

If this line passes through $(-1, 2)$ then substituting this point in the equation gives

$$2 = -\tfrac{1}{2}(-1) + c$$

> so
> $$c = \tfrac{3}{2}$$
> and the equation is $y = \tfrac{1}{2}(3 - x)$ or $x + 2y - 3 = 0$.

7.2.6 Intersecting lines

If two lines intersect, then their point of intersection must satisfy both of their equations. Thus, if the lines are:

$$ax + by = c$$
$$dx + ey = f$$

then at the point of intersection these equations must be satisfied simultaneously (48 ◀). Although one would find this intersection point by solving these equations, the geometrical picture provides a nice visual means of discussing the different possibilities for the solution of the equations. Thus, we find the following cases, with some examples you can check:

- if the lines are not parallel (i.e. do not have the same gradient) then they will intersect at some point – i.e. the equations will have a solution.

    Example
    $$x + 2y = 1$$
    $$x - y = 0$$

    have different gradients and intersect at $\left(\tfrac{1}{3}, \tfrac{1}{3}\right)$.

- If the lines are parallel (same gradient) then they will never intersect and there will be no solution to the equations.

    Example
    $$2x + y = 0$$
    $$4x + 2y = 3$$

    These are parallel, but through different points – the first passes through the origin, the second through $\left(0, \tfrac{3}{2}\right)$. These lines will never intersect and so the equations have no solution.

- If the lines are **identical**, i.e. coincide (same gradients, same intercepts), then they intersect at every point on the line(s) – the equations have an infinite number of solutions.

    Example
    $$x + 3y = 1$$
    $$2x + 6y = 2$$

The second equation is actually the same as the first – just cancel 2 throughout the equation. So these equations actually represent the same line and they therefore intersect at every point – there is an infinite number of solutions.

Solution to review question 7.1.6

You will save time by noting that the lines (i) $x + y = 3$ and (ii) $2x + 2y = -1$ are in fact parallel lines – they have the same gradient, -1. So they never intersect, and we have only to consider the intersection of each of them with the line (iii) $y = 3x - 1$.

(i) and (iii) intersect where

$$y = 3 - x = 3x - 1$$

Solving for x and then y gives

$$x = 1, y = 2$$

(ii) and (iii) intersect where

$$y = -\tfrac{1}{2} - x = 3x - 1$$

or

$$x = \tfrac{1}{8}, y = -\tfrac{5}{8}$$

7.2.7 Equation of a circle

◀ 205 222 ▶

Two things characterise a circle uniquely – its centre and its radius. Suppose therefore we refer a circle to a set of Cartesian axes in which its centre has coordinates (a, b) and its radius is r. Then we know that any point (x, y) on the circle is always a distance r from the centre (a, b). So, using Pythagoras' theorem (154 ◀ yet again) we can write the distance from (x, y) to the centre (a, b) as $\sqrt{(x - a)^2 + (y - b)^2}$. Since this is always equal to the radius r we can then write

$$(x - a)^2 + (y - b)^2 = r^2$$

This, with the square root sign removed by squaring, is the equation or **locus** for a circle with centre (a, b) and radius r, in Cartesian coordinates. This equation may be expanded (42 ◀) to give the form

$$x^2 + y^2 + 2fx + 2gy + h = 0$$

Problem
Express f, g, h in terms of a, b, r.

You should find

$$f = -a, \quad g = -b, \quad h = a^2 + b^2 - r^2$$

Often, it is the latter form of the equation of a circle that is given, and you are asked to find the centre and radius. To do this we complete the square (66 ◄) on x, y and return to the first form, from which centre and radius can be read off directly.

Note that in polar coordinates, the equation of a circle with radius a and centre at the origin takes the very simple form $r = a$. This illustrates how useful polar coordinates can be in some circumstances.

Solution to review question 7.1.7

A. The equation of the circle centre (a, b) with radius r is
$$(x - a)^2 + (y - b)^2 = r^2$$
Applying this to the data given produces:

(i) For centre $(0, 0)$ and radius 2 we get
$$(x - 0)^2 + (y - 0)^2 = 2^2, \text{ or } x^2 + y^2 = 4$$

(ii) For centre $(1, 2)$ radius 1 we get
$$(x - 1)^2 + (y - 2)^2 = 1 \text{ or } x^2 + y^2 - 2x - 4y + 4 = 0$$

(iii) For centre $(-1, 4)$ and radius 3 we get
$$(x + 1)^2 + (y - 4)^2 = 9 \text{ or } x^2 + y^2 + 2x - 8y + 8 = 0$$

B. To find the centre and radius of a circle whose equation is given in the form $x^2 + y^2 + 2fx + 2gy + h = 0$ we re-express it in the form
$$(x - a)^2 + (y - b)^2 = r^2$$
by completing the square in x and y (66 ◄).

(i) $x^2 + y^2 - 2x + 6y + 6 = x^2 - 2x + y^2 + 6y + 6$
$$= (x - 1)^2 - 1 + (y + 3)^2 - 9 + 6$$

on completing the square for $x^2 - 2x$ and $y^2 + 6y$
$$= (x - 1)^2 + (y + 3)^2 - 4$$

So the equation is equivalent to
$$(x - 1)^2 + (y + 3)^2 = 4 = 2^2$$
which gives a centre of $(1, -3)$ and radius 2.

(ii) $x^2 + y^2 + 4x + 4y - 1 = x^2 + 4x + y^2 + 4y - 1$
$$= (x + 2)^2 - 4 + (y + 2)^2 - 4 - 1$$
$$= (x + 2)^2 + (y + 2)^2 - 9$$

on completing the square. So the equation becomes
$$(x + 2)^2 + (y + 2)^2 = 9 = 3^2$$
giving a centre $(-2, -2)$ and a radius 3.

7.2.8 Parametric representation of curves

Sometimes, instead of writing an equation in the form $y = f(x)$, i.e. in terms of x, y, it is convenient to introduce some **parameter** t in terms of which x and y are jointly expressed in the form

$$x = x(t) \quad y = y(t)$$

For example the position of a projectile at time t may be expressed as

$$x = Vt\cos\alpha = x(t)$$
$$y = Vt\sin\alpha - \tfrac{1}{2}gt^2 = y(t)$$

where V, α are the initial projection velocity and angle respectively.

In principle, we can always return to the x, y form by eliminating t. In the projectile for example we have

$$t = \frac{x}{V\cos\alpha}$$

from the first equation, then substitution in the second gives the parabola:

$$y = V\frac{x}{V\cos\alpha}\sin\alpha - \frac{1}{2}g\left(\frac{x}{V\cos\alpha}\right)^2$$
$$= x\tan\alpha - \frac{1}{2}\frac{gx^2}{V^2}\sec^2\alpha$$

Solution to review question 7.1.8

(i) From the equations $x = 3t + 1$, $y = t + 2$ we have $t = y - 2$, so

$$x = 3t + 1 = 3(y - 2) + 1 = 3y - 5$$

The Cartesian equation is thus

$$x - 3y + 5 = 0$$

which is a straight line.

(ii) $x = 6\cos 2t$; $y = 6\sin 2t$

This is a good example of the need to have the elementary trig identities at your fingertips. Otherwise, it might not immediately occur to you to use the fact that $\cos^2\theta + \sin^2\theta = 1$ (185 ◄) to eliminate t, by squaring and adding

$$x^2 + y^2 = 36\cos^2 2t + 36\sin^2 2t = 36$$

which is a circle centre the origin and radius 6. This parametric form of the circle is very useful in the theory of oscillating systems.

(iii) In general, the parametric equations

$$x = r\cos\theta + a, \quad y = r\sin\theta + b$$

represent a circle with centre (a, b) and radius r, since:

$$(x - a)^2 + (y - b)^2 = r^2$$

The parameter θ can be regarded as the angle made by the radius with the x-axis, as shown in Figure 7.10.

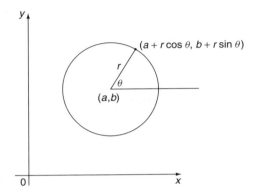

Figure 7.10 Parametric form of a circle.

So, for $x = 2\cos t - 1$ and $y = 2\sin t$ we get

$$2\cos t = x + 1$$

so

$$(x + 1)^2 + y^2 = 4\cos^2 t + 4\sin^2 t = 4$$

giving a circle centre $(-1, 0)$ and radius 2.

(iv) $x = \cos 2t$, $y = \cos t$
So from the double angle formula (188 ◄)

$$x = \cos 2t = 2\cos^2 t - 1 = 2y^2 - 1$$

which we may as well leave in the implicit form

$$x = 2y^2 - 1$$

7.3 Reinforcement

7.3.1 Coordinate systems in a plane   ◄◄ 204 205 ◄

A. Plot the points

(i) $(-1, -1)$ (ii) $(3, 2)$ (iii) $(-2, 3)$ (iv) $(0, 4)$

(v) $(4, 0)$ (vi) $(1, 1)$ (vii) $(3, -1)$ (viii) $(0, -2)$

B. Plot the points with polar coordinates

(i) $\left(\dfrac{3}{2}, \dfrac{3\pi}{4}\right)$ (ii) $(1, \pi/3)$ (iii) $(1, 120°)$ (iv) $(2, -60°)$

(v) $(2, \pi/2)$ (vi) $(3, \pi)$ (vii) $\left(2, \dfrac{5\pi}{4}\right)$ (viii) $(3, -150°)$

7.3.2 Distance between two points

◄◄ 204 208 ◄

Find the distance between each pair of points

(i) $(-1, 1), (2, 0)$ (ii) $(1, 0), (1, -2)$

(iii) $(2, 2), (3, 3)$ (iv) $(0, 1), (0, 3)$

(v) $(2, 1), (3, 1)$ (vi) $(-2, -1), (-1, -2)$

7.3.3 Midpoint and gradient of a line

◄◄ 204 209 ◄

A. Find the midpoint of the line segments joining the pairs of points in RE7.3.2.

B. Ditto for the gradients.

C. Ditto for the points dividing each line segment in the ratio 2 : 3.

7.3.4 Equation of a straight line

◄◄ 204 212 ◄

A. Find the equations of the straight lines through each pair of points in RE7.3.2.

B. Find the gradients and intercepts of the following lines

(i) $2x - 3y = 4$ (ii) $4x + 2y = 7$ (iii) $x - y + 1 = 0$

C. Show that the equations

$$\frac{y - y_1}{x - x_1} = \frac{y_2 - y_1}{x_2 - x_1} = m$$

may be rearranged to give the equation of the straight line as

$$y - y_1 = \left(\frac{y_2 - y_1}{x_2 - x_1}\right)(x - x_1)$$

or

$$y - y_1 = m(x - x_1)$$

Also show that it may be arranged into the 'symmetric form'

$$y = \frac{(x - x_1)}{(x_2 - x_1)} y_2 + \frac{(x - x_2)}{(x_1 - x_2)} y_1$$

7.3.5 Parallel and perpendicular lines

For the lines in RE7.3.4B determine:

    (a)   lines parallel to each of them through the origin,

    (b)   lines perpendicular to each of them through the point $(-1, 1)$.

7.3.6 Intersecting lines

Find all points where the following lines intersect

    (i) $x + y = 1$ (ii) $2x + 2y = 3$ (iii) $x - y = 1$

7.3.7 Equation of a circle

A. Write down the equations of the circles with centres and radii:

    (i) $(-1, 1)$, 4   (ii) $(2, -1)$, 1   (iii) $(4, 1)$, 2

B. Find the centre and radius of each of the circles:

    (i) $x^2 + y^2 - 2x - y = 4$ (ii) $x^2 + y^2 + 3x - 2y - 7 = 0$
    (iii) $x^2 + y^2 + y = 3$

C. A circle has the equation $x^2 + y^2 - 4y = 0$. Find its centre and radius, and the equation of the tangents at the points $(\pm\sqrt{2}, 2 + \sqrt{2})$, using only geometry and trig. Also find the point where the two tangents intersect.

D. A circle has the equation $x^2 + y^2 - 2x = 0$. Determine its centre and radius, and find the equation of the tangent at the point $\left(\dfrac{1}{2}, \dfrac{\sqrt{3}}{2}\right)$. Determine where this tangent cuts the axes.

7.3.8 Parametric representation of curves

Eliminate the parameters in the following pairs of equations

    (i) $x = 3\cos t,$ $y = 3\sin t$

    (ii) $x = 1 + 2\cos\theta,$ $y = 3 - \sin\theta$

    (iii) $x = 2u^2,$ $y = u - 2$

    (iv) $x = \dfrac{2}{t},$ $y = 3t$

    (v) $x = \cos 2t,$ $y = \sin t$

7.4 Applications

1. In simple **linear programming** we consider a set of linear relations that constrain two variables x, y in the form

   $$ax + by \leq c$$

   and look for the maximum or minimum value of some linear function

   $$f(x, y) = mx + ny$$

   as x and y vary subject to these constraints. By considering the equations of straight lines, show how to solve such problems graphically and by solving simultaneous equations. Find the values of x and y such that

   $$x \geq 0, y \geq 0$$
   $$4x - y - 24 \geq 0$$
   $$x - y + 7 \geq 0$$
   $$f(x, y) = x + y \text{ is a maximum}$$

2. This question considers a very simple model of the 'slingshot effect', in which a space module is transferred from an orbit around the earth to one around the moon, say. The trick is to eject the module from the earth orbit with just the right speed and direction for it to travel in a straight line to intercept the moon orbit at just the right point and speed to be captured and enter into orbit around the moon. Figure 7.11 shows how this can be modelled by finding the straight line that is tangent to the earth and moon orbits represented as circles centre the origin, radius R and centre $(0, D)$, radius r, respectively. Here D is the earth moon distance, R the earth orbit radius and r the moon orbit radius. Find the equation of this straight line, the distance from orbit to orbit, and the points where the orbits are left and entered all in terms of D, R, r. Try out your results on some realistic data.

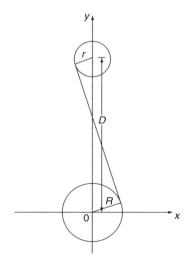

Figure 7.11

Answers to reinforcement exercises

7.3.1 Coordinate systems in a plane

A.

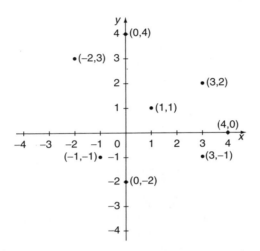

B.

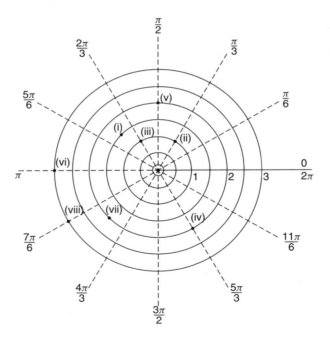

7.3.2 Distance between two points

(i) $\sqrt{10}$ (ii) 2   (iii) $\sqrt{2}$
(iv) 2   (v) 1   (vi) $\sqrt{2}$

7.3.3 Midpoint and gradient of a line

A. (i) $\left(\frac{1}{2}, \frac{1}{2}\right)$ (ii) $(1, -1)$ (iii) $\left(\frac{5}{2}, \frac{5}{2}\right)$

(iv) $(0, 2)$ (v) $\left(\frac{5}{2}, 1\right)$ (vi) $\left(\frac{-3}{2}, \frac{-3}{2}\right)$

B. (i) $-\frac{1}{3}$ (ii) 'infinite gradient' – vertical line

(iii) 1   (iv) vertical line   (v) 0

(vi) -1

C. (i) $\left(\frac{1}{5}, \frac{3}{5}\right)$ (ii) $\left(1, \frac{-4}{5}\right)$ (iii) $\left(\frac{12}{5}, \frac{12}{5}\right)$

(iv) $\left(0, \frac{9}{5}\right)$ (v) $\left(\frac{12}{5}, 1\right)$ (vi) $\left(\frac{-8}{5}, \frac{-7}{5}\right)$

7.3.4 Equation of a straight line

A. (i) $x + 3y - 2 = 0$ (ii) $x = 1$ (iii) $x - y = 0$
(iv) $x = 0$ (v) $y = 1$ (vi) $x + y + 3 = 0$

B. (i) $\frac{2}{3}, -\frac{4}{3}$ (ii) $-2, \frac{7}{2}$ (iii) $1, 1$

7.3.5 Parallel and perpendicular lines

(a) (i) $x + y = 0$ (ii) $2x + y = 0$ (iii) $x - y = 0$
(b) (i) $x - y + 2 = 0$ (ii) $x - 2y + 3 = 0$ (iii) $x + y = 0$

7.3.6 Intersecting lines

(i) and (ii) being parallel, never intersect
(i) and (iii) intersect at $(1, 0)$
(ii) and (iii) intersect at $\left(\frac{5}{4}, \frac{1}{4}\right)$

7.3.7 Equation of a circle

A. (i) $x^2 + y^2 + 2x - 2y - 14 = 0$ (ii) $x^2 + y^2 - 4x + 2y + 4 = 0$
(iii) $x^2 + y^2 - 8x - 2y + 13 = 0$

B. (i) $\left(1, \dfrac{1}{2}\right), \dfrac{\sqrt{21}}{2}$ (ii) $\left(-\dfrac{3}{2}, 1\right), \dfrac{\sqrt{41}}{2}$

(iii) $\left(0, \dfrac{-1}{2}\right), \dfrac{\sqrt{13}}{2}$.

C. $x + y - 2 - 2\sqrt{2} = 0, \quad x - y + 2 + 2\sqrt{2} = 0$
They intersect at $(0, 2 + \sqrt{2})$.

D. Centre $(1, 0)$, radius 1. Tangent is $x - \sqrt{3}y + 1 = 0$ with intercepts $(-1, 0)$ and $\left(0, \dfrac{1}{\sqrt{3}}\right)$.

7.3.8 Parametric representation of curves

(i) $x^2 + y^2 = 9$ (ii) $\dfrac{(x-1)^2}{4} + (y-3)^2 = 1$

(iii) $x = 2(y+2)^2$ (iv) $xy = 6$

(v) $x = 1 - 2y^2$

8

Techniques of Differentiation

Differentiation is a relatively straightforward side of calculus. We only have to remember a dozen or so standard derivatives (the elementary functions), a few rules (sum, product, quotient, function of a function) and the rest is routine. **However**, we do need to be highly proficient with differentiation, able to apply the rules quickly and accurately. One reason for this is to help you with **integration**. One of the difficulties with integration lies in recognising what was differentiated to give you what you have to integrate – you will more easily spot this (and check it) if your differentiation is good.

While it is perhaps not essential for the engineers, I have included **differentiation from first principles**. It is really not that difficult to follow, and the ideas and skills involved are useful elsewhere.

Prerequisites

It will be helpful if you know something about:

- the idea of a limit (126 ◄)
- basic algebra (Chapter 2 ◄)
- properties of the elementary functions (Chapter 3 ◄)
- the gradient of a line (210 ◄)
- function notation (90 ◄)
- composition of functions (97 ◄)
- implicit functions (91 ◄)
- parametric representation (91 ◄)

Objectives

In this chapter you will find:

- geometric interpretation of differentiation
- differentiation from first principles
- the standard derivatives
- rules of differentiation:
  - sum
  - product
  - quotient
  - function of a function
- implicit differentiation

Understanding Engineering Mathematics

- parametric differentiation
- higher order derivatives

Motivation

You may need the material of this chapter for:

- modelling rates of change, such as velocity, acceleration, etc.
- ordinary differential equations (Chapter 15)
- approximating the values of functions
- finding the maximum and minimum values of functions
- curve sketching

8.1 Review

8.1.1 Geometrical interpretation of differentiation   ▶ 230 243 ▶▶

Complete the following:

The derivative, $\dfrac{dy}{dx}$, is the [ ] of the curve $y = f(x)$ at the point x, which is defined as [ ] θ where θ is the angle made by the [ ] to the curve at x with the x-axis.

8.1.2 Differentiation from first principles   ▶ 230 243 ▶▶

Complete the following description of the evaluation of the derivative of x^2 from first principles:

If $y = f(x) = x^2$, then

$$y + \delta y = f(x + \delta x) = \boxed{} + (\delta x)^2$$

Hence

$$\delta y = f(x + \delta x) - f(x) = \boxed{}$$

and

$$\frac{\delta y}{\delta x} = \frac{f(x + \delta x) - f(x)}{\delta x} = \boxed{} + \delta x$$

So, 'in the limit', as $\delta x \to 0$

$$\frac{dy}{dx} = \frac{df}{dx} = \lim_{\delta x \to 0} \boxed{\dfrac{ - f(x)}{}}$$

$$= \lim_{\delta x \to 0} \boxed{} + \delta x$$

$$= \boxed{}$$

8.1.3 Standard derivatives   ➤ 232 243 ➤➤

Give the derivatives of the following functions

(i) 49 (ii) x^4 (iii) $\sqrt{x}$ (iv) $\dfrac{1}{x^2}$

(v) $\sin x$ (vi) e^x (vii) $\ln x$ (viii) 2^x

8.1.4 Rules of differentiation   ➤ 234 244 ➤➤

Differentiate and simplify

(i) $3x^4 - 2x^2 + 3x - 1$ (ii) $x^2 \cos x$

(iii) $\dfrac{x-1}{x+1}$ (iv) $\cos(x^2 + 1)$

(v) $\ln 3x$ (vi) e^{-2x}

(vii) $\sqrt{x^2 - 1}$

8.1.5 Implicit differentiation   ➤ 238 244 ➤➤

A. If $x^2 + 2xy + 2y^2 = 1$ obtain $\dfrac{dy}{dx}$ as a function of x and y.

B. Obtain the derivative of $\sin^{-1} x$ by using implicit differentiation.

C. Differentiate $y = 2^x$.

D. If $f(x) = \dfrac{x-1}{x+2}$ evaluate $f'(1)$.

8.1.6 Parametric differentiation   ➤ 240 245 ➤➤

If $x = 3t^2$, $y = \cos(t + 1)$ evaluate $\dfrac{dy}{dx}$ as a function of t.

8.1.7 Higher order derivatives   ➤ 241 245 ➤➤

A. Evaluate the second derivative of each of the following functions:

(i) $2x + 1$ (ii) $x^3 - 2x + 1$

(iii) $e^{-x} \cos x$ (iv) $\dfrac{x+1}{(x-1)(x+2)}$

B. If $x = t^2 + 1$, $y = t - 1$, evaluate $\dfrac{d^2y}{dx^2}$ as a function of t.

8.2 Revision

8.2.1 Geometrical interpretation of differentiation ◀ 228 243 ▶

If A and B are two points on a curve, the straight line AB is called a **chord** of the curve (157 ◀). A line which touches the curve at a single point (i.e. $A = B$) is called a **tangent** to the curve at that point (157 ◀). The **normal** to a curve at a point A is a line through A, perpendicular to the tangent at A. The **gradient** or **slope of a curve** at a point is the gradient or slope of the tangent to the curve at that point.

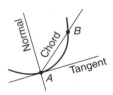

Figure 8.1

> Solution to review question 8.1.1
>
> The derivative, $\dfrac{dy}{dx}$, is the slope or gradient of the curve $y = f(x)$ at the point x, which is defined as $\tan\theta$ where θ is the angle made by the tangent to the curve at x with the positive x-axis.

8.2.2 Differentiation from first principles ◀ 228 243 ▶

As B gets closer to A, the gradient of the chord AB becomes closer to that of the tangent to the curve at A – that is, closer to the gradient of the curve at A. We say that in the limit, as B tends to A, the gradient of the chord tends to the gradient of the curve at A. We write this as

$$\lim_{B \to A} (\text{gradient chord } AB) = \text{gradient of curve at } A$$

where $\lim_{B \to A}$ means take the limiting value as B tends to A. We met the idea of a limit on page 126, in the context of letting a number become infinitely large. Here we are considering the situation where something – the distance between A and B – becomes 'infinitesimally small', i.e. tends to zero.

Let the equation of the curve be $y = f(x)$. The situation is illustrated in Figure 8.2.

Suppose δx ('delta x') denotes a small change in x (**not** 'δ times x'), so that x becomes $x + \delta x$. Then correspondingly $f(x)$ will change to $f(x + \delta x)$, producing a small change in y to $y + \delta y$. The change in y, δy, is then given by

$$\delta y = y + \delta y - y = f(x + \delta x) - f(x)$$

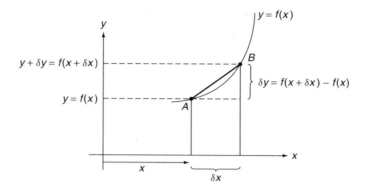

Figure 8.2

The gradient (210 ◀) of the extended chord AB is then given by

$$\text{Gradient } AB = \frac{f(x + \delta x) - f(x)}{\delta x}$$

As A gets closer to B this gradient will get closer to the gradient of the tangent at A, i.e. closer to the gradient of the curve at A, so:

$$\text{Gradient of curve at } A = \lim_{\delta x \to 0} \left[\frac{f(x + \delta x) - f(x)}{\delta x} \right]$$

where $\lim_{\delta x \to 0}$ means let δx tend to zero in the expression in square brackets.

Note that in all this discussion we are assuming that the curve $y = f(x)$ is continuous and smooth (i.e. has no breaks or sharp corners).

Taking limits is a sophisticated operation about which we will say more in Chapter 14 (see Section 14.5). For now we will only use very simple ideas.

We can repeat the above discussion at each point, i.e. each value of x, so the gradient is a **function of x** called the **derivative** of $f(x)$ (sometimes **differential coefficient**).

The standard forms of notation for the derivative of a function $y = f(x)$ with respect to x are:

$$\frac{dy}{dx} = \frac{df(x)}{dx} = f'(x) = \lim_{\delta x \to 0} \left[\frac{f(x + \delta x) - f(x)}{\delta x} \right]$$

The process of obtaining dy/dx is called **differentiation with respect to x**. $f'(a)$ means df/dx evaluated at $x = a$.

Evaluation of the derivative by calculating the limit is called **differentiation from first principles**. Using this one can construct a **table of standard derivatives of the elementary functions** and a number of **rules of differentiation** with which we may deduce many further derivatives.

The standard approach to such differentiation from first principles is as follows. For a function $y = f(x)$, let x increase by an 'infinitesimal' amount δx. Evaluate the corresponding increase in y:

$$\delta y = f(x + \delta x) - f(x)$$

and neglect any powers of δx greater than one to get something like

$$\delta y \simeq f'(x)\delta x$$

where $f'(x)$ denotes just a function of x alone. Then

$$\frac{\delta y}{\delta x} \simeq f'(x)$$

and letting $\delta x \to 0$ yields the derivative

$$\frac{dy}{dx} = f'(x)$$

as suggested by the notation. The solution to the review question illustrates this.

Solution to review question 8.1.2

If
$$y = f(x) = x^2, \text{ then}$$
$$y + \delta y = f(x + \delta x) = (x + \delta x)^2$$
$$= x^2 + 2x\delta x + (\delta x)^2$$

Hence
$$\delta y = f(x + \delta x) - f(x) = 2x\delta x + (\delta x)^2$$

and
$$\frac{\delta y}{\delta x} = \frac{f(x + \delta x) - f(x)}{\delta x} = 2x + \delta x$$

So, in the limit, as $\delta x \to 0$

$$\frac{dy}{dx} = \frac{df}{dx} = \lim_{\delta x \to 0} \frac{f(x + \delta x) - f(x)}{\delta x}$$
$$= \lim_{\delta x \to 0} 2x + \delta x$$
$$= 2x$$

8.2.3 Standard derivatives

By applying arguments such as that in the previous section we can build up a list of **standard derivatives**. These are basically derivatives of the 'elementary' functions such as powers of x, trigonometric functions, exponentials and logs. What you remember from these depends on the requirements of your course or programme, but I would certainly recommend that the minimal list given in Table 8.1 be not only remembered, but be second nature. Note, in anticipation of Chapter 9, that when you learn the standard **derivatives** you can also learn the **standard integrals** by reading the table from right to left (see Section 9.2.2).

Table 8.1

Function $f(x)$	Derivative $f'(x)$
$c = $ constant	0
x^n	$nx^{n-1} \quad n \neq 0$
$x^{n+1}/(n+1)$	$x^n (n \neq -1)$
$\ln|x| = \log_e |x|$	$1/x$
$\sin x$	$\cos x$
$\cos x$	$-\sin x$
$\tan x$	$\sec^2 x$
e^x	e^x
$\ln[u(x)]$	$\dfrac{u'(x)}{u(x)}$
$e^{u(x)}$	$u'(x)e^{u(x)}$
Integral $\int g(x)\,dx$	Function $g(x)$

Solution to review question 8.1.3

(i) $y = 49$ is a constant, so its derivative is 0. Think of it as a quantity whose rate of change or gradient is zero – its graph is always 'flat'.

(ii) $y = x^4$, so using $\dfrac{d(x^n)}{dx} = nx^{n-1}$ with $n = 4$ gives $\dfrac{dy}{dx} = 4x^3$

(iii) $y = \sqrt{x} = x^{\frac{1}{2}}$ so $\dfrac{dy}{dx} = \dfrac{1}{2} x^{-\frac{1}{2}} = \dfrac{1}{2\sqrt{x}}$

(iv) $y = \dfrac{1}{x^2} = x^{-2}$ so $\dfrac{dy}{dx} = -2x^{-3} = -\dfrac{2}{x^3}$

(v) $y = \sin x$, so $\dfrac{dy}{dx} = \cos x$ directly from Table 8.1. Note that x must be in radians here.

(vi) $y = e^x$ so $\dfrac{dy}{dx} = e^x$ again from the table. e^x is unique in being its own derivative and this is partly responsible for its immense importance.

(vii) For $y = \ln x$, $\dfrac{dy}{dx} = \dfrac{1}{x}$ from Table 8.1.

(viii) $y = 2^x$

This question is a bit naughty, since it **does not come from our list of standard integrals**, but beginners sometimes make the error of thinking it does and write

$$\text{`} \dfrac{d}{dx}(2^x) = x2^{x-1} \text{'}$$

> This is, of course, an incorrect application of the rule for the derivative of x^n. We will come back to 2^x later (Section 8.2.5) – it needs some more rules of differentiation, which takes us to the next section.

8.2.4 Rules of differentiation

In order to differentiate more complicated functions than those in the standard derivatives table we need to be able to differentiate sums, differences, products and quotients. Also we sometimes need to differentiate a composite function or 'function of a function' (97 ◄). These rules of differentiation can be proved by using the limit definition of differentiation (see Chapter 14). Here we will concentrate on making them at least plausible, and on getting used to using them.

The derivative of a sum (difference) is the sum (difference) of the derivatives.

$$\frac{d}{dx}(f(x) \pm g(x)) = \frac{df(x)}{dx} \pm \frac{dg(x)}{dx}$$

We say differentiation is a **linear operation**. An extension of this is:

$$\frac{d}{dx}(kf(x)) = k\frac{df(x)}{dx} \quad \text{for } k = \text{constant}$$

which simply says that to differentiate a constant multiple of a function you only have to take out the constant from the derivative.

It would be nice if the derivative of a product was a product – unfortunately it is not. To obtain the correct result, suppose we want to differentiate $y = uv$ where $u = u(x)$, $v = v(x)$ are functions of x.

Following our recipe of Section 8.2.2 let x increase to $x + \delta x$, say. Then u will increase to $u + \delta u$ and v to $v + \delta v$, say. So y increases by:

$$\delta y = (u + \delta u)(v + \delta v) - uv$$
$$= u\delta v + v\delta u + \delta u \delta v$$

Now we neglect $\delta u \delta v$ because it is of second degree (i.e. like $(\delta x)^2$ not δx) so we have

$$\delta y \simeq u\delta v + v\delta u$$

So, dividing through by δx

$$\frac{\delta y}{\delta x} = u\frac{\delta v}{\delta x} + v\frac{\delta u}{\delta x}$$

If we now let $\delta x \to 0$ we get the **product rule**:

$$\frac{dy}{dx} = \frac{d(uv)}{dx} = u\frac{dv}{dx} + v\frac{du}{dx}$$

A nice physical way to look at this is to think of how much the area of a rectangular plate increases when you heat it. Suppose the plate has sides u and v. Due to linear expansion

u and v will increase to $u + \delta u$ and $v + \delta v$ respectively, with δu and δv very small. The area therefore increases to

$$(u + \delta u)(v + \delta v) = uv + v\delta u + u\delta v + \delta u \delta v$$

So the increase in area will be

$$v\delta u + u\delta v + \delta u \delta v$$

Since δu, δv are very small, $\delta u \delta v$ is **very** very small, so we can neglect it and the increase in area is approximately

$$v\delta u + u\delta v$$

which is the form of the product rule.

The corresponding **quotient rule** is

$$\frac{d}{dx}\left(\frac{u}{v}\right) = \left(v\frac{du}{dx} - u\frac{dv}{dx}\right)\bigg/ v^2$$

We will see why below.

We can handle many functions by using the standard derivatives and the sum, product, quotient rules. We could not handle something like $\cos(e^x)$ however. This is an example of a **function** (cosine) **of a function** (exponential) or a composition of functions (97 ◄). In general we write:

$$y = f(u) \qquad \text{where } u = g(x)$$

e.g. $\qquad y = \cos u \qquad \text{where } u = e^x$

To differentiate such a function **with respect to** x we use the **function of a function** or the **chain rule**:

$$\frac{dy}{dx} = \frac{dy}{du}\frac{du}{dx}$$

So $\qquad \dfrac{dy}{dx} = -\sin u \times e^x = -e^x \sin(e^x)$

Think of this 'rule' of change of u w.r.t. x = rate of change of y w.r.t. u times rate of change of u w.r.t. x. Note that the chain rule suggests a useful result if we take the special case $y = x$:

$$\frac{dy}{du}\frac{du}{dx} = \frac{dx}{du}\frac{du}{dx} = \frac{dx}{dx} = 1$$

so $\qquad \dfrac{du}{dx} = 1 \bigg/ \dfrac{dx}{du}$

or $\qquad \dfrac{dy}{dx} = 1 \bigg/ \dfrac{dx}{dy}$

This result is useful in implicit differentiation for example (see Section 8.2.5). While it should **not** encourage you to think of $\dfrac{dy}{dx}$ as a **fraction** there are occasions when it is convenient to do this – with care. For example in differential equations we sometimes use such steps as:

$$\frac{dy}{dx}dx \equiv dy$$

which is really a bit of poetic license that is found to work most of the time.

We can combine the product and chain rules to obtain the rule for differentiating a quotient as follows:

$$\frac{d}{dx}\left(\frac{u}{v}\right) = \frac{d}{dx}(uv^{-1}) = v^{-1}\frac{du}{dx} + u\frac{d(v^{-1})}{dx}$$

by the product rule

$$= v^{-1}\frac{du}{dx} + u\left(-v^{-2}\frac{dv}{dx}\right)$$

by the chain rule

$$= \frac{1}{v}\frac{du}{dx} - \frac{u}{v^2}\frac{dv}{dx}$$

$$= \left(v\frac{du}{dx} - u\frac{dv}{dx}\right)\bigg/v^2$$

When first learning these rules you might use the given formulae, substituting $u = x$, $v = \sin x$, or whatever. However, the rules will be more useful to you if you practice and develop your skills to the point where you do not need to do this. That is, rather than actually use the formula, try to do the required procedure automatically. To help with this think of the rules in terms of words, describing what you actually do:

Sum rule: 'diff(+) = diff + diff'

Product: 'diff(×) = one × diff + diff × other'

Quotient diff(/) = $\dfrac{\text{diff top} \times \text{bottom - top} \times \text{diff bottom}}{(\text{bottom})^2}$

Function of a function 'diff$[f(g)]$ = (diff f w.r.t. g) × (diff g w.r.t. x)'

Solution to review question 8.1.4

(i) $y = 3x^4 - 2x^2 + 3x - 1$

This is a linear combination of terms of the form Ax^n. We differentiate each term in the polynomial and combine the results.

So $\dfrac{dy}{dx} = 3 \times 4x^3 - 2 \times 2x + 3$

$= 12x^3 - 4x + 3$

(ii) $y = x^2 \cos x$ is a product of the two elementary functions x^2 and $\cos x$.

So by product rule, with say $u = x^2$, $v = \cos x$,

$$\frac{dy}{dx} = \frac{d(uv)}{dx} = \frac{d}{dx}(x^2 \cos x)$$

$$= \frac{v\,du}{dx} + \frac{u\,dv}{dx} = \frac{d(x^2)}{dx}\cos x + x^2 \frac{d}{dx}(\cos x)$$

$$= 2x \cos x + x^2(-\sin x)$$

$$= 2x \cos x - x^2 \sin x$$

(iii) $y = \dfrac{x-1}{x+1}$

We can treat this as a quotient $y = \dfrac{u}{v}$ and use the quotient rule, **or** treat as a product:

$$y = (x-1) \times \left(\frac{1}{x+1}\right)$$

so $\dfrac{dy}{dx} = 1 \times \dfrac{1}{x+1} + (x-1)\left(-\dfrac{1}{(x+1)^2}\right)$

$$= \frac{x+1-(x-1)}{(x+1)^2} = \frac{2}{(x+1)^2}$$

(iv) $y = \cos(x^2 + 1)$

This is a function (cos) of a function ($x^2 + 1$) (97 ◄), so we can use the function of a function rule:

$$\frac{dy}{dx} = \frac{dy}{du}\frac{du}{dx} \quad \left(\begin{array}{l} u = x^2 + 1 \\ y = \cos u \end{array}\right)$$

giving

$$\frac{dy}{dx} = -\sin(x^2 + 1) \times (2x)$$

$$= -2x \sin(x^2 + 1)$$

(v) $y = \ln 3x$

Again, the function of a function rule can be used:

$$\frac{dy}{dx} = \frac{1}{3x} \times 3 = \frac{1}{x}$$

Or this can be simplified using rules of logarithms (131 ◄)

$$y = \ln 3x = \ln 3 + \ln x$$

Therefore

$$\frac{dy}{dx} = 0 + \frac{1}{x} \text{ as before}$$

(vi) $y = e^{-2x}$

This is a function of a function. Here we will practise a short hand approach to the function of a function rule.

$$\frac{dy}{dx} = \frac{d(e^{-2x})}{d(-2x)} \times \frac{d(-2x)}{dx} = e^{-2x}(-2) = -2e^{-2x}$$

(vii) $y = \sqrt{1 - x^2} = (1 - x^2)^{\frac{1}{2}}$

Again, function of a function – 'differentiate the ( )$^{\frac{1}{2}}$ and multiply by the derivative of the ( )'.

$$\frac{dy}{dx} = \frac{1}{2}(1 - x^2)^{-\frac{1}{2}} \times (-2x) = -\frac{x}{(1 - x^2)^{\frac{1}{2}}}$$

$$= -\frac{x}{\sqrt{1 - x^2}}$$

8.2.5 Implicit differentiation

◀ 229 244 ▶

If y is defined implicitly in terms of x by an equation of the form (91 ◀)

$$f(x, y) = 0$$

then $\dfrac{dy}{dx}$ may often be found by differentiating throughout using the function of a function rule. This is best illustrated by the example given below:

Example

If

$$x^2 + y^2 = 1$$

then differentiating through with respect to x gives

$$2x + \frac{d}{dx}(y^2) = \frac{d}{dx}(1)$$

so, using the function of a function rule on the left-hand side

$$2x + 2y\frac{dy}{dx} = 0$$

Hence

$$\frac{dy}{dx} = -\frac{x}{y}$$

Solution to review question 8.1.5

A. Differentiating through with respect to x, using the function of function and product rules we have

$$\frac{d}{dx}(x^2 + 2xy + 2y^2) = \frac{d}{dx}(1)$$

so
$$2x + 2y + 2x\frac{dy}{dx} + 4y\frac{dy}{dx} = 0$$

Hence $(x + y) + (x + 2y)\frac{dy}{dx} = 0$
and so
$$\frac{dy}{dx} = -\frac{x + y}{x + 2y}$$

B. This is a standard application of implicit differentiation. If $y = \sin^{-1} x$ then $x = \sin y$. But note that as usual with inverse trigonometric functions we must restrict the range of y to, say, $-\frac{\pi}{2} \leq y \leq \frac{\pi}{2}$ to obtain a single valued function (184 ◄).
Hence, on this range

$$\frac{dx}{dy} = \cos y$$

$$= \sqrt{1 - \sin^2 y}$$

where we have taken the positive root since the cosine is positive on $-\frac{\pi}{2} \leq y \leq \frac{\pi}{2}$

$$= \sqrt{1 - x^2}$$

So, using $\frac{dy}{dx} = 1 \Big/ \frac{dx}{dy}$ (see above) we get

$$\frac{dy}{dx} = \frac{1}{\sqrt{1 - x^2}}$$

C. $y = a^x$ is another case where we can use implicit differentiation to advantage. If we take logs to base e we have:

$$\ln y = \ln a^x = x \ln a$$

Now differentiate through with respect to x:

$$\frac{d}{dx}(\ln y) = \frac{d}{dx}(x \ln a)$$

or

$$\frac{1}{y}\frac{dy}{dx} = \ln a$$

by using the chain rule on the left-hand side.
So
$$\frac{dy}{dx} = y \ln a = a^x \ln a$$

You should now be able to repeat this argument with the special case of $a = 2$ to obtain

$$\frac{d}{dx}(2^x) = 2^x \ln 2$$

D. If $y = f(x) = \dfrac{x-1}{x+2}$, then we could use the quotient or product rule:

$$f'(x) = \frac{1}{x+2}\frac{d}{dx}(x-1) + (x-1)\frac{d}{dx}\left(\frac{1}{x+2}\right)$$

$$= \frac{1}{x+2} - \frac{x-1}{(x+2)^2}$$

So

$$f'(1) = \frac{1}{1+2} - \frac{0}{3^2} = \frac{1}{3}$$

However it is much easier by implicit differentiation. Rewrite as:

$$(x+2)y = x - 1$$

So, differentiating through with respect to x:

$$(1)y + (x+2)y' = 1$$

Substituting $x = 1$ and $y = 0$ gives

$$(1)(0) + 3y'(1) = 1$$

hence $\quad y'(1) = \dfrac{1}{3}$ as before.

8.2.6 Parametric differentiation

◀ 229 245 ▶

If x, y are defined in terms of a parameter, t, (91 ◀)

$$x = x(t) \qquad y = y(t)$$

then the function of a function rule gives

$$\frac{dy}{dx} = \frac{dy}{dt}\frac{dt}{dx}$$

Since $\dfrac{dt}{dx} = 1 \bigg/ \dfrac{dx}{dt}$ this can now be written as

$$\frac{dy}{dx} = \frac{dy/dt}{dx/dt}$$

Solution to review question 8.1.6

With $x = 3t^2$, $y = \cos(t+1)$ we have

$$\frac{dy}{dt} = -\sin(t+1) \text{ and } \frac{dx}{dt} = 6t, \text{ so:}$$

$$\frac{dy}{dx} = \frac{dy/dt}{dx/dt} = \frac{-\sin(t+1)}{6t}$$

8.2.7 Higher order derivatives

◀ 229 245 ▶

In general, $\dfrac{dy}{dx}$ will be a function of x. We can therefore differentiate it again with respect to x. We write this as

$$\frac{d}{dx}\left(\frac{dy}{dx}\right) = \frac{d^2y}{dx^2}$$

Note that $\dfrac{d^2y}{dx^2}$ does **not** mean $\left(\dfrac{dy}{dx}\right)^2$

We can, of course, differentiate yet again and write

$$\frac{d}{dx}\left(\frac{d^2y}{dx^2}\right) = \frac{d^3y}{dx^3}$$

and so on.

In general $\dfrac{d^n y}{dx^n}$ denotes the **nth order derivative** – differentiate y n times successively.

Solution to review question 8.1.7

A. (i) A first differentiation gives

$$\frac{d}{dx}(2x+1) = 2$$

So, differentiating again

$$\frac{d^2}{dx^2}(2x+1) = \frac{d}{dx}(2) = 0$$

(ii) A bit quicker this time:

$$\frac{d^2}{dx^2}(x^3 - 2x + 1) = \frac{d}{dx}(3x^2 - 2) = 6x$$

(iii) $\dfrac{d^2}{dx^2}(e^{-x}\cos x) = \dfrac{d}{dx}(-e^{-x}\cos x - e^{-x}\sin x)$

$= -\dfrac{d}{dx}(e^{-x}(\cos x + \sin x))$

$= -[-e^{-x}(\cos x + \sin x) + e^{-x}(-\sin x + \cos x)]$

$= -e^{-x}(-2\sin x)$

$= 2e^{-x}\sin x$

(iv) This one requires a bit of thought if you want to avoid a mess. It is easiest to use partial fractions in fact. Thus, to remind you of partial fractions (62 ◀), you can check that

$$\dfrac{x+1}{(x-1)(x+2)} \equiv \dfrac{2}{3(x-1)} + \dfrac{1}{3(x+2)}$$

Then

$\dfrac{d^2}{dx^2}\left[\dfrac{x+1}{(x-1)(x+2)}\right] = \dfrac{d^2}{dx^2}\left[\dfrac{2}{3(x-1)} + \dfrac{1}{3(x+2)}\right]$

$\dfrac{d}{dx}\left[-\dfrac{2}{3(x-1)^2} - \dfrac{1}{3(x+2)^2}\right]$

$= \dfrac{4}{3(x-1)^3} + \dfrac{2}{3(x+2)^3}$

B. This sort of example brings together a lot of what we have already done. First, note that

$$\dfrac{d^2y}{dx^2} \neq \dfrac{d^2y/dt^2}{d^2x/dt^2}$$

We must be more subtle. Start with

$$\dfrac{d^2y}{dx^2} = \dfrac{d}{dx}\left(\dfrac{dy}{dx}\right)$$

Now with parametric differentiation $\dfrac{dy}{dx}$ will be a function of t – in this case, with $x = t^2 + 1$, $y = t - 1$ we have

$$\dfrac{dy}{dx} = \dfrac{dy/dt}{dx/dt} = \dfrac{1}{2t}$$

If we were to differentiate now with respect to x we would have to express $\dfrac{1}{2t}$ in terms of x – possible but messy. Instead, we use the function of a function rule to change the differentiation to one with respect to t, and write

$$\frac{d^2y}{dx^2} = \frac{d}{dx}\left(\frac{dy}{dx}\right) = \frac{d}{dt}\left(\frac{dy}{dx}\right) \times \frac{dt}{dx}$$

$$= \frac{d}{dt}\left(\frac{dy}{dx}\right) \bigg/ \frac{dx}{dt}$$

on using $\dfrac{dt}{dx} = 1 \bigg/ \dfrac{dx}{dt}$

So:

$$\frac{d^2y}{dx^2} = \frac{d}{dt}\left(\frac{1}{2t}\right) \bigg/ 2t$$

$$-\frac{1}{2t^2} \bigg/ 2t = -\frac{1}{4t^3}$$

8.3 Reinforcement

8.3.1 Geometrical interpretation of differentiation ◀◀228 230◀

A. Evaluate the slopes of the following curves at the points specified:

(i) $y = x^3 - x$ $x = 1$
(ii) $y = \sin x$ $x = \pi$
(iii) $y = 2e^x$ $x = 0$
(iv) $y = \dfrac{3}{x}$ $x = 1$

B. Determine where the slope of the curve $y = 2x^3 + 3x^2 - 12x + 6$ is zero.

8.3.2 Differentiation from first principles ◀◀228 230◀

Differentiate from first principles:

(i) $3x$
(ii) $x^2 + 2x + 1$
(iii) x^3
(iv) $\cos x$

8.3.3 Standard derivatives

A. Differentiate without reference to a standard derivatives table:

(i) e^x
(ii) $\cos x$
(iii) x^{31}
(iv) $\ln x$
(v) $\sin x$
(vi) $x^{\frac{1}{3}}$
(vii) $\tan x$
(viii) $\dfrac{1}{x^3}$

B. What are the most general functions that you need to differentiate to obtain the following functions?

(i) x^4 (ii) $\cos x$ (iii) e^x (iv) $\sin x$

(v) $\dfrac{1}{x^4}$ (vi) $\sqrt{x}$ (vii) $\dfrac{1}{x}$ (viii) 0

(ix) $\dfrac{1}{\cos^2 x}$

8.3.4 Rules of differentiation

◄◄ 229 234 ◄

A. Using the definition of the functions and appropriate rules of differentiation obtain the derivatives of the following elementary functions (see Section 4.4 for hyperbolic functions).

(i) $\sec x$ (ii) $\operatorname{cosec} x$ (iii) $\cot x$ (iv) $\cosh x$

(v) $\sinh x$ (vi) $\tanh x$ (vii) $\operatorname{cosech} x$ (viii) $\operatorname{sech} x$

(ix) $\coth x$

B. Differentiate

(i) $\ln(\sec x)$ (ii) $\ln(\sin x)$ (iii) $\ln(\sec x + \tan x)$

(iv) $\ln(\operatorname{cosec} x + \cot x)$ (v) $\ln(\cosh x)$

(vi) $\ln(\sinh x)$

C. Differentiate

(i) $x^7 - 2x^5 + x^4 - x^2 + 2$ (ii) $(x^2 + 2)\tan x$

(iii) $\dfrac{\ln x}{x^2 + 1}$ (iv) $\exp(x^3 - 2x)$ (v) $\dfrac{x}{\sqrt{x^2 - 1}}$

(vi) $\ln(\cos x + 1)$ (vii) $\sin\left(\dfrac{x+1}{x}\right)$ (viii) $\sec x \tan x$

(ix) e^{6x} (x) xe^x (xi) e^{-x^2}

(xii) $\ln 5x$ (xiii) $e^x \ln x$ (xiv) $\ln e^{2x}$

8.3.5 Implicit differentiation

◄◄ 229 238 ◄

A. Use implicit differentiation to differentiate the functions

(i) $\cos^{-1} x$ (ii) $\tan^{-1} x$ (iii) π^x

B. Evaluate dy/dx at the points indicated.

(i) $x^2 + y^2 = 1$ $(0, 1)$ (ii) $x^3 - 2x^2 y + y^2 = 1$ $(1, 2)$

C. If $f(x) = \dfrac{x+1}{x-3}$, evaluate $f'(0)$.

8.3.6 Parametric differentiation

A. If $x = e^{2t}$, $y = e^t + 1$, evaluate $\dfrac{dy}{dx}$ and $\dfrac{d^2y}{dx^2}$ as functions of t by two different methods and compare your results.

B. Obtain $\dfrac{dy}{dx}$ and $\dfrac{d^2y}{dx^2}$ for each of the following parametric forms:

(i) $x = 3\cos t$, $y = 3\sin t$ (ii) $x = t^2 + 3$, $y = 2t + 1$ (iii) $x = e^t \sin t$, $y = e^t$
(iv) $x = 2\cosh t$, $y = 2\sinh t$

8.3.7 Higher order derivatives

A. Evaluate the second derivatives of each of the following functions:

(i) $x^2 + 2x + 1$ (ii) e^{x^2} (iii) $e^x \sin x$
(iv) $\dfrac{x-1}{(x+1)(x+2)}$

B. Evaluate the 20th derivative of each of the following functions:

(i) $x^{17} + 3x^{15} + 2x^5 + 3x^2 - x + 1$ (ii) e^{x-1}
(iii) e^{3x} (iv) $\dfrac{1}{x-1}$
(v) $\dfrac{x}{(x-1)(x+2)}$

8.4 Applications

1. Calculus was invented (by Newton and Liebnitz) partly to handle the mechanics of moving particles, culminating in Newton's laws of motion, which to this day forms one of the major foundations of engineering science. This question is an open invitation to link the topics of this chapter with what you know about dynamics. Things you might think about are:
   (i) the definitions of **time, position, displacement, speed** and **velocity, acceleration** and their mathematical expressions in terms of derivatives of each other, with respect to each other
   (ii) Newton's second law is $F = ma$ where F is force, m mass and a acceleration. Suppose F is given as a function of position and velocity. Then it would be convenient to have a derivative expression for acceleration that involves just these two variables. Use the function of a function rule to show that, with usual notation:

$$\frac{d^2s}{dt^2} = v\frac{dv}{ds}$$

(iii) Discuss the forms of motion described by the relations between position, s, and time, t, given below.

(a) $s = at + b$
(b) $s = at^2 + bt + c$
(c) $s = a \sin \omega t$
(d) $s = ae^{-bt}$

In each case a, b, c, ω are all constants which may be positive or negative.

2. Newton's second law gives us lots of examples of **differential equations**. You will study these in some detail in Chapter 15. For now, we just want to get used to what is meant by a **solution** to such an equation. A differential equation is any equation containing one or more derivatives. Examples are given below, (i)–(vi). A solution to such an equation is any **function** which when substituted into the equation, with its appropriate derivatives, makes the equation identically satisfied – i.e. it produces an identity in the independent variable. In (a)–(g) are listed a number of functions. By differentiating and substitution in the equations, determine which functions are solutions of which equations. Using these examples as inspiration, try to find other solutions of the equations, which you can then test.

Note that the full exercise of finding solutions to differential equations is at the very least a matter of integration (Chapter 9), and at its most advanced level may involve powerful mathematical tools such as transform theory (Chapter 17) or numerical methods.

(i) $\dfrac{dy}{dx} + 3y = 0$
(ii) $x\dfrac{dy}{dx} - y = x$
(iii) $\dfrac{d^2y}{dx^2} + 4y = 0$

(iv) $\dfrac{d^2y}{dx^2} + 2\dfrac{dy}{dx} - 3y = 0$
(v) $\dfrac{d^2y}{dx^2} + 6\dfrac{dy}{dx} + 9y = 0$

(vi) $\dfrac{d^2y}{dx^2} + 2\dfrac{dy}{dx} + 2y = 0$

(a) $3\cos 2x$
(b) $2e^x$
(c) $-2e^{-x} \sin x$
(d) $4e^{-3x}$
(e) $4x$
(f) $-6xe^{-3x}$
(g) $x \ln x$

Each one of the equations (i)–(vi) is an example of a general type of differential equation that plays a crucial role in one or more areas of engineering science.

3. There are many occasions in engineering when we need to know something about the tangent and normal to a curve at a given point (230 ◀). For example the reaction of a force at a smooth surface is normal, i.e. perpendicular, to the surface. In Chapter 5 we looked at tangents to circles – the normal at any point is of course along the radius (158 ◀). Using differentiation regarded as the slope of a curve, we can find the tangent at any point on a curve. Also, using the result that the gradients of perpendicular lines multiply to give -1 (214 ◀), we can then find the normal to a given surface, as the line perpendicular to the tangent. Investigate the tangents and normals to a general curve at a given point, using as examples the following:

(i) $y = x^2 + 3x - 4$ at $(0, -4)$
(ii) $y = \cos 3x$ at $x = \dfrac{\pi}{4}$
(iii) $y = e^{3x}$ at $x = 1$
(iv) $y = 2\ln x$ at $x = 1$

4. In a circuit with an inductor of inductance L and capacitor of capacitance C, the voltage across the inductance is given by

$$V = L\frac{dI}{dt}$$

where I is the current through the inductance, and the current through the capacitor is given by

$$I = C\frac{dV}{dt}$$

where V is the potential across the capacitor.
   (i) The time dependence of the current in a series RL circuit with a constant voltage source E can be accurately modelled by the expression

$$I = \frac{E}{R}\left[1 - \exp\left(-\frac{Rt}{L}\right)\right]$$

   What is the voltage across the inductor at time t?
   (ii) An RC circuit with a constant capacitor and constant voltage source includes a resistor whose resistance varies slowly according to a linear law $R = R_0(1 + \alpha t)$. The voltage across the capacitor in such circumstances can be modelled by the relation

$$V - V_0 = (E - V_0)[1 - (1 + \alpha t)^{-1/\alpha C R_0}]$$

   where V_0 is the voltage across the capacitor at $t = 0$. Determine the expression for the current through the capacitor at time t.
5. The potential in the junction region of a semiconductor pn junction is given by

$$V = \frac{A}{1 + \exp(-kx)}$$

where A and k are material-dependent constants. The force on an electron with charge e in such a potential region is given by

$$F = e\frac{dV}{dx}$$

Obtain an expression for this force in terms of hyperbolic functions (138 ◄).

Answers to reinforcement exercises

8.3.1 Geometrical interpretation of differentiation

A. (i) 2   (ii) −1   (iii) 2   (iv) −3
B. $x = 1$ and -2

8.3.2 Differentiation from first principles

(i) 3 (ii) $2x+2$ (iii) $3x^2$ (iv) $-\sin x$

8.3.3 Standard derivatives

A. (i) e^x (ii) $-\sin x$ (iii) $31x^{30}$ (iv) $\dfrac{1}{x}$ (v) $\cos x$
(vi) $\dfrac{1}{3}x^{-2/3}$ (vii) $\sec^2 x$ (viii) $-\dfrac{3}{x^4}$

B. (i) $\dfrac{x^5}{5}+C$ (ii) $\sin x + C$ (iii) $e^x + C$ (iv) $-\cos x + C$
(v) $C - \dfrac{1}{3x^3}$ (vi) $\dfrac{2}{3}x^{3/2}+C$ (vii) $\ln x + C$ (viii) C
(ix) $\tan x + C$

8.3.4 Rules of differentiation

A. (i) $\sec x \tan x$ (ii) $-\operatorname{cosec} x \cot x$ (iii) $-\operatorname{cosec}^2 x$
(iv) $\sinh x$ (v) $\cosh x$ (vi) $\operatorname{sech}^2 x$
(vii) $-\operatorname{cosech} x \coth x$ (viii) $-\operatorname{sech} x \tanh x$ (ix) $-\operatorname{cosech}^2 x$

B. (i) $\tan x$ (ii) $\cot x$ (iii) $\sec x$
(iv) $-\operatorname{cosec} x$ (v) $\tanh x$ (vi) $\coth x$

C. (i) $7x^6 - 10x^4 + 4x^3 - 2x$ (ii) $2x \tan x + (x^2+2)\sec^2 x$
(iii) $\dfrac{x^2 + 1 - 2x^2 \ln x}{x(x^2+1)^2}$ (iv) $(3x^2 - 2)e^{x^3 - 2x}$
(v) $-\dfrac{1}{(x^2-1)^{3/2}}$ (vi) $\dfrac{-\sin x}{\cos x + 1}$ (vii) $-\dfrac{1}{x^2}\cos\left(\dfrac{x+1}{x}\right)$
(viii) $\sec x \tan^2 x + \sec^3 x$ (ix) $6e^{6x}$ (x) $e^x(x+1)$
(xi) $-2xe^{-x^2}$ (xii) $\dfrac{1}{x}$ (xiii) $e^x(\ln x + \dfrac{1}{x})$
(xiv) 2

8.3.5 Implicit differentiation

A. (i) $-\dfrac{1}{\sqrt{1-x^2}}$ (ii) $\dfrac{1}{1+x^2}$ (iii) $\pi^x \ln \pi$

B. (i) 0 (ii) $\dfrac{5}{2}$

C. $-\dfrac{4}{9}$

8.3.6 Parametric differentiation

A. $\frac{1}{2}e^{-t}, -\frac{1}{4}e^{-3t}$

B. (i) $-\cot t, -\frac{1}{3}\csc^2 t$ (ii) $\frac{1}{t}, -\frac{1}{2t^3}$

(iii) $\dfrac{1}{\sin t + \cos t}, \dfrac{e^{-t}(\sin t - \cos t)}{(\sin t + \cos t)^3}$ (iv) $\coth t, -\frac{1}{2}\operatorname{cosech}^3 t$

8.3.7 Higher order derivatives

A. (i) 2   (ii) $e^{x^2}(4x^2 + 2)$ (iii) $2e^x \cos x$

(iv) $\dfrac{6}{(x+2)^3} - \dfrac{4}{(x+1)^3}$

B. (i) 0   (ii) e^{x-1} (iii) $3^{20}e^{3x}$

(iv) $\dfrac{20!}{(x-1)^{21}}$ (v) $\dfrac{1}{3}\dfrac{20!}{(x-1)^{21}} + \dfrac{2}{3}\dfrac{20!}{(x+2)^{21}}$

9

Techniques of Integration

Most of us find integration difficult – basically because we are always trying to 'undo' some differentiation, which might have been difficult in the first place! There are no short cuts to good integration skills – practice, practice, practice is the only way. Plenty of that is provided in this chapter. In this case the review test covers a wide range, which reflects the importance of the material.

Prerequisites

It will be helpful if you know something about:

- differentiation (Chapter 8 ◄)
- properties of the elementary functions (Chapters 4, 6)
- partial fractions (62 ◄)
- completing the square (66 ◄)

Objectives

In this chapter you will find:

- definition of integration
- standard integrals
- addition of integrals
- simplifying the integrand
- linear substitution in integration
- the $du = f'(x)\,dx$ substitution
- integration of rational functions
- use of trig identities in integration
- using trig substitutions in integration
- integration by parts
- choice of integration methods
- definite integrals

Motivation

You may need the material of this chapter for:

- solving differential equations (➤ Chapter 15)
- calculating areas and volumes (➤ Chapter 10)
- calculating mean and RMS values (➤ Chapter 10)
- calculating moments, centroids, etc. (➤ Chapter 10)
- calculating means, standard deviations, and probabilities

9.1 Review

9.1.1 Definition of integration
▶ 253 280 ▶▶

A. Differentiate

(i) x^3 (ii) $\sin 2x$ (iii) $2e^{3x}$

B. Find the integral of

(i) $3x^2$ (ii) $\cos 2x$ (iii) $2e^{3x}$

9.1.2 Standard integrals
▶ 255 281 ▶▶

A. Integrate the following functions (include integration constants)

(i) 3 (ii) $\dfrac{2}{x^2}$ (iii) $\sqrt{x}$ (iv) $x^{\frac{1}{3}}$

B. Give the integrals of

(i) $\sin x$ (ii) e^x (iii) $\dfrac{1}{x}$

9.1.3 Addition of integrals
▶ 257 281 ▶▶

Integrate $2x^2 + \dfrac{3}{x^2} + 4\sqrt{x} - 2e^x + 4\sin x$.

9.1.4 Simplifying the integrand
▶ 258 281 ▶▶

Find

(i) $\displaystyle\int \dfrac{dx}{\cos 2x + 2\sin^2 x}$ (ii) $\displaystyle\int xe^{2\ln x}\, dx$

(iii) $\displaystyle\int (\cos 2x \sin 3x - \cos 3x \sin 2x)\, dx$

Hint: think about what you are integrating!

9.1.5 Linear substitution in integration
▶ 260 282 ▶▶

Integrate

(i) $\dfrac{1}{x-1}$ (ii) $\cos(3x+2)$ (iii) e^{2x-1}

9.1.6 The $du = f'(x)\,dx$ substitution

Integrate the following functions by means of an appropriate substitution.

(i) $\dfrac{x+1}{x^2+2x+3}$ (ii) $x\sin(x^2+1)$ (iii) $\cos x\, e^{\sin x}$ (iv) $\sin x \cos x$

1. Compare the results of (iv) with that of Q9.1.8(iv). Are the answers the same? Explain.

9.1.7 Integrating rational functions

A. Find $\displaystyle\int \dfrac{dx}{x^2+x-2}$ using partial fractions.

B. Find $\displaystyle\int \dfrac{dx}{x^2+2x+2}$, given that $\displaystyle\int \dfrac{dx}{x^2+1} = \tan^{-1} x$.

C. Find $\displaystyle\int \dfrac{2x^2+5x+4}{x^2+2x+2}\,dx$

Hint: divide out first and think about Q9.1.6 and **B** immediately above.

D. Integrate

(i) $\dfrac{1}{x^2+x+1}$ (ii) $\dfrac{1}{x^2+3x+2}$ (iii) $\dfrac{2x+1}{x^2+x-1}$

(iv) $\dfrac{3x}{(x-1)(x+1)}$

9.1.8 Using trig identities in integration

Integrate the following functions using appropriate trig identities.

(i) $\sin^2 x$ (ii) $\cos^3 x$ (iii) $\sin 2x \cos 3x$

(iv) $\sin x \cos x$

1. Cf (iv) with Q9.1.6(iv) – are the answers the same?

9.1.9 Using trig substitutions in integration

Integrate

(i) $\dfrac{1}{\sqrt{9-x^2}}$ (ii) $\dfrac{1}{\sqrt{3-2x-x^2}}$

9.1.10 Integration by parts

Integrate by parts

(i) $x \sin x$ (ii) $x^2 e^x$ (iii) $e^x \sin x$

9.1.11 Choice of integration methods

Discuss the methods you would use to integrate the following – you will be asked to integrate them in RE9.3.11C.

(i) $\dfrac{x-3}{x^2-6x+4}$ (ii) $\sin^3 x$ (iii) $\dfrac{e^{\ln x}}{x}$

(iv) $\dfrac{x-1}{2x^2+x-3}$ (v) xe^{3x^2} (vi) $\dfrac{3}{\sqrt{3-2x-x^2}}$

(vii) $x\sin(x+1)$ (viii) $\cos^4 x$ (ix) $\ln\left(\dfrac{e^x}{x}\right)$

(x) $\dfrac{x+2}{x^2-5x+6}$ (xi) xe^{2x} (xii) $e^x \cos(e^x)$

(xiii) $\sin 2x \cos 2x$ (xiv) $\dfrac{x-1}{\sqrt{x^2-2x+3}}$ (xv) $\dfrac{x+3}{x^2+2x+2}$

(xvi) $\sin 4x \cos 5x$ (xvii) $x\cos(x^2+1)$

9.1.12 The definite integral

A. Evaluate

(i) $\displaystyle\int_0^1 (x^2+1)\,dx$ (ii) $\displaystyle\int_0^1 xe^x\,dx$ (iii) $\displaystyle\int_0^2 \dfrac{x^2}{x^3+1}\,dx$

(iv) $\displaystyle\int_0^{2\pi} \cos x \sin x\,dx$

B. What is wrong with

$$\int_0^2 \dfrac{dx}{x^2-1}\,?$$

9.2 Revision

9.2.1 Definition of integration

Integration (or more strictly **indefinite integration**) is the reverse of differentiation. Thus, if

$$\dfrac{dy}{dx} = f(x)$$

then

$$y = \int f(x)\,dx$$

is the **anti-derivative** or **indefinite integral** of $f(x)$ (also sometimes called the 'primitive'). $f(x)$ is called the **integrand**.

For example, because we know that the derivative of x^2 is $2x$, so we know that the integral of $2x$ is x^2. Actually, because the derivative of a constant is zero the most general integral of $2x$ is $x^2 + C$ where C is an **arbitrary constant of integration.** We write

$$\int 2x \, dx = x^2 + C$$

Solution to review question 9.1.1

A. Hopefully, this won't give you much trouble now:

(i) $\dfrac{d}{dx}(x^3) = 3x^2$

(ii) $\dfrac{d}{dx}(\sin(2x)) = 2\cos(2x)$

by the function of a function rule.

(iii) $\dfrac{d}{dx}(2e^{3x}) = 2e^{3x} \times 3 = 6e^{3x}$

B. Of course, these questions are not unrelated to **A**!

(i) Having just done it in **A**(i) we know that we can obtain $3x^2$ by differentiating x^3:

$$\dfrac{d}{dx}(x^3) = 3x^2$$

And in fact we get the same result if we add an arbitrary constant C to x^3:

$$\dfrac{d}{dx}(x^3 + C) = 3x^2$$

It follows that the most general integral of $3x^2$ is:

$$\int 3x^2 \, dx = x^3 + C$$

(ii) A little more thought is needed to integrate $\cos 2x$. Differentiating $\sin 2x$ gives us $2\cos 2x$ **not** $\cos 2x$. To get the latter we should differentiate $\tfrac{1}{2} \sin 2x$

$$\dfrac{d}{dx}\left(\tfrac{1}{2} \sin 2x\right) = \dfrac{1}{2} 2 \cos 2x = \cos 2x$$

Remembering the arbitrary constant we thus have:

$$\int \cos 2x \, dx = \dfrac{1}{2} \sin 2x + C$$

(iii) Again, we have to fiddle with the multiplier of the e^{3x}. Since

$$\dfrac{d}{dx}(e^{3x}) = 3e^{3x}$$

from **A**(iii), we see that

$$\frac{d}{dx}\left(\frac{2}{3}e^{3x}\right) = 2e^{3x}$$

So, again remembering the arbitrary constant:

$$\int 2e^{3x}\, dx = \frac{2}{3}e^{3x} + C$$

9.2.2 Standard integrals

◀ 251 281 ▶

We can compile a useful list of integrals just by reading a table of derivatives backwards. The simplest example is the derivative of x^α. For any power α (that is, α can be any positive or negative real number – check that the result does make sense when $\alpha = 0$) we have

$$\frac{d}{dx}(x^\alpha) = \alpha x^{\alpha-1}$$

from which, for any power α, **except $\alpha = -1$** (see below)

$$\frac{d}{dx}(x^{\alpha+1}) = (\alpha + 1)x^\alpha$$

and so

$$\int x^\alpha\, dx = \frac{x^{\alpha+1}}{\alpha + 1}$$

We must here exclude the case $\alpha = -1$ because otherwise we will have zero in the denominator. In fact, the result for $\alpha = -1$ is:

$$\int x^{-1}\, dx = \int \frac{dx}{x} = \ln|x| + C$$

which we know from $\dfrac{d}{dx}(\ln|x|) = \dfrac{1}{x}$.

Similarly, from the derivatives of other functions we can build up the table of standard integrals given below (from which the arbitrary constant has been omitted). Note that much of it is the table of standard derivatives given in Chapter 8 read 'backwards' (233 ◀). The term 'standard integral' is of course relative – an advanced calculus book might have many more such integrals that you are expected to know. Here however we confine ourselves to the most important elementary functions. Obviously, the more you do know, the easier integration will be, and in particular you will find it invaluable to commit all those in the table to memory.

Much of integration involves manipulating the integrand, or the whole integral, to make it expressible in terms of the standard integrals. Points to note about integration are:

- the better your differentiation, the better your integration
- know your basic algebra and trig skills
- practice and perseverance pay dividends
- trial and error may be needed

Table 9.1

$f(x)$	$\int f(x)\,dx$ (arbitrary constant omitted)
$x^\alpha \quad \alpha \neq -1$	$x^{\alpha+1}/(\alpha+1) \quad \alpha \neq -1$
$\dfrac{1}{x}$	$\ln|x|$
$\cos x$	$\sin x$
$\sin x$	$-\cos x$
$\sec^2 x$	$\tan x$
e^x	e^x
$1/\sqrt{a^2 - x^2}$	$\sin^{-1}(x/a)$
$a/(a^2 + x^2)$	$\tan^{-1}(x/a)$
$\dfrac{u'(x)}{u(x)}$	$\ln|u(x)|$
$u'(x)e^{u(x)}$	$e^{u(x)}$

- have the standard integrals (at least) at your fingertips
- check your answers by differentiation as often as possible – provides practice as well as confirmation
- there may be more than one way to do an integral – for example

$$\int \frac{x+1}{x^2+x-6}\,dx$$

can be done by partial fractions or by the substitution $u = x^2 + x - 6$
- there may be **no** way to do an integral – for example

$$\int e^{x^2}\,dx$$

simply cannot be integrated in terms of elementary functions
- integration skills build up sequentially – for example:
  (i)   first learn $\int \dfrac{dx}{x} = \ln|x|$
  (ii) then $\int \dfrac{dx}{x-1} = \ln|x-1|$
  (iii) then by partial fractions

$$\int \frac{dx}{x^2-1} = \int \left(\frac{1}{2(x-1)} - \frac{1}{2(x-1)}\right) dx$$

Solution to review question 9.1.2

A. These are all particular cases of

$$\int x^\alpha\,dx = \frac{x^{\alpha+1}}{\alpha+1}$$

As usual, the places where you might have trouble occur when negative signs and fractions are involved. In such cases, just take it steady and check each step.

(i) $\int 3\,dx = 3\int dx = 3x + C$

(ii) $\int \dfrac{2}{x^2}\,dx = 2\int x^{-2}\,dx = 2\dfrac{x^{-2+1}}{-2+1} + C = 2\dfrac{x^{-1}}{-1} + C = -\dfrac{2}{x} + C$

(iii) $\int \sqrt{x}\,dx = \int x^{\frac{1}{2}}\,dx = \dfrac{x^{\frac{1}{2}+1}}{\frac{1}{2}+1} + C = \dfrac{x^{3/2}}{3/2} + C = \dfrac{2}{3}x^{3/2} + C$

(iv) $\int x^{\frac{1}{3}}\,dx = \dfrac{x^{\frac{1}{3}+1}}{\frac{1}{3}+1} + C = \dfrac{x^{4/3}}{4/3} + C = \dfrac{3}{4}x^{4/3} + C$

B. These are all standard integrals, known because we know what to differentiate to get them.

(i) $\dfrac{d}{dx}(\cos x) = -\sin x$

so $\int \sin x\,dx = -\cos x + C$

(ii) $\dfrac{d}{dx}(e^x) = e^x$

so $\int e^x\,dx = e^x + C$

(iii) $\dfrac{d}{dx}(\ln |x|) = \dfrac{1}{x}$

so $\int \dfrac{1}{x}\,dx = \ln |x| + C$

NB Note that $\int \dfrac{1}{x}\,dx \neq \dfrac{x^0}{0}$! Also note the reminder about the modulus in $\ln |x|$.

These and others we can put together in the table of standard integrals on page 256.

9.2.3 Addition of integrals

◄ 251 281 ►

If you differentiate a sum then the result is the sum of the derivatives:

$$\dfrac{d}{dx}(f + g) = \dfrac{df}{dx} + \dfrac{dg}{dx}$$

Similarly, the integral of a sum is the sum of the integrals:

$$\int (f + g)\,dx = \int f\,dx + \int g\,dx$$

(not forgetting the arbitrary constant of course).

Applying the result a number of times we see for example that

$$\int 3f(x)\,dx = \int (f(x) + f(x) + f(x))\,dx$$
$$= \int f(x)\,dx + \int f(x)\,dx + \int f(x)\,dx$$
$$= 3\int f(x)\,dx$$

and in general for any numerical multiplier k:

$$\int kf(x)\,dx = k\int f(x)\,dx$$

In general, if k and l are constants then we have

$$\int (kf + lg)\,dx = k\int f\,dx + l\int g\,dx$$

For this reason integration is called a **linear operation**. This rule enables us to integrate any linear combination of standard integrals, including all polynomials, for example.

Solution to review question 9.1.3

The given function is a linear combination of standard integrals. We can therefore use the linearity of the integral operation:

$$\int \left(2x^2 + \frac{3}{x^2} + 4\sqrt{x} - 2e^x + 4\sin x\right) dx$$
$$= 2\int x^2\,dx + 3\int \frac{dx}{x^2} + 4\int x^{1/2}\,dx$$
$$- 2\int e^x\,dx + 4\int \sin x\,dx$$
$$= \frac{2x^3}{3} - \frac{3}{x} + \frac{8x^{3/2}}{3} - 2e^x - 4\cos x + C$$

Note that we only need one arbitrary constant for the overall integral and not one for each of the 'summands'.

9.2.4 Simplifying the integrand

◀ 251 281 ▶

When faced with a new integral, the first thing to do, after checking whether it is a standard integral, is to see if it is in the most convenient form for integration. Thus, in the integral $\int f(x)\,dx$, we have two things to play with:

- the integrand $f(x)$
- the variable x

We consider changes of variable, x, later under various substitution methods, but here we will have a first look at what we might be able to do with the integrand in some simple cases. In general $f(x)$ may be a polynomial or rational function, trig or hyperbolic, log or exponential, or any combination of these. We would first look at $f(x)$ to see if it can be simplified or rearranged to our advantage. This involves all that we have learnt so far, including such things as

- algebraic simplification
- partial fractions
- trig identities
- exponential and log properties

In this section we concentrate on the simplest kinds of rearrangements. For example, $\int e^{\ln x}\, dx$ looks awful – but it is simply $\int x\, dx$, a much easier proposition. The review question gives further examples.

Solution to review question 9.1.4

All the integrals look formidable (deliberately!) – but in fact, using the properties of exponentials and logs and trig identities the integrands can all be simplified to easily integrated functions.

(i) $\displaystyle\int \frac{dx}{\cos 2x + 2\sin^2 x} = \int \frac{dx}{1 - 2\sin^2 x + 2\sin^2 x}$

using $\cos 2x = 1 - 2\sin^2 x$

$\displaystyle = \int \frac{dx}{1} = x + C$

(ii) $\displaystyle\int xe^{2\ln x}\, dx = \int xe^{\ln x^2}\quad (\alpha \ln x = \ln x^\alpha)$

$\displaystyle = \int xx^2\, dx \quad (e^{\ln x} = x) = \int x^3\, dx$

$\displaystyle = \frac{x^4}{4} + C$

Note that here, as elsewhere, we have taken the modulus under the log for granted, as to include it would clutter up the expressions.

(iii) This is easy if you know your compound angle formulae backwards!

$\displaystyle\int (\cos 2x \sin 3x - \cos 3x \sin 2x)\, dx = \int \sin(3x - 2x)\, dx$

$\displaystyle = \int \sin x\, dx$

$\displaystyle = -\cos x + C$

9.2.5 Linear substitution in integration

It may be that $f(x)$ in $\int f(x)\,dx$ is inconvenient for integration because of the variable x, and it may be useful to change to a new variable. This entails changing dx as well as $f(x)$ of course. For example, consider the integral

$$\int (2x-4)^3\,dx$$

You may be tempted to expand the bracket out by the binomial theorem, and obtain a polynomial which is easily integrated. This is correct, but is an unnecessary complication. The way to go is to notice the similarity to x^3 and substitute

$$u = 2x - 4$$

We must of course also replace dx. To this end we note that

$$\frac{du}{dx} = 2$$

and so $du = 2\,dx$ and $dx = \dfrac{du}{2}$. This somewhat cavalier way of dealing with the dx and du is permissible provided we are careful. The upshot is that

$$\int (2x-4)^3\,dx = \int u^3 \frac{du}{2}$$
$$= \frac{1}{2}\int u^3\,du = \frac{u^4}{8} + C$$
$$= \frac{(2x-4)^4}{8} + C$$

on returning to the original variable.

In general, whenever we make a substitution of the form

$$u = ax + b$$

where a, b are constants, we call this a **linear substitution**. It is particularly simple because

$$\frac{du}{dx} = a$$

and therefore du and dx are simply proportional

$$du = \frac{du}{dx}\,dx = a\,dx$$

so

$$dx = \frac{du}{a}$$

With practice, you may be able to dispense with the formal '$u = x + 1$' substitution and do the above example as follows:

$$\int (2x-4)^3 \, dx = \frac{1}{2} \int (2x-4)^3 \, d(2x-4)$$
$$= \frac{1}{2}\frac{1}{4}(2x-4)^4 + C$$
$$= \frac{1}{8}(2x-4)^4 + C$$

$(d(2x-4) = 2\, dx)$

There is a useful 'intuitive' way of looking at linear substitution. When integrating a function of $ax + b$, $f(ax + b)$, we think 'Well, $ax + b$ is really little different to x. So treat it like x – if the integral of $f(x)$ is $F(x)$, take the integral of $f(ax + b)$ to be $F(ax + b)$. But if we differentiated this with respect to x we would get a multiplier "a" coming in from the function of a function rule. So we need to divide the integral by a to cancel this multiplier. The integral of $f(ax + b)$ is thus $\frac{1}{a}F(ax + b)$.' For the example above this goes through as follows: $(2x - 4)^3$ is like x^3, so take its integral to be $\frac{(2x-4)^4}{4}$. But if we differentiate this we get a 2 coming from differentiating the bracketed term by function of a function. So remove this by multiplying by $\frac{1}{2}$ so that the final form of the integral is $\frac{1}{2} \times \frac{(2x-4)^4}{4} = \frac{(2x-4)^4}{8}$, as we found above.

Another important point is that the linear substitution is very special – it is the **only case** in which du and dx are proportional. Suppose we substitute

$$u = g(x)$$

so

$$\frac{du}{dx} = g'(x)$$

and therefore

$$du = g'(x)\, dx$$

In the linear substitution case this gives

$$du = a\, dx$$

and we could simply divide by a to replace dx by

$$\frac{du}{a}$$

But in any other substitution we **can't do this**, we would get

$$dx = \frac{du}{g'(x)}$$

and we have the complication of substituting for x in terms of u in $g'(x)$. So, greater care is needed with other types of substitutions. In general, simple substitutions require special forms of the integrand – see Section 9.2.6.

Solution to review question 9.1.5

These questions all involve linear substitutions. In no case is it necessary to do anything to the function before substitution.

(i) If you did something like

$$\int \frac{dx}{x-1} = \int \left(\frac{1}{x} - 1\right) dx = \ldots\text{'}$$

stop now and go back to review your basic algebra. It is of course wrong. You cannot 'simplify' $\frac{1}{x-1}$ any further and must work with it as it is, and for this we use a substitution.

The closeness of $\frac{1}{x-1}$ to $\frac{1}{x}$ suggests making the substitution $u = x - 1$ so $du = dx$ and

$$\int \frac{dx}{x-1} = \int \frac{du}{u} = \ln u = \ln(x-1) + C$$

or, quicker

$$\int \frac{dx}{x-1} = \int \frac{d(x-1)}{x-1} = \ln(x-1) + C$$

(ii) Here you may have been tempted to use the compound angle formula to expand the cos and get simple sine and cosine integrals. Again, while correct, this is unnecessarily messy – simply substitute

$$u = 3x + 2, \quad dx = \frac{du}{3}$$

so

$$\int \cos(3x + 2)\,dx = \int \cos u\, \frac{du}{3}$$

$$= \frac{1}{3} \sin u + C$$

$$= \frac{1}{3} \sin(3x + 2) + C$$

or

$$\int \cos(3x + 2)\,dx = \frac{1}{3} \int \cos(3x + 2)\,d(3x + 2)$$

$$= \frac{1}{3} \sin(3x + 2) + C$$

(iii) By now, you will probably be happy with:

$$\int e^{2x-1}\,dx = \frac{1}{2} \int e^{2x-1}\,d(2x - 1)$$

$$= \frac{1}{2} e^{2x-1} + C$$

9.2.6 The $du = f'(x)\,dx$ substitution

Integration by substitution is often difficult. The substitution to use is not always obvious. One case when it is fairly easy to spot the substitution occurs when the integrand contains a **function f(x) and its derivative f'(x)**. One then tries substituting for the function, $u = f(x)$. This is the reverse of the function of a function rule of differentiation. This approach relies on 'noticing' the derivative – i.e. on good facility with differentiation. An important general example of this approach is the integral

$$\int \frac{f'(x)\,dx}{f(x)} = \ln|f(x)| + C$$

because if we put $u = f(x)$ we get $du = f'(x)\,dx$, which occurs on the top and the integral becomes

$$\int \frac{du}{u} = \ln|u| + C = \ln|f(x)| + C$$

Similarly, for example

$$\int \cos(f(x))f'(x)\,dx = \sin(f(x)) + C$$

by putting $u = f(x)$.

In general, to integrate $\int g(f(x))f'(x)\,dx$ put $u = f(x)$ and convert it to $\int g(u)\,du$, sometimes written $\int g(f(x))\,df(x)$, and hope that you can integrate this. We illustrate the process by a further example:

Given $\int xe^{x^2}\,dx$

Note that the derivative of x^2 is $2x$, and we have $\frac{1}{2}(2x) = x$ in the integrand. We know the derivative of e^x is e^x, so maybe the integral looks something like e^{x^2}? So differentiate e^{x^2} and see what we get

$$\frac{d}{dx}(e^{x^2}) = e^{x^2} \times (2x) = 2xe^{x^2}$$

Not quite – the factor 2 is the problem. But that is easily dealt with:

$$\frac{d}{dx}\left(\frac{1}{2}e^{x^2}\right) = xe^{x^2}$$

gives us what we want. So

$$\int xe^{x^2}\,dx = \frac{1}{2}e^{x^2} + C$$

You may need to try this a couple of times – and in the end it may not work, anyway. So, **BE FLEXIBLE AND KNOW YOUR DIFFERENTIATION.**

Solution to review question 9.1.6

(i) In the integral $\int \dfrac{x+1}{x^2+2x+3}\,dx$ the derivative of x^2+2x+3 is $2(x+1)$ and we have an $x+1$ on the top. This suggests substituting $u = x^2 + 2x + 3$. Then

$$\frac{du}{dx} = 2(x+1)$$

You may now be tempted to write

$$dx = \frac{du}{2(x+1)}$$

substitute in the integral and cancel the $x+1$. While correct, this is a bad habit – never mix variables in an integral. Instead, simply use the fact that $x+1$ is ready and waiting for us in the integrand and write

$$du = 2(x+1)\,dx$$

or

$$(x+1)\,dx = \frac{du}{2}$$

to get

$$\int \frac{(x+1)\,dx}{x^2+2x+3} = \int \frac{1}{u}\frac{du}{2}$$

$$= \frac{1}{2}\int \frac{du}{u} = \frac{1}{2}\ln u + C$$

$$= \frac{1}{2}\ln(x^2+2x+3) + C$$

(ii) Again, in $\int x\sin(x^2+1)\,dx$ the derivative of x^2+1 is $2x$ and we have an x multiplying the sine, which suggests putting $u = x^2 + 1$. Then

$$\int x\sin(x^2+1)\,dx \longrightarrow \int \sin u \frac{1}{2}\,du$$

$$= \frac{1}{2}\int \sin u\,du$$

$$= -\frac{1}{2}\cos u + C$$

$$= -\frac{1}{2}\cos(x^2+1) + C$$

Or:

$$\int x\sin(x^2+1)\,dx = \frac{1}{2}\int \sin(x^2+1)\,d(x^2+1)$$

$$= -\frac{1}{2}\cos(x^2+1) + C$$

(iii) Now you may be able to appreciate the following more easily:

$$\int \cos x\, e^{\sin x}\,dx = \int e^{\sin x}(\cos x\,dx)$$

$$= \int e^{\sin x}\,d(\sin x)$$

$$= e^{\sin x} + C$$

(iv) $\int \sin x \cos x\,dx = \int \sin x\,d(\sin x) = \frac{1}{2}\sin^2 x + C$

Compare this with the result that you might obtain for Q9.1.8(iv). In that case you may obtain

$$-\tfrac{1}{4}\cos 2x + C$$

Using the double angle formula for $\cos 2x$ (188 ◀) shows that this in fact only differs from the answer obtained above by a constant, which is unimportant because both forms of the answer contain an arbitrary constant.

9.2.7 Integrating rational functions ◀ 252 283 ▶

Any rational function has the form:

$$\frac{\text{polynomial}}{\text{polynomial}}$$

We can assume that the numerator has lower degree than the denominator, otherwise we could divide out to get a polynomial plus such a fraction. In many useful cases the polynomial in the denominator can be factorised into linear and/or quadratic factors with real coefficients. We could then use partial fractions and substitution to split the rational function up into integrals of the general form

$$\int \frac{ax+b}{cx^2+dx+e}\,dx$$

This form of integral is therefore very important, and we will study it in detail. How we approach it depends on whether or not the denominator factorises (45 ◀). If it does, we can use **partial fractions** (62 ◀). If it doesn't then **completing the square** (66 ◀) enables us to use a linear substitution and a standard integral.

Example

$$\int \frac{x}{x^2 - 3x + 2} dx = \int \frac{x}{(x-1)(x-2)} dx \quad \begin{pmatrix} \text{factorise} \\ \text{denominator} \end{pmatrix} \quad (45 \blacktriangleleft)$$

$$= \int \left[\frac{2}{x-2} - \frac{1}{x-1} \right] dx \quad \begin{pmatrix} \text{split into} \\ \text{partial fractions} \end{pmatrix} \quad (62 \blacktriangleleft).$$

$$= 2 \int \frac{dx}{x-2} - \int \frac{dx}{x-1} \quad \text{(isolate integrals)} \quad (258 \blacktriangleleft).$$

$$= 2 \ln(x-2) - \ln(x-1) + C \text{ (using substitutions)}$$

You can see the skill required at each step – if any steps still puzzle you go back to the appropriate paragraph. For a similar example using completing the square see the solution to the review question.

Given a rational function of the type

$$\frac{ax+b}{cx^2 + dx + e}$$

we may need to combine a number of methods – such as completing the square and substitution for example. This may require some manipulation of the integrand. Also we may not be told which method to use to integrate it, so we have to be able to spot the best approach. There are three methods we can employ, separately, or in combination:

- Substitution
- Completing the square
- Partial fractions

We can use substitution immediately if the top is proportional to the derivative of the bottom, using the result

$$\int \frac{f'(x) dx}{f(x)} = \ln f(x) + C$$

If it is not, then check whether the denominator factorises – if it does, then we can use partial fractions. If the bottom doesn't factorise, we can always rewrite the numerator to make it equal to the derivative and an additional constant, that is:

$$ax + b \equiv \frac{a(2cx + d)}{2c} - \frac{ad}{2c} + b$$

(Confirm this result to check that your algebra is up to scratch!) Then

$$\frac{ax+b}{cx^2 + dx + e} = \frac{a}{2c} \left(\frac{2cx + d}{cx^2 + dx + e} \right) + \left(b - \frac{ad}{2c} \right) \frac{1}{cx^2 + dx + c}$$

The first part can now be done by substituting $u = cx^2 + dx + e$, which leaves us with the problem of integrating something of the form

$$\int \frac{dx}{cx^2 + dx + e}$$

If the bottom doesn't factorise then we will have to use completing the square.

Solution to review question 9.1.7

If, in these questions, you have done something like:

$$\int \frac{dx}{x^2 + 2x + 2} = \frac{1}{2x+2} \ln(x^2 + 2x + 2)\text{'}$$

which is **incorrect**, then you have probably misunderstood the idea of substitution. 'Dividing by the derivative of the denominator' in this way will **only** work for a linear denominator, when the derivative is constant (261 ◄). You have only to differentiate the right-hand side to see that it cannot possibly be the integral of the left-hand side. The following solutions show the correct approach to such integrals.

A. In this case the denominator factorises and we don't need to complete the square – we can split into partial fractions:

$$\int \frac{dx}{x^2 + x - 2} = \int \frac{dx}{(x-1)(x+2)}$$

$$= \int \left[\frac{1}{3(x-1)} - \frac{1}{3(x+2)} \right] dx$$

$$= \frac{1}{3} \int \frac{dx}{x-1} - \frac{1}{3} \int \frac{dx}{x+2} + C$$

$$= \frac{1}{3} \ln(x-1) - \frac{1}{3} \ln(x+2) + C$$

using

$$\int \frac{dx}{x-a} = \ln(x-a) + C$$

We can tidy the answer up and write:

$$\int \frac{dx}{x^2 + x - 2} = \frac{1}{3} \ln\left(\frac{x-1}{x+2}\right) + C$$

B. In this case the denominator does not factorise, and the only option is to complete the square (66 ◄). We have:

$$\int \frac{dx}{x^2 + 2x + 2} = \int \frac{dx}{(x+1)^2 + 1}$$

Now this looks like the inverse tan integral (256 ◄), which we can get by substituting $u = x + 1$, $du = dx$:

$$\int \frac{dx}{(x+1)^2 + 1} = \int \frac{du}{u^2 + 1} = \tan^{-1} u$$

so

$$\int \frac{dx}{x^2 + 2x + 2} = \tan^{-1}(x+1)$$

C. This question illustrates that sometimes you may need to combine methods. Here, the numerator is of the same degree as the denominator, so we must first divide out:

$$\frac{2x^2 + 5x + 4}{x^2 + 2x + 2} = \frac{2(x^2 + 2x + 2) - 4x - 4 + 5x + 4}{x^2 + 2x + 2}$$

$$= 2 + \frac{x}{x^2 + 2x + 2}$$

So

$$\int \frac{2x^2 + 5x + 4}{x^2 + 2x + 2} dx = \int \left[2 + \frac{x}{x^2 + 2x + 2} \right] dx$$

$$= 2x + C + \int \frac{x \, dx}{x^2 + 2x + 2}$$

In the remaining integral the denominator doesn't factorise and the top is not the derivative of the bottom. We can remedy this by adding and subtracting 1 in the numerator, to manufacture the derivative of the bottom on the top!

$$\int \frac{x \, dx}{x^2 + 2x + 2} = \int \frac{x + 1 - 1}{x^2 + 2x + 2} dx$$

$$= \int \frac{(x + 1)}{x^2 + 2x + 2} dx - \int \frac{dx}{x^2 + 2x + 2}$$

$$= \frac{1}{2} \int \frac{d(x^2 + 2x + 2)}{x^2 + 2x + 2} - \int \frac{d(x + 1)}{(x + 1)^2 + 1}$$

$$= \frac{1}{2} \ln |x^2 + 2x + 2| - \tan^{-1}(x + 1) + C$$

So, finally, the integral is

$$2x + \frac{1}{2} \ln |x^2 + 2x + 2| - \tan^{-1}(x + 1) + C$$

Note that we don't need to duplicate the arbitrary constant C.

D. In these problems you have to decide which method to use for yourself. Also, (iv) illustrates that the 'obvious' method is not always the best, so you should always be on the look-out for easier alternatives.

(i) The denominator won't factorise, so complete the square

$$\int \frac{dx}{x^2 + x + 1} = \int \frac{dx}{\left(x + \frac{1}{2}\right)^2 + \frac{3}{4}}$$

$$= \frac{2}{\sqrt{3}} \tan^{-1} \left(\frac{x + \frac{1}{2}}{\sqrt{3}/2} \right) + C$$

$$= \frac{2}{\sqrt{3}} \tan^{-1} \left(\frac{2x + 1}{\sqrt{3}} \right) + C$$

(ii) The denominator factorises so we can use partial fractions

$$\int \frac{dx}{x^2 + 3x + 2} = \int \frac{dx}{(x+1)(x+2)}$$
$$= \int \left[\frac{1}{(x+1)} - \frac{1}{(x+2)} \right] dx$$
$$= \int \frac{dx}{x+1} - \int \frac{dx}{x+2}$$
$$= \ln\left(\frac{x+1}{x+2}\right) + C$$

(iii) The top is the derivative of bottom so use substitution

$$\int \frac{2x+1}{x^2+x-1} dx = \int \frac{d(x^2+x-1)}{x^2+x-1}$$
$$= \ln(x^2+x-1) + C$$

(iv) You may have tackled this by partial fractions. This is valid and will give the correct answer, but you might spot an easier way, if your algebra and substitution are on the ball.

$$\int \frac{3x}{(x-1)(x+1)} dx = 3 \int \frac{x}{x^2-1} dx$$
$$= \frac{3}{2} \int \frac{2x}{x^2-1} dx$$
$$= \frac{3}{2} \int \frac{d(x^2)}{x^2-1}$$
$$= \frac{3}{2} \ln(x^2-1) + C$$

9.2.8 Using trig identities in integration

◄ 252 283 ►

Using a simple linear substitution, $u = 3x$, we can integrate things such as:

$$\int \cos 3x \, dx = \frac{1}{3} \sin 3x + C$$

Using this and various trig identities we can perform some quite complicated integrations. The most common examples here include the use of the compound angle and double angle formulae

$$\sin(A \pm B) = \sin A \cos B \pm \sin B \cos A$$
$$\cos(A \pm B) = \cos A \cos B \mp \sin A \sin B$$

which enable us to integrate functions of the type

$$\sin mx \cos nx$$
$$\sin mx \sin mx$$
$$\cos mx \cos nx$$
$(m \neq n)$

Such integrals are fundamental to calculations in the theory of Fourier series (Chapter 17).

The double angle formulae can help us deal with integrals containing powers of $\sin x$ or $\cos x$. Odd powers of $\sin x$ and $\cos x$ can sometimes be dealt with using the Pythagorean identity $\cos^2 x + \sin^2 x = 1$ and substitution. In general, for the integral:

$$\int \sin^m x \cos^n x \, dx$$

if one of m or n is odd, try using the Pythagorean identity whilst if both are even, use double angle formula and other trig identities. The review questions illustrate this.

For integrals such as

$$\int \cos mx \cos nx \, dx, \int \sin mx \sin nx \, dx, \int \sin mx \cos nx \, dx$$

where $m \neq n$, use the compound angle formulae. Again, see the review questions. Hyperbolic functions can be dealt with similarly.

Solution to review question 9.1.8

The use of trig identities for the examples given relies on the fact that we can always integrate cos or sine of multiple angles:

$$\int \cos kx \, dx = \frac{1}{k} \sin kx + C$$

$$\int \sin kx \, dx = -\frac{1}{k} \cos kx + C$$

When faced with integrals such as those in this question, make a list of the identities you know involving products of sines and cosines:

$$\cos 2x = 2\cos^2 x - 1 = 1 - 2\sin^2 x$$
$$= \cos^2 x - \sin^2 x$$
$$\sin 2x = 2 \sin x \cos x$$
$$\sin(A + B) = \sin A \cos B + \sin B \cos A, \text{ etc.}$$

using these, the integrals are not so bad:

(i) $\displaystyle\int \sin^2 x \, dx = \int \frac{1}{2}(1 - \cos 2x) \, dx$

by the double angle formula for $\cos 2x$.

$$= \frac{1}{2} \int (1 - \cos 2x) \, dx$$

$$= \frac{1}{2} \left(x - \frac{\sin 2x}{2} \right) + C$$

NB: Be careful with functions such as $\cos 2x$, which, of course is not the same thing as $2 \cos x$!

(ii) We can write $\int \cos^3 x \, dx = \int \cos^2 x \cos x \, dx$

$$= \int (1 - \sin^2 x) \cos x \, dx$$

Now put $u = \sin x$, or equivalently:

$$= \int (1 - \sin^2 x) \, d(\sin x) = \sin x - \frac{\sin^3 x}{3} + C$$

(iii) We can express $\sin A \cos B$ as a sum of sines by using

$$\sin(A + B) = \sin A \cos B + \sin B \cos A$$
$$\sin(A - B) = \sin A \cos B - \sin B \cos A$$

Adding:

$$\sin A \cos B = \tfrac{1}{2}(\sin(A + B) + \sin(A - B))$$

So

$$\sin 2x \cos 3x = \tfrac{1}{2}(\sin(2x + 3x) + \sin(2x - 3x))$$
$$= \tfrac{1}{2}(\sin 5x + \sin(-x))$$
$$= \tfrac{1}{2}(\sin 5x - \sin x)$$

So we can write:

$$\int \sin 2x \cos 3x \, dx = \frac{1}{2} \int (\sin 5x - \sin x) \, dx$$

$$= \frac{1}{2} \left(-\frac{\cos 5x}{5} + \cos x \right) + C$$

$$= \frac{1}{2} \cos x - \frac{1}{10} \cos 5x + C$$

(iv) This is easy now, using the double angle formula for $\sin 2x$ backwards:

$$\sin x \cos x = \tfrac{1}{2} 2 \sin x \cos x$$
$$= \tfrac{1}{2} \sin 2x$$

> So
> $$\int \sin x \cos x \, dx = \frac{1}{2} \int \sin 2x \, dx$$
> $$= -\frac{1}{4} \cos 2x + C$$
>
> Refer back to Review Question 9.1.6(iv) for an alternative form of this.

9.2.9 Using trig substitutions in integration

Trig or hyperbolic substitutions may help you to deal with some rational and irrational functions. Thus if you have

$$\sqrt{a^2 - x^2} \quad \text{try } x = a \sin\theta \quad \text{or } x = \tanh\theta$$
$$\sqrt{x^2 - a^2} \quad \text{try } x = a \cosh\theta \quad \text{or } x = \sec\theta$$
$$\sqrt{a^2 + x^2} \quad \text{try } x = a \sinh\theta \quad \text{or } x = \tan\theta$$

The form of the substitution may depend on the range of values of x.
Another useful trig substitution is

$$t = \tan\theta/2$$

With this we have

$$\cos\theta = \frac{1 - t^2}{1 + t^2}, \quad \sin\theta = \frac{2t}{1 + t^2}$$

We also have to change between $d\theta$ and dt of course. We have

$$\frac{d\theta}{dt} = \frac{1}{2}\sec^2\frac{\theta}{2} = \frac{1}{2}\left(1 + \tan^2\frac{\theta}{2}\right) = \frac{1}{2}(1 + t^2)$$

from which

$$d\theta = \frac{2dt}{1 + t^2}$$

This substitution can sometimes be used to convert an integral of the form

$$\int f(\cos\theta, \sin\theta) \, d\theta$$

to an integral of a rational function in t, which may be easier to deal with.

Solution to review question 9.1.9

Whenever you see something like $\sqrt{a^2 - x^2}$ try a sine substitution $x = a \sin\theta$.

(i) For $\int \dfrac{dx}{\sqrt{9 - x^2}}$ the function $\sqrt{9 - x^2} = \sqrt{3^2 - x^2}$ appears and so we try a substitution $x = 3\sin\theta$. This gives $dx = 3\cos\theta\, d\theta$ and

$$\sqrt{9 - x^2} = \sqrt{9 - 9\sin^2\theta}$$
$$= 3\sqrt{\cos^2\theta} = 3\cos\theta$$

So

$$\int \dfrac{dx}{\sqrt{9 - x^2}} \longrightarrow \int \dfrac{3\cos\theta\, d\theta}{3\cos\theta}$$
$$= \int d\theta = \theta + C$$

But $\sin\theta = \dfrac{x}{3}$, so $\theta = \sin^{-1}\left(\dfrac{x}{3}\right)$ and therefore

$$\int \dfrac{dx}{\sqrt{9 - x^2}} = \sin^{-1}\left(\dfrac{x}{3}\right) + C$$

(ii) $\int \dfrac{dx}{\sqrt{3 - 2x - x^2}}$ doesn't seem to fit any of the simple forms given above. However, by completing the square and substituting we can make progress:

$$\int \dfrac{dx}{\sqrt{3 - 2x - x^2}} \equiv \int \dfrac{dx}{\sqrt{4 - (x + 1)^2}}$$
$$\equiv \int \dfrac{dx}{\sqrt{2^2 - (x + 1)^2}}$$
$$= \sin^{-1}\left(\dfrac{x + 1}{2}\right) + C$$

on putting $u = x + 1$.

9.2.10 Integration by parts

Useful for some products the integration by parts formula is derived by integrating the product rule of differentiation:

$$\dfrac{d(uv)}{dx} = u\dfrac{dv}{dx} + v\dfrac{du}{dx}$$

$$\int u\dfrac{dv}{dx}\,dx = uv - \int v\dfrac{du}{dx}\,dx$$

('integrate one, differentiate the other').

First, note that not all products can be integrated by parts, and that integration by parts can be useful for things that are not obvious products. We may also need to use it a number of times, or employ it in more cunning ways (see review solutions).

The first problem we face with integration by parts is which factor to integrate first. There are few hard and fast rules, and experience counts for a lot here. Still, we can get a lot from our knowledge of the elementary functions:

> Polynomials
> Rational functions
> Irrational functions
> Cosine and sine
> Exponentials

and their inverses. Rational and irrational (containing roots of algebraic functions) functions are difficult to deal with when combined with other functions, so we will focus on products of polynomials, cosine and sine and exponentials. By linearity, if we can handle a power function x^n then we can handle a polynomial. Also, cosine and sine are virtually equivalent so far as calculus is concerned, so we can concentrate on sine. Finally, $e^{\alpha x}$, $\sin(\alpha x)$ are little different to e^x, $\sin x$ so we can look at just products of the three functions x^n, e^x, $\sin x$. This gives us integrals such as:

$$I_1 = \int x^n e^x \, dx$$

$$I_2 = \int x^n \sin x \, dx$$

$$I_3 = \int e^x \sin x \, dx$$

In I_1, I_2 integrating or differentiating e^x or $\sin x$ is neither here nor there because essentially the same things results. However, integrating x^n will make things worse, while differentiating it will eventually remove it. So for I_1, I_2 we start off by integrating e^x or $\sin x$ so that the x^n will be differentiated. In I_3 it actually doesn't matter which we integrate first, we will always come back to where we started – and indeed this is the secret to this sort of integral (see review question).

$\ln x$ and inverse functions such as $\sin^{-1} x$ can be integrated by parts by treating them as a product with 1 and integrating the 1. For example,

$$\int \ln x \, dx = \int 1 \times \ln x \, dx$$

$$= x \ln x - \int x \times \frac{1}{x} \, dx = x \ln x - x + C$$

Solution to review question 9.1.10

(i) We will use the integration by parts formula

$$\int u \frac{dv}{dx} \, dx = uv - \int v \frac{du}{dx} \, dx$$

but only on this first example will we explicitly write

$$u = x \qquad \frac{dv}{dx} = \sin x$$

$$\frac{du}{dx} = 1 \qquad v = -\cos x$$

to help you on your way. Note that we chose $u = x$ so that when we differentiate the x becomes a much more amenable 1! We have

$$\int x \sin x \, dx = x(-\cos x) - \int \frac{d(x)}{dx}(-\cos x) \, dx$$

$$= -x \cos x + \int \cos x \, dx$$

$$= -x \cos x + \sin x + C$$

With practice, you should be able to dispense with explicit use of u, v, etc. – although continue with it until you feel confident enough to do without.

(ii) We may need to use integration by parts more than once (if you do, remember to keep integrating the same factor). Again, we are going to chose to differentiate the x^2 to reduce it, so we first integrate the e^x.

$$\int x^2 e^x \, dx = x^2(e^x) - \int 2x e^x \, dx$$

$$= x^2 e^x - 2\left[x e^x - \int e^x \, dx\right]$$

$$= x^2 e^x - 2x e^x + 2e^x + C$$

(iii) In integrating $\int e^x \sin x \, dx$ we sometimes feel like we are going round in circles with integration by parts! Write

$$I = \int e^x \sin x \, dx = e^x \sin x - \int e^x \cos x \, dx$$

(note that it doesn't matter which of e^x, $\sin x$ we integrate)

$$= e^x \sin x - \left[e^x \cos x - \int e^x(-\sin x) \, dx\right]$$

being careful to keep integrating the same factor

$$= e^x(\sin x - \cos x) + C - \int e^x \sin x \, dx$$

$$= e^x(\sin x - \cos x) - I + C$$

Note that we only need the one arbitrary constant. We can now transfer the I to the left-hand side to get

$$2I = e^x(\sin x - \cos x) + C$$

so $\quad I = \tfrac{1}{2} e^x(\sin x - \cos x) + C$

NB. Since C is arbitrary we have no need to keep changing it during the manipulations, or introducing new arbitrary constants.

9.2.11 Choice of integration methods

◀ 253 284 ▶

So far you have usually been told which method to use for a particular integral – a bit like someone handing you just the right tool for a particular job when you need it. By now you have developed a whole box of tools of integration, and there will not always be someone around to tell you which to choose for the given jobs – you need to know the tools and their uses well enough to choose the right ones for yourself. Efficient selection of the right tool from the box, as you will know, requires a lot of practice and experience, and even the best of us will make mistakes at times. Sometimes the tool we pick won't fit, so we have to try another. Sometimes different tools will do the same job, but one of them is the best to use in a particular instance. It helps to have a rough summary of the different types of tools, and for integration we have covered essentially:

- standard integrals
- simplifying the integrand
- linear substitution
- $du = f'(x) \, dx$ substitutions
- the use of trig identities and substitutions
- partial fractions
- completing the square
- integration by parts

Various strategies for choosing the best methods are discussed in other sections, and here we will simply use the review question to bring together more examples. Be prepared to go up a few blind alleys and don't be afraid to make mistakes in this topic.

Solution to review question 9.1.11

(i) We looked at rational functions such as $\dfrac{x-3}{x^2 - 6x + 4}$ in Section 9.2.7, and the common methods are

- simplification of the integrand (dividing out, etc.)
- substitution, $du = f'(x) \, dx$
- partial fractions
- completing the square

In this case we notice that the numerator is almost the derivative of the denominator, suggesting that we try the substitution $u = x^2 - 6x + 4$ here.

(ii) There are a number of ways we might tackle $\sin^3 x$, covered in Section 9.2.8, but perhaps the easiest is to use the Pythagorean identity to write $\sin^2 x = 1 - \cos^2 x$ and then use the substitution, $u = \cos x$ (271 ◄).

(iii) In $\dfrac{e^{\ln x}}{x}$ you may notice that $\dfrac{1}{x}$ is the derivative of $\ln x$, suggesting the substitution $u = \ln x$. But hopefully you are now happy enough with the exponential function to note the simplification $e^{\ln x} = x$ and take it from there (259 ◄)!

(iv) In $\dfrac{x-1}{2x^2 + x - 3}$ the top is not the derivative of the bottom, but we notice that the denominator factorises into $(2x + 3)(x - 1)$ and so the function can be simplified by cancelling the $x - 1$ (provided x is not equal to 1, of course), resulting in a simple linear substitution and log integral (263 ◄).

(v) No, xe^{3x^2} is not a prime candidate for integration by parts, despite being similar to products we have dealt with in that way. Instead, we notice that the x is almost the derivative of the x^2 in the exponent, and this suggests that the substitution $u = x^2$ will rescue us here (263 ◄).

(vi) Nasty looking functions such as $\dfrac{3}{\sqrt{3 - 4x - x^2}}$ containing roots of algebraic functions usually succumb only to some sort of trig or hyperbolic substitution as considered in Section 9.2.9 – though in this case we need to complete the square first. A similar integral is done in Review Question 9.1.9(ii).

(vii) $x \sin(x + 1)$ yields to a straightforward integration by parts, in which the $\sin(x + 1)$ is integrated first (273 ◄).

(viii) $\cos^4 x$ can be integrated by using the double angle formulae

$$\cos 2x = 2\cos^2 x - 1 = 1 - 2\sin^2 x$$

to replace $\cos^2 x$ and subsequently $\cos^2 2x$ to leave us with integrals of a constant, $\cos 2x$ and $\cos 4x$ (270 ◄).

(ix) By now you should see immediately that $\ln\left(\dfrac{e^x}{x}\right) = x - \ln x$ and we have only to integrate the $\ln x$, which can be done by parts, as shown in Section 9.2.10.

(x) The denominator of $\dfrac{x+2}{x^2 - 5x + 6}$ factorises to $(x - 2)(x - 3)$ and we have a straightforward partial fractions to deal with (266 ◄).

(xi) xe^{2x} provides a very typical integration by parts (273 ◄).

(xii) $\ln e^x \cos(e^x)$ we note that the derivative of e^x is of course e^x which tells us to substitute $u = e^x$ (263 ◄).

(xiii) $\sin 2x \cos 2x$ is, from the double angle formulae, $\frac{1}{2}\sin 4x$ and becomes a standard integral on substituting $u = 4x$ (271 ◄).

(xiv) $\dfrac{x-1}{\sqrt{x^2 - 2x + 3}}$ looks fearsome until we notice that the numerator is almost the derivative of the quadratic under the square root, prompting us to make the substitution $u = x^2 - 2x + 3$ (263 ◄).

(xv) $\dfrac{x+3}{x^2 + 2x + 2}$ can be rewritten as $\dfrac{x+1}{x^2 + 2x + 2} + \dfrac{2}{x^2 + 2x + 2}$ whence the first fraction can be integrated by the substitution $u = x^2 + 2x + 2$, and the second by completing the square (268 ◄).

(xvi) $\sin(4x)\cos(5x)$ looks like (xiii) but in this case we have to use the compound angle formulae to express it as a combination of $\sin(9x)$ and $\sin x$ – a similar integral is done in Review Question 9.1.8(iv) (271 ◄).

(xvii) $\ln x \cos(x^2 + 1)$ just notice that the x is almost the derivative of the $x^2 + 1$ and take it from there (263 ◄).

9.2.12 The definite integral   ◄ 253 284 ►

If
$$\int f(x)\,dx = F(x) + C$$

then the **definite integral of $f(x)$ between the limits $x = a$, $x = b$** is the **number**:

$$\int_a^b f(x)\,dx = [F(x)]_a^b = F(b) - F(a)$$

That is, substitute the upper limit b in the indefinite integral $F(x)$ and subtract the value of $F(x)$ with the lower limit substituted. Notice that even if we include the arbitrary constant in the actual integration, it will only cancel out when the difference between the upper and lower limits is taken – so we can discard it in the definite integral. The variable x in the definite integral is called a **dummy variable**, it being immaterial to the value of the integral – thus

$$\int_a^b f(x)\,dx = \int_a^b f(t)\,dt$$

In this respect the integration variable is like the summation index in the sigma notation (102 ◄). As we will see in Chapter 10, the definite integral can be interpreted as an area under a curve, and more rigorously as the limit of a sum – but here we are simply interested in the technicalities of **performing** the definite integral.

Obvious properties of the definite integral are:

$$\int_a^a f(x)\,dx = F(a) - F(a) = 0$$

$$\int_a^b f(x)\,dx = F(b) - F(a) = -(F(a) - F(b)) = -\int_b^a f(x)\,dx$$

Not so obvious is the fact that if $a < b < c$ then

$$\int_a^b f(x)\,dx + \int_b^c f(x)\,dx = F(b) - F(a) + F(c) - F(b) = F(c) - F(a)$$

$$= \int_a^c f(x)\,dx$$

Remember to change limits if you make a substitution – **and** to ensure that the substitution is consistent with the values of the limits.

When integrating a definite integral by parts you can substitute the limits in as you go along:

$$\int_a^b u \frac{dv}{dx}\,dx = [uv]_a^b - \int_a^b v \frac{du}{dx}\,dx$$

When putting limits on integrals we must be careful to avoid, in the integration interval, any points where the integral or integrand would not exist. For example

$$\int_0^1 \frac{dx}{2x - 1}$$

is an **improper integral** because the integrand is not defined at $x = \frac{1}{2}$, which is within the range of integration. Another type of 'improper integral' occurs when one or both of the limits involves infinity. We will see this specifically in Chapter 17, in the Laplace transform, but for now we simply state the definition:

$$\int_0^\infty f(x)\,dx = \lim_{a \to \infty} \int_0^a f(x)\,dx$$

That is, we first integrate using a finite limit a and then we let a tend to infinity in the result.

Solution to review question 9.1.12

A. (i) The indefinite integral of $f(x) = x^2 + 1$ is $F(x) = \frac{x^3}{3} + x$, ignoring the arbitrary constant. So

$$\int_0^1 (x^2 + 1)\,dx = [F(x)]_0^1 = F(1) - F(0) = \left[\frac{x^3}{3} + x\right]_0^1$$

$$= \frac{1}{3} + 1 - 0 = \frac{4}{3}$$

(ii) We can fit the limits in as we go along in integration by parts:

$$\int_0^1 xe^x\,dx = [xe^x]_0^1 - \int_0^1 e^x\,dx$$

$$= e - 0 - [e^x]_0^1$$

$$= e - (e - e^0) = 1 \text{ since } e^0 = 1$$

(iii) When we make a substitution, we can change the limits too – then we don't have to return to the original variables.

In $\int_0^1 \dfrac{x^2}{x^3+1} dx$ the numerator is the derivative of the denominator (almost), so put $u = x^3 + 1$.

Then $du = 3x^2 dx$ or $x^2 dx = \dfrac{du}{3}$.

The limits on u are obtained from the substitution

when $x = 0$, $u = 0^3 + 1 = 1$

when $x = 2$, $u = 2^3 + 1 = 9$

So
$$\int_0^2 \dfrac{x^2}{x^3+1} dx = \dfrac{1}{3}\int_1^9 \dfrac{du}{u} = \dfrac{1}{3}[\ln u]_1^9$$
$$= \dfrac{1}{3}(\ln 9 - \ln 1)$$
$$= \dfrac{1}{3}\ln 9$$

remembering that $\ln 1 = 0$.

(iv) This sort of integral is very important in the theory of Fourier series (see Chapter 17). Using the double angle formula we have

$$\int_0^{2\pi} \cos x \sin x \, dx = \dfrac{1}{2}\int_0^{2\pi} \sin 2x \, dx$$
$$= -\dfrac{1}{4}[\cos 2x]_0^{2\pi} = 0$$

B. $\int_0^2 \dfrac{dx}{x^2 - 1} = \int_0^2 \dfrac{dx}{(x-1)(x+1)}$

The integrand doesn't exist ('becomes singular') at $x = 1$, which is within the range of integration.

9.3 Reinforcement

9.3.1 Definition of integration

◀◀ 251 253 ◀

A. Differentiate the following functions:

(i) $3x^3$ (ii) $\sqrt{3x}$ (iii) $\dfrac{2}{x^4}$ (iv) $x^{2/5}$

(v) $\sin 3x$ (vi) $\cos x^3$ (vii) e^{5x} (viii) $\dfrac{1}{x+1}$

(ix) $\ln 2x$ (x) $\sqrt{x+1}$ (xi) $\ln(3x+1)$ (xii) $\tan^{-1} x$

B. Integrate the following functions:

(i) $\dfrac{2}{x}$ (ii) $3e^{5x}$ (iii) $\dfrac{2}{(1+x)^2}$ (iv) $\dfrac{1}{x^5}$

(v) $3x^2 \cos x^3$ (vi) $\dfrac{-3}{x^2+1}$ (vii) $x^{-\frac{3}{5}}$ (viii) $\dfrac{1}{\sqrt{x}}$

(ix) $2\cos 3x$ (x) x^2 (xi) $\dfrac{3}{\sqrt{x+1}}$ (xii) $\dfrac{1}{3x+1}$

9.3.2 Standard integrals

◀◀ 251 255 ◀

A. Integrate the following functions:

(i) 4   (ii) $3x^2 - 2x + 1$ (iii) $3/x^5$ (iv) $x^{2/3}$

(v) $\cos x$ (vi) $\sec^2 x$ (vii) $\ln x$ (viii) $\dfrac{1}{x^2+1}$

B. Integrate the following functions with respect to the appropriate variables:

(i) $4u$ (ii) $2s^2 - 3s + 2$ (iii) $6/t^7$ (iv) $x^{1/4}$

(v) $\sin \theta$ (vi) $\operatorname{cosec}^2 t$ (vii) e^{3t} (viii) $\dfrac{1}{s^2+1}$

9.3.3 Addition of integrals

◀◀ 251 257 ◀

Integrate:

(i) $2x^2 - \dfrac{3}{x^3}$ (ii) $2\cos x - \sin x$ (iii) $\cosh x = \dfrac{1}{2}(e^x + e^x)$

(iv) $\sin(x+3)$ by compound angle formulae

(v) The polynomial:
$$\sum_{r=0}^{n} a_r x^r$$

(vi) $\sin x \cos 2x$ (vii) $\sinh x$

9.3.4 Simplifying the integrand

◀◀ 251 258 ◀

A. Evaluate the following by simplifying the integrand – noting any special conditions needed.

281

(i) $\displaystyle\int \frac{x^2 - 3x + 2}{x - 2}\, dx$ (ii) $\displaystyle\int \ln e^{\cos x}\, dx$

(iii) $\displaystyle\int \frac{\cos 2x}{\cos x + \sin x}\, dx$ (iv) $\displaystyle\int \frac{e^{\cos^2 x} \ln e^{2x}}{e^{3\ln x} e^{-\sin^2 x}}\, dx$

(v) $\displaystyle\int (\cos x - \sin x)^2\, dx$

B. Sometimes you have to **complicate** a function to integrate it. Integrate $\sec x$ by multiplying and dividing by $\sec x + \tan x$. Integrate $\csc x$ in a similar way.

9.3.5 Linear substitution in integration   ◄◄ 251 260 ◄

Integrate:

(i) $(4x + 3)^7$ (ii) $(2x - 1)^{1/3}$ (iii) $\sin(3x - 1)$ (iv) $2e^{4x} + 1$

(v) $\dfrac{3}{2x + 1}$ (vi) $(2x - 1)^4$ (vii) $\sqrt{3x + 2}$ (viii) $\cos(2x + 1)$

(ix) $4e^{3x-1}$ (x) $\dfrac{1}{4x - 3}$

9.3.6 The $du = f'(x)\, dx$ substitution   ◄◄ 252 263 ◄

A. Write down the substitution $u = f(x)$ you would use to integrate the following and hence integrate them.

(i) $x^2 e^{x^3}$ (ii) $\sec^2 x \sin(\tan x + 2)$

(iii) $\cos x \sin^3 x$ (iv) $\dfrac{x + 1}{x^2 + 2x + 3}$ (v) $\tan x$

B. Integrate:

(i) $f'(x)\sin(f(x))$ (ii) $f(x)e^{(f(x))^2} f'(x)$ (iii) $x^2 \cos(x^3 + 1)$

(iv) $-\sin x \ln(\cos x)$ (v) $\dfrac{x - 1}{x^2 - 2x + 1}$ (vi) $\displaystyle\int f'(x) e^{f(x)}\, dx$

(vii) $\displaystyle\int f'(x) \cos(f(x) + 2)\, dx$

C. Integrate:

(i) $3x^2(x^3 + 2)^3$ (ii) $2x\sqrt{x^2 + 1}$ (iii) $\dfrac{2x - 1}{x^2 - x + 1}$

(iv) $(x + 1)\cos(x^2 + 2x + 1)$ (v) $\sin x\, e^{\cos x}$ (vi) $\dfrac{\sec^2 x}{\tan x}$

(vii) $\cos 2x(2\sin 2x + 1)^3$ (viii) $\dfrac{6x + 12}{\sqrt{x^2 + 4x + 4}}$

9.3.7 Integrating rational functions

A. Integrate the following by partial fractions:

(i) $\dfrac{2x}{(x-1)(x+3)}$ (ii) $\dfrac{x+1}{x^2+5x+6}$ (iii) $\dfrac{4}{2x^2-x-1}$

(iv) $\dfrac{3}{(x+1)(x^2+1)}$ (v) $\dfrac{x+1}{(x-1)^2(x-2)}$ (vi) $\dfrac{2x+1}{x^3+2x^2-x-2}$

B. Integrate the following by completing the square:

(i) $\dfrac{1}{x^2+2x+5}$ (ii) $\dfrac{3}{x^2-2x+2}$ (iii) $\dfrac{2}{2x^2+2x+1}$

(iv) $\dfrac{1}{x^2+6x+10}$ (v) $\dfrac{1}{2x^2+12x+27}$

C. Integrate:

(i) $\dfrac{x^2+1}{(x+1)(x+2)}$ (ii) $\dfrac{x^3}{x^2+2x+2}$ (iii) $\dfrac{3x^4}{(x-1)(x+1)}$

9.3.8 Using trig identities in integration

Integrate:

(i) $\cos^2 x \sin^3 x$ (ii) $\cos 2x \cos 3x$ (iii) $\cos^5 x$

(iv) $\cos 5x \sin 3x$ (v) $\sin 2x \sin 3x$

9.3.9 Using trig substitutions in integration

A. Integrate the following functions using appropriate substitutions:

(i) $\dfrac{1}{\sqrt{4-4x^2}}$ (ii) $\dfrac{2}{4+9x^2}$ (iii) $\dfrac{2}{\sqrt{1-9x^2}}$

(iv) $\dfrac{3}{1+4x^2}$ (v) $\dfrac{1}{\sqrt{8-2x-x^2}}$ (vi) $\dfrac{1}{\sqrt{6x-x^2}}$

B. Use the $t = \tan\dfrac{\theta}{2}$ substitution to integrate $\displaystyle\int \dfrac{1}{3+5\cos\theta}\, d\theta$.

9.3.10 Integration by parts

Integrate:

(i) $x \cos x$ (ii) $x^3 e^x$ (iii) $\sin^{-1} x$ (iv) $e^x \cos x$

(v) $x^2 \cos x$ (vi) $x \ln x$ (vii) $x^3 e^{x^2}$

9.3.11 Choice of integration methods

◀◀ 253 276 ◀

A. From the following three methods of integration:

    A standard integral
    B substitution
    C integration by parts

state which you would use to evaluate the following integrals:

(i) $\int 2e^x \, dx$ (ii) $\int xe^x \, dx$ (iii) $\int xe^{x^2} \, dx$

(iv) $\int x \cos(x^2) \, dx$ (v) $\int \sec x \tan x \, dx$ (vi) $\int \sec^2 x \tan x \, dx$

B. Choose from

    A partial fractions
    B completing the square
    C substitution

the methods (in the order in which you would use them) you would use to integrate

(i) $\dfrac{1}{x^2 + x + 1}$ (ii) $\dfrac{1}{x^2 + x - 2}$ (iii) $\dfrac{2x + 1}{x^2 + x - 1}$

(iv) $\dfrac{4x + 12}{\sqrt{x^2 + 6x + 6}}$ (v) $\dfrac{1}{\sqrt{7 - 6x - x^2}}$ (vi) $\dfrac{2}{x^2 + 2x + 1}$

C. Integrate the integrals in Review Question 9.1.11.

9.3.12 The definite integral

◀◀ 253 278 ◀

A. Evaluate

(i) $\displaystyle\int_0^1 x^2 \, dx$ (ii) $\displaystyle\int_0^{\pi/2} x \cos x \, dx$ (iii) $\displaystyle\int_0^1 \dfrac{x^2}{\sqrt{x^3 + 1}} \, dx$

(iv) $\displaystyle\int_0^1 \dfrac{x}{(x+1)(x+2)} \, dx$ (v) $\displaystyle\int_0^1 xe^x \, dx$

B. If $f(x)$ is an even function and $g(x)$ is an odd function, show that

(i) $\displaystyle\int_{-a}^{a} f(x) \, dx = 2 \int_0^a f(x) \, dx$ (ii) $\displaystyle\int_{-a}^{a} g(x) \, dx = 0$

9.4 Applications

We will be devoting a lot of Chapter 10 to applications of integration, and you will have ample opportunity to see applications of integration later in the book. For now we will apply the skills learnt in this chapter to look ahead to Laplace transforms and Fourier series, covered in Chapter?.

1. Suppose $f(t)$ is a function of t, defined for $t \geq 0$. Then the integral

$$\tilde{f}(s) = \int_0^\infty f(t)e^{-st}\, dt$$

is called the **Laplace transform of $f(t)$** (the use of t as the variable is conventional, since the Laplace transform is usually applied to functions of time). We will study this in detail in Chapter 17. We will want to evaluate this integral for all the simple elementary functions. In order to do this we will need to integrate finite integrals of the sort

$$\int_0^a f(t)e^{-st}\, dt$$

for a constant a. Do this for the functions $f(t) = 1, t, t^2, e^t, \sin t, \cos t$. Your results should be functions of a and s. If you feel confident enough take the limit as a tends to infinity, thereby obtaining the laplace transforms of the functions – see Chapter 17 for the results.

2. Let $f(t)$ be a function with period 2π. Then under certain conditions it can be expanded in a series of sines and cosines in the form

$$f(t) = \frac{a_0}{2} + \sum_{n=1}^{\infty} a_n \cos nt + \sum_{n=1}^{\infty} b_n \sin nt$$

This is called a **Fourier (series) expansion** for $f(t)$. We study such series in detail in Chapter 17, here we want to anticipate some of the results used there. The key task is to determine the coefficients a_n, b_n of the Fourier series for a given function $f(t)$. This involves the use of a number of integrals of products of sines and cosines, which we ask you to evaluate here.

If m, n are any integers, then:

$$\int_{-\pi}^{\pi} \sin mt \sin nt\, dt = 0 \quad \text{if } m \neq n$$

$$\int_{-\pi}^{\pi} \cos mt \cos nt\, dt = 0 \quad \text{if } m \neq n$$

$$\int_{-\pi}^{\pi} \cos^2 nt\, dt = \pi \qquad \int_{-\pi}^{\pi} \sin^2 nt\, dt = \pi$$

$$\int_{-\pi}^{\pi} \sin mt \cos nt\, dt = 0 \quad \text{for all } m, n$$

$$\int_{-\pi}^{\pi} \sin mt\, dt = \int_{-\pi}^{\pi} \cos mt\, dt = 0$$

These are called the **orthogonality relations for sine and cosine**. The limits $-\pi, \pi$ on the integrals may in fact be replaced by **any** integral of length 2π, or integer multiple of 2π.

The coefficients a_n, b_n of the Fourier series for a given function $f(t)$ can be determined by multiplying through by $\cos nx$ or $\sin nx$, integrating over a single period and using the above orthogonality relations to remove all but the desired coefficient (see Chapter 17). Show that in general

$$a_n = \frac{1}{\pi} \int_{-\pi}^{\pi} f(t) \cos nt\, dt \quad n = 0, 1, 2, \ldots$$

$$b_n = \frac{1}{\pi} \int_{-\pi}^{\pi} f(t) \sin nt\, dt \quad n = 1, 2, \ldots$$

Answers to reinforcement exercises

Note: an arbitrary constant C should be added to each indefinite integral.

9.3.1 Definition of integration

A. (i) $9x^2$ (ii) $\dfrac{\sqrt{3}}{2\sqrt{x}}$ (iii) $-\dfrac{8}{x^5}$ (iv) $\dfrac{2}{5}x^{-\frac{3}{5}}$

(v) $3\cos 3x$ (vi) $-3x^2 \sin x^3$ (vii) $5e^{5x}$ (viii) $\dfrac{-1}{(x+1)^2}$

(ix) $\dfrac{1}{x}$ (x) $\dfrac{1}{2\sqrt{x+1}}$ (xi) $\dfrac{3}{3x+1}$ (xii) $\dfrac{1}{x^2+1}$

B. (i) $2\ln x$ (ii) $\dfrac{3}{5}e^{5x}$ (iii) $\dfrac{-2}{(1+x)}$ (iv) $-\dfrac{1}{4x^4}$

(v) $\sin x^3$ (vi) $-3\tan^{-1} x$ (vii) $\dfrac{5}{2}x^{2/5}$ (viii) $2\sqrt{x}$

(ix) $\dfrac{2}{3}\sin 3x$ (x) $\dfrac{x^3}{3}$ (xi) $6\sqrt{x+1}$ (xii) $\dfrac{1}{3}\ln|3x+1|$

9.3.2 Standard integrals

A. (i) $4x$ (ii) $x^3 - x^2 + x$ (iii) $\dfrac{-3}{4x^4}$ (iv) $\dfrac{3}{5}x^{5/3}$

(v) $\sin x$ (vi) $\tan x$ (vii) $x\ln x - x$ (viii) $\tan^{-1} x$

B. (i) $2u^2$ (ii) $\dfrac{2s^3}{3} - \dfrac{3s^2}{2} + 2s$ (iii) $-\dfrac{1}{t^6}$ (iv) $\dfrac{4}{5}x^{5/4}$

(v) $-\cos\theta$ (vi) $-\cot t$ (vii) $\dfrac{e^{3t}}{3}$ (viii) $\tan^{-1} s$

9.3.3 Addition of integrals

(i) $\dfrac{2}{3}x^3 + \dfrac{3}{2}\dfrac{1}{x^2}$ (ii) $2\sin x + \cos x$ (iii) $\sinh x = \tfrac{1}{2}(e^x - e^x)$

(iv) $-\cos(x+3)$ (v) $\displaystyle\sum_{r=0}^{n} \dfrac{a_r x^{r+1}}{r+1}$ (vi) $\dfrac{1}{2}\left(\cos x - \dfrac{1}{3}\cos 3x\right)$

(vii) $\cosh x$

9.3.4 Simplifying the integrand

A. (i) $\dfrac{x^2}{2} - x \; (x \neq 2)$ (ii) $\sin x$ (iii) $\sin x + \cos x$, provided $\cos x - \sin x \neq 0$

(iv) $-\dfrac{2e}{x}\;(x>0)$ (v) $x + \dfrac{1}{2}\cos 2x$

B. $\ln|\sec x + \tan x|$, $\ln|\operatorname{cosec} x - \cot x|$

9.3.5 Linear substitution in integration

(i) $\dfrac{1}{32}(4x+3)^8$ (ii) $\dfrac{3}{8}(2x-1)^{4/3}$ (iii) $-\tfrac{1}{3}\cos(3x-1)$ (iv) $\tfrac{1}{2}e^{4x} + x$

(v) $\dfrac{3}{2}\ln(2x+1)$ (vi) $\dfrac{1}{10}(2x-1)^5$ (vii) $\dfrac{2}{9}(3x+2)^{3/2}$ (viii) $\tfrac{1}{2}\sin(2x+1)$

(ix) $\dfrac{4}{3}e^{3x-1}$ (x) $\dfrac{1}{4}\ln(4x-3)$

9.3.6 The $du = f'(x)\,dx$ substitution

A. (i) $u = x^3,\;\tfrac{1}{3}e^{x^3}$ (ii) $u = \tan x + 2,\; -\cos(\tan x + 2)$

(iii) $u = \sin x,\; \dfrac{\sin^4 x}{4}$ (iv) $u = x^2 + 2x + 3,\; \dfrac{1}{2}\ln(x^2 + 2x + 3)$

(v) $u = \cos x,\; \ln|\sec x|$

B. (i) $-\cos(f(x))$ (ii) $\tfrac{1}{2}e^{(f(x))^2}$ (iii) $\tfrac{1}{3}\sin(x^3 + 1)$

(iv) $\cos x \ln(\cos x) - \cos x$ (v) $\ln(x-1)$ (vi) $e^{f(x)}$

(vii) $\sin(f(x) + 2)$

C. (i) $\dfrac{1}{4}(x^3+2)^4$ (ii) $\dfrac{2}{3}(x^2+1)^{3/2}$ (iii) $\ln(x^2 - x + 1)$

(iv) $\dfrac{1}{2}\sin(x^2 + 2x + 1)$ (v) $-e^{\cos x}$ (vi) $\ln(\tan x)$

(vii) $\dfrac{1}{16}(2\sin 2x + 1)^4$ (viii) $6\sqrt{x^2 + 4x + 4}$

9.3.7 Integrating rational functions

A. (i) $\frac{1}{2}\ln|(x-1)(x+3)^3|$ (ii) $\ln\left|\frac{(x+3)^2}{(x+2)}\right|$ (iii) $\frac{4}{3}\ln\left|\frac{x-1}{2x+1}\right|$

(iv) $\frac{3}{4}\ln\left|\frac{(x+1)^2}{x^2+1}\right| + \frac{3}{2}\tan^{-1}x$ (v) $3\ln\left|\frac{x-2}{x-1}\right| + \frac{2}{x-1}$

(vi) $\frac{1}{2}\ln\left|\frac{x^2-1}{(x+2)^2}\right|$

B. (i) $\frac{1}{2}\tan^{-1}\left(\frac{x+1}{2}\right)$ (ii) $3\tan^{-1}(x-1)$ (iii) $2\tan^{-1}(2x+1)$

(iv) $\tan^{-1}(x+3)$ (v) $\frac{1}{3\sqrt{2}}\tan^{-1}\left[\frac{\sqrt{2}(x+3)}{3}\right]$

C. (i) $x + \frac{1}{3}\ln\left|\frac{(x-2)^5}{(x+1)^2}\right|$ (ii) $\frac{x^2}{2} - 2x + \ln(x^2+2x+2) + 2\tan^{-1}(x+1)$

(iii) $x^3 + 3x + \frac{3}{2}\ln\left|\frac{x-1}{x+1}\right|$

9.3.8 Using trig identities in integration

(i) $\frac{\cos^5 x}{5} - \frac{\cos^3 x}{3}$ (ii) $\frac{1}{10}\sin 5x + \frac{1}{2}\sin x$

(iii) $\sin x - \frac{2}{3}\sin^3 x + \frac{1}{5}\sin^5 x$ (iv) $-\frac{1}{16}(\cos 8x - 4\cos 2x)$

(v) $\frac{1}{10}(5\sin x - \sin 5x)$

9.3.9 Using trig substitutions in integration

A. (i) $\frac{1}{2}\sin^{-1}x$ (ii) $\frac{1}{3}\tan^{-1}\left(\frac{3x}{2}\right)$ (iii) $\frac{2}{3}\sin^{-1}3x$

(iv) $\frac{3}{2}\tan^{-1}2x$ (v) $\sin^{-1}\left(\frac{x+1}{3}\right)$ (vi) $\sin^{-1}\left(\frac{x-3}{3}\right)$

B. $\frac{1}{4}\ln\left|\frac{2+\tan\theta/2}{2-\tan\theta/2}\right|$

9.3.10 Integration by parts

(i) $x\sin x + \cos x$ (ii) $(x^3 - 3x^2 + 6x - 6)e^x$ (iii) $x\sin^{-1}x + \sqrt{1-x^2}$

(iv) $\dfrac{e^x}{2}(\cos x + \sin x)$ (v) $x^2 \sin x + 2x \cos x - 2\sin x$

(vi) $\dfrac{x^2}{2}\ln|x| - \dfrac{x^2}{4}$ (vii) $\dfrac{1}{2}(x^2 - 1)e^{x^2}$

9.3.11 Choice of integration methods

A. (i) A (ii) C (iii) B (iv) B (v) A (vi) B

B. (i) B, C (ii) A, C (iii) C (iv) C (v) B, C (vi) B, C

C. (i) $\ln|x^2 - 6x + 4|$ (ii) $\dfrac{\cos^3 x}{3} - \cos x$ (iii) x

(iv) $\dfrac{1}{2}\ln(2x + 3)$

(v) $\dfrac{1}{6}e^{3x^2}$ (vi) $3\sin^{-1}\left(\dfrac{x+1}{2}\right)$

(vii) $-x\cos(x + 1) + \sin(x + 1)$

(viii) $\dfrac{3}{8}x + \dfrac{1}{4}\sin 2x + \dfrac{1}{32}\sin 4x$ (ix) $\dfrac{x^2}{2} - x\ln x + x$

(x) $5\ln(x - 3) - 4\ln(x - 2)$ (xi) $\dfrac{1}{4}e^{2x}(2x - 1)$ (xii) $\sin(e^x)$

(xiii) $-\dfrac{1}{8}\cos 4x$ (xiv) $\sqrt{x^2 - 2x + 3}$

(xv) $\dfrac{1}{2}\ln(x^2 + 2x + 2) + 2\tan^{-1}(x + 1)$

(xvi) $\dfrac{1}{18}(9\cos x - \cos 9x)$ (xvii) $\dfrac{1}{2}\sin(x^2 + 1)$

9.3.12 The definite integral

A. (i) $\dfrac{1}{3}$ (ii) $\dfrac{\pi}{2} - 1$ (iii) $\dfrac{2}{3}(\sqrt{2} - 1)$ (iv) $2\ln 3 - 3\ln 2$ (v) 1

10

Applications of Differentiation and Integration

In general terms, the usefulness of calculus in elementary engineering maths lies in the application of differentiation as a rate of change or slope of a curve and integration as an area under a curve. These are essentially applications to very useful mathematical methods. There are also specific applications to various disciplines of engineering, such as centroids and moments of inertia in solid mechanics and rms values in electronic engineering. Also, there are examples in statistics. The applications to mathematical methods are the main subject of this chapter. Some engineering and statistics examples are covered under Applications.

Prerequisites

It will be helpful if you know something about:

- techniques of differentiation (Chapter 8 ◄)
- techniques of integration (Chapter 9 ◄)
- simple ideas of limits (230 ◄)
- coordinate geometry (Chapter 7 ◄)
- tangent and normal to a curve (230 ◄)
- equation of a line (212 ◄)
- inequalities (97 ◄)
- solving algebraic equations (Chapter 2 ◄)
- plotting graphs (91 ◄)

Objectives

In this chapter you will find

- differentiation as a gradient and a rate of change
- tangent and normal to a curve
- stationary points and points of inflection
- curve sketching
- the definite integral as the area under a curve
- volume of a solid of revolution

Motivation

You may need the material of this chapter for:

- finding and interpreting rates of change such as velocity, acceleration, etc.
- interpreting solutions of differential equations
- finding optimal values of various engineering variables
- sketching the graph of an engineering function
- finding areas, centroids, etc.
- evaluating volumes, moments of inertia, etc.
- evaluating mean and rms values of alternating currents
- evaluating probability, the mean and standard deviation in statistics

10.1 Review

10.1.1 The derivative as a gradient and rate of change ➤ 292 309 ➤➤

For each of the following functions find:

 (a) The gradient or slope of the graph of the function at the point specified
 (b) The rate of change of the function at the point specified
 (i) $y = x^2$, $x = 1$ (ii) $y = \cos x$, $x = \pi$

10.1.2 Tangent and normal to a curve ➤ 293 310 ➤➤

Find the equations of the (i) tangent and (ii) the normal to the curve $y = x^2 - x - 1$ at the point (2, 1).

10.1.3 Stationary points and points of inflection ➤ 294 310 ➤➤

Locate and classify any stationary points and points of inflection of the following functions

(i) $16x - 3x^3$ (ii) $x + \dfrac{1}{x}$ (iii) xe^x
(iv) $\sin x \cos x$ (v) x^4

10.1.4 Curve sketching in Cartesian coordinates ➤ 299 310 ➤➤

Sketch the functions given in Q10.1.3.

10.1.5 Applications of integration – area under a curve ➤ 304 310 ➤➤

A. Find the area enclosed between the curve, the x-axis and the limits stated for each of the following cases:

(i) $y = 4x^2 + 1$ $x = 0, 2$ (ii) $y = xe^x$ $x = 0, 1$

B. Find the area enclosed between the curves $y = x^2 - x$ and $y = 2x - x^2$.

10.1.6 Volume of a solid of revolution

▶ 308 311 ▶▶

Find the volume of the solid of revolution formed when the positive area enclosed under the following curve is rotated once about the x-axis.

$$y = 1 - x^2, \quad y = 0, \quad -1 \leq x \leq 1$$

10.2 Revise

10.2.1 The derivative as a gradient and rate of change

◀ 291 309 ▶

The derivative, $\dfrac{dy}{dx}$, of a function $y = f(x)$ is defined by a limiting process precisely to give the **gradient** or **slope** of the curve described by the function at a given point x (230 ◀). As such it describes the **rate of change** of the function – the 'steepness' of the curve. Thus, at a point where dy/dx is large and positive the curve of the function $y = f(x)$ is increasing steeply as x increases. Other cases are tabulated below, as examples – fill in the missing entries.

dy/dx	behaviour of $y = f(x)$
?	y increases slowly as x increases
large and negative	?
?	y decreases slowly as x increases

It is important to note that the derivative is defined **at a point**. Differentiation is therefore what is called a **local** operation. The derivative of a function will only tell us about the behaviour at a single point, and says nothing about the overall or global behaviour of the function.

Solution to review question 10.1.1

(i) For the function $y = x^2$ we have

$$\frac{dy}{dx} = 2x$$

At $x = 1$ this has the value 2 and so (a) the gradient at this point is 2, and (b) the rate of change is also 2.

(ii) For the function $y = \cos x$ the derivative is $\dfrac{dy}{dx} = -\sin x$ which is equal to zero at $x = \pi$. So in this case the (a) gradient or slope of the graph is zero at $x = \pi$, as is (b) the rate of change of the function. So, at this point the tangent to the curve is horizontal, its slope being zero, and the rate of change of the function is zero. We know, of course, from the properties of $\cos x$, that this point is a local minimum (see Section 10.2.3) (182 ◀).

10.2.2 Tangent and normal to a curve

The derivative at a point (a, b) on a curve $y = f(x)$ will give us the slope, m, of the **tangent** to the curve at that point. This tangent is a straight line with gradient m passing through the point (a, b) and therefore has an equation (212 ◀)

$$y - b = m(x - a)$$

The **normal** to the curve $y = f(x)$ at the point (a, b) is the line through (a, b) perpendicular to the tangent – see Figure 10.1.

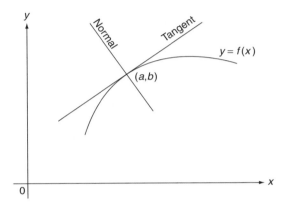

Figure 10.1 Tangent and normal to a curve.

In mechanics, the normal is important because it defines the direction of the reaction to a force applied to a smooth surface represented by the curve. If the gradient of $y = f(x)$ at (a, b) is m then the gradient of the normal through (a, b) will be $-\dfrac{1}{m}$ (214 ◀). So the equation of the normal is

$$y - b = -\frac{1}{m}(x - a)$$

Solution to review question 10.1.2

For the curve $y = x^2 - x - 1$ we have $\dfrac{dy}{dx} = 2x - 1$.

So at $(2, 1)$, the gradient is $m = 2 \times 2 - 1 = 3$.
The equation of the tangent at $(2, 1)$ is therefore

$$y - 1 = 3(x - 2)$$

or

$$y = 3x - 5$$

The gradient of the normal at $(2, 1)$ is

$$-\frac{1}{m} = -\frac{1}{3}$$

so the equation of the normal is

$$y - 1 = -\tfrac{1}{3}(x - 2)$$

or

$$x + 3y - 5 = 0$$

10.2.3 Stationary points and points of inflection

Since the derivative describes rate of change, or the slope, of a curve it can tell us a great deal about the shape of a curve, and the corresponding behaviour of the function. Figure 10.2 shows the range of possibilities one can meet.

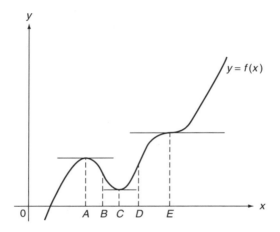

Figure 10.2 Stationary points and points of inflection.

In all cases we assume that the function is continuous and smooth (the graph has no breaks and no sharp points). At points A, C, E, where dy/dx is zero, y is not actually changing at all as x varies – the tangent to the curve is then **parallel to the x-axis** at such points as shown. These are called **stationary points**. So, at a stationary point of the function $y = f(x)$ we have

$$\frac{dy}{dx} = f'(x) = 0$$

The value of $f(x)$ at a stationary point is called a **stationary value** of $f(x)$. Note that we will usually use the 'dash' notation for derivatives in the rest of the book – it saves space!

There are a number of different types of stationary points illustrated in Figure 10.2, displaying the different possibilities – C is a **minimum** point, and A is a **maximum**. E is a stationary point where the tangent crosses the curve – this is an example of an important point called a **point of inflection**. B and D are also points of inflection of a different kind – the tangent crosses the curve, but is not horizontal – see below. The maximum and minimum points are called **turning points**, since the curve turns at such points and changes direction. As emphasised above, differentiation is a **local** procedure. So, for example any minimum identified by differentiation is only a **local** minimum, not necessarily an overall

global minimum of the function. Similarly for a maximum. Indeed, for a function such as $y = x + \dfrac{1}{x}$ the local minimum actually has a higher value than the local maximum (➤ 297).

While the graphical representation of the behaviour of functions is very suggestive, we need a means of distinguishing stationary points and points of inflection that depends only on the derivatives of the function in question. The graphical form provides a hint as to how this might work. As an example, consider a minimum point at say $x = x_0$ (see Figure 10.3).

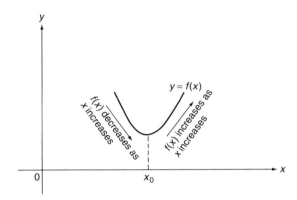

Figure 10.3 Behaviour at a minimum.

To the left of x_0, i.e. for $x < x_0$, $y = f(x)$ is decreasing as x increases, and $f'(x) < 0$. For $x > x_0$, y increases as x increases and so $f'(x) > 0$. And of course at $x = x_0$ we have $f'(x) = f'(x_0) = 0$. So, near to a minimum point we can summarise the situation as in the table below:

$x < x_0$	$x = x_0$	$x > x_0$
$f'(x) < 0$	$f'(x) = 0$	$f'(x) > 0$

This characterisation no longer relies on the graphical representation – it depends only on the values of the derivative at different points. Rather than use this table as a means of verifying a minimum point, we look at the implications it has for the second derivative of the function. In this case we see that as x passes through x_0, $f'(x)$ is steadily **increasing**. That is

$$\frac{d}{dx}\left(\frac{dy}{dx}\right) = \frac{d^2 y}{dx^2} = f''(x) > 0$$

So a **minimum point** $x = x_0$ on the curve $y = f(x)$ is characterised by

$$f'(x_0) = 0 \quad \text{and} \quad f''(x_0) > 0$$

Similarly, for a maximum point we obtain the table

$x < x_0$	$x = x_0$	$x > x_0$
$f'(x) > 0$	$f'(x) = 0$	$f'(x) < 0$

and so in this case the derivative is **decreasing** as we increase x through $x = x_0$ and thus a **maximum point** $x = x_0$ is characterised by

$$f'(x_0) = 0 \quad f''(x_0) < 0$$

Now what happens if $f''(x_0) = 0$? Then, it is not clear whether or not the **gradient**, $f'(x)$, is changing at $x = x_0$ and a more careful examination is necessary. An example of one possibility occurs at the point E in Figure 10.2, at which point

$$f'(x_0) = f''(x_0) = 0$$

This is a particular example of a **point of inflection**. B and D are other examples, but these are clearly not stationary points – the gradient is not zero. In these cases $f''(x_0) = 0$ but $f'(x_0) \neq 0$. So what precisely characterises a point of inflection?

Look more closely at the point B in Figure 10.2. To emphasise the point we will bend the curve somewhat near to B, as in Figure 10.4. Take $x = x_0$ as the coordinate of B here.

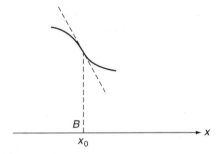

Figure 10.4 Detail of a point of inflection.

On either side of B the gradient is negative. However, on the left of B, $x < x_0$, the curve is **below the tangent**. We say the curve here is **concave down**. To the right of B, $x > x_0$, the curve is **above the tangent** and we say it is **concave up**. Looking at points D and E we see a similar change:

left of D – concave up
right of D – concave down

left of E – concave down
right of E – concave up

In general, any point where there is such a change of sense of concavity is called a **point of inflection**, meaning a change in the direction of bending of the curve. At such a point you will see that it is the rate of change of the **gradient** that is changing sign. For example:

left of B, gradient steepens as x increases – curve concave down
right of B, gradient decreases as x increases – curve concave up

So even though the gradient of $f(x)$ may not be zero at a point of inflection – i.e. we may not have a stationary value – the rate of change of the gradient, $f''(x_0)$, at such a point must be zero. However, note that the condition $f''(x_0) = 0$ does not itself guarantee a point of inflection. The Review Question 10.1.4(v) illustrates this.

Solution to review question 10.1.3

(i) For the function $f(x) = 16x - 3x^3$ we have $f'(x) = 16 - 9x^2$. Solving

$$f'(x) = 16 - 9x^2 = 0$$

gives stationary values at $x = \pm\frac{4}{3}$. To classify these consider the second derivative $f''(x) = -18x$. So, for $x = \frac{4}{3}$,

$$f''(\tfrac{4}{3}) = -24 < 0$$

and we have a **maximum**.
For $x = -\frac{4}{3}$ we have

$$f''(-\tfrac{4}{3}) = 24 > 0$$

and we have a **minimum**.
We also note that $f''(x) = 0$ at $x = 0$, so there is a possibility of a point of inflection at this point. To investigate this we have to consider how the gradient is changing on either side of $x = 0$. This is indicated by the sign of $f''(x)$:

$$x < 0 \quad f''(x) = -18x > 0$$
$$x > 0 \quad f''(x) = -18x < 0$$

So for $x < 0$, $f''(x) = (f'(x))'$ is positive and therefore the gradient $f'(x)$ is increasing as x increases and the curve is concave upwards. For $x > 0$, $f''(x) = (f'(x))'$ is negative, so the gradient $f'(x)$ is decreasing and x increases and the curve is concave downwards. Thus $x = 0$ is a point where the curve changes from concave up to concave down as x increases – it is a point of inflection. The curve is sketched in Figure 10.7.
Note that the complete story about the curve requires quite a careful study of its derivatives.

(ii) For $f(x) = x + \dfrac{1}{x}$ we have $f'(x) = 1 - \dfrac{1}{x^2}$. Solving

$$f'(x) = 1 - \frac{1}{x^2} = 0$$

gives us stationary values at $x = \pm 1$. We find

$$f''(x) = \frac{2}{x^3}$$

and so

$$f''(1) = 2 > 0$$

giving us a **minimum** at $x = 1$, while

$$f''(-1) = -2 < 0$$

giving us a **maximum** at $x = -1$.

In this case $f''(x)$ is never zero for finite x, so we have no points of inflection. Notice that in this case the local minimum value (2) is greater than the local maximum value (-2), emphasising the local nature of differentiation. The situation is sketched in Figure 10.8.

(iii) For $f(x) = xe^x$ we have $f'(x) = e^x + xe^x = 0$ when $x = -1$, since $e^x \neq 0$. At this stationary point we have

$$f''(x) = 2e^x + xe^x = e^{-1} > 0 \text{ at } x = -1$$

So $x = -1$ is a **minimum point** in this case.

Looking for points of inflection we note that $f''(x) = e^x(x+2) = 0$ at $x = -2$. For $x < -2$, $f''(x) < 0$, $f'(x)$ decreases as x increases and the curve is concave down. For $x > -2$, $f''(x) > 0$, $f'(x)$ increases as x increases and the curve is concave up. So $x = -2$ is a point of inflection. The curve is sketched in Figure 10.9.

(iv) We can write $f(x) = \sin x \cos x = \frac{1}{2} \sin 2x$.

Then $f'(x) = \cos 2x = 0$ when $2x = \dfrac{(2n+1)}{2}\pi$ where n is an integer.

So there are turning points where

$$x = \left(\frac{2n+1}{4}\right)\pi$$

Further,

$$f''(x) = -2\sin 2x = -2\sin\left(\left(\frac{2n+1}{2}\right)\pi\right)$$

For n even this is negative, while it is positive for n odd.

Thus we have maximum when $x = \left(\dfrac{2n+1}{4}\right)\pi$ for n even, and minimum when n is odd.

Points of inflection can occur when $f''(x) = -2\sin 2x = 0$, i.e.

$$2x = m\pi$$

where m is an integer, or $x = \dfrac{m\pi}{2}$.

(v) For $f(x) = x^4$ we have $f'(x) = 4x^3 = 0$ when $x = 0$. So there is a stationary point at $x = 0$. But

$$f''(x) = 12x^2 = 0 \text{ when } x = 0$$

So there may be a point of inflection at $x = 0$. However, we note that for $x < 0$, $f''(x) > 0$, so the curve is concave up, while for $x > 0$ we again have $f''(x) > 0$, again indicating that the curve is concave up. So the concavity does not change at $x = 0$ and therefore it is **not** a point of inflection. In fact, the graph is easy to sketch, and we clearly have a minimum at $x = 0$ (see Figure 10.11, Review Question 10.1.4(v). This illustrates that the vanishing of the second derivative, while essential for a point of inflection is not a guarantee of one.

10.2.4 Curve sketching in Cartesian coordinates

Curve sketching is exactly that – it does not mean **plotting** the curve, as in Chapter 3, although one might actually plot a few special points of the curve, such as where it intercepts the axes. A sketch of a curve shows its general shape and main features, it is not necessarily an accurate drawing. In sketching a curve we deduce what we can about it from quite general observations, such as where it increases, decreases, or remains bounded as x becomes large or small. Similarly we might look for stationary values, intercepts on the axes and so on.

With modern calculators capable of plotting almost any curve we might wonder why curve sketching is necessary at all – is it like shoeing horses, for example? Not at all. Apart from the fact that you might not always have a graphics calculator to hand, the main benefit of the skills of curve sketching is that it gives you an appreciation of how functions behave, and what the properties of the derivative can tell you about this. It gives you a 'feel' for the function. Being able to sketch the rough shape of a given rational function, for example, is a far more portable skill than being able to graph it on a calculator by pressing a few buttons.

Of course, we already know a large number of different graphs of the various 'elementary functions' we have considered, and the first step in sketching a curve is to see whether it is easily converted to one with which we are already familiar. The kinds of transformation we can make are listed below.

1. $y = f(x) + c$ translates the graph of $y = f(x)$ by c units in the direction $0y$.
2. $y = f(x + c)$ translates the graph of $y = f(x)$ by $-c$ units in the $0x$ direction.
3. $y = -f(x)$ reflects the graph of $y = f(x)$ in the x-axis.
4. $y = f(-x)$ reflects the graph of $y = f(x)$ in the y-axis.
5. $y = af(x)$ stretches $y = f(x)$ parallel to the y-axis by a factor a.
6. $y = f(ax)$ stretches $y = f(x)$ parallel to the x-axis by a factor $1/a$.

However, once these transformations are taken advantage of, we might still have quite complicated graphs to sketch. We can follow a systematic procedure for sketching graphs which can be nicely summarised in the acronym **S(ketch) GRAPH**:

- **S**ymmetry – does the curve have symmetry about the x- or y-axes, is the function odd or even?
- **G**ateways – where does the curve cross the axes?
- **R**estrictions – are there any limits on the variable or function values?
- **A**symptotes – any lines that the curve approaches as it goes to infinity?
- **P**oints – any points of special interest which are worth plotting?
- **H**umps and hollows – any stationary points or points of inflection?

(I am indebted to Peter Jack for most of this pretty acronym.)

Working through each of these (not necessarily in this order) should provide a good idea of the shape and major features of the curve. We describe each in turn.

S. We look for any **symmetry** of the curve. For example, if it is an even function, $f(x) = f(-x)$ then it is symmetric about the y-axis, and we only need to sketch half the curve. This may reduce considerably the amount of work we have to do in sketching the curve.

G. Look for points where the curve crosses the axes, or **'gateways'**. It crosses the y-axis at points where $x = 0$, i.e. at the value(s) $y = f(0)$. It crosses the x-axis at points where $y = 0$, i.e. at solutions of the equation $f(x) = 0$.

R. Consider any **restrictions** on either the domain or the range of the function – that is, forbidden regions where the curve cannot exist. The simplest such case occurs when we have **discontinuities**. Thus, for example

$$y = \frac{1}{x+1}$$

does not exist at the point $x = -1$ – there is a break in the curve at this point. See Figure 10.5.

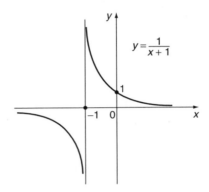

Figure 10.5 Asymptotes at a discontinuity.

Any rational function will have a number of such points equal to the number of real roots of the denominator. Irrational functions such as $\sqrt{1 - x^2}$ exhibit whole sets of values that x can't take, because for the square root to exist $1 - x^2$ must be positive, hence the curve $y = \sqrt{1 - x^2}$ does not exist for $|x| > 1$. Thus, the curve is confined to the region $|x| \leq 1$.

A. Look for **asymptotes**. These are lines to which the curve becomes infinitely close as x or y tend to infinity. Thus, the function $y = \dfrac{1}{x+1}$ has the x-axis as a horizontal asymptote, since $y \to 0$ as $x \to \pm\infty$. It also has the line $x = -1$ as an asymptote to which the curve tends as $y \to \pm\infty$. Of course the curve can never coincide with the line because of the restriction (see above) $x \neq -1$.

Also, look for the behaviour of the curve for very small and very large values of x or y. This is often very instructive. For example, the curve $y = x^3 + x$ behaves like the curve $y = x$ for small values of x, allowing us to approximate the curve near the origin by a straight line at 45° to the x-axis. For very large (positive or negative) values of x it behaves like the cubic curve $y = x^3$.

P. Consider any special **points**, apart from 'gateways' in the axes. They may literally be simply specific points you plot to determine which side of an asymptote the curve approaches from, for example.

H. Points of particular importance include maximum and minimum values (**humps and hollows**) and also points of inflection. While we do already know how to find these, it may not in fact be necessary. Information already available may hint strongly at certain turning values – for example a continuous curve which ends up going in the same direction for two different values of x must have passed through at least one turning value in between. Also, it is not always necessary to determine points of inflection, especially those with a non-zero gradient, unless we want a really accurate picture of the graph.

Solution to review question 10.1.4

(i) $y = 16x - 3x^3$

First check on the **symmetry**. The function is odd and so we need only sketch it for $x \geq 0$ and then obtain the whole curve by a rotation of $180°$ about the origin.

Now find the **gateways**. The curve crosses the y-axis ($x = 0$) when $y = 0$, i.e. it passes through the origin. It crosses the x-axis at

$$y = 16x - 3x^3 = 0$$

i.e. $\quad x = 0, \quad \pm\dfrac{4}{\sqrt{3}}$

There are no **restrictions** on the domain – the function, being a polynomial, exists for all values of x. It is continuous for all values of x too. There are no **asymptotes**. Near the origin the x^3 term is negligible, and $y \simeq 16x$, so it looks like a very steep straight line with positive slope. For very large values the curve behaves like $y \simeq -3x^3$, and clearly as $x \to \infty$, $y \to -\infty$.

This discussion alone is enough to tell us that there is likely to be at least one maximum point between 0 and $\dfrac{4}{\sqrt{3}}$ on the positive x-axis. Figure 10.6 illustrates this deduction.

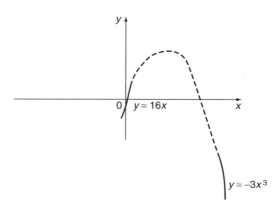

Figure 10.6 Going to extremes.

No particular special **points** spring to mind, except the **stationary points** and points of inflection. We know from Review Question 10.1.3(i) that there are two stationary points, at $x = \pm \frac{4}{3}$. The point at $x = \frac{4}{3}$ is a maximum and that at $x = -\frac{4}{3}$ is a minimum. There is also a point of inflection at $x = 0$ (this follows by symmetry anyway). This enables us to complete the sketch as shown in Figure 10.7.

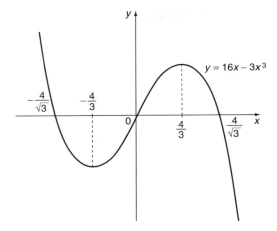

Figure 10.7 Sketch of $y = 16x - 3x^3$.

(ii) $y = x + \dfrac{1}{x}$

Again, this is an odd function and so is symmetric under a 180° rotation about the origin and we need only sketch it for $x \geq 0$.

Gateways: The curve does not cross the y-axis, since there is a discontinuity at $x = 0$. Also, it never crosses the x-axis, since

$$x + \frac{1}{x} = \frac{x^2 + 1}{x}$$

can never be zero.

Restrictions: For $x > 0$, $y > 0$, and for $x < 0$, $y < 0$, so the curve is confined to the upper half plane for $x > 0$ and to the lower half plane for $x < 0$.

Asymptotes: As $x \to 0$ from above (that is through positive values of x), $y \to +\infty$, so the y-axis is a vertical asymptote for $x > 0$. Also, for x very large, $y \sim x$, so the straight line $y = x$ is an asymptote as $x \to \infty$. Consideration of the asymptotes alone hints at a minimum for $x > 0$ (and a corresponding maximum for $x < 0$).

Special points: nothing, apart from:

Stationary points: From Review Question 10.1.3(ii) we have a minimum at $x = 1$, and a maximum at $x = -1$.

Putting all this together, the graph is as shown in Figure 10.8.

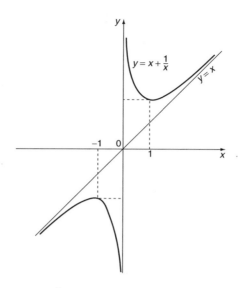

Figure 10.8 Sketch of $y = x + \dfrac{1}{x}$.

(iii) $y = xe^x$

Symmetry: none
Gateways: (0, 0) only
Restrictions: none – defined for all values of x, and no discontinuities
Asymptotes: As $x \to -\infty$, $y \to 0$ from below (i.e. from values less than zero)
Points: none special except the
Stationary value: which, from Review Question 10.1.3(iii) is a minimum at $x = -1$. We also have a **point of inflection** at $x = -2$
Putting all this together yields the graph shown in Figure 10.9.

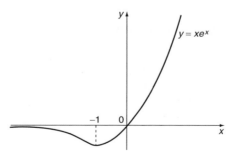

Figure 10.9 Sketch of $y = xe^x$.

(iv) $y = \sin x \cos x = \frac{1}{2} \sin 2x$

This is the graph of a standard elementary function, which we already know from Chapter 6.

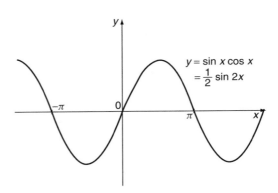

Figure 10.10 Sketch of $y = \sin x \cos x$.

(v) $y = x^4$

We have already discussed this relatively simple function in Review Question 10.1.3(v), and we know it has one minimum, at the origin. It is symmetric about the y-axis and clearly has shape similar to a parabola – but 'more squashed'. It is sketched in Figure 10.11.

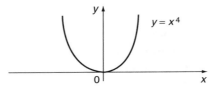

Figure 10.11 Sketch of $y = x^4$.

10.2.5 Applications of integration – area under a curve

There are many applications of integration in engineering and science, but for our purposes they can be grouped into the categories:

- Applications in further topics in engineering mathematics such as Laplace Transform and Fourier series (Chapter 17) and solutions of differential equations (Chapter 15).
- Applications in mechanics to such things as centre of mass and moment of inertia.
- Geometrical applications such as calculating areas, volumes, lengths of curves.
- Applications in probability and statistics, such as mean values, root mean square values, probability in the case of a continuous random variable.

The basic principle behind applications of integration is essentially that of obtaining a **total** of a quantity by regarding it as the sum of a very large (infinite) number of elementary quantities:

- an area split into strips
- a curve split into line segments
- a solid body split into very small particle like elements

The rest of this chapter will be concerned mainly with geometrical applications, although the Applications section contains some examples from mechanics. The main objective is to illustrate the general ideas and motivate the hard work necessary to master integration.

The area 'under' a curve is the, perhaps misleading, term used for the area enclosed between a given curve and the x-axis. Or, it may be the area enclosed between two quite general curves. Figure 10.12 illustrates the sorts of possibilities we can have.

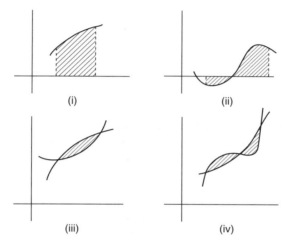

Figure 10.12 Areas and curves.

The case (i) is the easiest to deal with. In case (ii), note that areas **below** the axis are regarded as **negative**. This is to conform with such results as:

$$\int_{-\pi/2}^{0} \sin x \, dx = -1$$

In case (iii) we would need to find where the curves intersect and integrate the difference between the two functions over the appropriate region. Case (iv) also requires us to integrate the difference of the functions, but now care must be taken to allow for the difference in signs of the two areas.

Considering now the simple case (i) let us look at the connection between integration and the area under a curve.

At the elementary level there are two common viewpoints of integration –

- the integral as the inverse operation to differentiation, or **anti-derivative** (253 ◄):

$$\int 2x \, dx = x^2 + C$$

because

$$\frac{d}{dx}(x^2 + C) = 2x$$

- the definite integral interpreted as the **area under a curve** as illustrated in Figure 10.13.

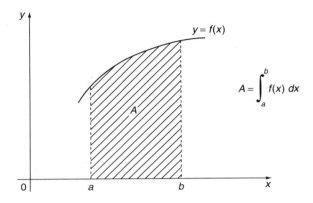

Figure 10.13 Area under a curve.

For theoretical discussion it is the latter viewpoint which is perhaps most useful – indeed, the indefinite integral can be viewed as the definite integral

$$\int_a^x f(u)\,du$$

First we will show, using only simple ideas of limits, the connection between the two viewpoints and then concentrate on the area viewpoint.

For a continuous curve, we can regard the area A under the curve as a function of x, as shown in Figure 10.14.

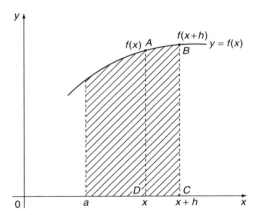

Figure 10.14 The indefinite integral as an area.

The area under the curve between a and $x + h$ is denoted $A(x + h)$. The difference between the area under the curve between a and $x + h$ and between a and x is $A(x + h) - A(x)$ and this is clearly the area of the thin strip above the interval $(x, x + h)$. The area of this thin strip is approximately $\frac{1}{2}[f(x + h) + f(x)] \times h$, the area of the thin trapezium $ABCD$, if h is very small. So we have

$$A(x + h) - A(x) \simeq \tfrac{1}{2}[f(x + h) + f(x)] \times h$$

or

$$\frac{A(x+h) - A(x)}{h} = \frac{1}{2}[f(x+h) + f(x)]$$

and this approximation improves as h gets smaller. Indeed, taking the limit $h \to 0$ we get

$$\lim_{h \to 0} \left[\frac{A(x+h) - A(x)}{h}\right] = \frac{1}{2}[f(x+h) + f(x)]$$
$$= f(x)$$

From Section 8.2.2 (◀) we recognise the left-hand side as dA/dx:

$$\lim_{h \to 0} \left[\frac{A(x+h) - A(x)}{h}\right] = \frac{dA}{dx} = f(x)$$

So, using the first viewpoint of the integral, as the inverse of differentiation, we have

$$A(x) = \int_a^x f(x)\, dx$$

for the area under the curve.

This argument for the equivalence of finding an area and reversing an integration can be extended to all other cases in Figure 10.12. The review questions illustrate what happens in practice.

Solution to review question 10.1.5

A. (i) The area under the curve $y = 4x^2 + 1$ between $x = 0$ and $x = 2$ is given by the definite integral

$$\int_0^2 (4x^2 + 1)\, dx = \left[\frac{4}{3}x^3 + x\right]_0^2 = \frac{38}{3}$$

(ii) For $y = xe^x$ between $x = 0$ and $x = 1$ we have the area

$$\int_0^1 xe^x\, dx = \left[xe^x - e^x\right]_0^1 = 1$$

B. A sketch is always useful in this kind of question. The curves $y = x^2 - x$ and $y = 2x - x^2$ intersect where

$$x^2 - x = 2x - x^2$$

or $x = 0, 3/2$. They are sketched in Figure 10.15.
The area required is shaded. You can now see the reason for the sketch. Between $x = 0$ and 1, the area for the curve $y = x^2 - x$ is negative.

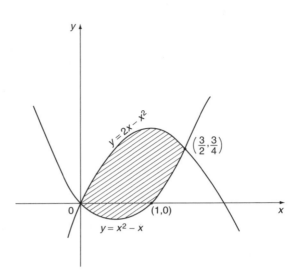

Figure 10.15 Area between $y = 2x - x^2$ and $y = x^2 - x$.

However, this will still be accommodated by integrating the difference of the two functions over the range $0 < x < 3/2$, as follows:

$$\int_0^{3/2} [2x - x^2 - (x^2 - x)] \, dx = \int_0^{3/2} (3x - 2x^2) \, dx$$

$$= \frac{9}{8} \text{ units}$$

10.2.6 Volume of a solid of revolution

When the area under a curve between the limits $x = a, b$ is rotated once about the x-axis a **solid of revolution** is formed. For the purposes of illustration we will assume that the curve is entirely above the x-axis, so that the area rotated is positive. See Figure 10.16.

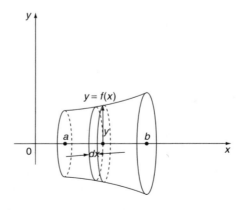

Figure 10.16 Volume of revolution.

If we rotate a thin strip of area, dx wide, about the axis, then a thin disc is formed, whose radius is $r = y$, and thickness is $h = dx$. The volume of this disc is approximately '$\pi r^2 h$' $= \pi y^2 \, dx$. So the total volume between the limits $x = a$ and $x = b$ is obtained by integration as

$$\int_a^b \pi y^2 \, dx = \pi \int_a^b y^2 \, dx$$

So, given a function $y = f(x)$ between values $x = a$ and $x = b$, entirely above the x-axis, by rotating the area under its curve we generate a volume of revolution determined by substituting the function and values of a and b in the above integral. Review Question 10.1.6 illustrates this.

Solution to review question 10.1.6

The volume of the solid of revolution obtained by rotating the area under the curve $y = f(x)$ once about the x-axis between the limits $x = a, b$ is given by

$$\pi \int_a^b y^2 \, dx = \pi \int_a^b [f(x)]^2 \, dx$$

The area bounded by the curves $y = 1 - x^2$, $y = 0$ for $-1 \leq x \leq 1$ is the positive area enclosed between the curve $y = 1 - x^2$ and the x-axis. The curve cuts the x-axis at $x = \pm 1$ and so the required volume is

$$\pi \int_{-1}^1 (1 - x^2)^2 \, dx = 2\pi \int_0^1 (x^4 - 2x^2 + 1) \, dx$$

$$= \frac{16\pi}{15} \text{ square units}$$

10.3 Reinforcement

10.3.1 The derivative as a gradient and rate of change ◄◄ 291 292 ◄

A. For the following functions find (a) the rate of change and (b) the slope of the graph at the points specified.

(i) $y = x^3 + 2x - 1$, $\quad x = 0, 2$

(ii) $y = \sin x$ $\quad\quad x = 0, \dfrac{\pi}{3}$

(iii) $y = e^x \cos x$ $\quad\quad x = 0, 1$

(iv) $y = \ln(x^2 + 1)$ $\quad\quad x = 0, 2$

B. Find the point on the curve

$$y = x^3 + 3x^2 - 9x + 1$$

where the gradient is -12.

10.3.2 Tangent and normal to a curve   ◄◄ 291 293 ◄

Find the equations of the tangents and the normals to the following curves at the points indicated.

(i) $y = x^2 + 2x - 3$ $x = 1$
(ii) $y = x^4 + 1$ $x = 1$
(iii) $y = \ln x$ $x = 1$
(iv) $y = e^x \sin x$ $x = 0$

10.3.3 Stationary points and points of inflection   ◄◄ 291 294 ◄

A. Locate and classify the stationary points and any points of inflection of the following functions:

(i) $x^2 - 4x + 3$
(ii) $x^3 - 12x + 2$
(iii) x^3
(iv) $\dfrac{x}{5} + \dfrac{5}{x}$
(v) $2x^3 - 15x^2 + 36x - 4$
(vi) $4x^3 + 3x^2 - 36x + 6$

B. Find the maximum and minimum values of the curve of the function $y = x(x^2 - 4)$, and also find the gradient of the curve at the point of inflection.

10.3.4 Curve sketching in Cartesian coordinates   ◄◄ 291 299 ◄

Sketch the graphs of the following functions.

(i) $x^3 - 2x^2 - x + 2$
(ii) $\dfrac{x}{x+1}$
(iii) $\dfrac{x+1}{x-1}$
(iv) $\dfrac{x-2}{x^2+1}$
(v) $\dfrac{x^2+4}{x^2+x-2}$
(vi) $3 + \cos\left(\dfrac{x}{2}\right)$
(vii) xe^{-x}

10.3.5 Applications of integration – area under a curve   ◄◄ 291 304 ◄

A. Find the area enclosed between the curve, the x-axis, and the limits stated for each of the following curves:

(i) $y = 2x^2 + x + 1$ $x = 0, 2$
(ii) $y = x - \dfrac{1}{x}$ $x = 1, 2$
(iii) $(x-1)e^x$ $x = 1, 2$
(iv) $y = \cos 2x$ $x = 0, \dfrac{\pi}{4}$
(v) $y = x^2 + \dfrac{1}{x^2}$ $x = 1, 3$
(vi) $y = \sin^2 x$ $x = 0, \pi/2$

B. Calculate the total signed area between the curves and the x-axis and the limits given by: (a) geometry, (b) integration.

(i) $y = 1 - x$, $x = 2, 4$ (ii) $y = x - 1$, $x = 0, 2$

C. Evaluate $\int_{-1}^{1} |x|\,dx$

D. Find the area enclosed by the curves $y = x^2 + 2$, and $y = 1 - x$, and the lines $x = 0$, $x = 1$.

10.3.6 Volume of a solid of revolution

Determine the volumes obtained by rotating the positive area under each of the following curves about the x-axis, between the given limits:

(i) $y = x(1 - x)$, $y = 0$
(ii) $xy = 1$, $y = 0$, $x = 2$, $x = 5$
(iii) $y = \sin x$, $x = 0$, $x = \pi$

10.4 Applications

1. The major application of differentiation in mechanics is in kinematics. In one-dimensional motion along the x-direction the displacement of a moving particle is a function of time, $x(t)$ and at any time t its velocity and acceleration are given respectively by

$$v = \frac{dx}{dt} = \dot{x} \quad a = \frac{d^2x}{dt^2} = \ddot{x}$$

(i)–(iv) give either the position $x(t)$ (metres) or velocity $v(t)$ (metres per second) of a particle at time t. In each case find (a) the velocity and (b) acceleration at the point $t = 1$ sec and (c) determine when the particle is stationary.

(i) $x(t) = 3 - 2t + t^2$
(ii) $v(t) = \dfrac{2t - 1}{t^2 + 1}$
(iii) $x(t) = e^{2t} \sin t$
(iv) $v(t) = \cos t$

2. Sometimes rates of change of a variable are not calculated directly but in terms of the rate of change of some other variable on which it depends. For example the rate of change of the area of a circle in terms of the rate of change of its radius. In such cases we use the function of a function rule:

$$A = \pi r^2 \quad \text{so} \quad \frac{dA}{dt} = 2\pi r \frac{dr}{dt}$$

and we can calculate dA/dt knowing the rate of change of r.

(i) An oil spill from a ruptured tanker in calm seas spreads out in a circular pattern with the radius increasing at a constant rate of 1 ms^{-1}. How fast is the area of the spill increasing when the radius is 300 m?

(ii) The radius of a spherical balloon is increasing at 0.001 ms^{-1}. At what rate is the (i) surface area (ii) volume increasing when the radius is 0.25 m?

3. The motion of a particle performing damped vibrations is given by

$$x = e^{-t} \sin 2t$$

where x is the displacement in metres from its mean position at time t secs. Determine the times at which x is a maximum and find the maximum distance for the least positive value of t. Determine the acceleration at this point.

4. The work done in an air compressor is given by

$$W = K \left[\left(\frac{p_1}{p}\right)^{\frac{n-1}{n}} + \left(\frac{p}{p_2}\right)^{\frac{n-1}{n}} - 2 \right]$$

where p_1, p_2, n, K are positive constants. Show that the work done is a minimum when $p = \sqrt{p_1 p_2}$.

5. One method of obtaining an estimate $\bar{x}$ of a quantity x from a set of n measurements of x (all of which may be subject to experimental error), $x_1, x_2, \ldots, x_n$ is to choose $\bar{x}$ to minimise the sum:

$$s = (x_1 - \bar{x})^2 + (x_2 - \bar{x})^2 + \cdots + (x_n - \bar{x})^2$$

This is called the **method of least squares**. It minimises the total squared deviation from $\bar{x}$.

Determine $\bar{x}$ according to this principle and comment on the result.

6. The resultant mass of a system of particles, masses m_i, situated at the points $P_i(x_i, y_i, z_i)$ acts at a fixed point $G(\bar{x}, \bar{y}, \bar{z})$ where

$$\bar{x} = \frac{\sum m_i x_i}{M} \quad \bar{y} = \frac{\sum m_i y_i}{M} \quad \bar{z} = \frac{\sum m_i z_i}{M} \quad \text{where } M = \sum m_i$$

G is called the **centre of mass**. The numerators in the above expressions are called the first moments of the system with respect to the yz-, zx-, xy-planes respectively. In the case of a continuous body the above particle system is replaced by the elements of the body and integrals used as the limits of the summations. For example:

$$\bar{x} = \frac{\int x \, dm}{\int dm}$$

The centre of mass G of a plane lamina lies in the plane of the lamina. If the lamina is uniform, G is called the **centroid** of the area of the lamina and if this area is symmetrical about any straight line, the centroid lies on this line. The coordinates of the centre of mass of a lamina with surface density $\rho(x, y)$ enclosed within the curve $y = f(x)$, the x-axis and the lines $x = a$, $x = b$, are given by

$$\bar{x} = \frac{\int_a^b \rho xy \, dx}{\int_a^b \rho y \, dx} \qquad \bar{y} = \frac{\frac{1}{2}\int_a^b \rho y^2 \, dx}{\int_a^b \rho y \, dx}$$

If the density is uniform, and such a lamina is rotated about the x-axis, then the centre of mass of the solid of revolution so formed will lie on the x-axis and the x-coordinate of the centroid will be given by

$$\bar{x} = \frac{\int_a^b xy^2 \, dx}{\int_a^b y^2 \, dx}$$

(i) Find the positions of the centroids of the following areas (taking $\rho = 1$):

(a) $y = \sin x \quad 0 \leq x \leq \pi, \quad y = 0$

(b) A semi-circle with radius a.

(ii) Calculate the positions of the centroids of the solids obtained by rotating the following curves about the given axes:

(a) $x^2 + y^2 = a^2$ about Ox, for $x \geq 0$, $y \geq 0$.

(b) $y^2 = 2x \quad x = 0, \quad x = 2$ about Ox.

7. In a rigid body of mass M rotating with angular velocity ω about a fixed axis, a particle P of mass δm at perpendicular distance r from the axis would have velocity ωr and kinetic energy $\frac{1}{2}\delta m(\omega r)^2$. By regarding a continuous body as the limit of a sum of such particles, its kinetic energy will be

$$\lim \sum \frac{1}{2}\omega^2 r^2 \delta m \to \frac{1}{2}\int \omega^2 r^2 \, dm = \frac{1}{2}\omega^2 \int r^2 \, dm$$

since ω is the same for all particles of the body. The expression

$$I = \int r^2 \, dm$$

is called the **moment of inertia** (MI) of the body about the axis of rotation. It is usually given in the form Mk^2 where k is called the **radius of gyration** about the axis. The MI is also called the **second moment** of the body.

(i) Find the moment of inertia and the radius of gyration of: (a) a uniform rod length a about perpendicular axis through centre; (b) a disc radius r about an axis perpendicular to the disc, through its centre.

(ii) Calculate the moment of inertia about the coordinate axes of

(i) $y = x^2$, $y = 0$, $x = 4$

(ii) $2y = x^3$, $x = 0$ $y = 4$

(iii) $xy = 4$, $x = 1$, $y = 0$, $x = 4$

8. The mean value of $y = f(x)$ between $x = a, b$ is given by $\dfrac{1}{b-a} \int_a^b f(x)\,dx$.

The RMS value is defined as $\sqrt{\dfrac{1}{b-a} \int_a^b y^2\,dx} = \sqrt{\dfrac{1}{b-a} \int_a^b (f(x))^2\,dx}$.

It is the square root of the mean value of the square of the function. The RMS value is very important in alternating current theory – applied to oscillatory currents it has the effect of averaging over the oscillations (power dissipated is proportional to the RMS of the current). The RMS is similar to the **standard deviation** of statistics.

Find the mean and RMS values of the following functions over the ranges indicated:

(i) $y = x$ $(x = 0, 1)$

(ii) $y = 3x$ $(x = 0, 1)$

(iii) $y = \sin x$ $(x = 0, \pi)$

(iv) $y = e^{-x}$ $(x = 0, 1)$

Answers to reinforcement exercises

10.3.1 The derivative as a gradient and rate of change

A. (i) (a) 2 at $x = 0$, 14 at $x = 2$ (b) 2 at $x = 0$, 14 at $x = 2$

(ii) (a) 1 at $x = 0$, $\dfrac{1}{2}$ at $x = \dfrac{\pi}{3}$ (b) 1 at $x = 0$, $\dfrac{1}{2}$ at $x = \dfrac{\pi}{3}$

(iii) (a) 1 at $x = 0$, $e(\cos 1 - \sin 1)$ at $x = 1$ (b) 1 at $x = 0$, $e(\cos 1 - \sin 1)$ at $x = 1$

(iv) (a) 0 at $x = 0$, $\dfrac{4}{5}$ at $x = 2$ (b) 0 at $x = 0$, $\dfrac{4}{5}$ at $x = 2$

B. $(-1, 12)$

10.3.2 Tangent and normal to a curve

(i) Tangent is $y = 4x - 4$, normal is $x + 4y - 1 = 0$

(ii) Tangent is $y = 4x - 2$, normal is $x + 4y - 9 = 0$

(iii) Tangent is $y = x - 1$, normal is $x + y - 1 = 0$

(iv) Tangent is $y = x$, normal is $y = -x$

10.3.3 Stationary points and points of inflection

A. (i) Min at $(2, -1)$

(ii) Min at $(2, -14)$, max at $(-2, 18)$, point of inflection at $(0, 2)$

(iii) Point of inflection at (0, 0)

(iv) Min at (5, 2), max at (−5, −2)

(v) Min at (3, 23), max at (2, 24), point of inflection at $x = \frac{5}{2}$

(vi) Min at $x = \frac{3}{2}$, min at $x = -2$, point of inflection at $x = -\frac{1}{4}$

B. Max of $\dfrac{16}{3\sqrt{3}}$ at $x = -\dfrac{2}{\sqrt{3}}$ and min of $-\dfrac{16}{3\sqrt{3}}$ at $x = \dfrac{2}{\sqrt{3}}$. The slope at the point of inflection, $x = 0$, is -4.

10.3.4 Curve sketching in Cartesian coordinates

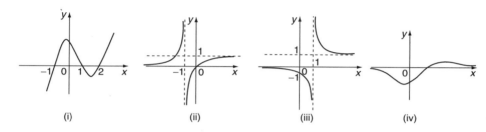

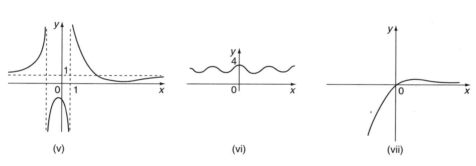

10.3.5 Applications of integration – area under a curve

A. (i) $\dfrac{28}{3}$

(ii) $\dfrac{3}{2} - \ln 2$

(iii) e

(iv) $\dfrac{1}{2}$

(v) $\dfrac{28}{3}$

(vi) $\dfrac{\pi}{4}$

B. (i) −4   (ii) 0

C. 1

D. $\dfrac{11}{6}$

10.3.6 Volume of a solid of revolution

(i) $\dfrac{\pi}{30}$

(ii) $\dfrac{3\pi}{10}$

(iii) $\dfrac{\pi^2}{2}$

11

Vectors

Vectors are the mathematical tools needed when we deal with engineering systems in more than one dimension. Vector quantities such as force have both a magnitude and direction associated with them, and so are represented by mathematical symbols that, similarly, have a magnitude and direction. We can construct an algebra of such quantities by which they may be added, subtracted, multiplied (in two different ways) – but not divided. These rules of combination reflect the ways in which the corresponding physical quantities behave – for example vectors add in the same way that forces do, the scalar product represents 'vector times vector = scalar' in the same way that 'velocity times velocity = energy'. By introducing a coordinate system, we can also represent vectors by arrays of ordinary numbers, in which form they are easier to combine and manipulate.

Prerequisites

It will be useful if you know something about

- ratio and proportion (14 ◀)
- Cartesian coordinate systems (205 ◀)
- Pythagoras' theorem (154 ◀)
- sines and cosines (175 ◀)
- cosine rule (197 ◀)
- simultaneous linear equations (48 ◀)
- function notation (90 ◀)
- differentiation (Chapter 8 ◀)

Objectives

In this chapter you will find

- definitions of scalars and vectors
- representation of a vector
- addition and subtraction of vectors
- multiplication of a vector by a scalar
- rectangular Cartesian coordinates in three dimensions
- distance in Cartesian coordinates
- direction cosines and ratios
- angle between two lines
- basis vectors (**i**, **j**, **k**)

- properties of vectors
- the scalar product of two vectors
- the vector product of two vectors
- vector functions
- differentiation of vectors

Motivation

You may need the material of this chapter for

- representing and manipulating directed physical quantities such as displacement, velocity, and force
- solving two and three dimensional problems
- working with phasors (➤ 371)
- describing and analysing structures

11.1 Introduction – representation of a vector quantity

Think of three examples of physical quantities: temperature, force, shear stress. Each of these is represented mathematically by a different type of object, reflecting the way in which these physical quantities behave and combine.

Temperature requires just one number for its specification – this is called a **scalar** quantity. Other examples of scalar quantities are mass, distance, speed.

Force requires a magnitude and a single direction for its specification – this is called a **vector** quantity. Other vector quantities are displacement and velocity.

Shear stress requires the relative motion of two parallel planes for its specification – this is called a **tensor** quantity. We do not consider tensors in this book, but they are very important in, for example, solid mechanics.

Scalars, vectors and tensors are all different types of mathematical objects that are defined and combined amongst themselves in such a way as to model the respective physical quantities.

The first conceptual hurdle that we have to overcome with vectors concerns how we represent them. A vector quantity can be represented at two levels:

- as a directed line segment drawn on a piece of paper, and represented by a symbol possessing both magnitude and direction, satisfying an algebra that reflects geometrical combinations – 'vectors as arrows'
- by a mathematical object consisting of some numbers that effectively describe the magnitude and the direction of the quantity, and combine in such a way as to represent combinations of the quantity – vectors as arrays of numbers, or 'components'. This requires an explicit coordinate system, and the numbers representing the vector will depend on this system.

Both of these representations are used at the elementary level, but the connection between them is not always easy to see. The situation is not helped by the fact that the terms 'scalar' and 'vector' are also used in matrix theory, and indeed matrices can be used to represent vectors – a 3×1 column matrix can be used to represent a vector for example. Again, the connections between these ideas are not always clear. However, we will usually use the representation of vectors by arrays, which is relatively straightforward to operate.

In this book we will usually denote vectors by bold lower case – in written work it is usual to denote them by under line (<u>a</u>) or over line ($\bar{a}$). To denote specific vector displacements from points A to B we use the notation

$$\overrightarrow{AB} \text{ or } \underline{AB}$$

Note that vectors have units of course, depending on the physical quantity they represent.

Whichever representation of vectors we use, we will eventually need a three dimensional coordinate system – simply so that we can state such things as length and direction unequivocally. There are a number of such coordinate systems in common use but we will use the simplest and most popular one – the generalisation of **rectangular Cartesian coordinates**.

Exercise on 11.1

Classify the following as scalar or vector quantities:

(i)   area   (ii)   gravitational force   (iii) electric field at a point
(iv) density   (v)   potential energy   (vi)   work
(vii) magnetic field (viii) time   (ix)   pressure
(x)   acceleration   (xi)   voltage   (xii) momentum

Answers

(i), (iv), (v), (vi), (viii), (xi) are all scalars. The rest are all vectors.

11.2 Vectors as arrows

This is the most common approach to vectors at the elementary level. We treat the line segment representing a vector **a** as an abstract geometrical 'free' object. If **a** is specifically fixed to an origin, then it is referred to as a **position vector** – its tip designating the position of the point at the tip. Pictorially, we might display vectors as in Figure 11.1.

Figure 11.1 Free vectors as arrows.

Note that the vector −**a** is simply **a** in the opposite sense.

An algebra of such objects can be constructed which reflects geometrically the way in which vector quantities combine. Naturally, two vectors are equal if and only if they have the same magnitude and direction. The **zero vector** has zero magnitude and undefined direction – it is denoted **0**.

The **length** or **magnitude** of a vector, **a**, is denoted |**a**| or just $\underline{a}$. The direction of **a** is represented by a vector in the same direction, but with unit magnitude, and is denoted by **â**. Thus, we may write **a** = |**a**|**â**.

Exercise on 11.2

Consider vectors in a plane, and use a two-dimensional Cartesian coordinate system with the usual rectangular x-, y-axes. Choosing suitable scales on the axes sketch vectors as follows:

(i) Three free vectors **a**, **b**, **c** all of the same length 2 units and making equal angles with the positive x- and y-axes.

(ii) A position vector **r** of length 3 units in the north-west direction (positive y-axis north).

Answers

(i)

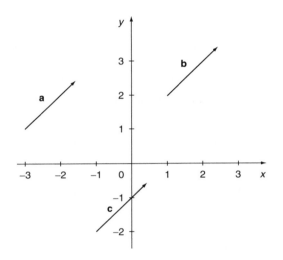

Figure 11(i)

(ii)

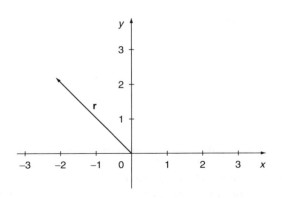

Figure 11(ii)

11.3 Addition and subtraction of vectors

Geometrically, vectors are **added** by the use of the **triangle law** in which **a** and **b** form two sides of a triangle in order, as shown in Figure 11.2. Then the resultant is the third side, denoted here with a double arrow. This representation of addition is in keeping with vectors describing displacements for example – following vector **a** and then continuing along vector **b** takes you to the same place that the resultant of vectors **a** and **b** takes you directly. In Figure 11.2 we represent this explicitly by showing displacements between three points A, B, C.

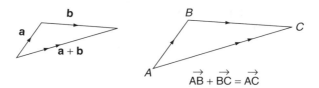

Figure 11.2 Triangle addition of vectors.

The resultant of **a** and **b** is denoted by the vector sum $\mathbf{a} + \mathbf{b}$ or by $\overrightarrow{AC}$. This can be regarded as '**a** followed by **b** gives $\mathbf{a} + \mathbf{b}$'.

An alternative means of geometrical addition, linked to the calculation of the resultant of two forces, is the **parallelogram law**. Thus to add vectors **a** and **b** we construct a parallelogram with **a** and **b** forming two sides and then their **sum** or **resultant**, in magnitude and direction, is represented by the diagonal of the parallelogram, as shown in Figure 11.3. This represents the way in which vector quantities such as force combine. For example, if you imagine two horses on either side of a canal pulling a barge, then the resultant force on the barge will be represented by the diagonal of the corresponding parallelogram.

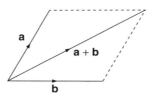

Figure 11.3 Parallelogram addition of vectors.

Using either viewpoint – triangle or parallelogram – by constructing appropriate diagrams you can convince yourself that

$\mathbf{a} + (\mathbf{b} + \mathbf{c}) = (\mathbf{a} + \mathbf{b}) + \mathbf{c}$ (vector addition is **associative**)

$\mathbf{a} + \mathbf{b} = \mathbf{b} + \mathbf{a}$ (vector addition is **commutative**)

Three or more vectors can be added by the **polygon law** as illustrated in Figure 11.4 – the resultant or sum 'closes', in the opposite direction, the polygon formed by following the vectors in the sum in turn.

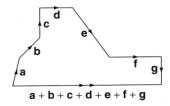

Figure 11.4 Polygon addition of vectors.

A particular example of this is **scalar multiplication** of a vector by a scalar – the vector $k\mathbf{a}$ is simply $\mathbf{a}$ with its magnitude scaled by k.

Vectors may be subtracted geometrically as shown in Figure 11.5. Thus, we can denote the vector $\mathbf{b} - \mathbf{a}$ by the 'arrow' that takes us from $\mathbf{a}$ to $\mathbf{b}$, since $\mathbf{a}$ followed by $\mathbf{b} - \mathbf{a}$ gives the same displacement as $\mathbf{b}$.

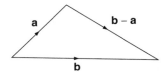

Figure 11.5 Subtraction of vectors.

Exercises on 11.3

1. Consider the pentagon below:

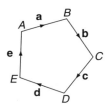

Express

    (i)   **e** in terms of **a**, **b**, **c**, **d**
    (ii) $\overrightarrow{CE}$ in terms of **e**, **a**, **b**
    (iii) $\overrightarrow{CE}$ in terms of **c** and **d**
    (iv) $\overrightarrow{EC}$ in terms of $\overrightarrow{EA}$ and $\overrightarrow{CA}$

2. On a two dimensional Cartesian coordinate system, **a** is the position vector from the origin to the point (1, 0), **b** is the position vector with length 2 at angle 60° (anticlockwise) to the positive x-axis. Describe the vector $\mathbf{b} - \mathbf{a}$.

Answers

1. (i) $\mathbf{e} = -(\mathbf{a} + \mathbf{b} + \mathbf{c} + \mathbf{d})$ (ii) $\overrightarrow{CE} = -(\mathbf{e} + \mathbf{a} + \mathbf{b})$ (iii) $\overrightarrow{CE} = \mathbf{c} + \mathbf{d}$
   (iv) $\overrightarrow{EC} = \overrightarrow{EA} - \overrightarrow{CA}$
2. $\mathbf{b} - \mathbf{a}$ is of length $\sqrt{3}$ parallel to the y-axis

11.4 Rectangular Cartesian coordinates in three dimensions

To represent points, geometrical objects, vectors, etc. in 3-dimensional space it is useful to have a **reference frame** or **coordinate system**. The simplest type is the Cartesian system of rectangular coordinates, which generalises the usual x-, y-axes of two dimensions – see Figure 11.6, which shows how a point P may be represented by coordinates (x_1, y_1, z_1) relative to a three-dimensional rectangular system of x-, y-, z-axes.

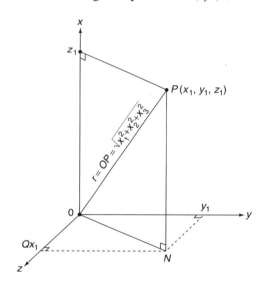

Figure 11.6 Rectangular Cartesian coordinates in three dimensions.

PN is the perpendicular to the x-y plane. The axes are shown arranged in the usual convention in which the Ox, Oy, Oz form a **right-handed set** – rotating Ox round to Oy drives a right-handed screw along Oz. The same applies if we replace $Ox \to Oy \to Oz \to Ox$, etc.

We say that the coordinates of P relative to the axes $Oxyz$ are (x_1, y_1, z_1) or 'P is the point (x_1, y_1, z_1)' where the coordinates are the distances of P along the x-, y-, z-axes respectively. We can label each point of 3-dimensional space with a set of 'coordinates' (x, y, z) relative to a particular coordinate system. For a given point, choosing a different set of axes gives a different set of coordinates.

Exercise on 11.4

Plot the points $(0, 0, 0)$, $(0, 0, 1)$, $(0, 1, 0)$, $(1, 0, 0)$ $(-1, 0, 0)$, $(1, 0, -1)$, $(2, -1, 1)$ on a perspective drawing of a Cartesian rectangular coordinate system.

Answer

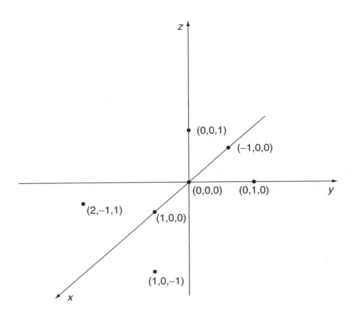

11.5 Distance in Cartesian coordinates

Look again at Figure 11.6. The distance, r, from the origin O to the point $P(x_1, y_1, z_1)$ is given by

$$r = OP = \sqrt{x_1^2 + y_1^2 + z_1^2}$$

This can be seen by applying Pythagoras' theorem (154 ◀) twice:

$$OP^2 = ON^2 + PN^2 = OQ^2 + QN^2 + PN^2 = x_1^2 + y_1^2 + z_1^2$$

In general the distance between two points $P(x, y, z)$, $P'(x', y', z')$ is given by:

$$PP' = \sqrt{(x'-x)^2 + (y'-y)^2 + (z'-z)^2}$$

which again follows from Pythagoras' theorem.

Problem 11.1

Calculate the distance between the two points $P(-1, 0, 2)$ and $P'(1, 2, 3)$.

The distance is given by substituting the coordinates into the above expression for PP':

$$PP' = \sqrt{(-1-1)^2 + (0-2)^2 + (2-3)^2}$$
$$= \sqrt{9} = 3$$

Exercise on 11.5

Calculate the distances of the points (i) (1, 0, 2), (ii) (−2, 1, −3), (iii) (−1, −1, −4) from the origin. Also calculate the distance between each pair of points.

Answer

(i) $\sqrt{5}$ (ii) $\sqrt{14}$ (iii) $\sqrt{18}$
Between (i) and (ii), $\sqrt{35}$. Between (i) and (iii), $\sqrt{41}$. Between (ii) and (iii), $\sqrt{6}$.

11.6 Direction cosines and ratios

NB. Sections 11.6–8 are not essential to what follows and may be safely omitted for the student who simply wants to be able to use vectors. However, the ideas are very important in more advanced engineering mathematics and occur in such topics as CAD and surveying.

While the coordinates of a point $P(x, y, z)$, tell us everything that we need to know about the line segment OP, they are not very convenient for indicating its direction, which is most easily done by means of angles. When soldiers aim a field gun they do not specify the coordinates of the tip of the barrel – they give the angle of elevation of the barrel, and the bearing of the target. Similarly, the best way to specify the direction of OP is to specify the angles it makes with the axes. But we still want to retain a link with the coordinates (x, y, z) and for this reason we use the cosines of these angles – the so-called **direction cosines**.

Let OP make angles α, β, γ with the axes Ox, Oy, Oz respectively – see Figure 11.7.

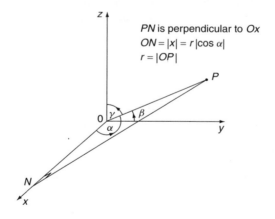

Figure 11.7 Direction angles in three dimensions.

The **direction cosines** of the line segment, or **radius vector**, OP are defined as $\cos \alpha$, $\cos \beta$, $\cos \gamma$ denoted by:

$$l = \cos \alpha, \quad m = \cos \beta, \quad n = \cos \gamma$$

Since
$$x = r \cos \alpha, \quad y = r \cos \beta, \quad z = r \cos \gamma$$

we have
$$l = \frac{x}{r}, \quad m = \frac{y}{r}, \quad n = \frac{z}{r}$$

from which, since
$$r^2 = x^2 + y^2 + z^2,$$
$$l^2 + m^2 + n^2 = 1$$

Note that the direction cosines of a line not passing through the origin are the same as those of a parallel line through origin.

The numbers l, m, n are not always the most convenient to use – they are all less than or equal to 1 in magnitude. Often all we need is simply the relation between them. Any three numbers a, b, c satisfying (14 ◀)

$$a : b : c = l : m : n$$

are called **direction ratios** of OP.

From $l^2 + m^2 + n^2 = 1$ we have

$$l = \frac{a}{d}, \quad m = \frac{b}{d}, \quad n = \frac{c}{d}$$

where
$$d = \pm\sqrt{a^2 + b^2 + c^2}$$

The $\pm$ denotes that to a given set of direction ratios there are two sets of direction cosines corresponding to oppositely directed parallel lines.

Problem 11.2

Calculate the direction cosines of each line to a vertex of a unit cube in the first octant with the lowest corner at the origin (Figure 11.8).

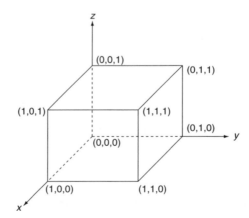

Figure 11.8

First note that the direction cosines of the origin are not defined, since $r = 0$ at the origin.

Consider the points on the axes – in this case the direction cosines are simply given by the points themselves (1, 0, 0), (0, 1, 0), (0, 0, 1).

Now consider the point (1, 0, 1). The position vector to this makes an angle 45° with the x- and z-axes, and 90° with the Oy-axis. The direction cosines are therefore directly obtained as

$$(\cos 45°, 0, \cos 45°) = \left(\frac{1}{\sqrt{2}}, 0, \frac{1}{\sqrt{2}}\right)$$

This is of course what we would obtain from the coordinates (1, 0, 1) themselves, taking the form

$$\left(\frac{x}{r}, \frac{y}{r}, \frac{z}{r}\right) = \left(\frac{1}{\sqrt{2}}, \frac{0}{\sqrt{2}}, \frac{1}{\sqrt{2}}\right) = \left(\frac{1}{\sqrt{2}}, 0, \frac{1}{\sqrt{2}}\right)$$

Similarly for:

(0, 1, 1) the direction cosines are $\left(0, \frac{1}{\sqrt{2}}, \frac{1}{\sqrt{2}}\right)$

(1, 1, 0) the direction cosines are $\left(\frac{1}{\sqrt{2}}, \frac{1}{\sqrt{2}}, 0\right)$

Finally, the far vertex with coordinates (1, 1, 1) has direction cosines

$$\left(\frac{x}{r}, \frac{y}{r}, \frac{z}{r}\right) = \left(\frac{1}{\sqrt{3}}, \frac{1}{\sqrt{3}}, \frac{1}{\sqrt{3}}\right)$$

The acute angle made with each axis is therefore

$$\cos^{-1}(\tfrac{1}{3}) = 70.53° \text{ to 2 dp}$$

Exercise on 11.6

Calculate the direction cosines of the position vectors defined by the points in the Exercise on 11.4.

Answer

In order, referring to the Exercise on 11.4, the direction cosines are:

not defined, (0, 0, 1), (0, 1, 0), (1, 0, 0), (−1, 0, 0), $\left(\frac{1}{\sqrt{2}}, 0, -\frac{1}{\sqrt{2}}\right)$, $\left(\frac{2}{\sqrt{14}}, -\frac{3}{\sqrt{14}}, \frac{1}{\sqrt{14}}\right)$

11.7 Angle between two lines through the origin

Using direction cosines we can obtain a simple and useful expression for the angle between two lines intersecting at the origin. In general, the angle between any two lines in space, not necessarily intersecting, is the angle between any pair of lines parallel to them. In particular we can always take a pair of lines through the origin. This expression for the angle between two lines will be useful in discussing the angle between two vectors. This comes into the different types of products that we can make with vectors (the scalar and the vector product – see Sections 11.10 and 11.11). Consider Figure 11.9.

Suppose two lines OA, OA' have direction cosines l, m, n and l', m', n' respectively with respect to axes $Oxyz$. Then we will show that

$$\cos\theta = ll' + mm' + nn'$$

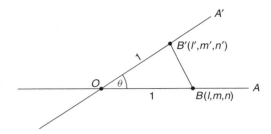

Figure 11.9 Angle between two lines.

This follows by applying the cosine rule to the triangle OBB' in Figure 11.10 where OB, OB' are of unit length. Then B, B' have the coordinates (l, m, n) (l', m', n') respectively.

We have, by the cosine rule (179 ◄)

$$BB'^2 = OB^2 + OB'^2 - 2OB\,OB'\cos\theta$$

or

$$(l' - l)^2 + (m' - m)^2 + (n' - n)^2 = 1^2 + 1^2 - 2\cos\theta = 2 - 2\cos\theta$$

Expanding the left-hand side and using $l^2 + m^2 + n^2 = 1$ for the direction cosines yields

$$l'^2 + m'^2 + n'^2 + l^2 + m^2 + n^2 - 2ll' - 2mm' - 2nn'$$
$$= 1 + 1 - 2(ll' + mm' + nn') = 2 - 2\cos\theta$$

This then gives the required result $\cos\theta = ll' + mm' + nn'$.

From this result it follows that if two lines are perpendicular then

$$ll' + mm' + nn' = 0$$

because $\cos(\pm 90°) = 0$

Exercise on 11.7

Using direction cosines find the acute angles between each pair of lines through the origin defined by the points in Exercises 11.5.

Answer

Between (i) and (ii), 17.02°; between (i) and (iii), 18.43°; between (ii) and (iii), 35.02° – all to two decimal places.

11.8 Basis vectors

We can use the coordinate system defined in Section 11.4 to represent vectors in terms of numerical components by choosing 'basis vectors' along the coordinate axes. The standard notation for this is to let $\mathbf{i}$, $\mathbf{j}$, $\mathbf{k}$ denote the unit vectors along the x-, y-, z-axes as shown in Figure 11.10.

In terms of these 'basis vectors' we can write any vector, $\mathbf{a}$, in the **component form**

$$\mathbf{a} = a_1\mathbf{i} + a_2\mathbf{j} + a_3\mathbf{k}$$

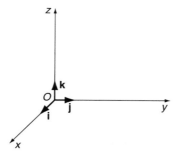

Figure 11.10 Basic vectors **i, j, k**.

and this provides the link with the 'arrows' representation of vectors given in earlier sections. It is common practice to represent $a_1\mathbf{i} + a_2\mathbf{j} + a_2\mathbf{k}$ by a triple (a_1, a_2, a_2) or a 'column matrix'

$$\mathbf{a} = \begin{bmatrix} a_1 \\ a_2 \\ a_3 \end{bmatrix}$$

but always remember that these only represent the vector **a** with respect to a particular set of basis vectors, and should not be confused with the notation for the coordinates of a point.

Any set of unit vectors, such as **i, j, k** which are mutually perpendicular to each other (and there can be at most three in 3-dimensional space) is called an **orthonormal set**. The **i, j, k** form an **orthonormal basis** for the set of all vectors, in the sense that any three dimensional vector can be expressed in terms of them.

Note that $a_1\mathbf{i} + a_2\mathbf{j} + a_3\mathbf{k} = \mathbf{0}$ can **only** mean $a_1 = a_2 = a_3 = 0$.

Problem 11.3

Find α, β, χ if $(\alpha + \beta + \chi)\mathbf{i} + (\beta - \chi + 1)\mathbf{j} + (\alpha + \chi)\mathbf{k} = \mathbf{0}$.

Don't let the Greek symbols put you off – as mathematics gets more advanced we soon run out of alphabets and so you will have to get used to such notation.

If $(\alpha + \beta + \chi)\mathbf{i} + (\beta - \chi + 1)\mathbf{j} + (\alpha + \chi)\mathbf{k} = \mathbf{0}$ then

$$\alpha + \beta + \chi = 0 \quad \text{(i)}$$
$$\beta - \chi + 1 = 0 \quad \text{(ii)}$$
$$\alpha + \chi = 0 \quad \text{(iii)}$$

So a single vector equation is equivalent to three 'scalar' equations. We will consider such systems of equations in detail in Chapter 13, but the above system is not difficult to solve. Substituting for β from (ii) into (i) gives

$$\alpha + 2\chi = 1$$
$$\alpha + \chi = 0$$

from which $\chi = 1$, $\alpha = -1$, then $\beta = \chi - 1 = 0$.

Exercises on 11.8

1. Express the position vectors with the given endpoints in terms of **i**, **j**, **k** vectors.

   (i) $(3, -1, 2)$ (ii) $(1, 0, 1)$ (iii) (a, b, c) (iv) $(-1, 1, -1)$

2. If $(\alpha - \beta)\mathbf{i} + (\beta + 2\chi)\mathbf{j} + (\alpha - \chi)\mathbf{k} = 2\mathbf{i} - \mathbf{k}$ determine α, β, χ.

Answers

1. (i) $3\mathbf{i} - \mathbf{j} + 2\mathbf{k}$ (ii) $\mathbf{i} + \mathbf{k}$ (iii) $a\mathbf{i} + b\mathbf{j} + c\mathbf{k}$ (iv) $-\mathbf{i} + \mathbf{j} - \mathbf{k}$
2. $\alpha = 0, \beta = -2, \chi = 1$

11.9 Properties of vectors

Two vectors $\mathbf{a} = a_1\mathbf{i} + a_2\mathbf{j} + a_3\mathbf{k}$ $\mathbf{b} = b_1\mathbf{i} + b_2\mathbf{j} + b_3\mathbf{k}$

referred to the same axes are **equal** if and only if

$$a_1 = b_1, \quad a_2 = b_2, \quad a_3 = b_3$$
$$(a_i = b_i \quad \text{for all } i)$$

A **zero vector** is one whose components are all zero:

$$\mathbf{0} = 0\mathbf{i} + 0\mathbf{j} + 0\mathbf{k}$$

Let A, B be points with coordinates $(a_1, a_2, a_3), (b_1, b_2, b_3)$ with respect to $Oxyz$ axes. The **position vector of B relative to A** is (see Figure 11.11)

$$\underline{\mathbf{AB}} = \overrightarrow{AB} = (b_1 - a_1)\mathbf{i} + (b_2 - a_2)\mathbf{j} + (b_3 - a_3)\mathbf{k}$$

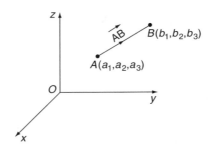

Figure 11.11 Position vector of B relative to A.

As a particular case let $A = (0, 0, 0)$, the origin, and $B = $ a point $P(x, y, z)$, as shown in Figure 11.12.

$\overrightarrow{OP}$ or $\underline{OP}$ or $\mathbf{r} = x\mathbf{i} + y\mathbf{j} + z\mathbf{k}$ is called the **position vector** of P (with respect to the axes $Oxyz$). We also refer to $\mathbf{r}$ as a **radius vector**.

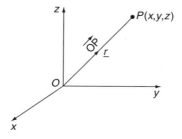

Figure 11.12 Position vector of a point P.

The **magnitude** or **length** of a vector $\mathbf{a} = a_1\mathbf{i} + a_2\mathbf{j} + a_3\mathbf{k}$ is given by

$$a = |\mathbf{a}| = \sqrt{a_1^2 + a_2^2 + a_3^2}$$

(also called **modulus** or **norm** of **a**).

A **unit vector** is one of unit norm or magnitude – usually denoted with a circumflex, for example $\hat{\mathbf{a}}$.

Problem 11.4

Show that $\mathbf{a} = \dfrac{\sqrt{3}}{2}\mathbf{i} + \dfrac{1}{2}\mathbf{j}$ is a unit vector.

We have $|\mathbf{a}| = \sqrt{\left(\dfrac{\sqrt{3}}{2}\right)^2 + \left(\dfrac{1}{2}\right)^2 + 0^2} = 1$.

Problem 11.5

Construct a unit vector parallel to $\mathbf{a} = \mathbf{i} + \mathbf{j} - 3\mathbf{k}$.

To construct a unit vector parallel to a given vector we have only to divide by its modulus, which for $\mathbf{a} = \mathbf{i} + \mathbf{j} - 3\mathbf{k}$ is

$$|\mathbf{a}| = \sqrt{1^2 + 1^2 + (-3)^2} = \sqrt{11}$$

so a unit vector parallel to **a** is

$$\hat{\mathbf{a}} = \frac{1}{\sqrt{11}}(\mathbf{i} + \mathbf{j} - 3\mathbf{k})$$

In general a vector $\mathbf{a} = a_1\mathbf{i} + a_2\mathbf{j} + a_3\mathbf{k}$ is **multiplied by a scalar** λ as follows:

$$\lambda \mathbf{a} = \lambda a_1 \mathbf{i} + \lambda a_2 \mathbf{j} + \lambda a_3 \mathbf{k}$$

i.e. all components are multiplied by λ.

Note that the magnitude of $\lambda \mathbf{a}$ is:

$$|\lambda \mathbf{a}| = \lambda |\mathbf{a}| = \lambda a$$

Multiplication by a positive (negative) λ leaves the direction of **a** unchanged (reversed). $-\mathbf{a}$ is defined by $(-1)\mathbf{a}$.

Vectors $\mathbf{a} = a_1\mathbf{i} + a_2\mathbf{j} + a_3\mathbf{k}$, $\mathbf{b} = b_1\mathbf{i} + b_2\mathbf{j} + b_3\mathbf{k}$ are added and subtracted 'componentwise':

$$\mathbf{a} + \mathbf{b} = (a_1 + b_1)\mathbf{i} + (a_2 + b_2)\mathbf{j} + (a_3 + b_3)\mathbf{k}$$

$$\mathbf{a} - \mathbf{b} = (a_1 - b_1)\mathbf{i} + (a_2 - b_2)\mathbf{j} + (a_3 - b_3)\mathbf{k}$$

It is obvious from this that, as noted in Section 11.3,

$$\mathbf{a} + \mathbf{b} = \mathbf{b} + \mathbf{a} \quad \text{(vector addition is commutative)}$$

$$\mathbf{a} + (\mathbf{b} + \mathbf{c}) = (\mathbf{a} + \mathbf{b}) + \mathbf{c} \quad \text{(vector addition is associative)}$$

Problem 11.6

For $\mathbf{a} = \mathbf{i} - \mathbf{j}$, $\mathbf{b} = 2\mathbf{i} + 3\mathbf{j} + 4\mathbf{k}$, $\mathbf{c} = -\mathbf{i} - 2\mathbf{j} - 4\mathbf{k}$, find the vectors representing

(i) $3\mathbf{a}$ (ii) $\mathbf{a} + \mathbf{b}$ (iii) $\mathbf{a} + 2\mathbf{b} - \mathbf{c}$ (iv) $2\mathbf{a} - 3\mathbf{c}$

Using the rules given above we have

(i) $3\mathbf{a} = 3(\mathbf{i} - \mathbf{j}) = 3\mathbf{i} - 3\mathbf{j}$

(ii) $\mathbf{a} + \mathbf{b} = \mathbf{i} - \mathbf{j} + 2\mathbf{i} + 3\mathbf{j} + 4\mathbf{k} = 3\mathbf{i} + 2\mathbf{j} + 4\mathbf{k}$

(iii) $\mathbf{a} + 2\mathbf{b} - \mathbf{c} = \mathbf{i} - \mathbf{j} + 2(2\mathbf{i} + 3\mathbf{j} + 4\mathbf{k}) - (-\mathbf{i} - 2\mathbf{j} - 4\mathbf{k})$

$\qquad\qquad\qquad = \mathbf{i} - \mathbf{j} + 4\mathbf{i} + 6\mathbf{j} + 8\mathbf{k} + \mathbf{i} + 2\mathbf{j} + 4\mathbf{k}$

$\qquad\qquad\qquad = 6\mathbf{i} + 7\mathbf{j} + 12\mathbf{k}$

(iv) $2\mathbf{a} - 3\mathbf{c} = 2(\mathbf{i} - \mathbf{j}) - 3(-\mathbf{i} - 2\mathbf{j} - 4\mathbf{k})$

$\qquad\qquad\qquad = 2\mathbf{i} - 2\mathbf{j} + 3\mathbf{i} + 6\mathbf{j} + 12\mathbf{k}$

$\qquad\qquad\qquad = 5\mathbf{i} + 4\mathbf{j} + 12\mathbf{k}$

Exercises on 11.9

1. If $\mathbf{a} = \mu\mathbf{i} + 2\mathbf{j} + (\lambda - \mu)\mathbf{k}$, $\mathbf{b} = 2\lambda\mathbf{i} + \upsilon\mathbf{j} + 2\mathbf{k}$

and

$$2\mathbf{a} + \mathbf{b} = \mathbf{0}$$

determine μ, λ, υ.

2. Referring to Q1, evaluate

(i) $|\mathbf{a}|$ (ii) $\hat{\mathbf{a}}$ (iii) $3\mathbf{b}$ (iv) $\mathbf{a} + 2\mathbf{k}$ (v) $3\mathbf{a} - \mathbf{b}$

Answers

1. $\mu = \dfrac{1}{2}$, $\lambda = -\dfrac{1}{2}$, $\upsilon = -4$

2. (i) $\dfrac{\sqrt{21}}{2}$ (ii) $\dfrac{1}{\sqrt{21}}(\mathbf{i}+4\mathbf{j}-2\mathbf{k})$ (iii) $-3\mathbf{i}-12\mathbf{j}+3\mathbf{k}$ (iv) $\dfrac{1}{2}\mathbf{i}+2\mathbf{j}+\mathbf{k}$
   (v) $\dfrac{5}{2}\mathbf{i}+10\mathbf{j}-4\mathbf{k}$

11.10 The scalar product of two vectors

So far we have simply added or subtracted vectors, or formed linear combinations of them. An obvious question is what sorts of 'products' can we define for two vectors. By 'product' we mean a generalisation of the multiplication of two real numbers. It is important to realise that we **are** free to **define** such products – vectors are new mathematical objects, completely different from anything we have considered before in this book, and we are at liberty to define the combinations of them in any way we wish **which is mathematically consistent**. When we invent new mathematical tools this is always the case – so long as our definitions do not lead to any silly contradictions (such as '1 = 0') we can define the rules exactly as we wish. **But** of all the possible rules, we are naturally going to choose the most useful. So if we do define products of vectors then we are going to do so in such a way that the result is useful to us in modelling reality. We have (at least) two possibilities for defining 'multiplication' of vectors:

$$\text{vector} * \text{vector} = \text{scalar}$$

$$\text{vector} * \text{vector} = \text{vector}$$

Not surprisingly these give rise to the 'scalar' and 'vector product' respectively.

The best way to introduce the scalar product is to do a little 'work'. Specifically, if a force of magnitude F acting in a fixed direction moves a particle through a displacement d in that direction, then we say that the 'work done' by the force on the particle is:

$$W = Fd \tag{1}$$

Now, both force and displacement are vector quantities – they require both magnitude and direction to specify them. This is hidden in the above result because both the force and the displacement have been taken as operating in the same fixed direction. So, the right-hand side of the above equation is essentially a vector times a vector.

On the other hand, work done, or energy expended, is not a vector quantity – energy is simply a scalar, numerical quantity, like temperature or heat. So the left-hand side of the equation is a scalar. Thus, the equation essentially says:

"vector '×' vector = scalar"

But, as we said, the vector content is hidden. To bring it out, suppose F no longer acts on the particle in its own direction, but at an angle to it. For example the particle may be a bead on a straight smooth wire, and the force applied by pulling a string at an angle θ to the wire, as shown in Figure 11.13.

Only the component of the force along the direction of the wire can do work on the bead (the component perpendicular to the wire will not result in any motion along the

333

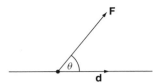

Figure 11.13 Force at an angle to the direction of motion.

wire). This component is $F\cos\theta$. So, if we move the bead a distance d along the wire then the work done is

$$W = (F\cos\theta)d$$
$$= Fd\cos\theta$$

This is the generalisation of equation (1) when the force and displacement make an angle θ with each other. So, let **F** be the vector representing the force and **d** represent the displacement. Then the work done is:

$$W = |\mathbf{F}|\,|\mathbf{d}|\cos\theta$$

i.e. the magnitude of the force **F** multiplied by the magnitude of the displacement **d**, times the cosine of the angle between the two vectors.

The result is, naturally, a scalar. But it has a magnitude component, $|\mathbf{F}|\,|\mathbf{d}|$, and a direction component θ, which suggest that it would be very useful to define this particular combination of vectors as a 'product' of vectors which yields a scalar. In general then the **scalar product** of two vectors **a**, **b** is denoted by $\mathbf{a}\cdot\mathbf{b}$ (sometimes called the **dot product**) and defined by

$$\mathbf{a}\cdot\mathbf{b} = ab\cos\theta$$

where $a = |\mathbf{a}|$, $b = |\mathbf{b}|$ and θ is the angle between **a** and **b** – see Figure 11.14.

Figure 11.14 The scalar product is $\mathbf{a}\cdot\mathbf{b} = ab\cos\theta$.

With this definition the work done by the force **F** moving a particle through a displacement **d** is given by

$$W = \mathbf{F}\cdot\mathbf{d}$$

This explains why we define the scalar product as above – it is very useful to do so!

Now, for the purposes of calculations it is invariably more convenient to have an expression for the scalar product in terms of components. The definition we have given above, as we will see shortly, is equivalent to the following expression in terms of components:

$$\mathbf{a}\cdot\mathbf{b} = a_1b_1 + a_2b_2 + a_3b_3 = \mathbf{b}\cdot\mathbf{a}$$

Note the special case:

$$\mathbf{a} \cdot \mathbf{a} = a_1^2 + a_2^2 + a_3^2 = a^2 \equiv |\mathbf{a}|^2$$
$$= \text{square of magnitude of } \mathbf{a}$$

In the above form of the scalar product it is not difficult to show that the scalar product has the following properties:

(i) $\mathbf{a} \cdot \mathbf{b}$ is a scalar
(ii) $\mathbf{a} \cdot \mathbf{b} = \mathbf{b} \cdot \mathbf{a}$
       i.e. the scalar product is commutative
(iii) $\mathbf{a} \cdot (\mathbf{b} + \mathbf{c}) = \mathbf{a} \cdot \mathbf{b} + \mathbf{a} \cdot \mathbf{c}$
       i.e. the scalar product is distributive over addition
(iv) $(k\mathbf{a}) \cdot \mathbf{b} = k(\mathbf{a} \cdot \mathbf{b}) = \mathbf{a} \cdot (k\mathbf{b})$ for any scalar k
(v)   If $\mathbf{a} \cdot \mathbf{b} = 0$ and $\mathbf{a}, \mathbf{b}$ are not zero then $\mathbf{a}$ is perpendicular to $\mathbf{b}$.

Problem 11.7

If $\mathbf{a} = 2\mathbf{i} - \mathbf{j} + 2\mathbf{k}$ and $\mathbf{b} = 2\mathbf{i} + 4\mathbf{j} - 3\mathbf{k}$ evaluate $\mathbf{a} \cdot \mathbf{b}, \mathbf{b} \cdot \mathbf{a}, a^2, b^2$

$$\mathbf{a} \cdot \mathbf{b} = 2(2) + (-1)(4) + 2(-3) = 4 - 4 = -6$$
$$\mathbf{b} \cdot \mathbf{a} = 2(2) + 4(-1) + (-3)(2) = -6, \text{ agreeing with property (ii)}$$
$$a^2 = |\mathbf{a}|^2 = \mathbf{a} \cdot \mathbf{a} = (2)^2 + (-1)^2 + 2^2 = 9$$
$$b^2 = 2^2 + 4^2 + (-3)^2 = 29$$

To see the connection of the above 'component expression' for the scalar product with the $\mathbf{a} \cdot \mathbf{b} = ab \cos \theta$ definition, we appeal to the work we did on direction cosines. You can omit this if you are prepared to take the result on trust. We have

$$\frac{\mathbf{a} \cdot \mathbf{b}}{ab} = \frac{a_1 b_1 + a_2 b_2 + a_3 b_3}{ab}$$
$$= \left(\frac{a_1}{a}\right)\left(\frac{b_1}{b}\right) + \left(\frac{a_2}{a}\right)\left(\frac{b_2}{b}\right) = \left(\frac{a_3}{a}\right)\left(\frac{b_3}{b}\right)$$
$$= l_a l_b + m_a m_b + n_a n_b$$
$$= \cos \theta$$

where (l_a, m_a, n_a) (l_b, m_b, n_b) are the direction cosines of $\mathbf{a}$ and $\mathbf{b}$ and θ is the angle between $\mathbf{a}$ and $\mathbf{b}$ (327 ◄). So $\mathbf{a} \cdot \mathbf{b} = ab \cos \theta = a_1 b_1 + a_2 b_2 + a_3 b_3$.

Problem 11.8

Find the scalar product of $\mathbf{a} = (-1, 2, 1)$ and $\mathbf{b} = (0, 2, 3)$ and hence find the acute angle between these vectors.

$$\mathbf{a} \cdot \mathbf{b} = (-1)(0) + 2(2) + 1(3) = 7$$
$$= ab \cos \theta = \sqrt{(-1)^2 + 2^2 + 1^2} \sqrt{0^2 + 2^2 + 3^2} \cos \theta$$
$$= \sqrt{6}\sqrt{13} \cos \theta$$

So
$$\cos \theta = \frac{7}{\sqrt{6}\sqrt{13}}$$

giving, to two decimal places,

$$\theta = 37.57°$$

for the acute angle between the vectors **a** and **b**.

Two non-zero vectors **a**, **b** are perpendicular to each other, or **mutually orthogonal** if

$$\mathbf{a} \cdot \mathbf{b} = 0$$

This follows directly from $\cos \theta = 0$ for $\theta = \pm \pi/2$.

For the **i**, **j**, **k** basis vectors we easily find: (Reinforcement Exercise 16)

$$\mathbf{i} \cdot \mathbf{i} = \mathbf{j} \cdot \mathbf{j} = \mathbf{k} \cdot \mathbf{k} = 1$$
$$\mathbf{i} \cdot \mathbf{j} = \mathbf{i} \cdot \mathbf{k} = \mathbf{j} \cdot \mathbf{k} = 0$$

Using these provides a direct derivation of the result:

$$\mathbf{a} \cdot \mathbf{b} = a_1 b_1 + a_2 b_2 + a_3 b_3$$

Exercises on 11.10

1. Using the component form for the scalar product prove the properties (i)–(v).
2. Find all possible scalar products between the vectors $\mathbf{a} = -\mathbf{i} + 2\mathbf{j}$, $\mathbf{b} = \mathbf{j} + 2\mathbf{k}$, $\mathbf{c} = \mathbf{i} + 2\mathbf{j} + 3\mathbf{k}$ and determine the angles between each pair of vectors.

Answers

2. $\mathbf{a} \cdot \mathbf{b} = 2$, $\mathbf{a} \cdot \mathbf{c} = 3$, $\mathbf{b} \cdot \mathbf{c} = 8$
   Angle between **a** and **b** is 66.42°, between **a** and **c** is 68.99°, between **b** and **c** is 17.02° all to two decimal places.

11.11 The vector product of two vectors

Now what about a vector valued 'product' of vectors? In this case we will simply define the vector product in terms of components first and give the connection with the a, b, θ form later. The mechanical application of the vector product as a moment of a force is given in the Applications section.

The **vector product** of $\mathbf{a} = a_1 \mathbf{i} + a_2 \mathbf{j} + a_3 \mathbf{k}$ with $\mathbf{b} = b_1 \mathbf{i} + b_2 \mathbf{j} + b_3 \mathbf{k}$ is denoted $\mathbf{a} \times \mathbf{b}$ ('**a** cross **b**') and defined by:

$$\mathbf{a} \times \mathbf{b} = (a_2 b_3 - a_3 b_2)\mathbf{i} + (a_3 b_1 - a_1 b_3)\mathbf{j} + (a_1 b_2 - a_2 b_1)\mathbf{k}$$

The notation $\mathbf{a} \wedge \mathbf{b}$ is sometimes used.

This vector product is often represented in the form of the array shown below, in order to aid memory.

$$\mathbf{a} \times \mathbf{b} \equiv \begin{vmatrix} \mathbf{i} & \mathbf{j} & \mathbf{k} \\ a_1 & a_2 & a_3 \\ b_1 & b_2 & b_3 \end{vmatrix}$$

$$\equiv \begin{vmatrix} a_2 & a_3 \\ b_2 & b_3 \end{vmatrix} \mathbf{i} - \begin{vmatrix} a_1 & a_3 \\ b_1 & b_3 \end{vmatrix} \mathbf{j} + \begin{vmatrix} a_1 & a_2 \\ b_1 & b_2 \end{vmatrix} \mathbf{k}$$

where we introduce the shorthand

$$\begin{vmatrix} a & b \\ c & d \end{vmatrix} = ad - bc$$

NB: This is just a mnemonic device for remembering the formula for the vector product and for calculating it – you don't need to know anything about determinants at this stage.

Problem 11.9

Find the vector product $\mathbf{a} \times \mathbf{b}$ between the vectors $\mathbf{a} = \mathbf{i} - \mathbf{j} + 2\mathbf{k}$ and $\mathbf{b} = 2\mathbf{i} + \mathbf{k}$. What is $\mathbf{b} \times \mathbf{a}$?

Using the component expression given we obtain (check these calculations against the symbolic expressions given above)

$$\mathbf{a} \times \mathbf{b} = \begin{vmatrix} \mathbf{i} & \mathbf{j} & \mathbf{k} \\ 1 & -1 & 2 \\ 2 & 0 & 1 \end{vmatrix}$$

$$= [(-1)(1) - 2(0)]\mathbf{i} - [1 \times 1 - 2 \times 2]\mathbf{j} + [1 \times 0 - (-1)(2)]\mathbf{k}$$

$$= -\mathbf{i} + 3\mathbf{j} + 2\mathbf{k}$$

You can also verify explicitly that

$$\mathbf{b} \times \mathbf{a} = \mathbf{i} - 3\mathbf{j} - 2\mathbf{k} = -\mathbf{a} \times \mathbf{b}$$

From the above definition it can be shown that the vector (or **cross**) product has the following properties:

(i) $\mathbf{a} \times \mathbf{b}$ is a vector
(ii) $\mathbf{a} \times (\mathbf{b} + \mathbf{c}) = \mathbf{a} \times \mathbf{b} + \mathbf{a} \times \mathbf{c}$ and $(\mathbf{a} + \mathbf{b}) \times \mathbf{c} = \mathbf{a} \times \mathbf{c} + \mathbf{b} \times \mathbf{c}$
i.e. the vector product is distributive over vector addition
(iii) $(k\mathbf{a}) \times \mathbf{b} = k(\mathbf{a} \times \mathbf{b}) = \mathbf{a} \times (k\mathbf{b})$ for any scalar k
(iv) If $\mathbf{a} \times \mathbf{b} = 0$ and $\mathbf{a}, \mathbf{b}$ are not zero vectors, then $\mathbf{a}$ and $\mathbf{b}$ are parallel
(v) $\mathbf{b} \times \mathbf{a} = -\mathbf{a} \times \mathbf{b}$ as illustrated by Problem 11.9
(vi) $\mathbf{a} \times \mathbf{a} = 0$, which follows from (v)
(vii) The vector product is not associative, i.e.

$$\mathbf{a} \times (\mathbf{b} \times \mathbf{c}) \neq (\mathbf{a} \times \mathbf{b}) \times \mathbf{c}$$

For the basis vectors **i**, **j**, **k** we find (Reinforcement Exercise 24)

$$\mathbf{i} \times \mathbf{i} = \mathbf{j} \times \mathbf{j} = \mathbf{k} \times \mathbf{k} = 0$$
$$\mathbf{i} \times \mathbf{j} = \mathbf{k}, \quad \mathbf{j} \times \mathbf{k} = \mathbf{i}, \quad \mathbf{k} \times \mathbf{i} = \mathbf{j}$$

Geometrically the vector product can be expressed as:

$$\mathbf{a} \times \mathbf{b} = ab \sin \theta \, \mathbf{n}$$

where **n** is a unit vector perpendicular to both **a** and **b** such that **a**, **b**, **n** form a right-handed set – see Figure 11.15.

Figure 11.15 The vector product is $\mathbf{a} \times \mathbf{b} = ab \sin \theta \, \mathbf{n}$.

The proof of this is an interesting exercise. Take the x-axis to be along **a** and **b** to be in the xy plane (we can always choose axes in this way). Then

$$\mathbf{a} = a\mathbf{i} \quad \mathbf{b} = b_1 \mathbf{i} + b_2 \mathbf{j}$$

so

$$\mathbf{a} \times \mathbf{b} = ab_2 \mathbf{i} \times \mathbf{j} = ab_2 \mathbf{k}$$

But also we have:

$$b_2 = b \sin \theta$$

so

$$\mathbf{a} \times \mathbf{b} = ab \sin \theta \, \mathbf{k}$$

which is equivalent to the above result.

The two vectors **a**, **b** are parallel or opposite and parallel if and only if $\mathbf{a} \times \mathbf{b} = 0$ (then $\theta = 0$ or $180°$).

Exercises on 11.11

1. Prove the properties (i) → (vi)
2. If $\mathbf{a} = \mathbf{i} + \mathbf{k}$ $\mathbf{b} = 2\mathbf{i} - \mathbf{j} + 3\mathbf{k}$ $\mathbf{c} = \mathbf{i} + 2\mathbf{j} + 3\mathbf{k}$, evaluate

  (i) $\mathbf{a} \times \mathbf{b}$ (ii) $\mathbf{a} \times (\mathbf{b} + \mathbf{c})$ (iii) $\mathbf{a} \times \mathbf{c}$ (iv) $\mathbf{b} \times \mathbf{b}$
  (v) $\mathbf{b} \times \mathbf{a}$ (vi) $\mathbf{a} \times (\mathbf{b} \times \mathbf{c})$ (vii) $(\mathbf{a} \times \mathbf{b}) \times \mathbf{c}$

Answers

2. (i) $\mathbf{i} - \mathbf{j} - \mathbf{k}$ (ii) $-\mathbf{i} - 3\mathbf{j} + \mathbf{k}$ (iii) $-2\mathbf{i} - 2\mathbf{j} + 2\mathbf{k}$ (iv) **0**
   (v) $-\mathbf{i} + \mathbf{j} + \mathbf{k}$ (vi) $3\mathbf{i} - 14\mathbf{j} - 3\mathbf{k}$ (vii) $-\mathbf{i} - 4\mathbf{j} + 3\mathbf{k}$

11.12 Vector functions

A **vector function** is a vector which is a function of some variable, say t. We write $\mathbf{f}(t)$ for a general vector function. For example the position of a projectile may be represented by a vector, $\mathbf{r}$ referred to some origin, and this position will vary with time, t. We can represent this by taking $\mathbf{r}$ to be a function of t, $\mathbf{r}(t)$. So $\mathbf{r}(t)$ is a vector function of time t. This is illustrated in Figure 11.16.

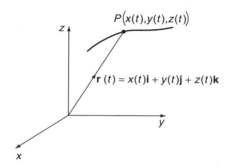

Figure 11.16 Position vector of a vector function.

The set of values of the variable t is called the **domain** of the vector function, while the set of possible vectors $\mathbf{f}(t)$ is called the **codomain** or **range** of the vector function.

A vector for which both the magnitude and direction are constant, i.e. each component is constant, is called a **constant vector**. Examples of constant vectors are $\mathbf{i}, \mathbf{j}, \mathbf{k}$.

It is useful to know the vector functions representing lines and planes. The position vector $\mathbf{r}$ of a general point P lying on a straight line passing through a point with position vector $\mathbf{a}$ is

$$\mathbf{r} = \mathbf{a} + t\mathbf{d}$$

where $\mathbf{d}$ is any constant vector parallel to the line – see Figure 11.17.

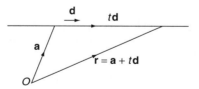

Figure 11.17 Equation of a line.

For the case of a plane, consider Figure 11.18.

Let P be any point in a plane, and let $\mathbf{n}$ be any vector normal to the plane. Let $\mathbf{a}$ be the position vector of a fixed point in the plane. If $\mathbf{r}$ is the position vector of P then $\mathbf{r} - \mathbf{a}$ will be a vector in the plane. As such it will be perpendicular to $\mathbf{n}$, from which we have

$$(\mathbf{r} - \mathbf{a}) \cdot \mathbf{n} = 0$$

So

$$\mathbf{r} \cdot \mathbf{n} = \mathbf{a} \cdot \mathbf{n} = \rho \text{ say}$$

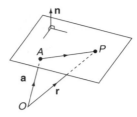

Figure 11.18 Vector equation of a plane.

is the general equation for a plane. If $\mathbf{n} = \alpha\mathbf{i} + \beta\mathbf{j} + \gamma\mathbf{k}$ then in terms of coordinates this is
$$\alpha x + \beta y + \gamma z = \rho$$

Exercise on 11.12

Sketch the path described by the position vector

$$\mathbf{f}(t) = \cos t\,\mathbf{i} + \sin t\,\mathbf{j} + \mathbf{k}$$

as t varies.

Answer

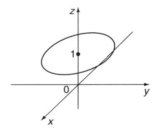

11.13 Differentiation of vector functions

The idea of a vector function $\mathbf{f}(t)$ of t raises the question of its rate of change as t varies. In the case of the projectile of Section 11.12, this rate of change would represent the velocity of the projectile at the time t. If you look back at the definition of the derivative on page 231, the following definition will probably now make sense to you. We are essentially applying the definition to each component of the vector function. The **derivative of a vector function $\mathbf{f}(t)$, with respect to t** is defined by

$$\frac{d\mathbf{f}(t)}{dt} = \lim_{h \to 0}\left(\frac{\mathbf{f}(t+h) - \mathbf{f}(t)}{h}\right)$$

Problem 11.10
Show that the derivative of a constant vector is zero.

Let $\mathbf{f}(t) = \mathbf{c}$ be a constant vector. Then $\mathbf{f}(t + h) = \mathbf{c}$, and so from the above definition

$$\frac{d\mathbf{f}}{dt} = \lim_{h \to 0} \left(\frac{\mathbf{f}(t+h) - \mathbf{f}(t)}{h} \right) = \lim_{h \to 0} \left(\frac{\mathbf{c} - \mathbf{c}}{h} \right) = 0$$

The properties of differentiation of a vector look very much like those of ordinary differentiation, although the notation may be a bit unnerving at first. We have three types of product to contend with of course – multiplication by a scalar, scalar product and vector product. On the other hand we don't have to worry about a quotient rule, since we have not defined division of vectors. The rules are as follows.

If $s(t)$ is a scalar function and $\mathbf{f}(t)$ a vector function, then

(i) $\quad \dfrac{d}{dt}(s(t)\mathbf{f}(t)) = \dfrac{ds(t)}{dt}\mathbf{f}(t) + s(t)\dfrac{d\mathbf{f}(t)}{dt}$

If $\mathbf{g}(t)$ is also a vector function, then

(ii) $\quad \dfrac{d}{dt}(\mathbf{f}(t) + \mathbf{g}(t)) = \dfrac{d\mathbf{f}(t)}{dt} + \dfrac{d\mathbf{g}(t)}{dt}$

(iii) $\quad \dfrac{d}{dt}(\mathbf{f} \cdot \mathbf{g}) = \mathbf{f} \cdot \dfrac{d\mathbf{g}}{dt} + \dfrac{d\mathbf{f}}{dt} \cdot \mathbf{g}$

(iv) $\quad \dfrac{d}{dt}(\mathbf{f} \times \mathbf{g}) = \mathbf{f} \times \dfrac{d\mathbf{g}}{dt} + \dfrac{d\mathbf{f}}{dt} \times \mathbf{g}$ where the order of vectors must be preserved

(i) and (ii) show that differentiation of the component form is simple:

$$\frac{d}{dt}(f_1(t)\mathbf{i} + f_2(t)\mathbf{j} + f_3(t)\mathbf{k}) = \frac{df_1(t)}{dt}\mathbf{i} + \frac{df_2(t)}{dt}\mathbf{j} + \frac{df_3(t)}{dt}\mathbf{k}$$

Problem 11.11

Differentiate the following vector functions with respect to t

(i) $\quad \cos 4t\,\mathbf{i} + 3\sin 4t\,\mathbf{j} + e^t\mathbf{k}$
(ii) $\quad t\cos t\,\mathbf{i} + e^{-t}\sin t\,\mathbf{j} + t^2\mathbf{k}$

(i) $\quad \dfrac{d}{dt}(\cos 4t\,\mathbf{i} + 3\sin 4t\,\mathbf{j} + e^t\mathbf{k})$

$\qquad = \dfrac{d}{dt}(\cos 4t)\mathbf{i} + \dfrac{d}{dt}(3\sin 4t)\mathbf{j} + \dfrac{d}{dt}(e^t)\mathbf{k}$

$\qquad = -4\sin 4t\,\mathbf{i} + 12\cos 4t\,\mathbf{j} + e^t\mathbf{k}$

(ii) $\quad \dfrac{d}{dt}(t\cos t\,\mathbf{i} + e^{-t}\sin t\,\mathbf{j} + t^2\mathbf{k})$

$\qquad = \dfrac{d}{dt}(t\cos t)\mathbf{i} + \dfrac{d}{dt}(e^{-t}\sin t)\mathbf{j} + \dfrac{d}{dt}(t^2)\mathbf{k}$

$\qquad = (\cos t - t\sin t)\mathbf{i} + (e^{-t}\cos t - e^{-t}\sin t)\mathbf{j} + 2t\mathbf{k}$

Second and higher derivatives may be obtained in the obvious way by repeated differentiation. Thus, for example:

$$\frac{d^2\mathbf{f}(t)}{dt^2} = \frac{d^2 f_1}{dt^2}\mathbf{i} + \frac{d^2 f_2}{dt^2}\mathbf{j} + \frac{d^2 f_3}{dt^2}\mathbf{k}$$

Problem 11.12

Evaluate $\dfrac{d^2\mathbf{f}}{dt^2}$ **and** $\dfrac{d^3\mathbf{f}}{dt^3}$ **for** $\mathbf{f} = t\mathbf{i} + e^{-t}\mathbf{j} + \cos t\,\mathbf{k}$

$$\frac{d\mathbf{f}}{dt} = \mathbf{i} - e^{-t}\mathbf{j} - \sin t\,\mathbf{k}$$

$$\frac{d^2\mathbf{f}}{dt^2} = e^{-t} - \cos t\,\mathbf{k}$$

$$\frac{d^3\mathbf{f}}{dt^3} = -e^{-t}\mathbf{j} + \sin t\,\mathbf{k}$$

When we introduced ordinary differentiation we referred it to the gradient or slope of a curve. We can do the same for differentiation of vector functions – but it is now a little more complicated. It is perhaps best to appeal to the example considered earlier of a vector function $\mathbf{r}(t)$ describing the position of a particle at time t. Thus, let $\mathbf{r}(t)$ be the position vector of a moving particle P. As t varies the particle moves along a curve in space – see Figure 11.19.

Figure 11.19 Definition of derivative of a vector.

Now for two points on the curve $\mathbf{r}(t)$, $\mathbf{r}(t+h)$, close to each other we have:

$$\frac{\mathbf{r}(t+h) - \mathbf{r}(t)}{h} = \frac{\overrightarrow{OP'} - \overrightarrow{OP}}{h} = \frac{\overrightarrow{PP'}}{h}$$

As $h \to 0$, $P' \to P$ and the vector $\overrightarrow{PP'}$ becomes tangential to the curve. Also the magnitude $\dfrac{|\overrightarrow{PP'}|}{h}$ is the average speed of the particle over the interval PP' and so as $h \to 0$ this becomes the velocity of the particle. Thus:

$$\lim_{h \to 0}\left(\frac{\mathbf{r}(t+h) - \mathbf{r}(t)}{h}\right) = \frac{d\mathbf{r}}{dt}$$

is a vector tangential to the curve, pointing in the direction of motion of the particle and with magnitude equal to the magnitude of the velocity of the particle. That is, $\dfrac{d\mathbf{r}}{dt}$ is the vector **velocity** of the particle:

$$\mathbf{v} = \frac{d\mathbf{r}}{dt}$$

Similarly $\mathbf{a} = \dfrac{d^2\mathbf{r}}{dt^2}$ is the **acceleration** of the particle.

Problem 11.13

Find the velocity, acceleration and kinetic energy of the particle mass m whose position vector at time t is:

$$\mathbf{r}(t) = 2\cos\omega t\,\mathbf{i} + 2\sin\omega t\,\mathbf{j}$$

Verify that for such a particle $\mathbf{v}\cdot\mathbf{r} = 0$.

The velocity is

$$\mathbf{v} = \frac{d\mathbf{r}}{dt} = -2\omega\sin\omega t\,\mathbf{i} + 2\omega\cos\omega t\,\mathbf{j}$$

From this we find

$$\mathbf{v}\cdot\mathbf{r} = 4\omega(-\sin\omega t\,\mathbf{i} + \cos\omega t\,\mathbf{j})\cdot(\cos\omega t\,\mathbf{i} + \sin\omega t\,\mathbf{j})$$
$$= 4\omega(-\sin\omega t\cos\omega t + \cos\omega t\sin\omega t) = 0$$

The acceleration is

$$\mathbf{a} = \frac{d^2\mathbf{r}}{dt^2} = \frac{d\mathbf{v}}{dt} = -2\omega^2\cos\omega t\,\mathbf{i} - 2\omega^2\sin\omega t\,\mathbf{j}$$
$$= -\omega^2\mathbf{r}$$

The kinetic energy is

$$\frac{1}{2}m\mathbf{v}^2 = \frac{1}{2}m((-2\omega\sin\omega t)^2 + (2\omega\cos\omega t)^2)$$
$$= \frac{1}{2}m\cdot 4\omega^2$$
$$= 2m\omega^2$$

Exercises on 11.13

1. Show that if $\mathbf{e}$ is a vector of constant magnitude, then

$$\mathbf{e}\cdot\frac{d\mathbf{e}}{dt} = 0$$

2. Verify $\dfrac{d}{dt}(\mathbf{f} \times \mathbf{g}) = \dfrac{d\mathbf{f}}{dt} \times \mathbf{g} + \mathbf{f} \times \dfrac{d\mathbf{g}}{dt}$

for the two vectors:

$$\mathbf{f} = t\mathbf{i} - 2t^2\mathbf{j} + e^t\mathbf{k} \qquad \mathbf{g} = \cos t\,\mathbf{i} - 2t\mathbf{j}$$

11.14 Reinforcement

1. The free vectors **a**, **b**, **c** are shown as arrows in the figure below. Sketch arrows representing the vectors

   (i) $\mathbf{a} + \mathbf{b}$ (ii) $\mathbf{a} - \mathbf{c}$ (iii) $\mathbf{a} + 2\mathbf{b}$ (iv) $\dfrac{1}{2}(\mathbf{a} + \mathbf{b})$ (v) $\dfrac{3}{2}\mathbf{a} - \mathbf{b}$ (vi) $\mathbf{a} - 2\mathbf{c}$

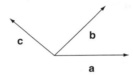

2. Determine the unknown vector **x** in terms of **a**, **b**, **c** in each case.
   (i) $3\mathbf{x} - \mathbf{a} + 2\mathbf{b} = 2\mathbf{a} - 4\mathbf{b} - \mathbf{x}$
   (ii) $2\mathbf{x} - 2\mathbf{a} + 3\mathbf{c} = 3\mathbf{c} - 2\mathbf{b} - 2\mathbf{x}$
   (iii) $4\mathbf{x} - 3\mathbf{a} + 2\mathbf{b} - \mathbf{c} = 4\mathbf{b} - 2\mathbf{x} - 3\mathbf{a} + 2\mathbf{c}$

3. Write down the coordinates of the six points which lie on the coordinate axes and are one unit from the origin.

4. Plot on a diagram the positions of

   (0, 0, 2);   (0, −1, 0);   (3, 0, 0);   (1, 1, 0);   (−2, 1, 0)
   (2, 0, 1);   (0, 1, 1);   (1, 1, 1);   (−2, 1, −1); (1, −2, 2)

5. Calculate the distance from the origin of each of the points:

   (i) (1, 1, 0)   (ii) (2, 0, −1)   (iii) (−2, 0, 1)   (iv) (−2, 3, 1)
   (v) (0, 4, 3)   (vi) ($\sqrt{2}$, 1, 1)

6. Find the distance between each of the following pairs of points:

   (i) (2, 0, 0); (1, 2, 3)
   (ii) (3, 0, −1); (1, 2, 1)
   (iii) (5, 4, 2); (0, 3, −1)
   (iv) (−2, −1, 1); (0, 2, −2)
   (v) $(a, -a, 0); (-a, 0, a)$

7. A straight rod is held with one end in the corner of a room. If it makes angles of 60° and 45° with the lines of intersection of the floor and the walls, find the angle that it makes with the vertical.

8. Find the direction cosines of the vectors OP for each of the points P in Q5.

9. Determine the acute angles between the position vectors defined by the pairs of points in Q6.

10. Express in terms of **i, j, k** the position vectors with endpoints

    (i) (2, −1, 2) (ii) (1, 2, 3) (iii) (−1, −2, −3) (iv) (2, −3, 4) (v) (2u, 3v, 4w)

11. Calculate the magnitude and direction for each of the position vectors:

    (i) **a** = 3**i**   (ii) **b** = **i** + **j**   (iii) **c** = 3**i** − 3**j**
    (iv) **d** = √3**i** + **j** (v) **e** = **i** + 2**j**   (vi) **f** = **i** + **j** + **k**
    (vii) **g** = −**i** − **j** + 2**k** (viii) **h** = 2**i** + 2**j** + 2**k**

    Find unit vectors in the direction of each vector.

12. For the vectors in Q11 find

    (i) **a** + **b**   (ii) **c** − **g**   (iii) **a** + **b** + **h**
    (iv) 2**a** − 3**b** − 3**f**   (v) 4**a** + 2**b** − 3**c** + **d** + **e** − **f** + **g** − 3**h**

13. For the vectors in Q11 determine the vector **x** such that:

    (i) 2**x** − 3**a** + **b** = 0   (ii) **a** + **b** − 2**c** + 3**x** = **d**
    (iii) 2**x** − 2**f** = 3**x** + 2**g**   (iv) 3**x** + 2**e** − **f** = **g** + **h** − **x**

14. The following describe position vectors with respect to the basis vectors **i, j, k**. Write down the component forms of the vectors.

    (i) **r** is in the xy plane, has length 3 units and bisects the angle between **i** and **j**.
    (ii) **r** is in the yz plane, has length 2 units and makes angle 30° with **j** and 60° with **k**.
    (iii) **r** is the diagonal of the unit cube in the first octant with sides **i, j, k**.

15. Vectors **a, b, c** have magnitudes 1, 2, 3 respectively and the angle between **a** and **b** is 60°, between **b** and **c** 45° and between **a** and **c** is 105°. Find all possible scalar products between **a, b, c**.

16. Prove the results

    $$\mathbf{i} \cdot \mathbf{i} = \mathbf{j} \cdot \mathbf{j} = \mathbf{k} \cdot \mathbf{k} = 1$$
    $$\mathbf{i} \cdot \mathbf{j} = \mathbf{i} \cdot \mathbf{k} = \mathbf{j} \cdot \mathbf{k} = 0$$

17. **a** and **b** are perpendicular to each other with magnitudes 2 and 4 respectively. If **x** = 2**a** − **b** and **y** = 3**a** − 4**b**, find **x** · **y**. If **z** = λ**a** − **b** determine λ such that **y** and **z** are perpendicular.

345

18. Find the scalar products of the following pairs of vectors and state if any of the three pairs are perpendicular

    (i) $\mathbf{i} - \mathbf{j}$, $3\mathbf{i} + 4\mathbf{j} + 5\mathbf{k}$
    (ii) $4\mathbf{i} + \mathbf{j} - 3\mathbf{k}$, $-\mathbf{i} + 3\mathbf{j} - 7\mathbf{k}$
    (iii) $3\mathbf{i} + \mathbf{j} + 4\mathbf{k}$, $2\mathbf{i} - 2\mathbf{j} - \mathbf{k}$

19. Find the angles between the pairs of vectors

    (i) $\mathbf{a} = -\mathbf{j} + \mathbf{k}$ $\mathbf{b} = 3\mathbf{i} + 4\mathbf{j} + 5\mathbf{k}$
    (ii) $\mathbf{a} = \mathbf{i} - \mathbf{j} - 2\mathbf{k}$ $\mathbf{b} = 2\mathbf{i} + \mathbf{j} + 3\mathbf{k}$
    (iii) $\mathbf{a} = 2\mathbf{i} + \mathbf{j} - \mathbf{k}$ $\mathbf{b} = \mathbf{i} + 2\mathbf{j} + \mathbf{k}$

20. Suppose the axes $Oxyz$ are such that Ox points East, Oy points North, Oz points vertically upwards. Evaluate the scalar products of the vectors **a** and **b** in each of the following cases:

    (i) **a** is of magnitude 3 and points SE. **b** has length 2 and points E.

    (ii) **a** is a unit vector pointing NE. **b** is of magnitude 2 and points vertically upwards.

    (iii) **a** is of unit magnitude and points NE. **b** is of magnitude 2 and points W.

21. If $\mathbf{a} = \mathbf{i} - \mathbf{j}$ $\mathbf{b} = -\mathbf{j} + 2\mathbf{k}$, show that

    $$(\mathbf{a} + \mathbf{b}) \cdot (\mathbf{a} - 2\mathbf{b}) = -9$$

22. Show that $\mathbf{i} + \mathbf{j} + \mathbf{k}$ and $a^2\mathbf{i} - 2a\mathbf{j} + \mathbf{k}$ are perpendicular if and only if $a = 1$.

23. Determine the vector product $\mathbf{a} \times \mathbf{b}$ for each of the pairs of vectors in Q18 and Q19. In each case find unit vectors perpendicular to the two vectors.

24. Prove the results

    $$\mathbf{i} \times \mathbf{j} = \mathbf{k} \quad \mathbf{j} \times \mathbf{k} = \mathbf{i} \quad \mathbf{k} \times \mathbf{i} = \mathbf{j}$$
    $$\mathbf{i} \times \mathbf{i} = \mathbf{j} \times \mathbf{j} = \mathbf{k} \times \mathbf{k} = 0$$

25. Find the vector products of the vectors **a**, **b**

    (i) $\mathbf{a} = 3\mathbf{i} + 7\mathbf{j} + 2\mathbf{k}$ $\mathbf{b} = \mathbf{i} + 3\mathbf{j} + \mathbf{k}$
    (ii) $\mathbf{a} = \mathbf{i} - 3\mathbf{j}$ $\mathbf{b} = -2\mathbf{i} + 5\mathbf{j}$
    (iii) $\mathbf{a} = 8\mathbf{i} + 8\mathbf{j} - \mathbf{k}$ $\mathbf{b} = 5\mathbf{i} + 5\mathbf{j} + 2\mathbf{k}$

26. If $\mathbf{a} = 3\mathbf{i} - 2\mathbf{j} + \mathbf{k}$, $\mathbf{b} = \mathbf{i} + \mathbf{j} + \mathbf{k}$ and $\mathbf{c} = 2\mathbf{i} + \mathbf{j} - 3\mathbf{k}$ evaluate, using any short-cuts you can:

    (i) the **triple scalar products**
       $\mathbf{a} \cdot \mathbf{b} \times \mathbf{c}$ $\mathbf{a} \cdot \mathbf{c} \times \mathbf{b}$ $\mathbf{b} \cdot \mathbf{a} \times \mathbf{b}$

    (ii) the **triple vector products**
       $\mathbf{a} \times (\mathbf{b} \times \mathbf{c})$ $(\mathbf{a} \times \mathbf{b}) \times \mathbf{c}$ $\mathbf{b} \times (\mathbf{b} \times \mathbf{a})$

27. For the pairs of vector functions $\mathbf{f}(t)$, $\mathbf{g}(t)$ verify the product rules:

$$\frac{d}{dt}(\mathbf{f} \cdot \mathbf{g}) = \mathbf{f} \cdot \frac{d\mathbf{g}}{dt} + \frac{d\mathbf{f}}{dt} \cdot \mathbf{g}$$

$$\frac{d}{dt}(\mathbf{f} \times \mathbf{g}) = \frac{d\mathbf{f}}{dt} \times \mathbf{g} + \mathbf{f} \times \frac{d\mathbf{g}}{dt}$$

(i) $\mathbf{f} = t^2\mathbf{i} + e^t\mathbf{j} + \sin t\,\mathbf{k}$ $\quad \mathbf{g} = t\mathbf{i} + t^2\mathbf{j} + 2\mathbf{k}$

(ii) $\mathbf{f} = \cos 2t\,\mathbf{i} + \sin 2t\,\mathbf{j} + 3\mathbf{k}$ $\quad \mathbf{g} = t\mathbf{i} + 2\mathbf{j} - 3t^2\mathbf{k}$

11.15 Applications

1. Vector methods can be used in geometrical applications. The following provide some examples for you to try.

   (i) The point P divides the line segment AB in the ratio $m : n$. If $\mathbf{a}$, $\mathbf{b}$ are respectively the position vectors of A and B with respect to an origin O, determine the position vector of the point P relative to O.

   (ii) Use the scalar product to prove Pythagoras' theorem

   (iii) Use the scalar product to prove the cosine rule

   (iv) Adjacent sides of a triangle represent vectors $\mathbf{a}$ and $\mathbf{b}$. Show that the area of the triangle is $\frac{1}{2}|\mathbf{a} \times \mathbf{b}|$.

2. There are countless examples of use of vectors in mechanics of course. The following give some typical examples.

   (i) Assuming the triangle rule for addition of vectors show that any set of forces represented by arrows forming a closed polygon with all sides like directed is in equilibrium (i.e. sums to zero).

   (ii) The **moment** or **torque** of a force $\mathbf{F}$ acting at the point P about the origin O is defined to be $\mathbf{M} = \mathbf{r} \times \mathbf{F}$ where $\mathbf{r}$ is the position vector of P. Describe the magnitude and direction of $\mathbf{M}$.

   Determine the torque of the force $\mathbf{F} = \mathbf{i} + 2\mathbf{j} + 3\mathbf{k}$ applied at the point $(-1, -1, 2)$ about the points

   (a) $(0, 0, 0)$ $\qquad$ (b) $(3, 2, -1)$ $\qquad$ (c) $(1, 1, 1)$

   (iii) The position vector of a particle at time t is given by

   $$\mathbf{r} = (t+1)\mathbf{i} - t^2\mathbf{j} + e^t\mathbf{k}$$

   Determine the velocity and acceleration at time t.

11.16 Answers to reinforcement exercises

1.

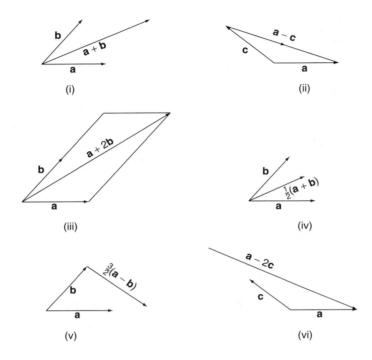

2. (i) $x = \dfrac{3}{4}a - \dfrac{3}{2}b$ (ii) $x = \dfrac{1}{2}a - \dfrac{1}{2}b$ (iii) $x = \dfrac{1}{3}b + \dfrac{1}{2}c$

3. $(\pm 1, 0, 0); (0, \pm 1, 0); (0, 0, \pm 1)$

4.

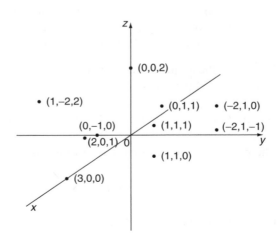

5. (i) $\sqrt{2}$ (ii) $\sqrt{5}$ (iii) $\sqrt{5}$ (iv) $\sqrt{14}$ (v) 5 (vi) 2
6. (i) $\sqrt{14}$ (ii) $2\sqrt{3}$ (iii) $\sqrt{35}$ (iv) $\sqrt{22}$ (v) $a\sqrt{6}$
7. 60°
8. (i) $\left(\frac{1}{\sqrt{2}}, \frac{1}{\sqrt{2}}, 0\right)$ (ii) $\left(\frac{2}{\sqrt{5}}, 0, -\frac{1}{\sqrt{5}}\right)$ (iii) $\left(-\frac{2}{\sqrt{5}}, 0, \frac{1}{\sqrt{5}}\right)$
   (iv) $\left(-\frac{2}{\sqrt{14}}, \frac{3}{\sqrt{14}}, \frac{1}{\sqrt{14}}\right)$ (v) $\left(0, \frac{4}{5}, \frac{3}{5}\right)$ (vi) $\left(\frac{1}{\sqrt{2}}, \frac{1}{2}, \frac{1}{2}\right)$
9. (i) 74.5° (ii) 75.04° (iii) 61.87° (iv) 54.74° (v) 60°
10. (i) $2\mathbf{i} - \mathbf{j} + 2\mathbf{k}$ (ii) $\mathbf{i} + 2\mathbf{j} + 3\mathbf{k}$ (iii) $-\mathbf{i} - 2\mathbf{j} - 3\mathbf{k}$ (iv) $2\mathbf{i} - 3\mathbf{j} + 4\mathbf{k}$
    (v) $2u\mathbf{i} + 3v\mathbf{j} + 4w\mathbf{k}$
11. (i) $|\mathbf{a}| = 3$, $\hat{\mathbf{a}} = \mathbf{i}$ (ii) $|\mathbf{b}| = \sqrt{2}$, $\hat{\mathbf{b}} = \frac{1}{\sqrt{2}}(\mathbf{i} + \mathbf{j})$
    (iii) $|\mathbf{c}| = 3\sqrt{2}$, $\hat{\mathbf{c}} = \frac{1}{\sqrt{2}}(\mathbf{i} - \mathbf{j})$ (iv) $|\mathbf{d}| = 2$, $\hat{\mathbf{d}} = \frac{\sqrt{3}}{2}\mathbf{i} + \frac{1}{2}\mathbf{j}$
    (v) $|\mathbf{e}| = 3$, $\hat{\mathbf{e}} = \frac{1}{\sqrt{5}}\mathbf{i} + \frac{2}{\sqrt{5}}\mathbf{j}$ (vi) $|\mathbf{f}| = \sqrt{3}$, $\hat{\mathbf{f}} = \frac{1}{\sqrt{3}}\mathbf{i} + \frac{1}{\sqrt{3}}\mathbf{j} + \frac{1}{\sqrt{3}}\mathbf{k}$
    (vii) $|\mathbf{g}| = \sqrt{6}$, $\hat{\mathbf{g}} = -\frac{1}{\sqrt{6}}\mathbf{i} - \frac{1}{\sqrt{6}}\mathbf{j} + \frac{2}{\sqrt{6}}\mathbf{k}$
    (viii) $|\mathbf{h}| = 2\sqrt{3}$, $\hat{\mathbf{h}} = \frac{1}{\sqrt{3}}\mathbf{i} + \frac{1}{\sqrt{3}}\mathbf{j} + \frac{1}{\sqrt{3}}\mathbf{k}$

    The coefficients of the unit vectors give the direction cosines that define the directions of the vectors.
12. (i) $4\mathbf{i} + \mathbf{j}$ (ii) $4\mathbf{i} - 2\mathbf{j} - 2\mathbf{k}$ (iii) $6\mathbf{i} + 3\mathbf{j} + 2\mathbf{k}$
    (iv) $-6\mathbf{j} - 3\mathbf{k}$ (v) $(\sqrt{3} - 2)\mathbf{i} + 6\mathbf{j} - 5\mathbf{k}$
13. (i) $\frac{1}{2}(8\mathbf{i} - \mathbf{j})$ (ii) $\frac{2+\sqrt{3}}{3}\mathbf{i} - 2\mathbf{j}$ (iii) $-6\mathbf{k}$ (iv) $-\frac{1}{2}\mathbf{j} + \frac{5}{4}\mathbf{k}$
14. (i) $\frac{3}{\sqrt{2}}\mathbf{i} + \frac{3}{\sqrt{2}}\mathbf{j}$ (ii) $\sqrt{3}\mathbf{j} + \mathbf{k}$ (iii) $\mathbf{i} + \mathbf{j} + \mathbf{k}$
15. Between **a** and **b** the scalar product is 1
    Between **b** and **c** the scalar product is $3\sqrt{2}$
    Between **a** and **c** the scalar products is -0.78 to two decimal places
17. $\mathbf{x} \cdot \mathbf{y} = 88$, $\lambda = -\frac{16}{3}$
18. (i) -1 (ii) 20 (iii) 0, so these are orthogonal
19. (i) 84.26° (ii) 123.06° (iii) 60°
20. (i) $3\sqrt{2}$ (ii) 0 (iii) $-\sqrt{2}$

23.
(i) $-5\mathbf{i} - 5\mathbf{j} + 7\mathbf{k}$, $\dfrac{1}{3\sqrt{11}}(-5\mathbf{i} - 5\mathbf{j} + 7\mathbf{k})$

(ii) $2\mathbf{i} + 31\mathbf{j} + 13\mathbf{k}$, $\dfrac{1}{\sqrt{1134}}(2\mathbf{i} + 31\mathbf{j} + 13\mathbf{k})$

(iii) $7\mathbf{i} + 11\mathbf{j} - 8\mathbf{k}$, $\dfrac{1}{\sqrt{234}}(7\mathbf{i} + 11\mathbf{j} - 8\mathbf{k})$

(iv) $-9\mathbf{i} + 3\mathbf{j} + 3\mathbf{k}$, $\dfrac{1}{\sqrt{11}}(-3\mathbf{i} + \mathbf{j} + \mathbf{k})$

(v) $-\mathbf{i} - 7\mathbf{j} + 3\mathbf{k}$, $\dfrac{1}{\sqrt{59}}(-\mathbf{i} - 7\mathbf{j} + 3\mathbf{k})$

(vi) $3\mathbf{i} - 3\mathbf{j} + 3\mathbf{k}$, $\dfrac{1}{\sqrt{3}}(\mathbf{i} - \mathbf{j} + \mathbf{k})$

25. (i) $\mathbf{i} - \mathbf{j} + 2\mathbf{k}$ (ii) $-\mathbf{k}$ (iii) $21\mathbf{i} - 21\mathbf{j}$

26. (i) $\mathbf{a} \cdot \mathbf{b} \times \mathbf{c} = -23$, $\mathbf{a} \cdot \mathbf{c} \times \mathbf{b} = 23$, $\mathbf{b} \cdot \mathbf{a} \times \mathbf{b} = 0$

(ii) $\mathbf{a} \times (\mathbf{b} \times \mathbf{c}) = -3\mathbf{i} - \mathbf{j} + 7\mathbf{k}$, $(\mathbf{a} \times \mathbf{b}) \times \mathbf{c} = \mathbf{i} + \mathbf{j} + \mathbf{k}$,

$\mathbf{b} \times (\mathbf{b} \times \mathbf{a}) = -7\mathbf{i} + 8\mathbf{j} - \mathbf{k}$

27. (i) $\dfrac{d}{dt}(\mathbf{f} \cdot \mathbf{g}) = 3t^2 + (t^2 + 2t)e^t + 2\cos t$

$\dfrac{d}{dt}(\mathbf{f} \times \mathbf{g}) = (2e^t - 2t\sin t - t^2 \cos t)\mathbf{i} - (4t - \sin t - t\cos t)\mathbf{j}$
$$+ (4t^3 - e^t - te^t)\mathbf{k}$$

(ii) $\dfrac{d}{dt}(\mathbf{f} \cdot \mathbf{g}) = 5\cos 2t - 2t\sin 2t - 18t$

$\dfrac{d}{dt}(\mathbf{f} \times \mathbf{g}) = -(6t\sin 2t + 6t^2 \cos 2t)\mathbf{i} + (3 - 6t^2 \sin 2t + 6t\cos 2t)\mathbf{j}$
$$- (5\sin 2t + 2t\cos 2t)\mathbf{k}$$

12

Complex Numbers

Complex numbers extend the notion of ordinary 'real numbers' to include a new kind of 'number', $j = \sqrt{-1}$. Such 'numbers' can be combined by the usual rules of algebra, except that we replace j^2 by -1 wherever it occurs. Complex numbers do not serve the usual role of numbers – we don't measure physical quantities with them – but they provide a very useful shorthand tool for dealing with certain combinations of real numbers. There is not much that is conceptually new in this chapter, but we have to become more practised at using basic, elementary mathematics.

Prerequisites

It will be helpful if you know something about:

- solving quadratic equations (65 ◄)
- elementary algebra (Chapter 2 ◄)
- polar coordinates (206 ◄)
- elementary trig, including compound angle formulae (187 ◄)
- the exponential function (124 ◄)

Objectives

In this chapter you will find:

- definition and representation of a complex number
- addition, subtraction, multiplication and division of complex numbers
- the Argand plane
- modulus and argument of a complex number
- polar form of a complex number
- multiplication and division in polar form
- the exponential form of a complex number
- powers and roots – De Moivre's theorem

Motivation

You may need the material of this chapter for:

- solving polynomial equations
- solving differential equations
- describing voltages, currents, etc. in AC electricity theory

12.1 What are complex numbers?

Numbers such as integers, rational numbers, etc. that we have been using so far are called **real numbers** – they can be used to count with, or measure distances, time, etc. From the rules of signs for combining negative numbers (11 ◄) we know that if we square two such real numbers, the result will always be positive:

$$2^2 = +4 \quad (-2)^2 = +4$$

An obvious question is, is there **any** sort of mathematical object that results in a negative number when squared? Why should such an object be of interest anyway? In real life we don't need such quantities – no one ever measured the length of a line to be, for example $\sqrt{-1}$ metres. True, we don't need such 'numbers' for measuring and representing real physical quantities. However, it does turn out that such 'numbers' are very useful as a mathematical tool in representing physical objects which have two parameters – for example the amplitude and phase of a signal waveform may be conveniently combined in 'complex' form. Also, such 'numbers' provide nice tools for calculational purposes – as for example in differential equations. So, let's accept that it's useful to look at the properties of such 'numbers' as $\sqrt{-1}$ and develop the tools necessary to use them.

Start by considering the quadratic equation

$$x^2 + 1 = 0 \quad \text{or} \quad x^2 = -1$$

This has no 'real' solution, but by introducing the symbol

$$j = \sqrt{-1} \quad \text{(alternative notation } i\text{)}$$

we can write

$$x = \pm j$$

The object j is sometimes called 'imaginary', and is an example of a 'complex number'. There is, however, nothing imaginary about it. While it certainly does not represent a 'quantity' in the normal sense that a 'real number' such as 2 does, it is still a perfectly proper symbol that serves a very useful purpose in mathematics. In particular, having defined it as above, it now allows us to write down the 'solution' of any quadratic equation.

Problem 12.1

Write the solution of the equation $x^2 + 2x + 2 = 0$, as given by the formula (66 ◄), in terms of $j = \sqrt{-1}$.

By the formula we have

$$x = \frac{-2 \pm \sqrt{-4}}{2} = \frac{-2 \pm \sqrt{-1}}{2} = -1 \pm j$$

Any number of this form

$$z = a + jb \quad j = \sqrt{-1}$$

where a, b are real numbers, is called a **complex number**. We will usually write z rather than x for such a number since z always signals that we are talking about complex numbers.

a is called the **real part** of z, denoted $\text{Re}\,z$ – it is a real number

b (**not** jb) is called the **imaginary part**, denoted $\text{Im}\,z$ – also a real number

Real numbers are special cases of complex numbers with zero imaginary part.

Exercise on 12.1

Solve the following equations in $a + jb$ form:

(i) $x^2 + 4 = 0$ (ii) $x^2 + x + 1 = 0$ (iii) $x^2 + 6x + 11 = 0$

(iv) $x^3 - 1 = 0$

Answer

(i) $\pm 2j$ (ii) $-\dfrac{1}{2} \pm \dfrac{\sqrt{3}}{2}j$ (iii) $-3 \pm \sqrt{2}j$ (iv) $1, -\dfrac{1}{2} \pm \dfrac{\sqrt{3}}{2}j$

12.2 The algebra of complex numbers

Complex numbers can be manipulated just like real numbers but using the property $j^2 = -1$ whenever appropriate. Many of the definitions and rules for doing this are simply common sense, and here we just summarise the main definitions.

Equality of complex numbers: $a + jb = c + jd$ means that $a = c$ and $b = d$.

To perform **addition** and **subtraction** of complex numbers we combine real parts together and imaginary parts separately:

$$(a + jb) + (c + jd) = (a + c) + j(b + d)$$
$$(a + jb) - (c + jd) = (a - c) + j(b - d)$$

Also note that $k(a + jb) = ka + jkb$ for any real number k.

At this point you may be noticing the similarity to our work on vectors in the previous chapter.

Problem 12.2

If $z = 1 + 2j$ and $w = 4 - j$ evaluate $2z - 3w$.

We have $2z - 3w = 2(1 + 2j) - 3(4 - j) = 2 + 4j - 12 + 3j = -10 + 7j$.

To **multiply** two complex numbers simply multiply out the brackets by ordinary algebra, use $j^2 = -1$ and gather terms:

$$(a + jb)(c + jd) = ac + ajd + jbc + j^2 bd = ac - bd + j(bc + ad)$$

353

Problem 12.3
Multiply $(3-j)$ and $(2+3j)$.

$$(3-j)(2+3j) = 6 + 9j - 2j - 3j^2 = 6 + 7j - 3(-1) = 9 + 7j$$

Given a complex number $z = a + jb$ there is a very useful operation we can perform, that converts the number to its 'conjugate'. Thus, the **complex conjugate** of a complex number $z = a + ib$ is defined as

$$\bar{z} = a + (-j)b \quad \text{(alternative notation } z^*\text{)}$$

NB: In general the complex conjugate of a complex expression is obtained by changing the sign of j everywhere it occurs. For examples of this see Reinforcement Exercise 5, and 9.

Problem 12.4
Show that $z\bar{z}$ is a positive real number which is only zero if $a = b = 0$, i.e. if $z = 0$. Also show that $z + \bar{z}$ is always a real number.

This is the crux of the importance of $\bar{z}$. We have

$$\bar{z}z = (a - jb)(a + jb) = a^2 - (jb)^2 = a^2 - j^2b^2 = a^2 + b^2$$

on using the difference of two squares (42 ◀). Now a^2 and b^2 are both non-negative numbers and so $z\bar{z}$ is clearly positive. Also, the result can only add up to zero if both a and b are zero. So $\bar{z}z = 0$ only if $a = 0$ and $b = 0$, i.e. if $z = 0$. We will see that these properties of $z\bar{z}$ are very important when we come to dividing complex numbers.

Also, we have directly

$$z + \bar{z} = a + jb + a - jb = 2a$$

which is a real number.

The complex conjugate is also useful for characterising the real and imaginary parts of a complex number. Thus, as you can easily check, if z is purely **real**, then $\bar{z} = z$ while if z is purely **imaginary**, then $\bar{z} = -z$. In general, you can see that (see Problem 12.4)

$$\operatorname{Re} z = \frac{z + \bar{z}}{2} \quad \operatorname{Im} z = \frac{z - \bar{z}}{2j}$$

Problem 12.5
For $z = 2 + 3j$ evaluate $z\bar{z}$, $\operatorname{Re} z$ and $\operatorname{Im} z$.

For $z = 2 + 3j$ we have $\bar{z} = 2 - 3j$ and so

$$z\bar{z} = 2^2 + 3^2 = 13$$

$$\operatorname{Re} z = \frac{2 + 3j + 2 - 3j}{2} = 2, \quad \operatorname{Im} z = \frac{2 + 3j - (2 - 3j)}{2j} = 3$$

which last results are obvious by inspection of z.

Division (or rationalization) of two complex numbers:

$$z = \frac{a + jb}{c + jd}$$

means converting into the standard form $A + jB$. We can do this by using the complex conjugate, $c - jd$, of $c + jd$ and the fact that $(c - jd)(c + jd) = c^2 + d^2$ which is real. So, if we multiply top and bottom of the above complex number z by $c - jd$ we get (see Applications, Chapter 2)

$$\frac{a + jb}{c + jd} = \frac{a + jb}{c + jd} \times \frac{c - jd}{c - jd} = \left(\frac{ac + bd}{c^2 + d^2}\right) + j\left(\frac{bc - ad}{c^2 + d^2}\right)$$

Problem 12.6
Divide $3 - 2j$ by $4 + j$.

We have

$$\frac{3 - 2j}{4 + j} = \frac{3 - 2j}{4 + j} \times \frac{4 - j}{4 - j} = \frac{10 - 11j}{17} = \frac{10}{17} - \frac{11}{17}j$$

Exercise on 12.2
For the complex numbers $z = 3 - j$ and $w = 1 + 2j$ evaluate
(i) $2z - 3w$ (ii) zw (iii) $z^2 w \bar{z}^2$ (iv) $\dfrac{z}{w}$

Answer
(i) $3 - 8j$ (ii) $5 + 5j$ (iii) $10 + 20j$ (iv) $\dfrac{1}{5} - \dfrac{7}{5}j$

12.3 Complex variables and the Argand plane

A general **complex variable** is denoted, in **Cartesian form**, by

$$z = x + jy$$

with x, y varying over real values. Such a variable can be represented by points in a plane called the **Argand plane** (or **complex plane**) – see Figure 12.1.

x is called the **real axis**; y the **imaginary axis**. In this representation the complex conjugate $\bar{z}$ is the mirror image of z in the x-axis. The distance of the complex number z from the origin is $r = \sqrt{x^2 + y^2}$, denoted $|z|$ and called the **modulus** of z – it is always taken to be the positive square root. The angle θ made by OP with the positive x-axis is called the **amplitude** or **argument** of z. r and θ are polar coordinates (206 ◀) of the point P defining the complex number z.

An alternative representation of a complex number – the **polar form** – can be obtained by using the polar coordinates (Figure 12.1) r, θ. Since $x = r\cos\theta$ and $y = r\sin\theta$ we have

$$z = x + jy = r(\cos\theta + j\sin\theta) \equiv r\operatorname{cis}\theta \equiv r\angle(\theta)$$

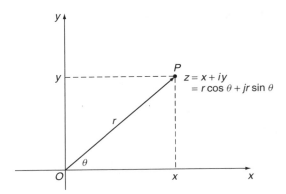

Figure 12.1 The Argand diagram.

cis θ is shorthand for $\cos\theta + i\sin\theta$, using the alternative notation i for j. $\angle(\theta)$ is the standard shorthand for $\cos\theta + j\sin\theta$ that we will use throughout this book.

Remember that θ is measured anti-clockwise positive from the x-axis, and, being an angle, it will represent the same point in the plane at an infinite number of different values, separated by integer multiples of 2π. To obtain a single valued representation of z, θ must be confined to a range of length 2π. The particular choice $-\pi < \theta \leq \pi$ is called the **principal value** of θ and is written Arg z. If the angle θ is not so confined we use a lower case notation arg z. The important point is that the argument of a complex number is not unique. Thus for $z = i$ we can take $\theta = \dfrac{\pi}{2}, \dfrac{5\pi}{2}, -\dfrac{3\pi}{2}, \ldots$, etc. In general we have

$$\angle(\theta) = \angle(\theta + 2k\pi)$$

for any integer k. This fact will play a crucial role in taking roots of a complex number (➤ 363). It is also worth noting that while it is not essential to work in radians in complex numbers, it tends to be the safest policy, particularly for theoretical work.

In terms of Cartesian coordinates we have Arg $z = \tan^{-1}\left(\dfrac{y}{x}\right)$, the sign and value of the angle being determined by x and y. Also note that $|z| = \sqrt{z\bar{z}}$.

You may find the above polar representation of complex numbers rather strange – what's wrong with the simple $x + jy$ form, you may ask? Well, as always there is method in mathematics – bear with me, it turns out that the polar form of a complex number is **much** easier to deal with in certain circumstances.

Exercises on 12.3

1. Plot the following complex numbers on the Argand plane and put them into polar form.

   (i) 1 (ii) j (iii) $-3j$ (iv) $1-j$

   (v) $2+j$ (vi) $-3-2j$ (vii) $-3+2j$

2. Convert into Cartesian form

   (i) $2\angle(0)$ (ii) $3\angle(\pi)$ (iii) $\angle(\pi/2)$

   (iv) $3\angle\left(\dfrac{3\pi}{4}\right)$ (v) $1\angle\left(-\dfrac{\pi}{3}\right)$ (vi) $2\angle\left(-\dfrac{\pi}{2}\right)$

Answers

1.

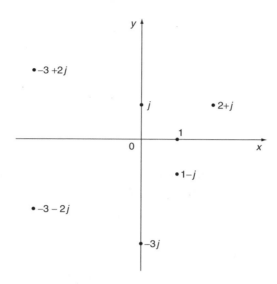

(i) $1\angle(0)$ (ii) $1\angle\left(\frac{\pi}{2}\right)$ (iii) $3\angle\left(-\frac{\pi}{2}\right)$

(iv) $\sqrt{2}\angle\left(-\frac{\pi}{4}\right)$ (v) $\sqrt{5}\angle\left(\text{Tan}^{-1}\left(\frac{1}{2}\right)\right)$

(vi) $\sqrt{13}\angle\left(\pi + \text{Tan}^{-1}\left(\frac{2}{3}\right)\right)$ (vii) $\sqrt{13}\angle\left(\pi - \text{Tan}^{-1}\left(\frac{2}{3}\right)\right)$

(Tan^{-1} denotes the principal value)

2. (i) 2   (ii) -3 (iii) j (iv) $-\frac{3}{\sqrt{2}} + \frac{3}{\sqrt{2}}j$

(v) $\frac{1}{2} - \frac{\sqrt{3}}{2}j$ (vi) $-2j$

12.4 Multiplication in polar form

Now comes the pay-off of polar forms – multiplication of complex numbers in such form is simplicity itself – indeed it reduces to **multiplication** of **m**odulii and **a**ddition of **a**rgument. See for yourself:

Problem 12.7

Multiply $\angle(\theta_1)$ and $\angle(\theta_2)$, use the compound angle formulae (187 ◀) and hence show that

$$r_1\angle(\theta_1)r_2\angle(\theta_2) = r_1r_2\angle(\theta_1 + \theta_2)$$

We have

$$\angle(\theta_1)\angle(\theta_2) = (\cos\theta_1 + j\sin\theta_1)(\cos\theta_2 + j\sin\theta_2)$$
$$= \cos\theta_1\cos\theta_2 - \sin\theta_1\sin\theta_2 + j(\sin\theta_1\cos\theta_2 + \sin\theta_2\cos\theta_1)$$
$$= \cos(\theta_1 + \theta_2) + j\sin(\theta_1 + \theta_2) = \angle(\theta_1 + \theta_2)$$

from which it follows directly that

$$r_1\angle(\theta_1)r_2\angle(\theta_2) = r_1r_2\angle(\theta_1 + \theta_2)$$

So, to multiply in polar form:

- Multiply **modulii** to obtain **modulus** of product
- Add **arguments** to obtain **argument** of product.

or:

$$|z_1z_2| = |z_1||z_2|$$
$$\arg(z_1z_2) = \arg(z_1) + \arg(z_2)$$
$$\mathrm{Arg}(z_1z_2) = \mathrm{Arg}(z_1) + \mathrm{Arg}(z_2) + 2k\pi \quad (k = \text{any integer})$$

Note carefully the last result. The $2k\pi$ is needed because the principal value of the arguments may add to give a value outside the principal value range.

Problem 12.8

Put $\sqrt{3} - j$ and $1 - \sqrt{3}j$ in polar form. Work the product of these both in Cartesian and polar form and compare the results.

It helps to plot the numbers on the Argand plane – see Figure 12.2.

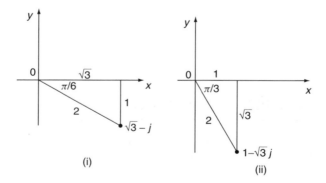

Figure 12.2 The complex numbers $\sqrt{3} - j$ and $1 - \sqrt{3}j$.

We have $|\sqrt{3} - j| = 2$ and $\mathrm{Arg}(\sqrt{3} - j) = -\dfrac{\pi}{6}$, so $\sqrt{3} - j = 2\angle(-\dfrac{\pi}{6})$.
Similarly, $|1 - \sqrt{3}j| = 2$ and $\mathrm{Arg}(1 - \sqrt{3}j) = -\dfrac{\pi}{3}$, so $1 - \sqrt{3}j = 2\angle\left(-\dfrac{\pi}{3}\right)$.

So in polar form the product becomes

$$(\sqrt{3} - j)(1 - \sqrt{3}j) = 2\angle\left(-\frac{\pi}{6}\right) 2\angle\left(-\frac{\pi}{3}\right)$$

$$= 4\angle\left(-\frac{\pi}{6} - \frac{\pi}{3}\right)$$

$$= 4\angle\left(-\frac{\pi}{2}\right) = -4j$$

on conversion back to Cartesian form. You can now check directly that

$$(\sqrt{3} - j)(1 - \sqrt{3}j) = \sqrt{3} - j - 3j + \sqrt{3}j^2$$

$$= -4j$$

You may feel that this is a rather roundabout approach to a simple multiplication – converting to polar form to multiply. However, in practice one already has the polar form – for example when using phasors to represent currents in alternating current theory. Also, the real power of conversion to polar form comes later, when considering powers and roots.

Exercises on 12.4

1. Evaluate all possible products, excluding powers, of the three complex numbers below and compare with the results obtained in $a + jb$ form (see Q2 Exercises on 12.3).

$$z_1 = \angle(\pi/2) \quad z_2 = 3\angle\left(\frac{3\pi}{4}\right) \quad z_3 = 1\angle\left(-\frac{\pi}{3}\right)$$

2. Show that if $z = r\angle(\theta)$, then

$$z^2 = r^2\angle(2\theta), \quad z^3 = r^3\angle(3\theta)$$

What do you think is the result for z^n where n is a positive integer? You will see more of this in Section 12.7.

Answers

1. $z_1 z_2 = 3\angle\left(-\frac{3\pi}{4}\right) = -\frac{3}{\sqrt{2}} - \frac{3}{\sqrt{2}}j \quad z_1 z_3 = \angle\left(\frac{\pi}{6}\right) = \frac{\sqrt{3}}{2} + \frac{1}{2}j$

$z_2 z_3 = 3\angle\left(\frac{5\pi}{12}\right) = \frac{3(\sqrt{3} - 1)}{2\sqrt{2}} + \frac{3(\sqrt{3} + 1)}{2\sqrt{2}}j$

$z_1 z_2 z_3 = 3\angle\left(\frac{11\pi}{12}\right) = -\frac{3(\sqrt{3} + 1)}{2\sqrt{2}} + \frac{3(\sqrt{3} - 1)}{2\sqrt{2}}j$

2. $z^n = r^n \angle(n\theta)$

12.5 Division in polar form

You can probably guess what happens with division in polar form now – divide modulii and subtract arguments. Again, check the details for yourself:

$$\frac{\angle(\theta_1)}{\angle(\theta_2)} = \frac{\cos\theta_1 + j\sin\theta_1}{\cos\theta_2 + j\sin\theta_2} = \frac{\cos\theta_1 + j\sin\theta_1}{\cos\theta_2 + j\sin\theta_2} \times \frac{\cos\theta_2 - j\sin\theta_2}{\cos\theta_2 - j\sin\theta_2}$$

$$= (\cos\theta_1\cos\theta_2 + \sin\theta_1\sin\theta_2) + j(\sin\theta_1\cos\theta_2 - \sin\theta_2\cos\theta_1)$$

(on using $\cos^2\theta_2 + \sin^2\theta_2 = 1$)

$$= \cos(\theta_1 - \theta_2) + j\sin(\theta_1 - \theta_2) = \angle(\theta_1 - \theta_2)$$

Hence

$$\frac{r_1\angle(\theta_1)}{r_2\angle(\theta_2)} = \frac{r_1}{r_2}\angle(\theta_1 - \theta_2)$$

So, to divide in polar form:

- Divide modulii to obtain the modulus of the quotient
- Subtract arguments to obtain the argument of the quotient.

In symbols

$$\left|\frac{z_1}{z_2}\right| = \frac{|z_1|}{|z_2|}$$

$$\arg\left(\frac{z_1}{z_2}\right) = \arg(z_1) - \arg(z_2)$$

$$\mathrm{Arg}\left(\frac{z_1}{z_2}\right) = \mathrm{Arg}\,z_1 - \mathrm{Arg}\,z_2 + 2k\pi \quad k \text{ an integer}$$

Problem 12.9

Work $\dfrac{\sqrt{3}-j}{1-\sqrt{3}j}$ in Cartesian and polar forms and compare the results.

We already know from Problem 12.8 that

$$\sqrt{3} - j = 2\angle\left(-\frac{\pi}{6}\right) \text{ and } 1 - \sqrt{3}j = 2\angle\left(-\frac{\pi}{3}\right)$$

So in polar form

$$\frac{\sqrt{3}-j}{1-\sqrt{3}j} = \frac{2\angle\left(-\frac{\pi}{6}\right)}{2\angle\left(-\frac{\pi}{3}\right)} = \angle\left(-\frac{\pi}{6} + \frac{\pi}{3}\right) = \angle\left(\frac{\pi}{6}\right)$$

$$= \cos\left(\frac{\pi}{6}\right) + j\sin\left(\frac{\pi}{6}\right)$$

$$= \frac{\sqrt{3}}{2} + \frac{1}{2}j$$

which we can check directly

$$\frac{\sqrt{3}-j}{1-\sqrt{3}j} = \frac{(\sqrt{3}-j)(1+\sqrt{3}j)}{1^2+(\sqrt{3})^2} = \frac{2\sqrt{3}+2j}{4}$$

$$= \frac{\sqrt{3}}{2} + \frac{1}{2}j$$

Exercise on 12.5

If $z_1 = \angle\left(\frac{\pi}{4}\right)$ and $z_2 = \angle\left(-\frac{\pi}{3}\right)$ evaluate $\frac{z_1}{z_2}$ in polar and Cartesian form.

Answer

$$\frac{z_1}{z_2} = \angle\left(\frac{7\pi}{12}\right) = \frac{(1-\sqrt{3})}{2\sqrt{2}} + \frac{(\sqrt{3}+1)}{2\sqrt{2}}j$$

12.6 Exponential form of a complex number

Having explained the usefulness of the polar form in multiplication and division of complex numbers, I will now introduce a result that is as pretty as it is powerful:

$$e^{j\theta} \equiv \cos\theta + j\sin\theta = \angle(\theta)$$

Remember that this result is unchanged if θ is replaced by $\theta + 2k\pi$ where k is an integer. This result for $e^{j\theta}$ was first stated explicitly by the Swiss mathematician Leonard Euler (1707–83) in 1748 – not so long ago really. It is known, along with countless other results, as **Euler's formula**. It can be proved by expanding the left-hand side in series, gathering together real and imaginary parts after using $j^2 = -1$ and summing the resulting series to $\cos\theta$ and $\sin\theta$ (see Reinforcement Exercise 18).

Problem 12.10

By taking particular values for θ show that
(i) $e^{j0} = e^{j2\pi} = 1$ (ii) $e^{j\pi} = -1$ (iii) $e^{j\frac{\pi}{2}} = j$

(i) $e^{j0} = \cos 0 + j\sin 0 = 1$ and similarly $e^{j2\pi} = \cos 2\pi + j\sin 2\pi = 1$
(ii) $e^{j\pi} = \cos\pi + j\sin\pi = -1$
(iii) $e^{j\frac{\pi}{2}} = \cos\frac{\pi}{2} + j\sin\frac{\pi}{2} = 0 + j1 = j$

It can be shown that $e^{j\theta}$ obeys all the usual indices laws:

$$e^{jx}e^{jy} = e^{j(x+y)} - \text{from multiplication in polar form}$$

$$\frac{e^{jx}}{e^{jy}} = e^{j(x-y)} - \text{from division in polar form}$$

$$(e^{jx})^y = e^{jxy} - \text{from an extension of Q2, Exercise 12.4}$$

Problem 12.11

Put into exponential form (i) 1   (ii) -1 (iii) $-2j$ (iv) $1+\sqrt{3}j$
(v) $\sqrt{3}-j$

Conversion to exponential form is of course no more than converting to polar form – i.e. finding r and θ.

(i) $z = 1$, $r = 1$, $\theta = 0$ so $z = 1e^{j0} = e^{j0}$
(ii) $z = -1$, $r = 1$, $\theta = \pi$ so $z = e^{j\pi}$
(iii) $z = -2j$, $r = 2$, $\theta = -\dfrac{\pi}{2}$ so $z = 2e^{-j\pi/2}$
(iv) $z = 1+\sqrt{3}j$, $r = 2$, $\theta = \dfrac{\pi}{3}$ so $z = 2e^{j\pi/3}$
(v) $z = \sqrt{3}-j$, $r = 2$, $\theta = -\dfrac{\pi}{6}$ so $z = 2e^{-j\pi/6}$

Exercises on 12.6

1. Show that the modulus of $e^{j\theta}$ is one
2. Show that $(e^{j\theta})^{-1} = e^{-j\theta}$
3. Show that $\cos\theta = \dfrac{e^{j\theta}+e^{j\theta}}{2}$ $\sin\theta = \dfrac{e^{j\theta}-e^{-j\theta}}{2j}$

12.7 De Moivre's theorem for integer powers

A remarkable result 'follows' from the above Euler formula, if we assume that the usual rules of indices apply. If n is an integer then

$$(\cos\theta + j\sin\theta)^n = (e^{j\theta})^n = e^{jn\theta} = \cos n\theta + j\sin n\theta$$

giving us **De Moivre's theorem for a positive integer**:

If n is an integer, then

$$(\cos\theta + j\sin\theta)^n = \cos n\theta + j\sin n\theta$$

Basically, this result is a generalization of the 'add arguments to form product' rule. As an example we have $(\cos\theta + j\sin\theta)^2 = \cos(\theta+\theta) + j\sin(\theta+\theta) = \cos 2\theta + j\sin 2\theta$. And if you were misguided enough to repeat this multiplication a few times you would find, for example, that $(\cos\theta + j\sin\theta)^5 = \cos 5\theta + j\sin 5\theta$. De Moivre's theorem gives this to us directly however.

Problem 12.12

Using De Moivre's theorem show that $(\sqrt{3}+j)^{-3} = -\dfrac{1}{8}j$.

First note that we can write $\sqrt{3}+j = 2\left(\dfrac{\sqrt{3}}{2} + \dfrac{1j}{2}\right) = 2\left(\cos\dfrac{\pi}{6} + j\sin\dfrac{\pi}{6}\right)$ and so

$$(\sqrt{3}+j)^{-3} = \left(2\left(\cos\frac{\pi}{6} + j\sin\frac{\pi}{6}\right)\right)^{-3} = \frac{1}{8}\left(\cos\frac{\pi}{6} + j\sin\frac{\pi}{6}\right)^{-3}$$
$$= \frac{1}{8}\left(\cos\left(-\frac{\pi}{2}\right) + j\sin\left(-\frac{\pi}{2}\right)\right) = -\frac{1}{8}j$$

Exercise on 12.7
By conversion to polar form and use of De Moivres' theorem evaluate

(i) j^7 (ii) $(1+j)^5$ (iii) $(\sqrt{3}-j)^{-4}$

Answer

(i) $-j$ (ii) $-4(1+j)$ (iii) $2^{-4}\left(-\frac{1}{2} + \frac{\sqrt{3}}{2}j\right)$

12.8 De Moivre's theorem for fractional powers

We have considered simple algebra of complex numbers – addition, subtraction, multiplication and 'division'. So far, apart from $\sqrt{-1}$ we have not however considered taking the roots of a complex number. We don't have to look far to see that this requires some care. A few examples will help:

$x^2 = 1$ implies

$$x = \pm(1)^{\frac{1}{2}} = \pm\sqrt{1} = \pm 1$$

since $(-1)^2 = 1^2 = 1$. Two values, as we might expect.

What about $x^3 = 1$? In this case 1 is again a root, $1^3 = 1$, but -1 won't do because $(-1)^3 = -1 \neq 1$. We would in any case expect **three** values of the cube root, since we expect three solutions to a cubic equation. In fact in this simple case we can actually find them by some algebra.

$$x^3 - 1 = (x-1)(x^2 + x + 1) = 0$$

So
$$x = 1$$

or
$$x^2 + x + 1 = 0$$

The latter equation gives two **complex** roots:

$$x = -\frac{1}{2} \pm j\frac{\sqrt{3}}{2}$$

So, in fact we **do** have three cube roots of $+1$:

$$1, -\frac{1}{2} + j\frac{\sqrt{3}}{2}, -\frac{1}{2} - j\frac{\sqrt{3}}{2}$$

but two of them are complex. The question is, how do we find such roots in general? Since the polar form is so useful in multiplication, and De Moivre's theorem is useful in raising to a power, we suspect that we can also use this form in taking roots. We can, but there is a subtlety that takes some getting used to.

Consider the cube root $1^{\frac{1}{3}}$. Written in polar form we have

$$1^{\frac{1}{3}} = (\cos 0 + j \sin 0)^{\frac{1}{3}} = (\angle(0))^{\frac{1}{3}}$$

Applying De Moivre's theorem as it would work for an integer gives

$$1^{\frac{1}{3}} = \angle\left(\frac{1}{3}0\right) = \angle(0) = 1$$

i.e. it only gives us the root 1. But we have missed a trick, namely that because over $\cos(\theta + 2k\pi) = \cos\theta$ and $\sin(\theta + 2k\pi) = \sin\theta$ for any integer k, we can always write

$$\angle(\theta) = \angle(\theta + 2k\pi) \quad k = \text{any integer}$$

as we noted in Section 12.3. That is, we can increase θ by any integer multiple of 2π, and we won't change $\angle(\theta)$. Normally, we wouldn't need to do this – we simply reproduce the same values of $\angle(\theta)$. However, when we take roots, the following happens:

$$(\angle(\theta))^{\frac{1}{3}} = (\angle(\theta + 2k\pi))^{\frac{1}{3}}$$
$$= \angle\left(\frac{\theta + 2k\pi}{3}\right)$$

when we apply De Moivre's theorem. **Now**

$$\angle\left(\frac{\theta}{3}\right) \neq \angle\left(\frac{\theta}{3} + \frac{2k\pi}{3}\right) \quad \text{for all integer values of } k$$

For example, with $k = 1$:

$$\angle\left(\frac{\theta}{3}\right) \neq \angle\left(\frac{\theta}{3} + \frac{2\pi}{3}\right)$$

Applying this extension of De Moivre's theorem to $1 = \angle(0)$ gives

$$(\angle(0))^{\frac{1}{3}} = [\angle(0 + 2k\pi)]^{\frac{1}{3}} \quad k = 0, \pm 1 \pm 2, \ldots$$
$$= \angle\left(\frac{0 + 2k\pi}{3}\right)$$
$$= \angle\left(\frac{2k\pi}{3}\right) \quad k = 0, \pm 1, \pm 2, \ldots$$

Now we appear to have **too many** values – one for each of the infinite number of values of k. However, you will find that these all reduce to just three different values:

Problem 12.13

Evaluate $\angle\left(\dfrac{2k\pi}{3}\right)$ **for** $k = 0, 1, 2, 3.$

Substituting for k we find:

$$k = 0, \angle\left(\dfrac{2k\pi}{3}\right) = \angle(0) = 1$$

$$k = 1, \angle\left(\dfrac{2\pi}{3}\right) = \cos\left(\dfrac{2\pi}{3}\right) + j\sin\left(\dfrac{2\pi}{3}\right) = -\dfrac{1}{2} + j\dfrac{\sqrt{3}}{2}$$

$$k = 2, \angle\left(\dfrac{4\pi}{3}\right) = \cos\left(\dfrac{4\pi}{3}\right) - j\sin\left(\dfrac{4\pi}{3}\right) = -\dfrac{1}{2} - j\dfrac{\sqrt{3}}{2}$$

$$k = 3, \angle\left(\dfrac{6\pi}{3}\right) = \angle(2\pi) = 1$$

We see that $k = 3$ brings us back to where we started. You can soon convince yourself that higher values of k simply reproduce the three roots in the order of $k = 0, 1, 2$.

So, we obtain all the different roots in polar form by first generalising the angle θ to $\theta + 2k\pi$, taking the root and then letting k take the appropriate number of values in turn, usually commencing with $k = 0$. Problem 12.13 should now help you to appreciate the general situation, which we now summarise.

We can obtain all q qth roots of $\cos\theta + j\sin\theta$ by writing it in its most general form:

$$\cos\theta + j\sin\theta \equiv \cos(\theta + 2k\pi) + j\sin(\theta + 2k\pi)$$

$$k = 0, \pm 1, \pm 2, \ldots$$

Then

$$[\cos(\theta + 2k\pi) + j\sin(\theta + 2k\pi)]^{1/q}$$
$$= \cos\left(\dfrac{\theta + 2k\pi}{q}\right) + j\sin\left(\dfrac{\theta + 2k\pi}{q}\right) \quad k = 0, 1, 2, \ldots, (q-1)$$

and by letting k take any q consecutive values, for example, $k = 0, 1, 2, \ldots (q-1)$, we obtain all of the q qth roots of $\cos\theta + j\sin\theta$.

The q qth roots of a complex number $z = x + jy$ can be obtained by converting to polar form and then using De Moivres theorem:

$$z^{1/q} \equiv [r(\cos\theta + j\sin\theta)]^{1/q}$$
$$= r^{1/q}\left[\cos\left(\dfrac{\theta + 2k\pi}{q}\right) + j\sin\left(\dfrac{\theta + 2k\pi}{q}\right)\right] \quad k = 0, 1, 2, \ldots, (q-1)$$

In geometrical terms these roots must all lie at equally spaced intervals of $\dfrac{2\pi}{q}$ around the circle radius $r^{1/q}$, centre the origin in the complex plane.

Exercises on 12.8

1. Find in $a + jb$ form and plot on Argand diagram:
   (i) The three values of $j^{1/3}$
   (ii) The four values of $(1 + j)^{1/4}$

2. Solve the following equations in $a + jb$ form and plot the roots in the Argand plane.
   (i) $x^6 - 1 = 0$ (ii) $x^3 + 8 = 0$

Answer

1. (i) $\dfrac{\sqrt{3}}{2} + \dfrac{1}{2}j, \quad -\dfrac{\sqrt{3}}{2} + \dfrac{1}{2}j, \quad -j$

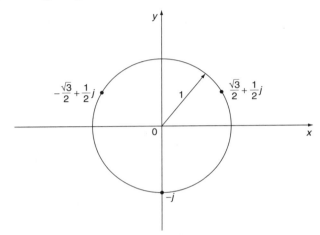

(ii) $2^{1/8}\left(\cos\left(\dfrac{\pi}{16}\right) + j\sin\left(\dfrac{\pi}{16}\right)\right) \quad 2^{1/8}\left(\cos\left(\dfrac{9\pi}{16}\right) + j\sin\left(\dfrac{9\pi}{16}\right)\right)$

$2^{1/8}\left(\cos\left(\dfrac{17\pi}{16}\right) + j\sin\left(\dfrac{17\pi}{16}\right)\right) \quad 2^{1/8}\left(\cos\left(\dfrac{25\pi}{16}\right) + j\sin\left(\dfrac{25\pi}{16}\right)\right)$

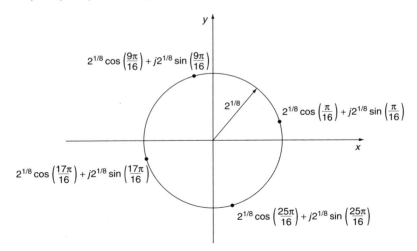

2. (i) ± 1 $\quad\quad\quad\dfrac{1}{2} \pm \dfrac{\sqrt{3}}{2}j$ $\quad\quad\quad -\dfrac{1}{2} \pm \dfrac{\sqrt{3}}{2}j$

   (ii) -2 $\quad\quad\quad 1 \pm \sqrt{3}j$

12.9 Reinforcement

1. Write the following in simplest form in terms of real numbers and j.

   (i) $\sqrt{-1}$ $\quad\quad$ (ii) $\sqrt{9}$ $\quad\quad$ (iii) $\sqrt{-9}$ $\quad\quad$ (iv) j^2

   (v) $-j^2$ $\quad\quad$ (vi) $\dfrac{1}{j}$ $\quad\quad$ (vii) $(-j)^2$

2. Solve the following equations, writing the answer in $z = a + jb$ form:

   (i) $z^2 + 25 = 0$ $\quad\quad\quad\quad$ (ii) $z^2 + 4z + 5 = 0$

   (iii) $z^4 - 3z^2 - 4 = 0$ $\quad\quad\quad\quad$ (iv) $z^3 + z - 2 = 0$

   (v) $z^3 + 1 = 0$ $\quad\quad\quad\quad$ (vi) $z^2 + 2jz + 1 = 0$

   Using equations (i)–(v) verify the result that the equations with real coefficients have real roots and/or complex roots occurring in conjugate pairs.

3. Express in the form $a + jb$:

   (i) j^3 $\quad\quad$ (ii) j^{27} $\quad\quad$ (iii) $3(1 + j) - 2(1 - j)$

   (iv) $(-2j)^6$ $\quad\quad$ (v) $j(j + 2)$ $\quad\quad$ (vi) j^3/j

   (vii) $2j(j - 1) + j^3(2 + j)$

4. Find the real and imaginary parts of:

   (i) $(1 - j)(1 + j)$ $\quad\quad$ (ii) $(3 - 4j)(1 + j)$ $\quad\quad$ (iii) $(4 + 3j)^2$

   (iv) $(4 + 3j)^3$

5. Write down the complex conjugates of:

   (i) $5 + 3j$ $\quad\quad$ (ii) $\dfrac{1}{3 - 4j}$ $\quad\quad$ (iii) $\dfrac{j}{j + 2}$ $\quad\quad$ (iv) $\dfrac{j - 2}{3 + 2j}$

   (v) $\left(\dfrac{j - 2}{3 + 2j}\right)^4$ $\quad\quad$ (vi) $\dfrac{1}{j(2 - 5j)^2}$

   Where appropriate put both the original complex number and its conjugate into $a + jb$ form and check your results.

6. Put into $a + jb$ form

   (i) $(2+5j)(4-3j)$ (ii) $(4-j)(1+j)(3+4j)$ (iii) $\dfrac{1}{3-4j}$

   (iv) $\dfrac{2-5j}{1+4j}$ (v) $\dfrac{-1+3j}{(3-2j)(2+j)}$

7. Evaluate

   (i) $\dfrac{1}{4-3j} + \dfrac{1}{4+3j}$ (ii) $\dfrac{2+j}{4-3j} - \dfrac{2-j}{4+3j}$ (iii) $\dfrac{1}{(5+3j)(5-3j)}$

   and explain why each is either purely real or purely imaginary.

8. Simplify the complex number $\dfrac{2-j}{3+j} + \dfrac{1+j}{1-j}$. Find the modulus and argument of the result.

9. State by inspection only (no arithmetic is necessary) whether each of the following numbers is purely real, purely imaginary or complex.

   (i) $\dfrac{4+j}{5-2j} - \dfrac{4-j}{5+2j}$ (ii) $\dfrac{j}{5+4j} - \dfrac{j}{5-4j}$ (iii) $\left(\dfrac{1+2j}{2-3j}\right)^3 \left(\dfrac{1-2j}{2+3j}\right)^4$

10. Mark each of the following numbers on an Argand diagram and find the modulus and the principal value of the argument of each:

    (i) 2 (ii) -1 (iii) $3j$

    (iv) $-j$ (v) $1+j\sqrt{3}$ (vi) $-\sqrt{3}-j$

    (vii) $-2+2j$ (viii) $-3-3j$

    Write down the numbers in polar form.

11. Convert to Cartesian form

    (i) $4\angle(0)$ (ii) $3\angle\left(-\dfrac{\pi}{2}\right)$ (iii) $2\angle(\pi)$

    (iv) $10\angle(\pi)$ (v) $10\angle\left(\dfrac{\pi}{2}\right)$ (vi) $2\angle\left(\dfrac{\pi}{4}\right)$

    (vii) $3\angle\left(-\dfrac{\pi}{4}\right)$ (viii) $2\angle\left(-\dfrac{3\pi}{4}\right)$ (ix) $\sqrt{3}\angle\left(\dfrac{\pi}{3}\right)$

    (x) $3\angle\left(-\dfrac{2\pi}{3}\right)$ (xi) $\angle\left(\dfrac{\pi}{6}\right)$ (xii) $3\angle\left(-\dfrac{5\pi}{6}\right)$

12. If $|z_1| = 5$, $\text{Arg } z_1 = \pi/3$, $|z_2| = 3$, $\text{Arg } z_2 = \pi/4$, find the Cartesian forms of z_1 and z_2 and the values of:

    (i) $|z_1 z_2|$ (ii) $\left|\dfrac{z_1}{z_2}\right|$ (iii) $|z_1^2|$

    (iv) $\text{Arg}(z_1 z_2)$ (v) $\text{Arg}\left(\dfrac{z_1}{z_2}\right)$ (vi) $\text{Arg } \bar{z}_1$

13. Show that multiplication by j rotates a complex number through $\dfrac{\pi}{2}$ in the anticlockwise direction and division by j rotates it $\dfrac{\pi}{2}$ in the clockwise direction.

14. If $z_1 = 3\angle\left(\dfrac{\pi}{6}\right)$, $z_2 = 2\angle\left(\dfrac{\pi}{18}\right)$, $z_3 = \angle\left(\dfrac{\pi}{3}\right)$ find the polar form of:

    (i) $z_1 z_2$ (ii) $\dfrac{z_1}{z_2}$ (iii) $z_1 z_2 z_3$

    (iv) $\dfrac{z_2}{z_3^2}$ (v) $\dfrac{z_2 z_3}{z_1}$ (vi) $\dfrac{z_1^2 z_3}{z_2}$

15. Evaluate the powers indicated by use of the polar form.

    (i) $(1 - j)^8$ (ii) $(\sqrt{3} + j)^6$ (iii) $(2 + 2j)^4$

16. Plot the complex numbers

    (i) $1 - j$ (ii) $2j$ (iii) $\dfrac{\sqrt{3}}{2} + \dfrac{1}{2}j$ (iv) $-3j$

    on the Argand diagram and put them in the form $e^{j\theta}$.

17. Express the following numbers in the form $a + jb$:

    (i) $e^{j\pi/3}$ (ii) $e^{-j\pi/6}$ (iii) $e^{-(1+j\pi)/3}$ (iv) $\dfrac{e^{j\pi/3}}{j}$

    (v) $\dfrac{e^{-j\pi/4}}{1 + 2j}$ (vi) $\dfrac{e^{-j\pi/6}}{2 - j}$

18. Use the power series for e^x with $x = j\theta$ to find $e^{j\theta}$ in the form $A + jB$ where A and B are real power series in θ. Hence show that:

    $$e^{j\theta} = \cos\theta + j\sin\theta$$

    Use this result to prove that

    $$(\cos\theta + j\sin\theta)^n = \cos n\theta + j\sin n\theta$$

    Hence evaluate $\left(\dfrac{1}{2} + j\dfrac{\sqrt{3}}{2}\right)^{43}$

19. Simplify (i) $(1+j\sqrt{3})^6 + (1-j\sqrt{3})^6$ and (ii) $(\sqrt{3}-j)^{15}$ by using De Moivre's theorem.

20. Simplify
$$\frac{(\cos 3\theta - j\sin 3\theta)^{-3}(\cos 2\theta + j\sin 2\theta)^4}{(\cos 7\theta + j\sin 7\theta)^5(\cos 5\theta - j\sin 5\theta)^{-4}}$$

21. Determine all the roots in each case:

   (i) Square roots of 1
   (ii) Square roots of -1
   (iii) Square roots of j
   (iv) Cube roots of $-j$
   (v) Square roots of $(1+j)$
   (vi) Square roots of $1+\sqrt{3}j$
   (vii) Fourth roots of $1-j$
   (viii) Fourth roots of $\sqrt{3}+j$

22. Find the values of z satisfying
$$z^4 + 1 = 0$$

23. Simplify each of the following numbers to the $a+jb$ form:

   (i) $\dfrac{\cos 4\alpha + j\sin 4\alpha}{\cos 3\alpha - j\sin 3\alpha}$
   (ii) $\dfrac{1}{\sin 2\alpha + j\cos 2\alpha}$
   (iii) $\dfrac{\cos 2\alpha - j\sin 2\alpha}{\sin 3\alpha + j\cos 3\alpha}$

24. Find the polar forms of the fourth roots of -16 and indicate the results on the Argand diagram.

12.10 Applications

1. If $y = Ae^{j\alpha t}$ where A and α are constants, show that $\dfrac{d^2y}{dt^2} + \alpha^2 y = 0$. By assuming a solution of the form $y = e^{\lambda t}$, find solutions for the differential equation

$$\frac{d^2y}{dt^2} + 2\frac{dy}{dt} + 2y = 0$$

by obtaining an equation for λ. This is the approach we will follow in Chapter 15.

2. (i) By equating the real parts of $\cos 5\theta + j\sin 5\theta = (\cos\theta + j\sin\theta)^5$ show that $\cos 5\theta = 16\cos^5\theta - 20\cos^3\theta + 5\cos\theta$. Also express $\sin 5\theta$ as powers of $\sin\theta$.

   (ii) By using $\sin\theta = \dfrac{e^{j\theta} - (e^{j\theta})^{-1}}{2j}$ and $\cos\theta = \dfrac{e^{j\theta} + (e^{j\theta})^{-1}}{2}$ express in terms of sines and cosines of multiple angles (a) $\sin^4\theta$, (b) $\cos^5\theta$. Use (a) to evaluate $\int_0^{\pi/3} \sin^4\theta\, d\theta$.

   (iii) Write $\cos\theta = R(e^{j\theta})$ and hence sum the series $\cos\theta + \cos 2\theta + \cdots + \cos n\theta$.

3. In the relationship

$$(R + jpL)\left(S - j\frac{1}{pC}\right) = \frac{P}{Q}$$

all the quantities are real except j. Show that

$$p = \sqrt{\frac{R}{LSC}}$$

and find R in terms of C, L, P, Q, S.

4. In electrical circuit theory impedance is represented by a complex variable in which the resistance R is the real part and the reactance X is the imaginary part:

$$Z = R + jX$$

The combined impedance, $Z = R + jX$ of two impedance's $Z_1 = R_1 + jX_1$, $Z_2 = R_2 + jX_2$ in parallel is given by the usual result:

$$\frac{1}{Z} = \frac{1}{Z_1} + \frac{1}{Z_2}$$

Obtain explicit expressions for R and X in terms of R_1, R_2, X_1, X_2.

5. A typical sinusoidal function

$$x(t) = A \sin(\omega t + \phi)$$

describing an oscillating signal with period $2\pi/\omega$, amplitude A and phase angle ϕ may be represented, with the frequency ω understood, by a line of length A making angle ϕ with a horizontal axis in an Argand plane. This is called a **phasor diagram**, the line being the **phasor** of the signal $x(t)$.

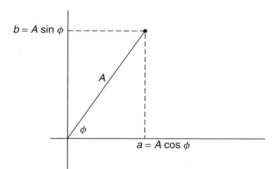

The phasor X of $x(t)$ is then represented mathematically by a complex number of magnitude A and argument ϕ:

$$X = a + jb = A \cos\phi + jA \sin\phi = Ae^{i\phi}$$

In terms of X we can write $x(t)$ as:

$$x(t) = \text{Im}(Xe^{j\omega t})$$

wherein lies the usefulness of phasors – the frequency behaviour of $x(t)$ is separated out.

Phasors representing sinusoids of the same frequency can be added and subtracted by complex algebra. Thus, if the phasors $X_1 = A_1 e^{j\phi_1}$, $X_2 = A_2 e^{j\phi_2}$ represent signals $x_1(t) = A_1 \sin(\omega t + \phi_1)$, $x_2(t) = A_2 \sin(\omega t + \phi_2)$ respectively then the phasor $X_1 + X_2$ represents the signal $x_1(t) + x_2(t)$ and the phasor $X_1 - X_2$ represents the signal $x_1(t) - x_2(t)$.

1. Sketch and write down the phasors corresponding to the sinusoids $2\sin(\omega t + \pi/4)$, $\sin(\omega t + 4\pi/3)$, $3\cos(\omega t + \pi/3)$, $4\sin(\omega t - \pi/6)$ – write these in Cartesian form $a + jb$.

2. Sketch and write down the phasors corresponding to the sinusoids:

   (i) $3\sin(\omega t + \pi/6)$ (ii) $2\sin(\omega t + \pi/2)$

   (iii) $\sin(\omega t - \pi/4)$ (iv) $3\cos(\omega t - \pi/3)$

   Express the phasors in $a + jb$ form.

Answers:

(i) $3e^{j\pi/6}$, $\dfrac{3\sqrt{3}}{2} + j\dfrac{3}{2}$ (ii) $2e^{j\pi/2}$, $2j$

(iii) $e^{-j\pi/4}$, $\dfrac{1}{\sqrt{2}} - j\dfrac{1}{\sqrt{2}}$ (iv) $3e^{j\pi/6}$, $\dfrac{3\sqrt{3}}{2} + j\dfrac{3}{2}$

3. Find the sinusoidal functions (assume frequency ω) corresponding to the following phasors in $a + jb$ form

   (i) $3 + 3j$ (ii) $-4 + 2\sqrt{3}j$ (iii) $-4 - 3j$ (iv) $3 - 5.2j$

Answers:

(i) $3\sqrt{2}\sin(\omega t + \pi/4)$ (ii) $2\sqrt{7}\sin(\omega t + 2.43)$

(iii) $5\sin(\omega t + 3.79)$ (iv) $6\sin(\omega t - 1.05)$

NB. All angles in radians.

4. Express $2\cos \omega t$ and $\sqrt{2}\cos(\omega t + \pi/4)$ in phasor form and hence determine the amplitude and phase of

$$2\cos \omega t + \sqrt{2}\cos(\omega t + \pi/4)$$

12.11 Answers to reinforcement exercises

1. (i) j (ii) 3   (iii) $3j$
   (iv) -1 (v) 1   (vi) $-j$
   (vii) -1

2. (i) $\pm 5j$ (ii) $-2+j, -2-j$ (iii) $2, -2, j, -j$
   (iv) $1, -\frac{1}{2}+\frac{\sqrt{7}}{2}j, -\frac{1}{2}-\frac{\sqrt{7}}{2}j$ (v) $-1, \frac{1}{2} \pm j\frac{\sqrt{3}}{2}$
   (vi) $-(1 \pm \sqrt{2})j$

3. (i) $-j$ (ii) $-j$ (iii) $1+5j$ (iv) -64
   (v) $-1+2j$ (vi) -1 (vii) $-1-4j$

4. (i) 2, 0   (ii) 7, -1 (iii) 7, 24   (iv) -44, 117

5. (i) $5-3j$ (ii) $\dfrac{1}{3+4j}$ (iii) $\dfrac{j}{j-2}$
   (iv) $\dfrac{j+2}{2j-3}$ (v) $\left(\dfrac{j+2}{2j-3}\right)^4$ (vi) $\dfrac{j}{(2+5j)^2}$

6. (i) $23+14j$ (ii) $3+29j$ (iii) $\dfrac{3}{24}+\dfrac{4j}{25}$
   (iv) $-\dfrac{18}{17}-\dfrac{13j}{17}$ (v) $-\dfrac{11}{65}+\dfrac{23j}{65}$

7. (i) 8/25, real, sum of complex conjugates.
   (ii) $4j/5$, imaginary, difference of complex conjugates.
   (iii) 1/34, real, product of complex conjugates.

8. $\dfrac{1}{2}+\dfrac{1}{2}j$ Modulus $=\dfrac{1}{\sqrt{2}}$ Argument $=\dfrac{\pi}{4}$

9. (i) imaginary   (ii) real   (iii) complex

10. (i) 2, 0   (ii) 1, π (iii) 3, $\pi/2$
    (iv) 1, $-\pi/2$ (v) 2, $\pi/3$ (vi) 2, $-5\pi/6$
    (vii) $2\sqrt{2}, 3\pi/4$ (viii) $3\sqrt{2}, -3\pi/4$

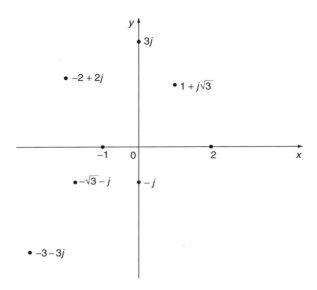

11. (i) 4   (ii) $-3j$ (iii) -2

(iv) -10 (v) $10j$ (vi) $\sqrt{2} + \sqrt{2}j$

(vii) $\dfrac{3}{\sqrt{2}} - \dfrac{3}{\sqrt{2}}j$ (viii) $-\sqrt{2} - \sqrt{2}j$ (ix) $\dfrac{\sqrt{3}}{2} + \dfrac{3}{2}j$

(x) $-\dfrac{3}{2} - \dfrac{3\sqrt{3}}{2}j$ (xi) $\dfrac{\sqrt{3}}{2} + \dfrac{1}{2}j$ (xii) $-\dfrac{3\sqrt{3}}{2} - \dfrac{3}{2}j$

12. $z_1 = 5(1 + j\sqrt{3})/2$ $z_2 = 3(1 + j)/\sqrt{2}$

(i) 15   (ii) $5/3$ (iii) 25

(iv) $7\pi/12$ (v) $\pi/12$ (vi) $2\pi/3$

14. (i) $6\angle\left(\dfrac{2\pi}{9}\right)$ (ii) $\dfrac{3}{2}\angle\left(\dfrac{\pi}{9}\right)$ (iii) $6\angle\left(\dfrac{5\pi}{9}\right)$

(iv) $2\angle\left(-\dfrac{11\pi}{18}\right)$ (v) $\dfrac{2}{3}\angle\left(\dfrac{2\pi}{9}\right)$ (vi) $\dfrac{9}{2}\angle\left(\dfrac{11\pi}{18}\right)$

15. (i) 16   (ii) -64 (iii) -64

16. (i) $\sqrt{2}e^{-j\pi/4}$ (ii) $2e^{j\pi/2}$ (iii) $2e^{j\pi/6}$ (iv) $3e^{-j\pi/2}$

17. (i) $\dfrac{1}{2} + \dfrac{\sqrt{3}}{2}j$ (ii) $\dfrac{\sqrt{3}}{2} - \dfrac{1}{2}j$ (iii) $\dfrac{e^{-1/3}}{2} - \dfrac{\sqrt{3}e^{-1/3}}{2}j$

(iv) $\dfrac{\sqrt{3}}{2} - \dfrac{1}{2}j$ (v) $\dfrac{3}{5\sqrt{2}} - \dfrac{1}{5\sqrt{2}}j$ (vi) $\dfrac{2\sqrt{3}+1}{10} - \dfrac{2-\sqrt{3}}{10}j$

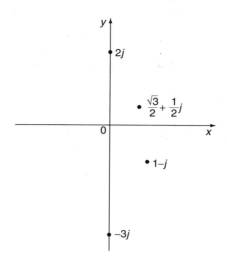

18. $\dfrac{1}{2} + j\dfrac{\sqrt{3}}{2}$

19. (i) 2^7 (ii) $-2^{15}j$

20. $\cos(38\theta) - j\sin(38\theta)$

21. (i) ± 1

(ii) $\pm j$

(iii) $\pm\left(\dfrac{1}{\sqrt{2}} + \dfrac{1}{\sqrt{2}}j\right)$

(iv) $\dfrac{\sqrt{3}}{2} - \dfrac{1}{2}j,\ j,\ -\dfrac{\sqrt{3}}{2} - \dfrac{1}{2}j$

(v) $2^{1/4}\left(\cos\dfrac{\pi}{8} + j\sin\dfrac{\pi}{8}\right),\ 2^{1/4}\left(\cos\left(-\dfrac{7\pi}{8}\right) + j\sin\left(-\dfrac{7\pi}{8}\right)\right)$

(vi) $\pm\sqrt{2}\left(\dfrac{\sqrt{3}}{2} + \dfrac{1}{2}j\right)$

(vii) $2^{1/8}\angle\left(-\dfrac{\pi}{16}\right),\ 2^{1/8}\angle\left(\dfrac{7\pi}{16}\right),\ 2^{1/8}\angle\left(\dfrac{15\pi}{16}\right),\ 2^{1/8}\angle\left(-\dfrac{9\pi}{16}\right)$

(viii) $2^{1/4}\angle\left(\dfrac{\pi}{24}\right),\ 2^{1/4}\angle\left(\dfrac{13\pi}{24}\right),\ 2^{1/4}\angle\left(-\dfrac{23\pi}{24}\right),\ 2^{1/4}\angle\left(-\dfrac{11\pi}{24}\right)$

22. $\dfrac{1+j}{\sqrt{2}},\ \dfrac{1-j}{\sqrt{2}},\ \dfrac{-1+j}{\sqrt{2}},\ \dfrac{-1-j}{\sqrt{2}}$

23. (i) $\cos 7\alpha + j \sin 7\alpha$ (ii) $\sin 2\alpha - j \cos 2\alpha$ (iii) $\sin \alpha + j \cos \alpha$

24. $2 \angle \left(\dfrac{\pi + 2k\pi}{4} \right)$ or $\pm \sqrt{2}(1 \pm j)$

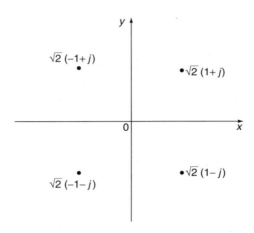

13

Matrices and Determinants

Matrices are rectangular arrays of numbers treated as mathematical entities in themselves and satisfying an algebra that suits their particular application. In this respect they are similar to vectors (Chapter 11), but whereas vectors are designed to reflect the behaviour of physical quantities that have a direction associated with them, matrices are designed to enable us to handle systems of linear equations of the sort introduced in Section 2.2.4. Matrices have simple rules of addition and multiplication, but are complicated to invert, for which it is convenient to introduce certain numbers, called determinants, that are associated with matrices. Determinants also help us to define the eigenvalue problem, which is one of immense importance in engineering and science.

Prerequisites

It will be helpful if you know something about:

- solving simultaneous linear equations (48 ◄)
- the principles and practice of elementary algebra (Chapter 12 ◄)
- the sigma notation (102 ◄)

Objectives

In this chapter you will find:

- definitions of matrices and determinants
- addition and subtraction of matrices
- multiplication of matrices
- zero and unit matrices
- properties of determinants
- the adjoint and inverse of a matrix
- Cramer's rule for solving systems of linear equations
- eigenvalues and eigenvectors

Motivation

You may need the material of this chapter for:

- solving systems of linear equations
- frequency analysis of coupled engineering systems
- solving distribution and scheduling problems in operational research
- solving systems of differential equations

13.1 An overview of matrices and determinants

Everywhere in engineering and science we are constantly having to deal with situations where there are many variables. Matrices provide the mathematical shorthand for dealing with a number of quantities at the same time – basically a generalisation of vectors (indeed the language is similar in the two topics). Matrices are used in modelling complex mechanical systems, control systems, electronic circuits, optical systems, economic and financial systems, network problems, robotics, and much more besides.

A **matrix** is essentially a rectangular array of numbers, each independent from each other – only their position in the array is significant. In matrix algebra the array itself is treated as a single mathematical object – just like a vector. Rules have been invented for combining such arrays, much like the rules of arithmetic, but with some significant differences. Thus, matrices can be added and subtracted. They can also be multiplied, but the commutative rule $xy = yx$ no longer holds in general – if A, B are two matrices then AB and BA are not necessarily the same thing. Matrices can also be 'divided', but care is needed, as always with division. The rules of multiplication and division of matrices are constructed so that we can use matrices in such things as solving systems of equations.

Remember that in complex numbers we found it useful to introduce a real number – the modulus, $|z|$ – associated with a complex number, z, in order to define division by z. We do a similar, but more general thing with matrices. Thus, a **determinant** is a number associated with a given matrix or array. It is evaluated in a certain way from the elements of the matrix and gives us useful information about it. Determinants enable us to define a series of numbers associated with a (square) matrix, called its **eigenvalues**. These are extremely important both mathematically and for applications. If for example the matrix describes a particular control system then the eigenvalues of that matrix tell us important things about the stability of the control system.

13.2 Definition of a matrix and its elements

Consider the system of equations

$$2x + y = 3$$
$$x - 2y = -1 \qquad (13.1)$$

In a sense the important information here is the set of coefficients, which we can arrange in two arrays:

$$A = \begin{bmatrix} 2 & 1 \\ 1 & -2 \end{bmatrix} \quad b = \begin{bmatrix} 3 \\ -1 \end{bmatrix}$$

Such arrays are called **matrices**. A is referred to as a **2 × 2 matrix**, because it has two rows and two columns. Similarly b is a 2 × 1 matrix, or **column vector**. The numbers in the array are called its **elements**. Their positions are important, and note that signs must be properly included. The elements can be complex numbers. What we want to do is construct an 'algebra' for such arrays that enables us to treat A and b as much as possible like ordinary algebraic objects. The system of equation (13.1) already suggests some ideas here.

We put the quantities x, y in a 2×1 matrix of their own:

$$u = \begin{bmatrix} x \\ y \end{bmatrix}$$

and write (13.1) as

$$\begin{bmatrix} 2 & 1 \\ 1 & -2 \end{bmatrix} \begin{bmatrix} x \\ y \end{bmatrix} = \begin{bmatrix} 3 \\ -1 \end{bmatrix}$$

or

$$Au = b$$

If this is to be a symbolic shorthand for (13.1) then this essentially defines the 'product' Au of the two matrices A and u. Have a go at determining the rule for multiplying matrices yourself. The formal definition is given in Section 13.3

Problem 13.1

Write the system of equations

$$a + 3b = 1$$
$$2a + b = 0$$

in matrix form $Au = b$.

The arrays are:

$$A = \begin{bmatrix} 1 & 3 \\ 2 & 1 \end{bmatrix} \quad b = \begin{bmatrix} 1 \\ 0 \end{bmatrix} \quad u = \begin{bmatrix} a \\ b \end{bmatrix}$$

so the equations become

$$\begin{bmatrix} 1 & 3 \\ 2 & 1 \end{bmatrix} \begin{bmatrix} a \\ b \end{bmatrix} = \begin{bmatrix} 1 \\ 0 \end{bmatrix}$$

Problem 13.2

'Multiply' the following matrix products

(i) $\begin{bmatrix} -1 & 0 \\ 1 & 2 \end{bmatrix} \begin{bmatrix} x \\ y \end{bmatrix}$ (ii) $\begin{bmatrix} 2 & 3 \\ 4 & 1 \end{bmatrix} \begin{bmatrix} 2 \\ 1 \end{bmatrix}$

(iii) $\begin{bmatrix} 3 & -1 \\ 4 & 2 \end{bmatrix} \begin{bmatrix} 1 & 2 \\ -1 & 1 \end{bmatrix}$

(i) By analogy with how we have rewritten the pairs of equations we 'multiply' by plugging rows of the first matrix into columns of the second matrix in the product. So:

$$\begin{bmatrix} -1 & 0 \\ 1 & 2 \end{bmatrix} \begin{bmatrix} x \\ y \end{bmatrix} = \begin{bmatrix} (-1)x + 0y \\ (1)x + 2y \end{bmatrix} = \begin{bmatrix} -x \\ x + 2y \end{bmatrix}$$

The quantities $-x$, $x + 2y$ are the elements of the new 2×1 matrix on the right-hand side.

(ii)
$$\begin{bmatrix} 2 & 3 \\ 4 & 1 \end{bmatrix} \begin{bmatrix} 2 \\ 1 \end{bmatrix} = \begin{bmatrix} 2 \times 2 + 3 \times 1 \\ 4 \times 2 + 1 \times 1 \end{bmatrix} = \begin{bmatrix} 7 \\ 9 \end{bmatrix}$$

(iii) In this case the second factor has two columns. No problem, the pattern is

$$\begin{bmatrix} 3 & -1 \\ 4 & 2 \end{bmatrix} \begin{bmatrix} 1 & 2 \\ -1 & 1 \end{bmatrix} = \begin{bmatrix} 3 \times 1 + (-1)(-1) & 3 \times 2 + (-1)(1) \\ 4 \times 1 + 2(-1) & 4 \times 2 + 2 \times 1 \end{bmatrix}$$

$$= \begin{bmatrix} 4 & 5 \\ 2 & 10 \end{bmatrix}$$

a 2×2 matrix. Essentially we are treating each of the two columns in the second factor as we did the single columns in (i) and (ii). Take care with the signs in these calculations!

It is common to use the notation a_{ij} for the elements of a matrix, the subscripts i, j, denoting the ith row and jth column. We now give the general definition of a matrix.

An ***mxn* matrix** A is an array of mn numbers with m rows and n columns denoted by:

$$A = \begin{bmatrix} a_{11} & a_{12} & \ldots & a_{1n} \\ a_{21} & a_{22} & & a_{2n} \\ \vdots & & & \vdots \\ a_{m1} & a_{m2} & \ldots & a_{mn} \end{bmatrix}$$

The number a_{ij} is the ijth element of A, read as 'a, i, j'

      row column

We sometimes write

$$A = [a_{ij}]$$

The case when the number of rows and columns is equal, i.e. $m = n$, is very important. Such matrices are called **square**, and we will have a lot to say about them later on.

A matrix does not have a numerical value (whereas a **determinant** – see Section 13.4 – does). Note that round brackets are sometimes used for matrices. This is not advised since when handwritten such brackets can be confused with determinant lines.

We will see more on the algebra of matrices, but for now just note that:

Addition of matrices is **associative**:

$$A + (B + C) = (A + B) + C$$

and **commutative**:

$$A + B = B + A$$

Multiplication of matrices, considered in Section 13.3, is **associative**:

$$A(BC) = (AB)C$$

but **not commutative**:

$$AB \neq BA$$

We also define some special matrices:

Zero or **null matrix** – all elements zero. It satisfies the obvious rules $A + 0 = A$, $0A = 0$, etc.

Unit matrix, I – all elements zero except for one's on the leading diagonal. It must therefore be a square matrix, and it satisfies

$$IA = AI' = A \quad (I' \text{ and } I \text{ may be different size})$$

for any matrix A.

Inverse matrix – if A is a square matrix then it may be possible to find another matrix A^{-1} such that:

$$AA^{-1} = A^{-1}A = I$$

Such a matrix is called the **inverse** of A. We show how to evaluate the inverse matrix in Section 13.6.

Transpose – the transpose of a matrix A, denoted A^T is the matrix obtained by changing the rows of A into columns, thus, if A is $m \times n$, A^T is $n \times m$.

Exercise on 13.2

Write the following system of equations in matrix form $Ax = b$. In each case, specify the location of the coefficient -3 in the matrix A in the form a_{ij}

(i) $x - 3y = 1$
     $2x + 4y = 4$

(ii) $3x + 2y - z = 1$
     $-x - 3y + 2z = 1$
     $x + y + z = 2$

Answer

(i) $\begin{bmatrix} 1 & -3 \\ 2 & 4 \end{bmatrix} \begin{bmatrix} x \\ y \end{bmatrix} = \begin{bmatrix} 1 \\ 4 \end{bmatrix} \quad a_{12} = -3$

(ii) $\begin{bmatrix} 3 & 2 & -1 \\ -1 & -3 & 2 \\ 1 & 1 & 1 \end{bmatrix} \begin{bmatrix} x \\ y \\ z \end{bmatrix} = \begin{bmatrix} 1 \\ 1 \\ 2 \end{bmatrix} \quad a_{22} = -3$

13.3 Adding and multiplying matrices

Two matrices are **equal** if and only if corresponding elements are identical. Formally, two matrices are equal if and only if they are of the same size and the corresponding elements are the same, i.e.

$$[b_{ij}] = [a_{ij}] \text{ if and only if } b_{ij} = a_{ij}$$

for every value of i and j.

We **add/subtract** matrices by adding/subtracting corresponding elements. So we can only add or subtract matrices of the same size:

$$[a_{ij}] \pm [b_{ij}] = [a_{ij} \pm b_{ij}]$$

Generalising the idea of repeated addition we can define **multiplication of a matrix by a scalar**, k, according to:

$$k[a_{ij}] = [ka_{ij}]$$

i.e. we multiply each element by the scalar k. This is just like multiplication by a scalar in vector algebra.

Problem 13.3

Add all possible pairs of the following matrices:

$$A = \begin{bmatrix} 1 & -1 \\ 0 & 1 \end{bmatrix} \quad B = \begin{bmatrix} 1 & 3 & -1 \\ -1 & 0 & 2 \\ 2 & 1 & 0 \end{bmatrix}$$

$$C = \begin{bmatrix} 2 & 3 \\ -1 & 2 \end{bmatrix} \quad D = \begin{bmatrix} 1 & 2 & 3 \\ 1 & 2 & 3 \\ 1 & 2 & 3 \end{bmatrix}$$

$$E = \begin{bmatrix} 2 & -1 \\ 1 & 0 \end{bmatrix}$$

Since only matrices of the same size can be added we can only add A, C, E together and B, D together. The possibilities are therefore

$$A + C = \begin{bmatrix} 1 & -1 \\ 0 & 1 \end{bmatrix} + \begin{bmatrix} 2 & 3 \\ -1 & 2 \end{bmatrix} = \begin{bmatrix} 1+2 & -1+3 \\ 0-1 & 1+2 \end{bmatrix}$$

$$= \begin{bmatrix} 3 & 2 \\ -1 & 3 \end{bmatrix}$$

$$C + E = \begin{bmatrix} 2 & 3 \\ -1 & 2 \end{bmatrix} + \begin{bmatrix} 2 & -1 \\ 1 & 0 \end{bmatrix} = \begin{bmatrix} 4 & 2 \\ 0 & 2 \end{bmatrix}$$

$$A + E = \begin{bmatrix} 1 & -1 \\ 0 & 1 \end{bmatrix} + \begin{bmatrix} 2 & -1 \\ 1 & 0 \end{bmatrix} = \begin{bmatrix} 3 & -2 \\ 1 & 1 \end{bmatrix}$$

$$B + D = \begin{bmatrix} 1 & 3 & -1 \\ -1 & 0 & 2 \\ 2 & 1 & 0 \end{bmatrix} + \begin{bmatrix} 1 & 2 & 3 \\ 1 & 2 & 3 \\ 1 & 2 & 3 \end{bmatrix} = \begin{bmatrix} 2 & 5 & 2 \\ 0 & 2 & 5 \\ 3 & 3 & 3 \end{bmatrix}$$

Problem 13.4

If $\begin{bmatrix} a & b \\ c & d \end{bmatrix} + \begin{bmatrix} 3 & -1 \\ 1 & 0 \end{bmatrix} = \begin{bmatrix} 1 & 0 \\ 0 & 1 \end{bmatrix}$

find a, b, c, d.

We have

$$\begin{bmatrix} a & b \\ c & d \end{bmatrix} + \begin{bmatrix} 3 & -1 \\ 1 & 0 \end{bmatrix} = \begin{bmatrix} a+3 & b-1 \\ c+1 & d+0 \end{bmatrix} = \begin{bmatrix} 1 & 0 \\ 0 & 1 \end{bmatrix}$$

Equating corresponding elements gives

$a + 3 = 1$ so $a = -2$
$b - 1 = 0$ so $b = 1$
$c + 1 = 0$ so $c = -1$
$d = 1$ so $d = 1$

Note that an equation in 2×2 matrices is equivalent to **four** equations in terms of the elements. Alternatively we can simply subtract the matrices:

$$\begin{bmatrix} a & b \\ c & d \end{bmatrix} = \begin{bmatrix} 1 & 0 \\ 0 & 1 \end{bmatrix} - \begin{bmatrix} 3 & -1 \\ 1 & 0 \end{bmatrix} = \begin{bmatrix} -2 & 1 \\ -1 & 1 \end{bmatrix}$$

I think you will agree that most of the above is fairly sensible and you might expect that an obvious generalisation of addition of matrices would be to define multiplication of matrices as multiplying corresponding elements. However, this is not the most useful definition. As noted above, we define multiplication to fit in with the structure of linear equations – 'plugging rows into columns'. For this to be possible, without trying to fit a 3-pin plug into a 2-pin socket, for example, we can only form AB if A has the same number of columns as B has rows. We then say A and B are **conformable** and we form the product AB by multiplying each row of A into each column of B, multiplying elements in pairs and adding up the results. The formal definition looks more fearsome than it actually is – have a look at Problem 13.5 to make sense of it.

Thus let:

$A = [a_{ij}]$ $(m \times n)$
$B = [b_{ij}]$ $(n \times r)$

Then the elements of the product matrix:

$C = AB = [c_{ij}]$ $(m \times r)$

are given by

$$c_{ij} = \sum_{k=1}^{n} a_{ik} b_{kj}$$

That is, the element in the ith row and jth column of the product matrix C is obtained from the 'scalar' product of the ith row of A with the jth column of B.

As an example:

$$A = \begin{bmatrix} a_{11} & a_{12} & a_{13} \\ a_{21} & a_{22} & a_{23} \end{bmatrix} \quad (2 \times 3)$$

$$B = \begin{bmatrix} b_{11} & b_{12} \\ b_{21} & b_{22} \\ b_{31} & b_{32} \end{bmatrix} \quad (3 \times 2)$$

$$C = AB = \begin{bmatrix} a_{11} & a_{12} & a_{13} \\ a_{21} & a_{22} & a_{23} \end{bmatrix} \begin{bmatrix} b_{11} & b_{12} \\ b_{21} & b_{22} \\ b_{31} & b_{32} \end{bmatrix}$$

$$= \begin{bmatrix} a_{11}b_{11} + a_{12}b_{21} + a_{13}b_{31} & a_{11}b_{12} + a_{12}b_{22} + a_{13}b_{32} \\ a_{21}b_{11} + a_{22}b_{21} + a_{23}b_{31} & a_{21}b_{12} + a_{22}b_{22} + a_{23}b_{32} \end{bmatrix}$$

Note:

(i) AB and BA need not be the same – in fact, they are only both defined if A is $m \times n$ and B is $n \times m$ (then AB is $m \times m$ and BA $n \times n$)

(ii) $AB = 0$ is possible even if both A and B have non zero elements.

Problem 13.5

Evaluate all possible products of the following matrices, excluding powers

$$A = \begin{bmatrix} 1 & -2 \\ 3 & 1 \end{bmatrix} \quad B = \begin{bmatrix} 2 & 1 & 0 \\ 1 & 0 & -1 \end{bmatrix} \quad C = \begin{bmatrix} 0 & -1 & 2 \\ -1 & 0 & -4 \\ 3 & 1 & 2 \end{bmatrix}$$

Looking at the numbers of rows and columns we see that we can only form the following products:

$$AB = \begin{bmatrix} 1 & -2 \\ 3 & 1 \end{bmatrix} \begin{bmatrix} 2 & 1 & 0 \\ 1 & 0 & -1 \end{bmatrix} = \begin{bmatrix} 0 & 1 & 2 \\ 7 & 3 & -1 \end{bmatrix}$$

$$BC = \begin{bmatrix} 2 & 1 & 0 \\ 1 & 0 & -1 \end{bmatrix} \begin{bmatrix} 0 & -1 & 2 \\ -1 & 0 & -4 \\ 3 & 1 & 2 \end{bmatrix} = \begin{bmatrix} -1 & -2 & 0 \\ -3 & -2 & 0 \end{bmatrix}$$

No higher order products are possible without resulting in powers of A, B or C.

Problem 13.6

Evaluate

$$\begin{bmatrix} -1 & 2 & 3 \\ 0 & 1 & 1 \\ -1 & 0 & 2 \end{bmatrix} \begin{bmatrix} 2 & -4 & -1 \\ -1 & 1 & 1 \\ 1 & -2 & -1 \end{bmatrix}$$

and interpret the result.

$$\begin{bmatrix} -1 & 2 & 3 \\ 0 & 1 & 1 \\ -1 & 0 & 2 \end{bmatrix} \begin{bmatrix} 2 & -4 & -1 \\ -1 & 1 & 1 \\ 1 & -2 & -1 \end{bmatrix}$$

$$= \begin{bmatrix} -1 & 0 & 0 \\ 0 & -1 & 0 \\ 0 & 0 & -1 \end{bmatrix} = - \begin{bmatrix} 1 & 0 & 0 \\ 0 & 1 & 0 \\ 0 & 0 & 1 \end{bmatrix}$$

$$= -I$$

where I is called the 3×3 **unit matrix**. It follows that if

$$A = \begin{bmatrix} -1 & 2 & 3 \\ 0 & 1 & 1 \\ -1 & 0 & 2 \end{bmatrix}$$

and

$$B = \begin{bmatrix} -2 & 4 & 1 \\ 1 & -1 & -1 \\ -1 & 2 & 1 \end{bmatrix}$$

(note the overall change of sign) then

$$AB = I$$

You can also check that

$$BA = I$$

We say in these circumstances that B is the **inverse** of A – and then of course A is the inverse of B. We will see how to find the inverse matrix in Section 13.6.

Problem 13.7

Evaluate

$$\begin{bmatrix} 1 & -1 \\ 0 & 0 \end{bmatrix} \begin{bmatrix} 1 & 2 \\ 1 & 2 \end{bmatrix}$$

The product is the 2×2 zero matrix

$$\begin{bmatrix} 0 & 0 \\ 0 & 0 \end{bmatrix}$$

This is an example of $AB = 0$ even though neither A nor B is zero.

Exercises on 13.3

1. If
$$\begin{bmatrix} x+y & 3 \\ -1 & x-y \end{bmatrix} = \begin{bmatrix} 2 & X+Y \\ Y-X & 1 \end{bmatrix}$$
find x, y, X, Y.

2. For the matrices

$$A = \begin{bmatrix} -1 & 0 & 1 \\ 2 & 1 & 0 \\ 3 & 2 & -1 \end{bmatrix} \quad B = \begin{bmatrix} 4 & 2 & 3 \\ -2 & 4 & 1 \\ 3 & 2 & 1 \end{bmatrix}$$

evaluate

(i) $A + B$ (ii) $A - B$ (iii) $3A + 2B$

(iv) AB (v) BA (vi) A^2

Answers

1. $x = \frac{3}{2}$, $y = \frac{1}{2}$, $X = 2$, $Y = 1$

2. (i) $\begin{bmatrix} 3 & 2 & 4 \\ 0 & 5 & 1 \\ 6 & 4 & 0 \end{bmatrix}$ (ii) $\begin{bmatrix} -5 & -2 & -2 \\ 4 & -3 & -1 \\ 0 & 0 & -2 \end{bmatrix}$ (iii) $\begin{bmatrix} 5 & 4 & 9 \\ 2 & 11 & 2 \\ 15 & 10 & -1 \end{bmatrix}$

(iv) $\begin{bmatrix} -1 & 0 & -2 \\ 6 & 8 & 7 \\ 5 & 12 & 10 \end{bmatrix}$ (v) $\begin{bmatrix} 9 & 8 & 1 \\ 13 & 6 & -3 \\ 4 & 4 & 2 \end{bmatrix}$ (vi) $\begin{bmatrix} 4 & 2 & -2 \\ 0 & 1 & 2 \\ -2 & 0 & 4 \end{bmatrix}$

13.4 Determinants

First, an admission. The full theory of determinants is conceptually quite straightforward, but it is very intricate. I will not, therefore, be going into much depth, and I will ask you to make a leap of faith on occasions. However, determinants have great theoretical importance – in defining the inverse matrix, and in eigenvalue theory. It may appear therefore that I am selling you short in skimming over determinants. However, in **practice** no one uses determinants to invert matrices or find eigenvalues for real problems – there are far more effective means studied at more advanced level. Another point is that the definition of a determinant appears a little complicated and so I'll work up to it gently, so that you can see where it comes from. To help you interpret the symbols, a numerical example is done in parallel.

Consider the simple system of equations:

$$a_{11}x_1 + a_{12}x_2 = b_1 \quad (i)$$
$$a_{21}x_1 + a_{22}x_2 = b_2 \quad (ii)$$

Multiply (i) by a_{21}, (ii) by a_{11} and subtract

$$(a_{11}a_{22} - a_{21}a_{12})x_2 = a_{11}b_2 - a_{21}b_1$$

Problem 13.8

Solve the pair of equations

$$3x_1 + 2x_2 = 1 \quad (i)$$
$$x_1 - 2x_2 = 2 \quad (ii)$$

Multiply (i) by 1, (ii) by 3 and subtract

$$(3 \times (-2) - 1 \times 2)x_2 = 3 \times 2 - 1 \times 1$$

So

$$x_2 = \frac{a_{11}b_2 - a_{21}b_1}{a_{11}a_{22} - a_{21}a_{12}}$$

Similarly we find

$$x_1 = \frac{b_1 a_{22} - a_{12} b_2}{a_{11}a_{22} - a_{21}a_{12}}$$

Introduce the notation (337 ◀)

$$\begin{vmatrix} a & b \\ c & d \end{vmatrix} = ad - bc$$

then we can write

$$x_1 = \frac{\begin{vmatrix} b_1 & a_{12} \\ b_2 & a_{22} \end{vmatrix}}{\begin{vmatrix} a_{11} & a_{12} \\ a_{21} & a_{22} \end{vmatrix}}$$

$$x_2 = \frac{\begin{vmatrix} a_{11} & b_1 \\ a_{21} & b_2 \end{vmatrix}}{\begin{vmatrix} a_{11} & a_{12} \\ a_{21} & a_{22} \end{vmatrix}}$$

So

$$x_2 = \frac{3 \times 2 - 1 \times 1}{3 \times (-2) - 1 \times 2} = -\frac{5}{8}$$

Similarly we find

$$x_1 = \frac{1(-2) - 2 \times 2}{3(-2) - 1 \times 2} = \frac{3}{4}$$

Introduce the notation

$$\begin{vmatrix} 3 & 2 \\ 1 & -2 \end{vmatrix} = 3(-2) - 1 \times 2$$

then we can write

$$x_1 = \frac{\begin{vmatrix} 1 & 2 \\ 2 & -2 \end{vmatrix}}{\begin{vmatrix} 3 & 2 \\ 1 & -2 \end{vmatrix}} = \frac{3}{4}$$

$$x_2 = \frac{\begin{vmatrix} 3 & 1 \\ 1 & 2 \end{vmatrix}}{\begin{vmatrix} 3 & 2 \\ 1 & -2 \end{vmatrix}} = -\frac{5}{8}$$

Notice the patterns in the last steps:

(i) The denominator is the same for x_1 and x_2 and clearly obtained from the coefficients of the left-hand side of the equations.

(ii) The numerator for x_1 is given by replacing the x_1 column of the array of coefficients by the right-hand side column of b's. Similarly, the numerator for x_2 is obtained by replacing the x_2 column by the right-hand side column.

This rule, capitalising on the notation introduced, is called **Cramer's rule** – we will say more on this in Section 13.5.

The quantity $\begin{vmatrix} a & b \\ c & d \end{vmatrix}$ is called a **2 × 2** or **second order determinant**. It can be thought of as associated with the matrix $\begin{bmatrix} a & b \\ c & d \end{bmatrix}$. We denote the determinant of a matrix A by $|A|$ or $\det(A)$. By solving systems of 3, 4, ... equations, we are led to definitions of 3, 4, ...

order determinants to fit in with the above pattern. The simplest way to define these higher order determinants is in terms of second order determinants. Thus, we define a 3×3 determinant by:

$$\begin{vmatrix} a_{11} & a_{12} & a_{13} \\ a_{21} & a_{22} & a_{23} \\ a_{31} & a_{32} & a_{33} \end{vmatrix} = a_{11} \begin{vmatrix} a_{22} & a_{23} \\ a_{32} & a_{33} \end{vmatrix} - a_{12} \begin{vmatrix} a_{21} & a_{23} \\ a_{31} & a_{33} \end{vmatrix} + a_{13} \begin{vmatrix} a_{21} & a_{22} \\ a_{31} & a_{32} \end{vmatrix}$$

$$= a_{11}a_{22}a_{33} - a_{11}a_{23}a_{32} - a_{12}a_{21}a_{33} + a_{12}a_{23}a_{31}$$
$$+ a_{13}a_{21}a_{32} - a_{13}a_{22}a_{31}$$

The elements of the first row are multiplied by the determinant obtained by crossing out the row and column of the element, with an alternating sign. This is called the **expansion of the determinant by the first row**. In fact, one can expand by any row or column and we now show how to do this for general order determinants. Let

$$\Delta_n = \begin{vmatrix} a_{11} & a_{12} & \ldots & a_{1n} \\ a_{21} & a_{22} & \ldots & a_{2n} \\ \vdots & \vdots & & \vdots \\ a_{n1} & a_{n2} & \ldots & a_{nn} \end{vmatrix}$$

be a general nth order determinant. Choose any element a_{ij}. The **cofactor** of a_{ij}, denoted A_{ij}, is the $(n-1)$th order determinant $(-1)^{i+j} D_{ij}$ where D_{ij} is the determinant obtained by deleting the i row and j column of Δ_n (D_{ij} is called the **minor** of a_{ij}).

Then the general nth order determinant can be expressed in terms of $(n-1)$th order determinants (and hence in terms of $(n-2)$th order, etc., down to second order) by **an expansion in the cofactors of any row or column**. For example, for expansion of Δ_n by the first row we have

$$\Delta_n = a_{11}A_{11} + a_{12}A_{12} + a_{13}A_{13} + \ldots + a_{1n}A_{1n}$$

Take the 3×3 case as an example:

$$\Delta_3 = \begin{vmatrix} a_{11} & a_{12} & a_{13} \\ a_{21} & a_{22} & a_{23} \\ a_{31} & a_{32} & a_{33} \end{vmatrix}$$

The cofactors of the second row are

$$A_{21} = (-1)^{2+1} \begin{vmatrix} a_{12} & a_{13} \\ a_{32} & a_{33} \end{vmatrix}$$

$$A_{22} = (-1)^{2+2} \begin{vmatrix} a_{11} & a_{13} \\ a_{31} & a_{33} \end{vmatrix}$$

$$A_{23} = (-1)^{2+3} \begin{vmatrix} a_{11} & a_{12} \\ a_{31} & a_{32} \end{vmatrix}$$

Then we can easily check that

$$\Delta_3 = a_{21}A_{21} + a_{22}A_{22} + a_{23}A_{23}$$

Similarly, we can check that

$$\Delta_3 = a_{13}A_{13} + a_{23}A_{23} + a_{33}A_{33}$$

(expansion by the third column).

You may have spotted that the example equations in Problem 13.8 can be solved much more simply than by employing determinants. Adding them gives

$$4x_1 = 3$$

yielding x_1 immediately and then x_2 can be obtained from one of the equations. So this already appears to be a sledgehammer to crack a nut – why bother with determinants at all? Because they provide a systematic, 'easily' remembered, and automatic means of writing down the solution of any system of linear equations no matter how complicated. The determinantal form does not depend on accidental properties of the coefficients.

Problem 13.9

Evaluate $\begin{vmatrix} 3 & 1 & -2 \\ 1 & 0 & 2 \\ -3 & -1 & 4 \end{vmatrix}$

Expand by the second row because it contains a zero:

$$= 1(-1) \begin{vmatrix} 1 & -2 \\ -1 & 4 \end{vmatrix} + 0 \begin{vmatrix} 3 & -2 \\ -3 & 4 \end{vmatrix} + 2(-1) \begin{vmatrix} 3 & 1 \\ -3 & -1 \end{vmatrix}$$
$$= -(4 - 2) + 0 = -2$$

Note that the final determinant is zero because the two rows (or columns) are proportional to each other – see below.

Above we have both defined determinants and given a method for evaluating them. We now need their properties. These are best illustrated using 2×2 examples.

(i) A determinant and its transpose (i.e. the determinant obtained from it by transposing rows into columns – cf: transpose of a matrix (381 ◄)) have the same value.

(ii) Interchanging two rows or two columns changes the sign of the determinant.

(iii) If two rows or two columns are identical the determinant is zero (follows from (ii)).

(iv) Any factors common to a row (column) may be factored out from the determinant.

(v) A row may be increased or decreased by equal multiples of another row (similarly, for columns) without changing the determinant.

(vi) A determinant in which a row or a column consists of the sum or difference of two or more terms, may be expanded as the sum or difference of two or more determinants.

(vii) The determinant of the product of two (square) matrices is the product of the determinants of the factors.

Problem 13.10

Illustrate the above properties of determinants by applying them to general 2×2 determinants.

(i) $\begin{vmatrix} a & b \\ c & d \end{vmatrix} = ad - bc = \begin{vmatrix} a & c \\ b & d \end{vmatrix}$

(ii) $\begin{vmatrix} a & b \\ c & d \end{vmatrix} = ad - bc = -(bc - ad) = -\begin{vmatrix} b & a \\ d & c \end{vmatrix}$

(iii) If two rows (columns) are the same then exchanging them will not change the determinant at all, but by (ii) it will change the sign. So the determinant will be its own negative and must therefore be zero. Or, directly:

$$\begin{vmatrix} a & a \\ b & b \end{vmatrix} = ab - ab = 0$$

(iv) $\begin{vmatrix} ka & b \\ kc & d \end{vmatrix} = kad - kbc = k(ad - bc) = k\begin{vmatrix} a & b \\ c & d \end{vmatrix}$

(v) $\begin{vmatrix} a+kb & b \\ c+kd & d \end{vmatrix} = (a+kb)d - b(c+kd)$

$$= ad - bc = \begin{vmatrix} a & c \\ b & d \end{vmatrix}$$

(vi) $\begin{vmatrix} a+e & b \\ c+f & d \end{vmatrix} = (a+e)d - b(c+f)$

$$= ad - bc + ed - bf$$

$$= \begin{vmatrix} a & b \\ c & d \end{vmatrix} + \begin{vmatrix} e & b \\ f & d \end{vmatrix}$$

(vii) $\begin{bmatrix} a & b \\ c & d \end{bmatrix} \begin{bmatrix} e & f \\ g & h \end{bmatrix} = \begin{bmatrix} ae+bg & af+bh \\ ce+dg & cf+dh \end{bmatrix}$

Now

$$\begin{vmatrix} ae+bg & af+bh \\ ce+dg & cf+dh \end{vmatrix} = (ae+bg)(cf+dh) - (af+bh)(ce+dg)$$

$$= aecf + aedh + bgcf + bgdh - afce - afdg$$

$$- bhce - bhdg$$

$$= adeh + bcfg - bceh - adfg$$
$$= (ad - bc)(eh - fg) \qquad (41 \blacktriangleleft)$$
$$= \begin{vmatrix} a & b \\ c & d \end{vmatrix} \begin{vmatrix} e & f \\ g & h \end{vmatrix}$$

as required.

Exercises on 13.4

1. Evaluate (i) $\begin{vmatrix} 3 & 2 \\ -1 & 1 \end{vmatrix}$ (ii) $\begin{vmatrix} 3 & 3 \\ 1 & 1 \end{vmatrix}$

2. Expand by (i) first row (ii) second column (iii) second row

$$\begin{vmatrix} 3 & 2 & 0 \\ 1 & 0 & 7 \\ 2 & 4 & 1 \end{vmatrix}$$

3. Simplify and evaluate $\begin{vmatrix} 9 & 10 & 10 \\ 2 & 3 & -1 \\ 3 & 3 & 1 \end{vmatrix}$

Answers

1. (i) 5 (ii) 0   2. -58 in each case   3. -26

13.5 Cramer's rule for solving a system of linear equations

We will now show how to generalise the 2×2 version of Cramer's rule given in Section 13.4 by looking at the case of three equations in three unknowns.

$$a_{11}x_1 + a_{12}x_2 + a_{13}x_3 = b_1$$
$$a_{23}x_1 + a_{22}x_2 + a_{23}x_3 = b_2$$
$$a_{31}x_1 + a_{32}x_2 + a_{33}x_3 = b_3$$

If we solve this system by elimination then it is found (believe me!) that the solution can be expressed in a simple determinant form that generalises the 2×2 case of Section 13.4. This is **Cramer's rule**. Cramer's rule gives the solution in terms of 3×3 determinants as:

$$x_1 = \frac{\begin{vmatrix} b_1 & a_{12} & a_{13} \\ b_2 & a_{22} & a_{23} \\ b_3 & a_{32} & a_{33} \end{vmatrix}}{\Delta} \qquad x_2 = \frac{\begin{vmatrix} a_{11} & b_1 & a_{13} \\ a_{21} & b_2 & a_{23} \\ a_{31} & b_3 & a_{33} \end{vmatrix}}{\Delta}$$

$$x_3 = \frac{\begin{vmatrix} a_{11} & a_{12} & b_1 \\ a_{21} & a_{22} & b_2 \\ a_{31} & a_{32} & b_3 \end{vmatrix}}{\Delta}$$

where

$$\Delta = \begin{vmatrix} a_{11} & a_{12} & a_{13} \\ a_{21} & a_{22} & a_{23} \\ a_{31} & a_{32} & a_{33} \end{vmatrix}$$

Again, note the pattern – denominators are always the same and equal to the determinant of the coefficients, Δ. The numerator of x_i is obtained by replacing the column of x_i coefficients in Δ by the column of right-hand sides.

Problem 13.11

Solve the system of linear equations

$$3x + 2y - z = 0$$
$$2x - y + z = 1$$
$$x - y + 2z = -1$$

Using Cramer's rule we have

$$x = \frac{\begin{vmatrix} 0 & 2 & -1 \\ 1 & -1 & 1 \\ -1 & -1 & 2 \end{vmatrix}}{\begin{vmatrix} 3 & 2 & -1 \\ 2 & -1 & 1 \\ 1 & -1 & 2 \end{vmatrix}}$$

$$= \frac{0\begin{vmatrix} -1 & 1 \\ -1 & 2 \end{vmatrix} - 2\begin{vmatrix} 1 & 1 \\ -1 & 2 \end{vmatrix} - 1\begin{vmatrix} 1 & -1 \\ -1 & -1 \end{vmatrix}}{3\begin{vmatrix} -1 & 1 \\ -1 & 2 \end{vmatrix} - 2\begin{vmatrix} 2 & 1 \\ 1 & 2 \end{vmatrix} - 1\begin{vmatrix} 2 & -1 \\ 1 & -1 \end{vmatrix}}$$

$$= \frac{-2(2+1) - (-1-1)}{3(-2+1) - 2(4-1) - (-2+1)}$$

$$= \frac{-4}{-8} = \frac{1}{2}$$

$$y = \frac{\begin{vmatrix} 3 & 0 & -1 \\ 2 & 1 & 1 \\ 1 & -1 & 2 \end{vmatrix}}{-8}$$

$$= \frac{3(2+1) - 1(-2-1)}{-8}$$

$$= \frac{9+3}{-8} = -\frac{3}{2}$$

$$z = \frac{\begin{vmatrix} 3 & 2 & 0 \\ 2 & -1 & 1 \\ 1 & -1 & -1 \end{vmatrix}}{-8}$$

$$= \frac{3(1+1) - 2(-2-1)}{-8}$$

$$= \frac{6+6}{-8} = -\frac{3}{2}$$

Note that in the case when at least one of the values b_i is non-zero the system of equations only has a unique solution if the determinant of coefficients Δ is non-zero. This provides a simple test for the **consistency** of a system. If **all** b_i are zero we say that the system is **homogeneous**. In this case, if the determinant of coefficients is non-zero then we simply get a zero solution for all the variables – the so-called 'trivial solution'. On the other hand, if the determinant of coefficients is zero then we get an indeterminate result $\frac{0}{0}$. In this case it may be possible to get non-zero solutions for some or all of the variables. We consider homogeneous systems further in Section 13.7.

Geometrically, an equation of the form

$$ax + by + cz = d$$

represents a plane in three dimensions (340 ◀). Three equations of this form would therefore be expected to yield the intersection of three planes and usually this would be a single unique point corresponding to the unique solution of the system. However, there are some exceptional cases:

(i) 2 or 3 of the planes parallel, in which case the determinant of coefficients $\Delta = 0$ and x, y, z are not uniquely defined.

(ii) 3 planes coincide or have a line in common in which case we again have $\Delta = 0$, the numerators for x, y, z, are all zero, and in this case there are an infinite number of solutions.

Exercise on 13.5

Solve the system of equations where possible

(i) $3x - 2y + z = 1$ (ii) $x + 2y - 3z = 0$

        $x + y + z = 0$ $2x - y + z = 1$

        $2x - y + 2z = 1$ $x - 3y + 4z = 2$

Answer

(i) $x = 0$, $y = -\frac{1}{3}$, $z = \frac{1}{3}$ (ii) No solution

13.6 The inverse matrix

Before defining the inverse matrix A^{-1} of a square matrix A we define another matrix, **Adj$A$**, associated with A. The definition seems a bit strange, but all will be revealed shortly. The **adjoint** of a square matrix A, denoted by **Adj$A$**, is the transpose of the matrix in which each element of A is replaced by its corresponding cofactors. Try the problem quickly to get a proper grasp of this!

Problem 13.12
Find the determinant and adjoint of

$$A = \begin{bmatrix} -1 & 2 & 3 \\ 0 & 1 & 1 \\ -1 & 0 & 2 \end{bmatrix}$$

and evaluate $A \, \text{Adj} A$

$$|A| = \begin{vmatrix} -1 & 2 & 3 \\ 0 & 1 & 1 \\ -1 & 0 & 2 \end{vmatrix} = -1(2) - 2(1) + 3(1)$$

$$= -1$$

$$\text{Adj} A = \begin{bmatrix} \begin{vmatrix} 1 & 1 \\ 0 & 2 \end{vmatrix} & -\begin{vmatrix} 0 & 1 \\ -1 & 2 \end{vmatrix} & \begin{vmatrix} 0 & 1 \\ -1 & 0 \end{vmatrix} \\ -\begin{vmatrix} 2 & 3 \\ 0 & 2 \end{vmatrix} & \begin{vmatrix} -1 & 3 \\ -1 & 2 \end{vmatrix} & -\begin{vmatrix} -1 & 2 \\ -1 & 0 \end{vmatrix} \\ \begin{vmatrix} 2 & 3 \\ 1 & 1 \end{vmatrix} & -\begin{vmatrix} -1 & 3 \\ 0 & 1 \end{vmatrix} & \begin{vmatrix} -1 & 2 \\ 0 & 1 \end{vmatrix} \end{bmatrix}^T$$

Note, for example, that the element $-\begin{vmatrix} -1 & 2 \\ -1 & 0 \end{vmatrix}$ in the 2, 3 position of the matrix about to be transposed is indeed the cofactor of the element a_{23} in A and check the other elements similarly.

$$= \begin{bmatrix} 2 & -1 & 1 \\ -4 & 1 & -2 \\ -1 & 1 & -1 \end{bmatrix}^T$$

$$= \begin{bmatrix} 2 & -4 & -1 \\ -1 & 1 & 1 \\ 1 & -2 & -1 \end{bmatrix}$$

We then find

$$A \, \text{Adj} A = \begin{bmatrix} -1 & 2 & 3 \\ 0 & 1 & 1 \\ -1 & 0 & 2 \end{bmatrix} \begin{bmatrix} 2 & -4 & -1 \\ -1 & 1 & 1 \\ 1 & -2 & -1 \end{bmatrix}$$

$$= \begin{bmatrix} -1 & 0 & 0 \\ 0 & -1 & 0 \\ 0 & 0 & -1 \end{bmatrix}$$

$$= -1 \begin{bmatrix} 1 & 0 & 0 \\ 0 & 1 & 0 \\ 0 & 0 & 1 \end{bmatrix}$$

$$= -1I$$

$$= |A|I$$

This is an example of the general result

$$A \frac{\text{Adj} A}{|A|} = I$$

which we will return to below.

Before pressing on to the inverse matrix, let's think about why we might need it and what it might look like. Remember that we introduced the idea of a matrix so that we could write a system of equations in the single form

$$A u = b$$

This is of little use unless we can push the new notation to help us to solve such a system. Think of a simple equation

$$3x = 6$$

Of course, the solution is $x = 2$. But, spelling out the details, we first multiply by $1/3 = 3^{-1}$,

$$3^{-1}(3x) = (3^{-1}3)x = 1x$$
$$= x = 3^{-1}6 = 2$$

In order to repeat this for the matrix equation we need A^{-1}, the equivalent of 3^{-1}. This motivates the following definition.

The **inverse**, A^{-1}, of a square matrix A is defined by

$$A^{-1}A = I = AA^{-1}$$

where I is the appropriate unit matrix. The inverse of a matrix exists only if the matrix is **non-singular** (i.e. its determinant is non-zero). This follows because, since the determinant of a product is the product of the determinants (390 ◄):

$$|A^{-1}A| = |A^{-1}||A| = |I| = 1$$

and so we must have $|A| \neq 0$. Also, from this equation follows:

$$|A^{-1}| = 1/|A| = |A|^{-1}$$

So
$$\det(A^{-1}) = [\det(A)]^{-1}$$

In Problem 13.12 above we saw that

$$A \text{Adj} A = -I = |A|I$$

This is, in fact, a general result and leads to the following expression for A^{-1}:

$$A^{-1} = \frac{\text{Adj}A}{|A|}$$

Note that in finding the inverse matrix it is wise to calculate $|A|$ first, because if this is zero, the inverse does not exist and it is a waste of time finding **Adj**A.

Problem 13.13

Find, if possible, the inverse of each of the matrices

(i) $A = \begin{bmatrix} 1 & -1 & 3 \\ 2 & 2 & 1 \\ 0 & 4 & -5 \end{bmatrix}$ (ii) $B = \begin{bmatrix} 2 & -1 & 1 \\ 2 & 0 & 3 \\ 1 & -1 & 0 \end{bmatrix}$

(i) First check the determinant of A:

$$A = \begin{vmatrix} 1 & -1 & 3 \\ 2 & 2 & 1 \\ 0 & 4 & -5 \end{vmatrix}$$

$$= 1 \begin{vmatrix} 2 & 1 \\ 4 & -5 \end{vmatrix} - (-1) \begin{vmatrix} 2 & 1 \\ 0 & -5 \end{vmatrix} + 3 \begin{vmatrix} 2 & 2 \\ 0 & 4 \end{vmatrix}$$

$$= -10 - 4 + 2(-5) - 0 \times 1 + 3(2 \times 4 - 0 \times 2)$$

$$= 0$$

So A is **singular** – its determinant is zero, and so its inverse does not exist.

(ii) Hoping for better luck with B we have

$$|B| = \begin{vmatrix} 2 & -1 & 1 \\ 2 & 0 & 3 \\ 1 & -1 & 0 \end{vmatrix}$$

$$= 2(3) + (-3) + (-2)$$

$$= 1$$

So the inverse of B exists and we are not wasting our time finding **Adj$B$**.

$$\text{Adj}B = \begin{bmatrix} \begin{vmatrix} 0 & 3 \\ -1 & 0 \end{vmatrix} & -\begin{vmatrix} 2 & 3 \\ 1 & 0 \end{vmatrix} & \begin{vmatrix} 2 & 0 \\ 1 & -1 \end{vmatrix} \\ -\begin{vmatrix} -1 & 1 \\ -1 & 0 \end{vmatrix} & \begin{vmatrix} 2 & 1 \\ 1 & 0 \end{vmatrix} & -\begin{vmatrix} 2 & -1 \\ 1 & -1 \end{vmatrix} \\ \begin{vmatrix} -1 & 1 \\ 0 & 3 \end{vmatrix} & -\begin{vmatrix} 2 & 1 \\ 2 & 3 \end{vmatrix} & \begin{vmatrix} 2 & -1 \\ 2 & 0 \end{vmatrix} \end{bmatrix}^T$$

$$= \begin{bmatrix} 3 & 3 & -2 \\ -1 & -1 & 1 \\ -3 & -4 & -2 \end{bmatrix}^T$$

$$= \begin{bmatrix} 3 & -1 & -3 \\ 3 & -1 & -4 \\ -2 & 1 & 2 \end{bmatrix}$$

As an exercise you might like to check that $B\,\text{Adj}B = |B|I$ at this stage. So we now have

$$B^{-1} = \frac{\text{Adj}B}{|B|} = \text{Adj}B$$

$$= \begin{bmatrix} 3 & -1 & -3 \\ 3 & -1 & -4 \\ -2 & 1 & 2 \end{bmatrix}$$

Exercise on 13.6

Where possible, solve the systems of Exercise 13.5 using the inverse matrix.

Answer

See answer to Exercise 13.5.

13.7 Eigenvalues and eigenvectors

Consider the system of equations

$$3x + 2y = 0$$
$$6x + 4y = 0$$

Note that the determinant of coefficients of the left-hand side is

$$\begin{vmatrix} 3 & 2 \\ 6 & 4 \end{vmatrix} = 0$$

If either of the right-hand sides was non-zero this would mean the system has no solution. However, when the right-hand sides are all zero, we say the system is **homogeneous**, and in this case there **is** the possibility of a solution. In fact, cancelling 2 from the second equation reveals that the equations are the same – we really only have one equation:

$$3x + 2y = 0$$

This has an infinite number of solutions – choose any value of y, $y = s$, then take x as $-\frac{2}{3}s$.

In fact, in the homogeneous case it is essential for a non-trivial solution that the determinant of the coefficients is zero. We can generalise this. Returning to the original motivation for the inverse matrix, remember that we wanted to solve

$$Au = b$$

by multiplying by A^{-1}:

$$A^{-1}(Au) = (A^{-1}A)u = Iu = u$$
$$= A^{-1}b.$$

So we have to insist in this case on $|A| \neq 0$, if $b \neq 0$. On the other hand, if $b = 0$, and we have the **homogeneous system**

$$Au = 0$$

then if A is non-singular we only get the **trivial solution** $u = 0$. It is called trivial because if it were, say, to correspond to a physical dynamical system then it would represent the state of zero motion – no activity. In the case of homogeneous systems, we **only** get non-trivial solutions **if** $|A| = 0$. All this would be of little interest were it not for the fact that homogeneous systems are very important, especially in engineering mathematics. When a complicated structure such as an aircraft wing is being studied to examine the kinds of vibrations it might experience it turns out that the mathematical modelling relies heavily on homogeneous systems of equations of a particular type that leads to quantities called **eigenvalues**. These quantities are actually related to the frequencies of vibrations of the various components of the aircraft wing, so their importance is obvious. They have countless other applications too, but here we will concentrate on their mathematical significance and properties. Once again, the definitions may be a little complicated and you may have to take a few things on trust, but the results are worth it, and the technical manipulations involved are really not that difficult.

Eigenvalues are numerical quantities related to the determinant of a certain type of square matrix. They arise in, for example, vibrating systems because the mathematical models used involve systems of equations that take the matrix form

$$Au = \lambda u$$

Here A is some square matrix describing the system, λ (Greek, l, 'lambda' – the standard notation for eigenvalues) is some parameter depending on the characteristics of the vibrating system and u is some 'solution vector' that might, for example, represent the displacements of the vibrating components being modelled. Now this system of equations is a particular example of a homogeneous system of the form

$$(A - \lambda I)u = 0$$

where I is a unit matrix of the same size as A. It will therefore only have a non-trivial solution for u if the determinant of coefficients is zero, which gives the **eigenvalue equation**

$$|A - \lambda I| = 0$$

This is in fact a polynomial equation for λ. For a given matrix A only certain values of λ will satisfy this equation, and only for these 'eigenvalues' will the original equation for u have a non-trivial solution. Such non-trivial solutions for u are called the **eigenvectors** corresponding to the eigenvalues λ.

The above is a lot to take in but the objective is to motivate the rather strange definitions we are introducing. The mechanics of the actual mathematical manipulations are, as noted above, not too bad. Try the problem quickly!

Problem 13.14
Find the eigenvalues and corresponding eigenvectors of the matrix

$$A = \begin{bmatrix} 6 & -3 \\ 2 & 1 \end{bmatrix}$$

If λ is an eigenvalue of A, then it satisfies

$$|A - \lambda I| = \left| \begin{bmatrix} 6 & -3 \\ 2 & 1 \end{bmatrix} - \lambda \begin{bmatrix} 1 & 0 \\ 0 & 1 \end{bmatrix} \right|$$

$$= \begin{vmatrix} 6 - \lambda & -3 \\ 2 & 1 - \lambda \end{vmatrix}$$

$$= \lambda^2 - 7\lambda + 12$$
$$= (\lambda - 3)(\lambda - 4) = 0$$

So the eigenvalues are $\lambda = 3$, $\lambda = 4$.

To find corresponding eigenvectors we must solve the system of homogeneous equations $A\boldsymbol{u} = \lambda \boldsymbol{u}$ with each of the eigenvalues substituted. For $\lambda = 3$ the equations become

$$[A - 3I]\boldsymbol{u} = \begin{bmatrix} 3 & -3 \\ 2 & -2 \end{bmatrix} \begin{bmatrix} u_1 \\ u_2 \end{bmatrix} = 0 \quad \left(\boldsymbol{u} = \begin{bmatrix} u_1 \\ u_2 \end{bmatrix}\right)$$

This gives two equations for u_1, u_2 which actually reduce to one, with solution

$$u_1 = u_2$$

In general then we get an eigenvector corresponding to $\lambda = 3$ of the form

$$\boldsymbol{u} = \begin{bmatrix} u \\ u \end{bmatrix} = u \begin{bmatrix} 1 \\ 1 \end{bmatrix}$$

where u is an arbitrary number, roughly analogous to the arbitrary constant in integration – we choose it to suit our purposes, usually to give a column vector with unit magnitude (the magnitude of a column vector, like that of a vector (331 ◄), is the square root of the sum of the squares of its components – if that puts you in mind of Pythagoras again then so it should, the magnitude of a column vector is in a sense the 'length' of a vector it might represent). In the above case choosing u to make $\boldsymbol{u}$ of unit magnitude gives

$$\boldsymbol{u} = \frac{1}{\sqrt{2}} \begin{bmatrix} 1 \\ 1 \end{bmatrix}$$

Repeating the above arguments for the other eigenvalue $\lambda = 4$ you should find a corresponding eigenvector

$$\boldsymbol{v} = v \begin{bmatrix} 3 \\ 2 \end{bmatrix}$$

where v is again arbitrary. Taking $\boldsymbol{v}$ to be of unit magnitude gives

$$\boldsymbol{v} = \frac{1}{\sqrt{13}} \begin{bmatrix} 3 \\ 2 \end{bmatrix}$$

Hopefully, you will agree that actually finding the eigenvalues and eigenvectors is relatively routine – in this case simply expanding a determinant and solving a quadratic equation, then solving a system of linear equations. In real engineering problems one might have hundreds of components or **degrees of freedom**, resulting in matrices with thousands of entries, and things are then not so simple – we have no hope of expanding the corresponding determinants or solving the resulting equations so easily. In such cases we have to resort to completely different (numerical) methods for finding eigenvalues – but the basic principles are still the same.

Understanding Engineering Mathematics

In the Applications section (➤ 403) you are led through an application of eigenvalues and eigenvectors to the simplest vibrating system there is – a single simple pendulum. Full details are given on the book website, although it involves some material we have not covered yet. But be assured, this topic is one of the most important in engineering mathematics and is well worth the effort to master.

Exercise on 13.7

Find the eigenvalues and corresponding eigenvectors for the matrix

$$A = \begin{bmatrix} 4 & 0 & 1 \\ -2 & 1 & 0 \\ -2 & 0 & 1 \end{bmatrix}$$

Answer

$$\lambda = 1, \begin{bmatrix} 0 \\ 1 \\ 0 \end{bmatrix}; \quad \lambda = 2, \begin{bmatrix} -1 \\ 2 \\ 2 \end{bmatrix}; \quad \lambda = 3, \begin{bmatrix} -1 \\ 1 \\ 1 \end{bmatrix}$$

Remember that each of the eigenvectors could be multiplied by a different arbitrary constant.

13.8 Reinforcement

1. Express the following in matrix form

   (i) $2x - y = -1$
   $x + 2y = 0$

   (ii) $3x - y + 2z = 1$
   $2x + 2y + 3z = 2$
   $-3x - z = -3$

   (iii) $a + 2b + 3c = 1$
   $a - b - c = 3$

   (iv) $4u - 2v = 1$
   $3u + 3v = 2$
   $u - v = -1$

2. If $A = \begin{bmatrix} -1 & 2 & 3 \\ 4 & 0 & 2 \\ -1 & 1 & 2 \end{bmatrix}$ $B = \begin{bmatrix} 0 & 1 & 3 \\ 2 & -1 & 4 \\ 3 & -1 & 2 \end{bmatrix}$

   (i) Write down $a_{12}, a_{31}, a_{33}, b_{11}, b_{21}, b_{32}$

   (ii) Evaluate $\sum_{k=1}^{3} a_{1k}b_{k3}$, $\sum_{k=1}^{3} a_{3k}b_{k2}$

   (iii) Evaluate (a) $2A - 3B$ (b) A^2 (c) AB (d) BA

3. If $\begin{bmatrix} x & -1 & y \\ 2 & 0 & 3 \\ z & 3 & 2 \end{bmatrix} = \begin{bmatrix} 3 & a & x \\ b & 0 & c \\ 2 & 3 & d \end{bmatrix}$

   evaluate a, b, c, d, x, y, z.

4. If $\begin{bmatrix} \cos\theta & \sin\theta \\ -\sin\theta & \cos\theta \end{bmatrix} = \begin{bmatrix} \frac{\sqrt{3}}{2} & \frac{1}{2} \\ -\frac{1}{2} & \frac{\sqrt{3}}{2} \end{bmatrix}$

determine θ in the first quadrant, i.e. $0 < \theta < 90°$.

5. Using the following matrices evaluate every possible sum and product of pairs of the matrices (repetitions such as A^2 allowed):

$$A = \begin{bmatrix} 2 & -1 \\ 3 & 0 \end{bmatrix} \qquad B = \begin{bmatrix} 3 \\ -1 \\ 4 \end{bmatrix} \qquad C = [-2 \ \ 0]$$

$$D = \begin{bmatrix} 0 & 1 & -1 \\ 2 & 0 & 3 \\ -1 & -3 & 2 \end{bmatrix} \qquad E = \begin{bmatrix} 3 & 4 \\ -1 & 2 \\ 0 & 3 \end{bmatrix}$$

$$F = \begin{bmatrix} 1 & 1 & 1 \\ 2 & 2 & 2 \end{bmatrix} \qquad G = [2 \ \ 1 \ \ 0]$$

6. $A = \begin{bmatrix} 2 & -1 \\ 3 & 0 \\ -1 & 1 \end{bmatrix} \qquad B = \begin{bmatrix} 3 & 2 & -1 \\ 0 & 1 & 2 \\ 1 & 1 & 1 \end{bmatrix} \qquad C = \begin{bmatrix} 4 & 2 \\ 3 & 1 \\ 0 & -1 \end{bmatrix}$

$D = \begin{bmatrix} 2 & 1 & 4 \\ 3 & 0 & -1 \end{bmatrix}$

$u = \begin{bmatrix} -2 \\ 1 \\ -1 \end{bmatrix} \qquad v = \begin{bmatrix} 1 \\ 2 \\ 3 \end{bmatrix} \qquad w = \begin{bmatrix} -1 \\ 3 \\ 2 \end{bmatrix}$

Find, where possible,

(i) $3A + B$ (ii) $4A + 2C$ (iii) $3D - 2A$

(iv) $2AD + 3B$ (v) $3u - 2v + B$ (vi) $2u + 3v - w$

(vii) $u - 2w + Bv$

7. A, B, C are the matrices

$$A = \begin{bmatrix} 3 & 1 \\ -1 & 2 \end{bmatrix} \qquad B = \begin{bmatrix} 2 & -1 & 0 \\ 3 & 1 & 2 \end{bmatrix} \qquad C = \begin{bmatrix} 1 & 1 & 2 \\ -2 & 2 & 3 \end{bmatrix}$$

Verify that $A(B + C) = AB + AC$. Is it true that $(B + C)A = BA + CA$?

8. Evaluate

(i) $\begin{vmatrix} 2 & 3 \\ -2 & 4 \end{vmatrix}$ (ii) $\begin{vmatrix} 1 & 0 & 2 \\ 3 & 4 & 5 \\ 5 & 6 & 7 \end{vmatrix}$

(iii) $\begin{vmatrix} 1 & 0 & 6 \\ 3 & 4 & 15 \\ 5 & 6 & 21 \end{vmatrix}$ (iv) $\begin{vmatrix} 1 & 0 & 0 \\ 2 & 3 & 5 \\ 4 & 1 & 3 \end{vmatrix}$

(v) $\begin{vmatrix} 0 & 2 & 3 \\ -2 & 0 & 4 \\ -3 & -4 & 0 \end{vmatrix}$ (vi) $\begin{vmatrix} \cos\theta & \sin\theta \\ -\sin\theta & \cos\theta \end{vmatrix}$

9. Write down the following expansions of the determinant

$$|A| = \begin{vmatrix} 3 & -1 & 2 \\ 0 & 1 & 2 \\ 4 & -1 & 2 \end{vmatrix}$$

(i) By first row
(ii) By second row
(iii) By 3rd column
(iv) By last row

and check that they all lead to the same result.

10. Invert the following matrices

(i) $\begin{bmatrix} 2 & 3 \\ 1 & 4 \end{bmatrix}$
(ii) $\begin{bmatrix} 2 & 3 & 1 \\ 1 & 2 & 3 \\ 3 & 1 & 2 \end{bmatrix}$
(iii) $\begin{bmatrix} 1 & 2 & -1 \\ -1 & 1 & 2 \\ 2 & -1 & 1 \end{bmatrix}$

(iv) $\begin{bmatrix} 2 & 3 & 4 \\ 4 & 3 & 1 \\ 1 & 2 & 4 \end{bmatrix}$
(v) $\begin{bmatrix} 1 & 2 & 0 \\ 3 & -1 & 4 \\ 2 & 0 & 6 \end{bmatrix}$

11. Solve the equations below by matrix inversion where possible:

(i) $x + y + z = 4$
     $2x + 5y - 2z = 3$
     $x + 7y - 7z = 5$

(ii) $2x + 3y - 4z = -15$
     $3x - 2y + 3z = 15$
     $5x + 7y + 5z = -6$

(iii) $2x + 3y - z = 5$
      $x - y + 3z = 8$
      $3x + 4y - 2z = 5$

12. Solve the equation

$$\begin{vmatrix} x+1 & 6 & 2 \\ 1-x & -5 & x-1 \\ x-1 & 4 & 0 \end{vmatrix} = 0$$

13. Solve the following systems of equations by Cramer's rule – when possible:

(i) $x + y = 1$
     $2x - y = 2$

(ii) $4x - 12y = 3$
     $11x - 2y = 1$

(iii) $x + y + z = 1$
      $x + 2y + 3z = 0$
      $x - y - z = 0$

(iv) $x - y + 2z = 1$
     $x + 2y - 3z = 0$
     $2x + y - z = 2$

14. Decide which of the following systems of equations have non-trivial solutions:

(i) $3x - 2y = 0$
     $x + y = 0$

(ii) $x + 4y = 0$
     $2x + 8y = 0$

(iii) $x + y + z = 0$
$x - y + 2z = 0$
$2x + y - 3z = 0$

(iv) $x + y + z = 0$
$x - y - z = 0$
$3x + y + z = 0$

(v) $2x - 2y + z = 0$
$3x - y + z = 0$
$x + y = 0$

(vi) $2x - y + z = 0$
$x + y - z = 0$
$3x + 2y + 2z = 0$

15. Construct a system of three linear homogeneous equations in three unknowns that has a non-trivial solution

16. Determine the eigenvalues of the following:

(i) $\begin{bmatrix} 1 & 0 \\ 0 & 2 \end{bmatrix}$

(ii) $\begin{bmatrix} 6 & -3 \\ 2 & 1 \end{bmatrix}$

(iii) $\begin{bmatrix} 2 & 3 \\ -1 & -2 \end{bmatrix}$

(iv) $\begin{bmatrix} 8 & -4 \\ 2 & 2 \end{bmatrix}$

(v) $\begin{bmatrix} -1 & 1 & 0 \\ 0 & -2 & 2 \\ 3 & -9 & 6 \end{bmatrix}$

(vi) $\begin{bmatrix} 1 & 0 \\ 2 & 1 \end{bmatrix}$

(vii) $\begin{bmatrix} 2 & 0 & -1 \\ 0 & 2 & 0 \\ -1 & 0 & 2 \end{bmatrix}$

(viii) $\begin{bmatrix} 1 & 2 & -1 \\ 2 & 1 & 1 \\ -1 & 1 & 0 \end{bmatrix}$

(ix) $\begin{bmatrix} 1 & 0 & 0 \\ 2 & 1 & -2 \\ 3 & 2 & 1 \end{bmatrix}$

17. Determine the eigenvectors for each of the matrices in Q16.

13.9 Applications

1. A standard example in which systems of linear equations in more than one unknown arise is that of the modelling of electrical circuits or **networks**. A system of electrical components and connections might be very complicated, but can be analysed by regarding it as comprised of simple loops and branches to which **Kirchoff's Laws** can be applied. A typical problem might be to determine the current in some part of a circuit given the resistances, emfs, etc. in the network components. Such a problem resulted in the following system of equations for currents i_1, i_2, i_3 (amps) in a network:

$$i_1 + i_2 + i_3 = 0$$
$$-8i_2 + 10i_3 = 0$$
$$4i_1 - 8i_2 = 6$$

Solve the system for i_1, i_2, i_3 by (i) Cramer's rule (ii) matrix inversion.

Notes:
- For practical circuits of any size some standard software package would be used, probably involving numerical methods.
- Useful mathematical analysis of the network can be done prior to solution, using the techniques of **graph theory**, to optimise the solution process
- Cramer's rule and matrix inversion are essentially equivalent.

2. Systems of forces acting on a body in equilibrium can also lead to systems of linear equations. For example, resolution of forces and balancing of moments leads to the following equations for three forces F_1, F_2, F_3 (Newtons) acting on one of the struts in an aircraft wing.

$$F_1 - F_2 = 0$$
$$2F_1 + F_2 - 2F_3 = 20$$
$$F_2 - F_3 = 4$$

Find the forces by (i) Cramer's rule (ii) matrix inversion.

3. **The simple harmonic oscillator**
The position, x, of a particle performing simple harmonic motion about the origin at a time t can be described through Newton's laws of motion by the differential equation:

$$\frac{d^2x}{dt^2} + \omega^2 x = 0$$

where ω^2 is a positive quantity. It is standard practice to introduce a new variable:

$$\frac{dx}{dt} = y$$

to give the **second order system**:

$$\frac{dx}{dt} = y$$
$$\frac{dy}{dt} = -\omega^2 x$$

which may be written, in matrix form as,

$$\frac{d}{dt}\mathbf{x} = \begin{bmatrix} 0 & 1 \\ -\omega^2 & 0 \end{bmatrix} \mathbf{x} \quad \mathbf{x} = \begin{bmatrix} x \\ y \end{bmatrix}$$
$$= A\mathbf{x}$$

For various reasons arising from the theory of differential equations (Chapter 15) we now try to solve such systems by assuming solutions of the form

$$x = x_0 e^{\lambda t}$$
$$y = y_0 e^{\lambda t}$$

where λ is some constant to be determined, and x_0, y_0 will depend on the initial conditions. Substituting in the equations gives the matrix form

$$\begin{bmatrix} -\lambda & 1 \\ -\omega^2 & -\lambda \end{bmatrix} \begin{bmatrix} x_0 \\ y_0 \end{bmatrix} = 0$$

We only get non-trivial values for (x_0, y_0) if the determinant of coefficients is zero:

$$\begin{vmatrix} -\lambda & 1 \\ -\omega^2 & -\lambda \end{vmatrix} = \left\| \begin{bmatrix} 0 & 1 \\ -\omega^2 & 0 \end{bmatrix} - \lambda \begin{bmatrix} 1 & 0 \\ 0 & 1 \end{bmatrix} \right\| = 0$$

This is precisely the eigenvalue equation for the **matrix of coefficients** of the original system:

$$A = \begin{bmatrix} 0 & 1 \\ -\omega^2 & 0 \end{bmatrix}$$

Convert the equation

$$\frac{d^2x}{dt^2} + 4x = 0$$

to a first order system as described above. Try a solution of the form $x = x_0 e^{\lambda t}$, $y = y_0 e^{\lambda t}$ in this system and find the possible values of λ that will yield a non-trivial solution – they will in fact be complex. Hence solve the equations and express the solutions without complex numbers.

13.10 Answers to reinforcement exercises

1. (i) $\begin{bmatrix} 2 & -1 \\ 1 & 2 \end{bmatrix} \begin{bmatrix} x \\ y \end{bmatrix} = \begin{bmatrix} -1 \\ 0 \end{bmatrix}$

   (ii) $\begin{bmatrix} 3 & -1 & 2 \\ 2 & 2 & 3 \\ -3 & 0 & -1 \end{bmatrix} \begin{bmatrix} x \\ y \\ z \end{bmatrix} = \begin{bmatrix} 1 \\ 2 \\ -3 \end{bmatrix}$

   (iii) $\begin{bmatrix} 1 & 2 & 3 \\ 1 & -1 & -1 \end{bmatrix} \begin{bmatrix} a \\ b \\ c \end{bmatrix} = \begin{bmatrix} 1 \\ 3 \end{bmatrix}$

   (iv) $\begin{bmatrix} 4 & -2 \\ 3 & 3 \\ 1 & -1 \end{bmatrix} \begin{bmatrix} u \\ v \end{bmatrix} = \begin{bmatrix} 1 \\ 2 \\ -1 \end{bmatrix}$

2. If $A = \begin{bmatrix} -1 & 2 & 3 \\ 4 & 0 & 2 \\ -1 & 1 & 2 \end{bmatrix}$ $B = \begin{bmatrix} 0 & 1 & 3 \\ 2 & -1 & 4 \\ 3 & -1 & 2 \end{bmatrix}$

   (i) $2, -1, 2, 0, 2, -1$

   (ii) $11, -4$

   (iii) (a) $\begin{bmatrix} -2 & 1 & -3 \\ 2 & 3 & -8 \\ -11 & 5 & -2 \end{bmatrix}$ (b) $\begin{bmatrix} 6 & 1 & 7 \\ -6 & 10 & 16 \\ 3 & 0 & 3 \end{bmatrix}$ (c) $\begin{bmatrix} 13 & -6 & 11 \\ 6 & 2 & 16 \\ 8 & -4 & 5 \end{bmatrix}$

   (d) $\begin{bmatrix} 1 & 3 & 8 \\ -10 & 8 & 12 \\ -9 & 8 & 11 \end{bmatrix}$

405

3. $a = -1, b = 2, c = 3, d = 2, x = y = 3, z = 2$

4. $\theta = 30°$

5. Additions

$$2A = \begin{bmatrix} 4 & -2 \\ 6 & 0 \end{bmatrix} \qquad 2B = \begin{bmatrix} 6 \\ -2 \\ 8 \end{bmatrix} \qquad 2C = [-4 \quad 0]$$

$$2D = \begin{bmatrix} 0 & 2 & -2 \\ 4 & 0 & 6 \\ -2 & -6 & 4 \end{bmatrix} \qquad 2E = \begin{bmatrix} 6 & 8 \\ -2 & 4 \\ 0 & 6 \end{bmatrix}$$

$$2F = \begin{bmatrix} 2 & 2 & 2 \\ 4 & 4 & 4 \end{bmatrix} \qquad 2G = [4 \quad 2 \quad 0]$$

Multiplications

$$AF = \begin{bmatrix} 0 & 0 & 0 \\ 3 & 3 & 3 \end{bmatrix} \qquad A^2 = \begin{bmatrix} 1 & -2 \\ 6 & -3 \end{bmatrix} \qquad BG = \begin{bmatrix} 6 & 3 & 0 \\ -2 & -1 & 0 \\ 8 & 4 & 0 \end{bmatrix}$$

$$CA = [-4 \quad 2] \qquad CF = [-2 \quad -2 \quad -2] \qquad D^2 = \begin{bmatrix} 3 & 3 & 1 \\ -3 & -7 & 4 \\ -8 & -7 & -4 \end{bmatrix}$$

$$DB = \begin{bmatrix} -5 \\ 18 \\ 8 \end{bmatrix} \qquad DE = \begin{bmatrix} -1 & -1 \\ 6 & 17 \\ 0 & -4 \end{bmatrix} \qquad EA = \begin{bmatrix} 18 & -3 \\ 4 & 1 \\ 9 & 0 \end{bmatrix}$$

$$EF = \begin{bmatrix} 11 & 11 & 11 \\ 3 & 3 & 3 \\ 6 & 6 & 6 \end{bmatrix} \qquad FB = \begin{bmatrix} 6 \\ 12 \end{bmatrix} \qquad FD = \begin{bmatrix} 1 & -2 & 4 \\ 2 & -4 & 8 \end{bmatrix}$$

$$FE = \begin{bmatrix} 2 & 9 \\ 4 & 18 \end{bmatrix} \qquad GB = 5 \qquad GD = [4 \quad 2 \quad 1]$$

$$GE = [5 \quad 10]$$

6. (i) Not possible (ii) $\begin{bmatrix} 16 & 0 \\ 18 & 2 \\ -4 & 2 \end{bmatrix}$ (iii) $\begin{bmatrix} 8 & 8 \\ 3 & 3 \\ 2 & -5 \end{bmatrix}$

(iv) $\begin{bmatrix} 11 & 10 & 15 \\ 12 & 9 & 30 \\ 5 & 1 & -7 \end{bmatrix}$ (v) u, v, B not all same size

(vi) $\begin{bmatrix} 0 \\ 5 \\ 5 \end{bmatrix}$ (vii) $\begin{bmatrix} 4 \\ 3 \\ 1 \end{bmatrix}$

7. A, B, C are the matrices

$$A = \begin{bmatrix} 3 & 1 \\ -1 & 2 \end{bmatrix} \qquad B = \begin{bmatrix} 2 & -1 & 0 \\ 3 & 1 & 2 \end{bmatrix} \qquad C = \begin{bmatrix} 1 & 1 & 2 \\ -2 & 2 & 3 \end{bmatrix}$$

$(B + C)A \neq BA + CA$, indeed neither of $(B + C)A$, BA, or CA exist

8. (i) 14 (ii) −6 (iii) −18 (iv) 4
   (v) 0 (vi) 1

9. (i) $3\begin{vmatrix}1&2\\-1&2\end{vmatrix}+\begin{vmatrix}0&2\\4&2\end{vmatrix}+2\begin{vmatrix}0&1\\4&-1\end{vmatrix}$

   (ii) $-0\begin{vmatrix}-1&2\\-1&2\end{vmatrix}+1\begin{vmatrix}3&2\\4&2\end{vmatrix}-2\begin{vmatrix}3&-1\\4&-1\end{vmatrix}$

   (iii) $2\begin{vmatrix}0&1\\4&-1\end{vmatrix}-2\begin{vmatrix}3&-1\\4&-1\end{vmatrix}+2\begin{vmatrix}3&-1\\0&1\end{vmatrix}$

   (iv) $4\begin{vmatrix}-1&2\\1&2\end{vmatrix}+1\begin{vmatrix}3&2\\0&2\end{vmatrix}+2\begin{vmatrix}3&-1\\0&1\end{vmatrix}$

   All equal to −4

10. (i) $\begin{bmatrix}4/5&-3/5\\-1/5&2/5\end{bmatrix}$ (ii) $\dfrac{1}{18}\begin{bmatrix}1&-5&7\\7&1&-5\\-5&7&1\end{bmatrix}$

    (iii) $\dfrac{1}{14}\begin{bmatrix}3&-1&5\\5&3&-1\\-1&5&3\end{bmatrix}$ (iv) $\dfrac{1}{5}\begin{bmatrix}-10&4&9\\15&-4&-14\\-5&1&6\end{bmatrix}$

    (v) $\dfrac{1}{26}\begin{bmatrix}6&12&-8\\10&-6&4\\-2&-4&7\end{bmatrix}$

11. (i) No solution (ii) $x=1, y=-3, z=2$ (iii) $x=1, y=2, z=3$

12. $x = 1$ or 4

13. (i) $x=1, y=0$ (ii) $x=\dfrac{3}{62}, y=-\dfrac{29}{124}$ (iii) $x=\dfrac{1}{2}, y=2, z=-\dfrac{3}{2}$
    (iv) No solution

14. (i) Trivial solution only (ii) Non-trivial solution
    (iii) Trivial solution only (iv) Non-trivial solution
    (v) Non-trivial solution (vi) Trivial solution only

15. Example:
    $$x+y+z=0$$
    $$x+2y+3z=0$$
    $$2x+3y+4z=0$$

16. (i) 1, 2 (ii) 3, 4 (iii) 1, −1 (iv) 4, 6
    (v) 0, 1, 2 (vi) 1 (twice) (vii) 1, 2, 3 (viii) 1, −2, 3
    (ix) $1, 1\pm 2j$

17. (i) $1, \begin{bmatrix} 1 \\ 0 \end{bmatrix}; 2, \begin{bmatrix} 0 \\ 1 \end{bmatrix}$

(ii) $3, \dfrac{1}{\sqrt{2}}\begin{bmatrix} 1 \\ 1 \end{bmatrix}; 4, \dfrac{1}{\sqrt{13}}\begin{bmatrix} 3 \\ 2 \end{bmatrix}$

(iii) $1, \dfrac{1}{\sqrt{10}}\begin{bmatrix} -3 \\ 1 \end{bmatrix}; -1, \dfrac{1}{\sqrt{2}}\begin{bmatrix} 1 \\ -1 \end{bmatrix}$

(iv) $4, \dfrac{1}{\sqrt{2}}\begin{bmatrix} 1 \\ 1 \end{bmatrix}; 6, \dfrac{1}{\sqrt{5}}\begin{bmatrix} 2 \\ 1 \end{bmatrix}$

(v) $0, \dfrac{1}{\sqrt{3}}\begin{bmatrix} 1 \\ 1 \\ 1 \end{bmatrix}; 1, \dfrac{1}{\sqrt{14}}\begin{bmatrix} 1 \\ 2 \\ 3 \end{bmatrix}; 2, \dfrac{1}{\sqrt{46}}\begin{bmatrix} 1 \\ 3 \\ 6 \end{bmatrix}$

(vi) $1 \text{ (twice)}, \begin{bmatrix} 0 \\ 1 \end{bmatrix}$

(vii) $1, \dfrac{1}{\sqrt{2}}\begin{bmatrix} 1 \\ 0 \\ 1 \end{bmatrix}; 2, \begin{bmatrix} 0 \\ 1 \\ 0 \end{bmatrix}; 3, \dfrac{1}{\sqrt{2}}\begin{bmatrix} 1 \\ 0 \\ -1 \end{bmatrix}$

(viii) $1, \dfrac{1}{3}\begin{bmatrix} 1 \\ 2 \\ -2 \end{bmatrix}; -2, \dfrac{1}{\sqrt{3}}\begin{bmatrix} 1 \\ -1 \\ 1 \end{bmatrix}; 3, \dfrac{1}{\sqrt{2}}\begin{bmatrix} 1 \\ 1 \\ 0 \end{bmatrix}$

(ix) $1, \dfrac{1}{\sqrt{17}}\begin{bmatrix} 2 \\ -3 \\ 2 \end{bmatrix}; 1+2j, \begin{bmatrix} 0 \\ j \\ 1 \end{bmatrix}; 1-2j, \begin{bmatrix} 0 \\ 1 \\ j \end{bmatrix}$

14

Analysis for Engineers – Limits, Sequences, Iteration, Series and All That

The topics considered in this chapter are often regarded as something of a luxury for engineers and scientists, who are normally more concerned with using techniques rather than worrying too much about the underlying theory. However, the ideas are not really that difficult, and there are in fact many engineering situations where it is necessary to pay particular attention to things such as continuity and differentiability. Also, even though some of the techniques may not be of immediate practical use, they are important in applications of numerical methods in engineering. Thus, before engaging in costly computational approaches, it is important to check that the problem is mathematically well defined. Does a 'solution' exist? Is it unique? Will the computational scheme converge to it? How long will it take? What are the error bounds? ... etc. These are the sorts of questions that analysis addresses.

Prerequisites

It will be helpful if you know something about:

- different types of number (5 ◀)
- elementary notion of a limit (231 ◀)
- properties of zero (6 ◀)
- sketching graphs (91 ◀)
- the binomial theorem (71 ◀)
- properties of rational functions (56 ◀)
- slope of a curve (230 ◀)
- inequalities (97 ◀)
- sequences and series (105 ◀)
- the geometric series (105 ◀)
- differentiation (Chapter 8 ◀)
- the modulus sign (96 ◀)
- elementary functions – trig, exponential, etc. (Chapters 4 and 6 ◀)

Objectives

In this chapter you will find:

- more on irrational numbers

- evaluation and properties of limits
- continuity
- slope of a curve and theory of differentiation
- infinite sequences
- iteration and Newton's method
- infinite series
- infinite power series
- convergence of sequences and series

Motivation

You may need the material of this chapter for:

- understanding when particular mathematical methods are applicable
- solving equations by iteration
- testing convergence of numerical methods

14.1 Continuity and irrational numbers

What do we mean by a 'continuous curve'? Plotting the result of an experiment on a graph will lead to a series of points separated by gaps (which may be small, but will always be there). We can never **plot** a **continuous curve**, although we usually draw one through the points. In fact, the idea of a continuous curve is a convenient mathematical abstraction which allows us to use geometry to talk about slopes and rates of change. When we draw a continuous curve we tacitly assume that the curve passes through every point on it (which it doesn't – there will at least be gaps between the molecules in the chemicals of the ink!).

To **plot** a point on a curve we must **measure** its distance from some origin. This can only be done to a certain level of accuracy and so can only be done using a terminating decimal, e.g. 3.412. As noted in Chapter 1 such numbers can always be written as **rational numbers** – those which can be expressed in the form m/n where m and n are **integers**.

Problem 14.1
Write the decimal 3.412 as a rational number.

Quite simply in this case we have

$$3.412 = \frac{3412}{1000}$$

However, there are numbers which cannot be written in this form as fractions – i.e. there exist quantities which we can **never** measure exactly and yet which do have a real existence! These can never be plotted on a graph and so represent 'holes' in the apparently continuous curve. An elementary example of such a number is the diagonal of the unit square. By Pythagoras' theorem (154 ◄) we know that the unit square has diagonal $\sqrt{2}$ units. Clearly, this number 'exists', otherwise we couldn't cross a square diagonally.

However, we will show that it cannot in fact be represented by a decimal with a finite number of places – i.e. it is **not** a rational number, and so therefore it cannot be 'measured' or 'plotted' exactly. The method of proof is **by contradiction**, that is, we **assume** that $\sqrt{2}$ can be written as a rational number and then derive a **contradiction**. The proof is one of the prettiest in elementary pure mathematics and it has great physical significance, demonstrating that there is a real physical quantity of great importance that we cannot actually measure. It also gives good practice in elementary algebra, so try to work through it.

Assume that we can write $\sqrt{2}$ as a rational number:

$$\sqrt{2} = \frac{m}{n}$$

where m, n are integers, each cancelled down to their lowest form (i.e. **m and n have no factors in common** (12 ◄)). Then

$$2 = \frac{m^2}{n^2}$$

or

$$m^2 = 2n^2$$

This implies that m^2 is even. Since the square of an odd number is odd m must also be even. Let's write it $m = 2k$ where k is an integer. Then:

$$m^2 = 4k^2 = 2n^2$$

or

$$n^2 = 2k^2$$

By the same argument as for m, this implies that n is even.

We have therefore shown that both m and n are even and therefore contain (at least) the factor 2 in common. This is a contradiction since we assumed that m and n have no factors in common. So we have to conclude that $\sqrt{2}$ cannot be written as a rational number.

Numbers such as $\sqrt{2}$, which cannot be written in the form m/n, are called **irrational numbers**. They are clearly of more than academic interest because we cannot draw a continuous curve without them. So, if we definitely cannot plot all points on a continuous curve, how do we define such a curve mathematically? To define continuity precisely we have to introduce the idea of a **limit** of a function at a point. This is the value of the function as we approach closer and closer to the point indefinitely, without actually reaching the point. It might or might not exist.

Exercises on 14.1

1. π is often approximated crudely by $\frac{22}{7}$. Explain why this expression cannot be exactly equal to π.
2. Using a calculator or computer evaluate $\sqrt{2}$ to as many decimal places as you can. Square your result – do you retrieve 2?

Answers

1. $\frac{22}{7}$ is rational, whereas it can be shown that π is not.

2. On my calculator $\sqrt{2} = 1.41421356237$. My calculator squares this to 1.9999999998, so it is clearly not the square root of 2.

14.2 Limits

Suppose we are trying to find the velocity of a particle at a time t. As velocity is distance moved divided by time taken we have to measure a small distance moved and divide by the time taken for this. But this will only give us the **average velocity** over this small interval. Actually, at the point t there is zero movement in zero time – is the velocity 0/0? Remember that in Chapter 1 we said this is not defined (7 ◀). The way out of this is to consider smaller and smaller intervals and find the average velocity over these intervals. As the interval decreases, the average velocity over it should be nearer to the actual velocity at t. We say that we find the **limit of the velocity as the interval goes to zero**. Notice that we never actually evaluate the velocity **at** the point t – because this just gives 0/0, as seen above – we just get nearer and nearer to it without actually reaching it. This is formalised in the following definition.

The **limit of $f(x)$ as x tends to a** is defined as the value of $f(x)$ as x approaches closer and closer to a without actually reaching it and is denoted by:

$$\lim_{x \to a} f(x)$$

There are some points to note:

(i) We do not evaluate the limit by actually substituting $x = a$ in $f(x)$ in general, although in some simple cases it is possible.
(ii) The value of the limit can depend on 'which side it is approached' – from 'above' or 'below', i.e. through values of x less than a or through values of x greater than a respectively. The two possible values may not be the same – in which case the limit does not exist.
(iii) The limit may not exist at all – and even if it does it may not be equal to $f(a)$.

The sort of limits that cause most difficulty and which are probably the most important are those arising from so called 'indeterminate forms'. Any expression that yields results of the form 0/0, ∞/∞, or $0 \times \infty$ is called an **indeterminate form**. Examples are

$$\frac{x^2 - 1}{x - 1} = \frac{0}{0} \text{ at } x = 1 \quad \text{and} \quad \frac{\sin x}{x} = \frac{0}{0} \text{ at } x = 0$$

Even though the function does not exist at such points, its **limit** at the point may exist. For example as we will see below,

$$\lim_{x \to 1} \left(\frac{x^2 - 1}{x - 1} \right) \quad \text{and} \quad \lim_{x \to 0} \left(\frac{\sin x}{x} \right)$$

both exist.

Problem 14.2

Evaluate $f(x) = x^2 + 2x + 3$ for $x = 1.1, 1.01, 1.001, 1.0001$. What do you think $f(x)$ approaches as x gets closer and closer to 1?

We obtain the values in the table below

x	$f(x)$
1.1	6.41
1.01	6.0401
1.001	6.004001
1.0001	6.00040001

So as x gets closer and closer to 1, $f(x)$ gets closer and closer to 6 – we say that the limit as x tends to 1 is 6. Similar results are obtained if we consider values of x less than and increasingly closer to 1. The results get increasingly closer to 6. This is also the value of $f(x)$ at $x = 1$ of course. Graphically, the situation is illustrated in Figure 14.1.

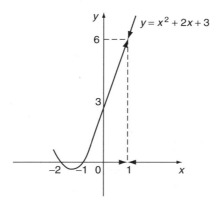

Figure 14.1 The graph of $y = x^2 + 2x + 3$.

Problem 14.3

Evaluate the function $f(x) = \dfrac{x^2 - 1}{x - 1}$ for $x = 1.2, 1.02, 1.002, 1.0002$ – what do you think is the limit as x tends to 1?

We obtain the table below

x	$f(x)$
1.2	2.2
1.02	2.02
1.002	2.002
1.0002	2.0002

The table of values suggests that as $x \to 1$, $f(x) \to 2$ (the arrow represents 'tends to'). However, of course,

$$f(1) = \frac{1^2 - 1}{1 - 1} = \frac{0}{0}$$

which is **indeterminate** indicating that in this case we cannot evaluate the limit simply by substituting the value of $x = 1$. Instead, we proceed as follows:

$$\lim_{x \to 1} \frac{x^2 - 1}{x - 1} = \lim_{x \to 1} \frac{(x - 1)(x + 1)}{(x - 1)} = \lim_{x \to 1} (x + 1) = 2$$

The cancellation of the $x - 1$ factor under the limit **is** permissible because in the limit we only consider values of x **arbitrarily close** but never **equal** to 1. So $x - 1$ is **never** zero and so can be cancelled under the limit. See Figure 14.2.

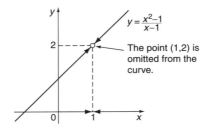

Figure 14.2 The graph of $y = \dfrac{x^2 - 1}{x - 1}$.

Problem 14.4

Investigate the limit of $f(x) = \dfrac{1}{x - 1}$ as x tends to 1.

The graph of this function is shown in Figure 14.3 and it is clear that the limit at $x = 1$ does not exist. If we approach $x = 1$ from below the limit tends to $-\infty$. If we approach from above it tends to $+\infty$. We say there is a **discontinuity** at $x = 1$.

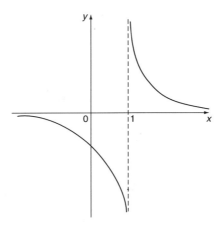

Figure 14.3 The graph of $y = \dfrac{1}{x - 1}$.

The properties of limits are fairly well what we might expect. Thus, if

$$\lim_{x \to a} f(x) = b \quad \lim_{x \to a} g(x) = c$$

then

L1. $\lim_{x \to a} kf(x) = kb$ for any fixed k

L2. $\lim_{x \to a}[f(x) \pm g(x)] = \lim_{x \to a} f(x) \pm \lim_{x \to a} g(x) = b \pm c$

L3. $\lim_{x \to a}[f(x)g(x)] = \lim_{x \to a} f(x) \lim_{x \to a} g(x) = bc$

L4. $\lim_{x \to a}\left[\dfrac{f(x)}{g(x)}\right] = \dfrac{\lim_{x \to a} f(x)}{\lim_{x \to a} g(x)} = \dfrac{b}{c}$ if $c \neq 0$

L5. $\lim_{x \to a}[f(x)]^{\frac{1}{n}} = b^{\frac{1}{n}}$ provided $[f(x)]^{\frac{1}{n}}$ and $b^{\frac{1}{n}}$ are real.

NB. We must always check that $\lim_{x \to a} f(x)$, $\lim_{x \to a} g(x)$ exist before applying the above results, all of which are proved rigorously in pure maths books, but are fairly 'obvious'.

Problem 14.5

Evaluate the limits

(i) $\lim_{x \to 1}\left[\dfrac{x^2 - 1}{x - 1} \times (x^2 + 2x + 3)\right]$ (ii) $\lim_{x \to 1}\left[\dfrac{x^2 - 1}{(x - 1)\sqrt{x^2 + 2x + 3}}\right]$

We know from Problems 14.2 and 14.3 that

$$\lim_{x \to 1} \dfrac{x^2 - 1}{x - 1} = 2 \quad \text{and} \quad \lim_{x \to 1}(x^2 + 2x + 3) = 6$$

(i) Using rule L3, we have

$$\lim_{x \to 1}\left[\dfrac{x^2 - 1}{x - 1} \times (x^2 + 2x + 3)\right] = \lim_{x \to 1}\dfrac{x^2 - 1}{x - 1} \times \lim_{x \to 1}(x^2 + 2x + 3)$$
$$= 2 \times 6 = 12$$

which you can see we also get from:

$$\lim_{x \to 1}[(x + 1) \times (x^2 + 2x + 3)] = 12$$

(ii) Using rules L4 and L5, we have

$$\lim_{x \to 1}\left[\dfrac{x^2 - 1}{(x - 1)\sqrt{x^2 + 2x + 3}}\right] = \lim_{x \to 1}\dfrac{x^2 - 1}{x - 1} \div \lim_{x \to 1}\sqrt{x^2 + 2x + 3}$$
$$= 2 \div \sqrt{\lim_{x \to 1} x^2 + 2x + 3}$$
$$= 2 \div \sqrt{6} = \dfrac{2}{\sqrt{6}} = \dfrac{\sqrt{6}}{3}$$

and again you can see we get this directly from

$$\lim_{x \to 1} \left[\frac{x+1}{\sqrt{x^2 + 2x + 3}} \right] = \frac{2}{\sqrt{6}}$$

Note that it would **not** be correct to write

$$\lim_{x \to 1} \left[\frac{x^2 - 1}{(x-1)\sqrt{x^2 + 2x + 3}} \right] = \lim_{x \to 1}(x^2 - 1) \div \lim_{x \to 1}((x-1)\sqrt{x^2 + 2x + 3})$$

since the last limit, which we are trying to divide by, is zero.

Exercises on 14.2

1. Investigate the (very useful) limit of $\frac{1}{x}$ as x tends to infinity, i.e. gets infinitely large. Evaluate

   (i) $\lim\limits_{x \to \infty} \dfrac{1}{x-1}$ (ii) $\lim\limits_{x \to \infty} \dfrac{2x}{x-4}$

2. Consider the values of the following functions as x gets closer and closer (but not equal) to $x = 2$:

   (i) $x - 2$
   (ii) $x + 2$
   (iii) $x^2 - 4$
   (iv) $\dfrac{1}{x^2 - 4}$
   (v) $\dfrac{x-2}{x^2 - 4}$

   In each case evaluate the limit as $x \to 2$ using the techniques of this section.

3. Evaluate the limits

   (i) $\lim\limits_{x \to 0} x^2$ (ii) $\lim\limits_{x \to 0} \dfrac{x}{x^2 - x}$ (iii) $\lim\limits_{x \to 1} \dfrac{x - 1}{x^2 - 3x + 2}$

Answers

1. $\lim\limits_{x \to \infty} \dfrac{1}{x} = 0$ (i) 0   (ii) 2
2. (i) 0   (ii) 4   (iii) 0   (iv) doesn't exist   (v) $\frac{1}{4}$
3. (i) 0   (ii) -1 (iii) -1

14.3 Some important limits

There are further examples of limits that are very important in calculus and approximations. If you wish, simply remember the results and how to use them, but working through the proofs will give you good practice in basic mathematics as well as enhancing your understanding of the results.

(i) $\lim\limits_{\theta \to 0} \dfrac{\sin \theta}{\theta} = 1$

When $\theta = 0$ this yields the indeterminate form 0/0.

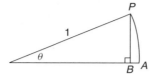

Figure 14.4

From Figure 14.4 we see that if PA is an arc of a unit circle subtending an angle θ at the centre (173 ◄), then we have:

$$\frac{PB}{1} = \sin\theta$$

and $PA = 1 \times \theta$ (rads).

So $\dfrac{\sin\theta}{\theta} = \dfrac{PB}{PA}$ and clearly as $\theta \to 0$, $PB \to PA$, so $\dfrac{\sin\theta}{\theta} \to 1$

Thus we obtain the given result. This limit implies that for θ small we have $\sin\theta \simeq \theta$. The same sort of reasoning can also be used to show that for very small θ, $\cos\theta \simeq 1$. It is also useful to note that provided $k \neq 0$,

$$\lim_{\theta \to 0} \frac{\sin k\theta}{k\theta} = 1$$

So, for example

$$\lim_{\theta \to 0} \frac{\sin 6\theta}{6\theta} = 1$$

Once again, remember that in all of this discussion, and in use of the limits defined, **θ must be in radians**.

(ii) $\lim\limits_{x \to a} \left(\dfrac{x^n - a^n}{x - a} \right) = na^{n-1}$

When $x = a$ this has the form 0/0. To find the value of the limit as $x \to a$ we consider values of x close to a and therefore take $x = a + h$ so as $x \to a$, $h \to 0$. Using the binomial theorem (71 ◄), we have

$$\lim_{h \to 0}\left[\frac{(a+h)^n - a^n}{a+h-a}\right] = \lim_{h \to 0}\left[\frac{a^n + na^{n-1}h + \dfrac{n(n-1)}{2}a^{n-2}h^2 + \cdots - a^n}{h}\right]$$

$$= \lim_{h \to 0}\left[na^{n-1} + \frac{n(n-1)}{2}a^{n-2}h + \cdots\right] = na^{n-1}$$

This result is essentially the differentiation of x^n from first principles (231 ◄).

(iii) $\lim\limits_{x \to \infty} \dfrac{e^x}{x^n} = \lim\limits_{x \to \infty} \left[\dfrac{1}{x^n}\left(1 + x + \dfrac{x^2}{2!} + \cdots + \dfrac{x^n}{n!} + \cdots \right)\right]$
(126 ◄)

$$= \lim_{x \to \infty}\left(\frac{1}{x^n} + \frac{1}{x^{n-1}} + \cdots + \frac{1}{n!} + \frac{x}{(n+1)!} + \cdots\right)$$

$$= \infty$$

So as $x \to \infty$, $\dfrac{e^x}{x^n} \to \infty$ – i.e. e^x is 'stronger' than x^n; for any positive integer n. So for example:

$$\frac{x^n}{e^x} = x^n e^{-x} \to 0 \quad \text{as } x \to \infty$$

In fact, it is not difficult to extend the above proof to apply for any value of n, not necessarily integer. This limit is very useful in Laplace transforms and should be thoroughly understood (➤ 502).

Exercise on 14.3

Evaluate the following limits

(i) $\displaystyle\lim_{x \to 0} \frac{\sin 2x}{x}$ (ii) $\displaystyle\lim_{x \to 3} \frac{x^3 - 27}{x - 3}$ (iii) $\displaystyle\lim_{x \to \infty} \frac{2x^3 - x}{e^{3x}}$

Answer

(i) 2 (ii) 27 (iii) 0

14.4 Continuity

We can now use the idea of a limit to put the concept of continuity into mathematical form. If we focus on a particular point $x = a$ on the curve of a function $f(x)$, then clearly if the curve is continuous at that point then we would want the limit at $x = a$ to be the same 'from both sides' and to be equal to $f(a)$. Only in this way can the two parts of the curve on either side of $x = a$ 'join up' without leaving a hole in the curve. We express this formally in the following definition:

A function $f(x)$ is **continuous at $x = a$** if $\lim_{x \to a} f(x) = f(a)$.

The graphs of continuous functions are continuous curves, and some examples of continuous and discontinuous curves are illustrated in Figure 14.5.

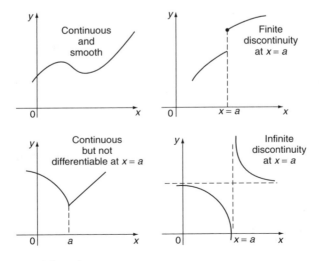

Figure 14.5 Continuous and discontinuous curves.

A function $f(x)$ such that

$$\lim_{x \to a} f(x) \neq f(a)$$

is said to be **discontinuous at $x = a$**.

For example, $f(x) = \dfrac{x+3}{(x-1)(x+5)}$ is discontinuous at $x = 1$ and $x = -5$. Another example of a discontinuous function is the $\tan x$ function (see Figure 6.10).

In general, discontinuous functions are a nuisance in mathematics, but various techniques have been devised to deal with them. They are sometimes useful to approximate rapid continuous variations, as for example the use of the **step function**, $H(t) = 0$ for $t < 0$ and $= 1$ for $t > 0$, in representing the 'instantaneous' throwing of a switch (Figure 14.6).

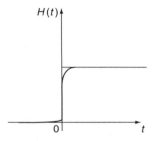

Figure 14.6 The step function $H(t)$.

Although we have used limits to define continuity, in practice we often know that a given function $f(x)$ is continuous, and then we can simply put $x = a$ to find the 'limit' or value of $f(x)$ at $x = a$. For example, the general rational function $P(x)/Q(x)$ can only have discontinuities where $Q(x) = 0$ and these are infinite if $P(x)$ is non-zero. If however, at $x = a$, $P(a) = Q(a) = 0$, then we have an indeterminate form 0/0 (the same discussion applies for ∞/∞) and a more detailed study is required. This can take the form of a substitution such as $z = a + h$ and studying the limit as $h \to 0$. This is equivalent to focusing attention on the point $x = a$ and looking closely at the behaviour near this point. We can perform any algebraic operations on $P(x)/Q(x)$ to simplify it, or rearrange it to a more convenient form for taking the limit – we can for example, cancel factors like $x - a$, because we never actually consider what happens at $x = a$. Another approach is to expand the function in a power series. In fact our main concern with limits and continuity lies in the theory of differentiation, where we need to consider indeterminate expressions such as 0/0.

Problem 14.6

Investigate the continuity of the function

$$f(x) = \frac{x-3}{x^2 - 4x + 3}$$

Sketch the function and define a new function, $g(x)$, that is continuous and such that $g(x) = f(x)$ for all $x > 2$, except at $x = 3$.

This is a rather sophisticated problem, but is worth working through because it brings together a number of the key subtleties of limits and continuity, and is very instructive. We have

$$f(x) = \frac{x-3}{x^2 - 4x + 3} = \frac{x-3}{(x-3)(x-1)}$$

This has two discontinuities, at $x = 1$ and $x = 3$. At $x = 3$ the function is indeterminate, but otherwise we can cancel the $x - 3$ and write

$$f(x) = \frac{1}{x-1} \quad \text{provided } x \neq 3$$

At $x = 1$ the discontinuity is infinite, of the same kind as that of $1/x$ at $x = 0$ (cf: Problem 14.4), **except** that the point $x = 3$ must be omitted, see Figure 14.7.

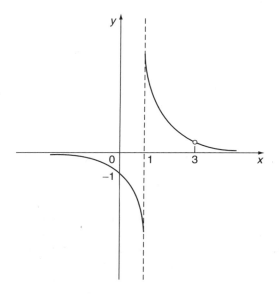

Figure 14.7 The function $f(x) = \dfrac{(x-3)}{(x-3)(x-1)}$.

While $f(x)$ does not exist at $x = 3$, the limit of the function does:

$$\lim_{x \to 3} f(x) = \lim_{x \to 3} \frac{x-3}{(x-3)(x-1)} = \lim_{x \to 3} \frac{1}{(x-1)}$$
$$= \tfrac{1}{2}$$

So we can 'plug the gap' and form a continuous function from $f(x)$ for $x > 2$ by defining a new function $g(x)$ that is equal to $f(x)$ for $x > 2$, $x \neq 3$, but is equal to $\tfrac{1}{2}$ at $x = 3$:

$$g(x) = f(x) \quad x > 2, x \neq 3$$
$$= \tfrac{1}{2} \quad \text{at } x = 3$$

The function $g(x)$ **is continuous** and indeed is equivalent to

$$g(x) = \frac{1}{x-1} \text{ for } x > 2$$

Exercise on 14.4

Consider the following functions for $0 < x < \infty$, and discuss whether or not they are continuous for these values of x.

(i) $x + 1$ (ii) $\frac{1}{x}$ (iii) $\frac{1}{x-1}$

(iv) $\sin x$ (v) $\ln x$ (vi) $\frac{x^2 - 1}{x + 1}$

(vii) $\frac{x+1}{x^2-1}$ (viii) $\frac{x^2-4}{x+2}$

(ix) $f(x) = \frac{x^2 - 4}{x - 2}$ $x \neq 2$
$= 4$ $x = 2$

Answer

(i) C   (ii) C   (iii) D   (iv) C   (v) C
(vi) C   (vii) D   (viii) C   (ix) C

14.5 The slope of a curve

The slope of a curve at a point is defined as the slope of the tangent at that point. It can be evaluated by a limiting process illustrated in Figure 14.8 (We adopted a similar approach in Chapter 8, using a different but equivalent notation (230 ◄).)

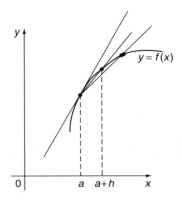

Figure 14.8 Definition of the derivative.

The value of the function at $x = a$ is $f(a)$, and the value at $x = a + h$ is $f(a + h)$, so the slope of the extended chord (also called the **secant**) joining these two points on the curve is

$$\frac{f(a+h) - f(a)}{a+h-a} = \frac{f(a+h) - f(a)}{h}$$

As $h \to 0$, the chord approaches the tangent to the curve at $x = a$, so the slope of the curve at $x = a$ is

$$\lim_{h \to 0} \frac{f(a+h) - f(a)}{h}$$

An equivalent and possibly more useful form is

$$\lim_{x \to a} \frac{f(x) - f(a)}{x - a}$$

Note that when $h = 0$, or $x = a$, this is of the indeterminate form 0/0. It is called the **derivative of $f(x)$** at $x = a$ (231 ◀). We use the notation

$$\lim_{x \to a} \left(\frac{f(x) - f(a)}{x - a} \right) = \frac{df(x)}{dx} \text{ at } x = a$$

Problem 14.7

Determine the slope of the curve $f(x) = x^n$ at the point $x = a$.

From above, the slope of $f(x) = x^n$ at $x = a$ is given by

$$\lim_{x \to a} \frac{f(x) - f(a)}{x - a} = \lim_{h \to 0} \left(\frac{x^n - a^n}{x - a} \right) = na^{n-1}$$

from Section 14.3.

Exercise on 14.5

Evaluate the slope of the function $f(x) = 3x^3$ using the limit definition.

Answer

$9x^2$

14.6 Introduction to infinite series

The purpose of this section is to ease you gently into the topic of infinite series, before we embark on a more formal treatment. How would you evaluate $\cos x$ for $x = 0.1$ radians? Probably tap it out on your calculator. But suppose you wanted greater accuracy than your calculator can give? Writing a short computer programme might be the answer. But how do you instruct the computer? You can't tell it to evaluate $\cos x$ from its geometrical definition 'adjacent over hypotenuse'. The logic circuits of computers can only do very simple arithmetic operations such as addition and multiplication. Fortunately, we can express $\cos x$ (x in radians!) in terms of just such operations by an **infinite power series** (107 ◄):

$$\cos x = 1 - \frac{x^2}{2!} + \frac{x^4}{4!} - \frac{x^6}{6!} + \cdots$$

If this is new to you, don't worry where it comes from just now. Because the series is infinite – the terms go on forever – we can't even total all of the terms and so can never write down the exact value of, for example, $\cos 0.1$:

$$\cos 0.1 = 1 - \frac{(0.1)^2}{2!} + \frac{(0.1)^4}{4!} - \frac{(0.1)^6}{6!} + \cdots$$

However, by taking a sufficient number of terms we can obtain the value of $\cos 0.1$ with as much accuracy as we desire. The terms in the series only involve multiplication and division, which a computer can do easily. So such series are just the thing for evaluating complicated functions. They are also useful in mathematical methods too – for example if x is small enough then we might be able to neglect terms such as $x^4, x^6, \ldots$ and take the **approximation**:

$$\cos x \simeq 1 - \frac{x^2}{2}$$

The advantage here is that the right-hand side is much easier to work with than the left-hand side.

We will look at how we find such series later, in Section 14.11. Here we will focus on issues arising from the fact that we have an **infinite** series. Because it is infinite, we can never see directly what is going to happen when we 'gather up' all the terms. For the $\cos x$ example above, the terms $\frac{(0.1)^r}{r!}$ got smaller and smaller, so maybe we can feel confident that we are all right to neglect the remaining infinity of terms after we cut the series off. Well, look at another series,

$$S = 1 + \tfrac{1}{2} + \tfrac{1}{3} + \tfrac{1}{4} + \cdots$$

The terms get smaller and smaller – maybe we can get a good approximation by 'truncating' the series? In fact, we can't. This can be seen by a very pretty argument. Because S is the sum of only positive terms, we can obviously get a quantity **less** than S by replacing some of its terms by smaller values. Suppose we did this as follows

$$S > 1 + \tfrac{1}{2} + (\tfrac{1}{4} + \tfrac{1}{4}) + (\tfrac{1}{8} + \tfrac{1}{8} + \tfrac{1}{8} + \tfrac{1}{8}) + \cdots$$

You can see the pattern:

$$= 1 + \tfrac{1}{2} + \tfrac{1}{2} + \tfrac{1}{2} + \cdots$$

But this clearly adds up to an infinite total and yet we know by construction that this is less than S – so S must also be infinite, showing that the series diverges.

This is typical of the sort of 'strange' results and pretty arguments one meets in the theory of series. From a practical point of view it means we have to be very careful when dealing with infinite series.

Exercise on 14.6

Adapt the argument of this section to show that the series

$$1 + \frac{1}{\sqrt{2}} + \frac{1}{\sqrt{3}} + \frac{1}{\sqrt{4}} + \cdots + \frac{1}{\sqrt{r}} + \cdots$$

does not add up to a finite quantity, i.e. it diverges.

14.7 Infinite sequences

As a prelude to infinite series we look at infinite **sequences** (102 ◄), which are simply lists that continue indefinitely – adding terms of an infinite sequence produces an infinite series. Such sequences arise naturally in such infinite processes as **iteration**, where we continually improve an approximation by recycling it until we get the accuracy we want.

An ordered set of numbers:

$$u_1, u_2, u_3, \ldots, u_n, \ldots$$

is called a **sequence** or **infinite sequence**. u_n is called the **nth term** (105 ◄). The nth term may be defined by a formula:

$$u_n = f(n)$$

or by a **recurrence relation** (difference equation), such as:

$$u_n = u_{n-1} + u_{n-2} \quad u_1 = a, \quad u_2 = b$$

Problem 14.8

Write down an expression or a recurrence relation for the nth term u_n in each of the following cases:

(i) $1, 4, 9, 16, 25, \ldots$

(ii) $1, \dfrac{1}{2!}, \dfrac{1}{3!}, \dfrac{1}{4!}, \dfrac{1}{5!}, \ldots$

(iii) $1, 4, 7, 10, \ldots$

(iv) $2, 4, 10, 24, 58 \ldots$

This is really a matter of educated guesswork – find a result that fits the first few terms of the sequence, check that it holds for the other terms and assume that it holds in general for all terms, even those that are not explicitly written down.

(i) We note that the terms can be written as $1^2, 2^2, 3^2, 4^2, 5^2, \ldots$ which suggests that in general we can take

$$u_n = n^2$$

(ii) The same idea suggests $u_n = \dfrac{1}{n!}$ in this case.

(iii) In this case $u_n = 1 + (n-1)3$ is pretty obvious.

(iv) In this case it is perhaps not quite so easy to spot a pattern for an explicit expression for u_n. You might spot more easily that there is a relation between u_n and u_{n-1}, u_{n-2}. Explicitly we find the recurrence relation $u_n = 2u_{n-1} + u_{n-2}$; $u_1 = 2, u_2 = 4$.

In case you are wondering what the explicit expression for u_n would be in (iv) that is equivalent to the recurrence relation, it is actually

$$u_n = \frac{1}{\sqrt{2}}(1+\sqrt{2})^n - \frac{1}{\sqrt{2}}(1-\sqrt{2})^n$$

Not something that readily springs to mind, as do (i) – (iii)! This illustrates why recurrence relations are sometimes to be preferred. As a useful exercise in surds you might like to check that this explicit result does generate the given sequence. Also, you may like to check, using the binomial expansion that an equivalent form to the above result, not involving surds, is

$$u_n = \sum_{r=1}^{n} C_r 2^{(r+1)/2}$$

where the sum ranges only over odd values of r. Again, not a form that springs immediately to mind!

An obvious question is, as we continue along an infinite sequence, does the nth term tend to a definite limit? That is, what is the behaviour of u_n as $n \to \infty$? Symbolically, what is $\lim_{n\to\infty} u_n$? Clearly this depends on the form of the nth term.

Problem 14.9

Examine the limits of the following sequences as n increases indefinitely:

(i) $1, \dfrac{1}{2}, \dfrac{1}{3}, \dfrac{1}{4}, \ldots, \dfrac{1}{n}, \ldots$

(ii) $\dfrac{1}{2}, \dfrac{3}{4}, \dfrac{7}{8}, \ldots, 1-\left(\dfrac{1}{2}\right)^n, \ldots$

(iii) $0, 3, 8, \ldots, n^2 - 1, \ldots$

Essentially we have to examine the limit as n tends to infinity.

(i) In this case we have $\lim_{n\to\infty} u_n = \lim_{n\to\infty} \dfrac{1}{n} = 0$

(ii) In this case we have $\lim_{n\to\infty} u_n = \lim_{n\to\infty} \left(1 - \left(\tfrac{1}{2}\right)^n\right)$

$$= \lim_{n\to\infty} 1 - \lim_{n\to\infty} \left(\tfrac{1}{2}\right)^n = 1 - 0 = 1$$

where we have used the fact that an 'infinite power' of any number between 0 and 1 is zero.

(iii) Fairly straightforward in this case

$$\lim_{n\to\infty} u_n = \lim_{n\to\infty} (n^2 - 1) = \infty$$

So this sequence 'diverges' – the terms get larger and larger indefinitely.

Understanding Engineering Mathematics

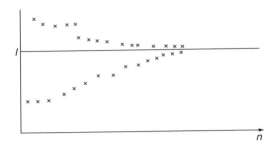

Figure 14.9 A sequence approaching a limit l.

If a sequence has a finite limit, we say it **converges**, if not, it **diverges**. Graphically, a sequence converges if the points plotted for values of n approach a definite line (Figure 14.9).

Note that the behaviour of the sequence depends not at all on the first terms of the sequence, but rather on the 'infinite tail' – i.e. on the last infinity of terms!

Exercise on 14.7

Investigate the convergence of the sequences

$u_n = $ (i) $(-1)^n$ (ii) 2^n (iii) $\left(\dfrac{1}{3}\right)^n$ (iv) $-\dfrac{1}{n}$

(v) $\dfrac{1}{r^n}$ $|r| > 1$ (vi) $\dfrac{1}{r^n}$ $|r| < 1$

Answer

(i) D ('oscillating')   (ii) D   (iii) C   (iv) C   (v) C   (vi) D

14.8 Iteration

One of the most important practical applications of the theory of sequences is in the discussion of **iterative methods** used for example in numerical methods for solving equations. In an iterative method for solving an equation we assume an initial guess and use it in some rearrangement of the equation to calculate a (hopefully) improved approximation to the solution. We will describe two methods for this, but mainly as examples of convergence of sequences.

One way of finding an approximate solution to the equation

$$f(x) = 0$$

is to write it in the form

$$f(x) \equiv F(x) - x = 0$$

and solve this equation by iteration:

$$x_{n+1} = F(x_n)$$

where x_n is the **nth approximation**. Starting with a first approximation x_1 we find another by:

$$x_2 = F(x_1)$$

and another

$$x_3 = F(x_2)$$

and so on.

This generates a sequence of approximate solutions:

$$x_1, x_2, x_3, \ldots, x_n, \ldots$$

and under favourable circumstances the limit of this sequence is the solution of the equation. To get a good enough approximation we then take n large enough. Clearly, whether this is possible will depend on the form of the function F.

Perhaps the most popular iteration method is **Newton's method**. In this method the $(n+1)$th approximation, x_{n+1}, to the solution of an equation $f(x) = 0$ is calculated from the nth approximation x_n, subject to certain conditions, using the relation

$$x_{n+1} = x_n - \frac{f(x_n)}{f'(x_n)}$$

Again we generate an infinite sequence of approximate solutions that (we hope) will eventually converge on the actual solution.

Exercise on 14.8

Rewriting the equation $x^3 - 5x - 3 = 0$ in the form

$$x = \tfrac{1}{5}(x^3 - 3)$$

and starting with the value $x_0 = -1$, use your calculator to generate a sequence of approximate solutions to the equation. Do you think this sequence converges to a solution? What happens if you try $x_0 = -2$?

Now rewrite the equation in the form

$$x = \frac{5x+3}{x^2}$$

and again start with $x_0 = -2$. What happens?

Answer

Starting with $x_0 = -1$ the first rearrangement leads quickly to the approximation $x = -0.657$. Starting with $x_0 = -2$ leads to a divergent sequence and no solution. Starting with $x_0 = -2$ in the second rearrangement however leads quickly to an approximate solution $x = -1.83$. Note that some other starting values x_0 might not lead to a root at all.

14.9 Infinite series

An infinite series (105 ◄) is one of the form

$$S = u_1 + u_2 + u_3 + \cdots + u_n + \cdots$$

i.e. the sum of the terms of an infinite sequence. To consider the convergence of such a series, i.e. whether it adds up to a finite quantity, we introduce some definitions. Firstly, we pretend that the series is finite and define the nth **partial sum of the series** as the sum of the first n terms:

$$S_n = \text{sum to } n \text{ terms} = u_1 + u_2 + u_3 + \cdots + u_n$$

Then the infinite series can be regarded as this nth partial sum as n tends to infinity, i.e.

$$S = \text{sum to infinity} = \lim_{n \to \infty} S_n$$

If $S = \lim_{n \to \infty} S_n = l$, a definite value, then we say that the series **converges** to the value l. Otherwise the series is **divergent**.

In terms of sigma notation (102 ◄) we write:

$$S_n = \sum_{r=1}^{n} u_r$$

$$S = \sum_{r=1}^{\infty} u_r$$

Clearly, whether or not a series converges depends on the terms u_r and also on their relation to each other – i.e. on u_r/u_{r-1}. This last quantity appears in the **ratio test** for convergence.

One way of testing for convergence, of course, is to investigate the limit of the nth partial sum directly.

Problem 14.10

Investigate the convergence of the geometric series

$$S = a + ar + ar^2 + \cdots + ar^{n-1} + \cdots$$

(105 ◄)

We looked at this in Section 3.2.10. We have

$$S_n = a + ar + ar^2 + \cdots + ar^{n-1}$$
$$= \frac{a(1-r^n)}{1-r} \quad (\text{Section } 3.2.9)$$
$$= \frac{a}{1-r} - \left(\frac{a}{1-r}\right)r^n$$

Now, provided $-1 < r < 1$ we have $r^n \to 0$ as $n \to \infty$, so

$$\lim_{n \to \infty} S_n = \frac{a}{1-r}$$

Thus, the sum to infinity of the geometric series is

$$S = \frac{a}{1-r}$$

provide $|r| < 1$. On the other hand, if $|r| \geq 1$ then $r^n \to \infty$ as $n \to \infty$ and so the series diverges.

Problem 14.11

Show that the infinite arithmetic series

$$S = a + (a+d) + (a+2d) \cdots$$

diverges for any value of the common difference d (104 ➤).

Again, we found in Section 3.2.9 that

$$S_n = a + (a+d) + (a+2d) + \cdots + (a+(n-1)d)$$
$$= \tfrac{1}{2}n(2a + (n-1)d)$$

Now, as $n \to \infty$, $S_n \to \infty$, whatever the value of d, so

$$\lim_{n \to \infty} S_n = \infty$$

and therefore the arithmetic series diverges.

It is not always possible to use the nth partial sum in this way. For example the nth partial sum of the harmonic series

$$S = 1 + \frac{1}{2} + \frac{1}{3} + \frac{1}{4} + \cdots + \frac{1}{n} + \cdots$$

is

$$S_n = 1 + \frac{1}{2} + \frac{1}{3} + \cdots + \frac{1}{n}$$

In this case there is no simple formula for S_n, and so we have to find other ways to test convergence. We did actually do this in Section 14.6 where it was shown that the harmonic series diverges. We will look at other tests of convergence in the next section.

Exercise on 14.9

Find S_n for the series

$$1 + \tfrac{1}{3} + \left(\tfrac{1}{3}\right)^2 + \cdots + \left(\tfrac{1}{3}\right)^r + \cdots$$

and hence evaluate the sum to infinity.

Answer

$\frac{3}{2}\left[1-\left(\frac{1}{3}\right)^n\right]$, $\frac{3}{2}$

14.10 Tests for convergence

As noted in the previous section, if we can calculate the general form for S_n then we can find $\lim_{n\to\infty} S_n$ and check convergence directly, but this is not always possible. Fortunately, a number of different tests have been developed to determine whether a series converges by considering its general form.

Before we even start to discuss the convergence of a series, we must first check that $\lim_{n\to\infty} u_n = 0$, because if this is not so then the series cannot converge. To see this suppose that the series converges. Then from:

$$u_n = S_n - S_{n-1}$$

we have

$$\lim_{n\to\infty}(S_n - S_{n-1}) = 0 = \lim_{n\to\infty} u_n$$

So, if the series converges, $\lim_{n\to\infty} u_n = 0$. However, the converse is not true, i.e. $\lim_{n\to\infty} u_n = 0$, does not imply that the series converges. Thus, we saw in Section 14.6 that the harmonic series actually diverges and yet it clearly satisfies $\lim_{n\to\infty} u_n = \lim_{n\to\infty} \frac{1}{n} = 0$. On the other hand, for the arithmetic series (Problem 14.11) $\lim_{n\to\infty} u_n = \infty$ and we do know that this series always diverges.

Having verified that $\lim_{n\to\infty} u_n = 0$, we can continue to apply the tests for convergence given below. First, a couple of general points:

(1) In studying the convergence of a series it is only the 'infinite tail' which is important, not an initial finite number of terms.
(2) Some of the tests given below apply only to series whose terms all eventually become positive.

The comparison test

This is essentially a generalisation of the method we used for the harmonic series. Suppose we wish to study the convergence of the series of positive terms:

$$U = u_1 + u_2 + \cdots + u_n + \cdots$$

Then in the comparison test we **compare** this with a known series. Thus, let

$$V = v_1 + v_2 + \cdots + v_n + \cdots$$

be a known standard series. Then if $u_r \leq cv_r$, $c =$ a constant and V is convergent, then so is U. If $u_r \geq cv_r$ and V diverges, then so does U. This test is intuitively obvious.

Problem 14.12

Apply the comparison test to the series

(i) $U = 1 + \dfrac{1}{2!} + \dfrac{1}{3!} + \cdots (= e^1 - 1)$ (ii) $U = \dfrac{1}{2} + \dfrac{1}{4} + \dfrac{1}{6} + \cdots$

(i) As standard series take the geometric series

$$V = 1 + \frac{1}{2} + \frac{1}{2^2} + \cdots$$

Clearly, $u_r \leq 1 v_r$ and we know that V converges because its common ratio is less than 1, and so U also converges.

(ii) $U = \frac{1}{2} + \frac{1}{4} + \frac{1}{6} + \cdots$

Take $V = 1 + \frac{1}{2} + \frac{1}{3} + \frac{1}{4} + \cdots$, the harmonic series
Then $u_r = \frac{1}{2} v_r$ and since V diverges, so does U.

Alternating series

In the particular case of an **alternating series**, such as:

$$1 - \tfrac{1}{2} + \tfrac{1}{3} - \tfrac{1}{4} + \cdots$$

where the signs alternate, there is a very simple test for convergence. Thus, let

$$S = u_1 - u_2 + u_3 - u_4 + \cdots$$

(all $u_i \geq 0$). Then if

$$\lim_{n \to \infty} u_n = 0 \text{ and } u_n \leq u_{n-1} \text{ for all } n > N$$

(N a certain finite number) then the series converges.

Problem 14.13

Examine the convergence of the series

$$S = 1 - \tfrac{1}{2} + \tfrac{1}{3} - \tfrac{1}{4} + \cdots$$

In this case it is clear that

(i) $\lim_{n \to \infty} u_n = 0$ (ii) $u_n < u_{n-1}$ for all n

So the required conditions are satisfied and therefore the series converges. Notice the difference a sign or two makes – if all the signs are positive, then we have the harmonic series, which we know diverges.

D'Alembert's ratio test

Lest, as an engineer, you doubt the relevance of all this pure mathematics to your studies, a few words about Jean Le Rond D'Alembert (1717–1783). Thought to be the illegitimate son of a Parisian noble who paid for his education, he was found as a baby on the steps of the church of St Jean Le Rond. He is most famous for his work in the field of dynamics (**d'Alembert's Principle**) where he made great contributions to applying Newton's Laws to the motion of complicated systems of planets. In this he helped lay the foundations for the approximation methods now used routinely in the analysis of complex mechanical systems occurring in all aspects of engineering. His convergence test described here is just one small example of the sorts of mathematical tool one needs to apply and justify such approximation methods.

This useful test, the **d'Alembert ratio test**, deals with the positive values of the ratio of successive terms.

For the series

$$S = u_1 + u_2 + \cdots + u_n + \cdots$$

let

$$\lim_{n \to \infty} \left| \frac{u_{n+1}}{u_n} \right| = l$$

Then the series is **divergent if $l > 1$**
convergent if $l < 1$

and if $l = 1$ the test fails and we have to find some other test, for example, the comparison test.

We can see roughly why the ratio test works. We note that $\lim_{n \to \infty} \left| \frac{u_{n+1}}{u_n} \right|$ gives the behaviour of the 'last terms of the series' – after a large number of terms the series will have terms $u_n, u_{n+1} \ldots$ which we can compare with a geometric series with common ratio $\left| \frac{u_{n+1}}{u_n} \right|$. If the common ratio is less than 1 the series will converge, otherwise it will diverge (unless possibly it is equal to 1, in which case further investigation is needed).

Problem 14.14

Examine the convergence of the following series by the ratio test:

(i) $1 + \dfrac{1}{2!} + \dfrac{1}{3!} + \dfrac{1}{4!} + \cdots$

(ii) $1 + \dfrac{2}{1} + \dfrac{2^2}{2!} + \dfrac{2^3}{3!} + \dfrac{2^4}{4!} + \dfrac{2^5}{5!} + \cdots$

(iii) $1 + \dfrac{1}{2^2} + \dfrac{1}{3^2} + \dfrac{1}{4^2} + \dfrac{1}{5^2} + \cdots$

(i) We have

$$u_n = \frac{1}{n!}, \quad u_{n+1} = \frac{1}{(n+1)!}$$

So

$$\lim_{n \to \infty} \left| \frac{u_{n+1}}{u_n} \right| = \lim_{n \to \infty} \left| \frac{1}{(n+1)!} n! \right| = 0 < 1$$

So the series converges, as we already know from Problem 14.12. It is in fact the series for $e^1 - 1$.

(ii) For $1 + \dfrac{2}{1} + \dfrac{2^2}{2!} + \dfrac{2^3}{3!} + \dfrac{2^4}{4!} + \dfrac{2^5}{5!} + \cdots$ we have $u_n = \dfrac{2^n}{n!}$.

So
$$\lim_{n\to\infty}\left|\frac{u_{n+1}}{u_n}\right| = \lim_{n\to\infty}\left|\frac{2^{n+1}}{(n+1)!}\frac{n!}{2^n}\right| = \lim_{n\to\infty} 2\left|\frac{1}{n+1}\right| = 0 < 1$$

Therefore the series converges.

(iii) For the series $1 + \dfrac{1}{2^2} + \dfrac{1}{3^2} + \dfrac{1}{4^2} + \dfrac{1}{5^2} + \cdots$ we have

$$\lim_{n\to\infty}\left|\frac{u_{n+1}}{u_n}\right| = \lim_{n\to\infty}\left|\frac{1}{(n+1)^2}n^2\right| = 1$$

So in this case the ratio test is **inconclusive**. However, it is possible to extend the method used for the harmonic series to prove that the above series **converges** and in fact more generally, the series

$$1 + \frac{1}{2^p} + \frac{1}{3^p} + \frac{1}{4^p} + \frac{1}{5^p} + \cdots$$

converges if $p > 1$ and diverges if $p \leq 1$. This series provides a good standard series for use in a comparison test.

Absolute convergence

The infinite series $u_1 + u_2 + u_3 + \cdots$ is said to be **absolutely convergent** if the corresponding series of positive terms $|u_1| + |u_2| + |u_3| + \cdots$ is convergent. If the series $u_1 + u_2 + u_3 + \cdots$ is convergent, but the series $|u_1| + |u_2| + |u_3| + \cdots$ is divergent, we say the series is **conditionally convergent**. For example:

$$1 - \frac{1}{2^2} + \frac{1}{3^2} - \cdots \text{ is absolutely convergent (see Problem 14.14(iii)).}$$

$1 - \frac{1}{2} + \frac{1}{3} - \frac{1}{4} + \cdots$ is conditionally convergent (by the harmonic series).

The real power of absolute convergence comes in when we consider infinite power series. If a power series in x is absolutely convergent, then we can differentiate or integrate it with respect to x – term by term, which may not be possible otherwise.

Convergence testing summary

Below we summarise a systematic approach to testing the convergence of a series – not foolproof, but a good start.

When testing the series:

$$S = u_1 + u_2 + u_3 + u_4 + \cdots$$

follow the steps below:

(i) Check that $u_n \to 0$ as $n \to \infty$. If u_n does not tend to zero then the series diverges.

(ii) Apply the ratio test:

if
$$l = \lim_{n \to \infty} \left| \frac{u_{n+1}}{u_n} \right|$$

then the series converges if $l < 1$ and diverges if $l > 1$.

(iii) If $l = 1$, use a comparison test with a known standard series.

(iv) If the series is alternating then it is convergent if $u_n \to 0$, as $n \to \infty$.

Exercise on 14.10

Test the following series for convergence

(i) $\dfrac{1}{2} + \dfrac{1}{3}\left(\dfrac{1}{2}\right)^3 + \dfrac{1}{5}\left(\dfrac{1}{2}\right)^5 + \dfrac{1}{7}\left(\dfrac{1}{2}\right)^7 + \cdots + \dfrac{1}{2r-1}\left(\dfrac{1}{2}\right)^{2r-1} + \cdots$

(ii) $\dfrac{1}{2} + \dfrac{1}{3}\left(-\dfrac{1}{2}\right)^3 + \dfrac{1}{5}\left(-\dfrac{1}{2}\right)^5 + \dfrac{1}{7}\left(-\dfrac{1}{2}\right)^7 + \cdots + \dfrac{1}{2r-1}\left(-\dfrac{1}{2}\right)^{2r-1} \cdots$

(iii) $2 + \dfrac{1}{3}2^3 + \dfrac{1}{5}2^5 + \cdots + \dfrac{1}{2n-1}2^{2n-1} + \cdots$

Answer

(i) C (ii) C (iii) D

14.11 Infinite power series

A series of form

$$a_0 + a_1 x + a_2 x^2 + a_3 x^3 + \cdots$$

where the coefficients are constants, is called an **infinite power series in x** (107 ◀).
Such series are used to provide useful alternative forms for functions – for example

$$e^x = 1 + x + \frac{x^2}{2!} + \frac{x^3}{3!} + \cdots$$

$$\cos x = 1 - \frac{x^2}{2!} + \frac{x^4}{4!} - \cdots \quad (x \text{ in radians})$$

The right-hand sides have the advantage that they can, subject to convergence, be used to evaluate the function to any required accuracy. In general such series are obviously most helpful for small values of x. We say they give approximations **near to the origin** $x = 0$ and they are also called **power series about the origin**, or **Maclaurin's series**. If we require series that approximate a function near to some particular non–zero value of x, say $x = a$, then we use a **Taylor series about $x = a$**:

$$f(x) = a_0 + a_1(x - a) + a_2(x - a)^2 + \cdots$$

A Maclaurin's series is thus just a Taylor series about the origin. Note that by changing the variable, $X = x - a$ we can always convert a Taylor series to a Maclaurin's series, so we will confine our attention to the latter here.

The coefficients in a Maclaurin's series can be found by a nice argument, which also illustrates the condition that a Maclaurin's series for a function $f(x)$ only exists if $f(x)$ is differentiable an infinite number of times.

We start with the series:

$$f(x) = a_0 + a_1 x + a_2 x^2 + \cdots$$

The job is to find the coefficients a_r. a_0 is easy, just put $x = 0$:

$$a_0 = f(0)$$

We now get the $a_1, a_2 \ldots$ by differentiating enough times to isolate each of them as the constant term, and then put $x = 0$. Thus:

$$f'(x) = a_1 + 2a_2 x + 3a_3 x^2 + \cdots$$

so

$$f'(0) = a_1$$
$$f''(x) = 2a_2 + 3 \times 2a_3 x + \cdots$$

so

$$f''(0) = 2a_2$$

and

$$a_2 = \tfrac{1}{2} f''(0)$$
$$f'''(x) = 3 \times 2a_3 + \cdots$$

so

$$f'''(0) = 3 \times 2a_3$$

and

$$a_3 = \frac{1}{3 \times 2} f'''(0) = \frac{1}{3!} f'''(0)$$

You should now be able to see that in general

$$a_r = \frac{1}{r!} f^{(r)}(0)$$

where $f^{(r)}(0)$ denotes the rth derivative of $f(x)$ at $x = 0$. So the Maclaurin's series for $f(x)$ may be written as:

$$f(x) = f(0) + f'(0)x + \frac{f''(0)}{2!} x^2 + \cdots + \frac{f^{(r)}(0)}{r!} x^r + \cdots$$

$$= \sum_{r=0}^{\infty} \frac{f^{(r)}(0) x^r}{r!}$$

where $f^{(0)}(0)$ means $f(0)$.

Problem 14.15

Obtain the Maclaurins series for the functions (i) e^x (ii) $\cos x$

(i) e^x is the easy one – it just keeps repeating on differentiation and we have

$$f^{(r)}(x) = e^x \quad \text{for all } r$$

so

$$f^{(r)}(0) = e^0 = 1$$

So the series is

$$e^x = 1 + x + \frac{x^2}{2!} + \frac{x^3}{3!} + \cdots$$

(ii) $\cos x$ is almost as easy:

$$f(0) = \cos 0 = 1$$
$$f'(0) = -\sin 0 = 0$$
$$f''(0) = -\cos 0 = -1$$
$$f'''(0) = \sin 0 = 0$$
$$f^{(iv)}(0) = \cos 0 = 1$$

and so on, giving the series

$$\cos x = 1 - \frac{x^2}{2!} + \frac{x^4}{4!} - \cdots$$

So far as the convergence of power series is concerned we can use tests such as the ratio test described in Section 14.10. The results of such tests will usually depend on x and so the series may only converge for certain (if any) values of x. The values of x for which the series converges is usually called the **radius of convergence**. For example, the binomial series for $(1-x)^{-1}$ only converges for $|x| < 1$. Clearly, if a series does not converge for a particular value of x, then it cannot represent a sensible function at that value.

Problem 14.16

Investigate the convergence of

(i) $e^x = 1 + x + \dfrac{x^2}{2!} + \dfrac{x^3}{3!} + \dfrac{x^4}{4!} + \cdots$ (ii) $x - \dfrac{x^2}{2} + \dfrac{x^3}{3} - \dfrac{x^4}{4} + \cdots$

(i) For $1 + x + \dfrac{x^2}{2!} + \dfrac{x^3}{3!} + \dfrac{x^4}{4!} + \cdots$ we have $u_n = \dfrac{x^n}{n!}$ and so

$$\lim_{n \to \infty} \left| \frac{u_{n+1}}{u_n} \right| = \lim_{n \to \infty} \left| \frac{x^{n+1}}{(n+1)!} \frac{n!}{x^n} \right| = \lim_{n \to \infty} \left| \frac{x}{n+1} \right| = 0 \text{ for all finite } x$$

So the ratio test tells us that this series converges for all finite x. It is, of course, the series for e^x.

(ii) For $x - \dfrac{x^2}{2} + \dfrac{x^3}{3} - \dfrac{x^4}{4} + \cdots$, we have $|u_n| = \left|\dfrac{x^n}{n}\right|$ and so

$$\lim_{n\to\infty}\left|\dfrac{u_{n+1}}{u_n}\right| = \lim_{n\to\infty}\left|\dfrac{x^{n+1}}{(n+1)}\dfrac{n}{x^n}\right| = \lim_{n\to\infty}\left|\dfrac{nx}{n+1}\right| = |x|$$

So by the ratio test the series is convergent if $|x| < 1$ and divergent if $|x| > 1$. If $|x| = 1$ the ratio test is inconclusive and we have to consider this case separately. There are two cases to consider, $x = 1$ and $x = -1$.

If $x = 1$ the series is

$$S = 1 - \tfrac{1}{2} + \tfrac{1}{3} - \tfrac{1}{4} + \cdots$$

and this is an alternating series with decreasing terms that tend to zero, and so it **converges**.
If $x = -1$:

$$S = -1 - \tfrac{1}{2} - \tfrac{1}{3} - \tfrac{1}{4} + \cdots$$

and this is the negative of the harmonic series and so **diverges.**
Thus we can summarize our results:

$$x - \dfrac{x^2}{2} + \dfrac{x^3}{3} - \cdots$$

converges if $-1 < x \leq 1$ and diverges if $x > 1$ or $x \leq -1$. This is in fact the series for $\log(1 + x)$.

Note that the ratio test ensures **absolute convergence**, since it involves taking the modulus of the term of the series. Thus the series for $\tan^{-1} x$ (see Exercise on 14.11) is absolutely convergent for $x^2 < 1$ and in particular can be differentiated term by term to give

$$\dfrac{1}{1 + x^2} = 1 - x^2 + x^4 - x^6 + \cdots$$

which can be checked by the binomial theorem (71 ◀).

Exercise on 14.11

Derive the series $\tan^{-1} x = x - \dfrac{x^3}{3} + \dfrac{x^5}{5} - \dfrac{x^7}{7} + \cdots + (-1)^{n+1}\dfrac{x^{2n-1}}{2n-1} + \cdots$ and discuss its convergence.

Answer

The Maclaurin's series for $\tan^{-1} x$ is valid for $-1 \leq x \leq 1$.

14.12 Reinforcement

1. (i) Find the values of the function $(3x + 1)/x$ when x has the values 10, 100, 1000, 1,000,000.
   (ii) What limit does the function approach as x becomes infinite?

2. Find $\lim_{x \to 2} \dfrac{x^2 - 4}{x^2 - 2x}$

3. Find $\lim_{x \to \infty} \dfrac{4x^2 + x - 1}{3x^2 + 2x + 1}$

4. The distance fallen by a particle from rest is given by $s = 16t^2 m$. Representing an increase in time by δt and the corresponding increase in s by δs, find an expression for δs in terms of δt and hence find $\delta s/\delta t$. Using this result find the average velocity for the following intervals:

   (i) 2 secs. to 2.2 secs.   (ii) 2 secs. to 2.1 secs.
   (iii) 2 secs. to 2.01 secs.   (iv) 2 secs. to 2.001 secs.

   From these results infer the velocity at the end of 2 secs.

5. Which of the following functions exist at the stated values of x? If the functions do not exist, find the limit of the function at the values concerned, if it exists.

   (i) $x^2 |_{x=4}$
   (ii) $\dfrac{(x - 1)(x + 2)}{x - 1}\Big|_{x=1}$
   (iii) $\dfrac{(x + 4)^4}{(x + 4)}\Big|_{x=-4}$
   (iv) $\dfrac{1}{x}\Big|_{x=0}$
   (v) $\dfrac{x^2 - 4}{x + 2}\Big|_{x=-2}$
   (vi) $\dfrac{x^2 + 1}{x^4 + 1}\Big|_{x=-1}$
   (vii) $\dfrac{x^4 - 16}{x - 2}\Big|_{x=2}$
   (viii) $\dfrac{x + 1}{x - 1}\Big|_{x=1}$

6. Given $\lim_{x \to 0} \dfrac{\sin x}{x} = 1$ find the following limits:

   (i) $\lim_{x \to 0} \dfrac{\sin 2x}{2x}$
   (ii) $\lim_{x \to 0} \dfrac{\sin 3x}{x}$
   (iii) $\lim_{x \to 0} \dfrac{\sin \alpha x}{\beta x} \quad \alpha \neq 0, \beta \neq 0$
   (iv) $\lim_{x \to 0} \left(\dfrac{\sin x}{x}\right)^3$
   (v) $\lim_{x \to 0} \left(\dfrac{\sin \alpha x}{\beta x}\right)^\gamma$
   (vi) $\lim_{x \to 0} \dfrac{\sin^2 x}{x}$
   (vii) $\lim_{x \to 0} \dfrac{\sin x}{\pi - x}$

7. Which of the functions in Q5 are continuous for all x? Give any points of discontinuity and where possible give a function which is continuous everywhere and is equal to the given function away from the points of discontinuity.

8. Find the limit of $\dfrac{(x + h)^3 - x^3}{h}$ as $h \to 0$. Hence find the slope of the curve $y = x^3$ at the point $x = 1$.

9. Write down the first four terms of the sequence whose nth term is:

   (i) n (ii) $(n^2 + 3n)/5n$ (iii) $\cos \tfrac{1}{2}n\pi$ (iv) $(1 + (-1)^n)/n$
   (v) $1/\sqrt{n}$ (vi) $(n + \sin \tfrac{1}{2}n\pi)/(2n + 1)$ (vii) ar^{n-1}
   (viii) $(-1)^n x^{2n+1}/(2n + 1)$

   Investigate the limits of these sequences.
   Hints   (vii) limit depends on r (viii) limit depends on x

10. Write down the nth terms of the sequences

    (i) $\dfrac{1}{2}, \dfrac{1}{4}, \dfrac{1}{8}, \dfrac{1}{16}, \ldots$ (ii) $0, \dfrac{1}{2}, \dfrac{2}{3}, \dfrac{3}{4}, \dfrac{4}{5}, \ldots$ (iii) $\dfrac{1}{1.2}, \dfrac{1}{2.3}, \dfrac{1}{3.4}, \ldots$

    (iv) $\dfrac{3}{2}, \dfrac{5}{4}, \dfrac{9}{8}, \dfrac{17}{16}, \ldots$ (v) $-1, \dfrac{1}{2}, -\dfrac{1}{6}, \dfrac{1}{24}, -\dfrac{1}{120}, \ldots$

    (vi) $\dfrac{x}{2.3}, \dfrac{2x^2}{3.4}, \dfrac{3x^3}{4.5}, \ldots$ (vii) $-\dfrac{x^3}{3}, \dfrac{x^5}{5}, -\dfrac{x^7}{7}, \ldots$ (viii) $\dfrac{x}{2}, \dfrac{2x}{3}, \dfrac{3x}{4}, \dfrac{4x}{5}, \ldots$

    Investigate the limits of these sequences.
    Hints (vi), (vii), and (viii) depend on x.

11. Evaluate the following limits:

    (i) $\lim\limits_{x \to \infty} x^2 e^{-x}$ (ii) $\lim\limits_{n \to \infty} \dfrac{1}{n}$ (iii) $\lim\limits_{n \to \infty} \dfrac{1}{n!}$

    (iv) $\lim\limits_{n \to \infty} \dfrac{n}{n + 2}$ (v) $\lim\limits_{n \to \infty} \left| \dfrac{x}{n + 1} \right|$ (vi) $\lim\limits_{n \to \infty} \left| \dfrac{xn}{n + 1} \right|$

    (vii) $\lim\limits_{n \to \infty} \dfrac{a(l - r^n)}{l - r}$

12. If $l = \lim_{n \to \infty} \left| \dfrac{u_{n+1}}{u_n} \right|$ find l in the following cases:

    (i) $u_n = \dfrac{1}{n^2}$ (ii) $u_n = \dfrac{1}{n!}$ (iii) $u_n = \dfrac{(-1)^n (x/2)^n}{(n + 1)}$

13. Plot or sketch the function $f(x) = 3x^3 - 4x + 5$ and verify that it has a root near $x = -1.5$. Use the Newton–Raphson method to obtain this root to 2 decimal places.

14. Using the Newton–Raphson method find to 2 decimal places the value of

    (i) $\sqrt{291.7}$ (ii) $\sqrt[3]{3.074}$

    Hint (i) Treat as the equation $x^2 - 291.7 = 0$. Similarly for (ii).

15. Find the sequence of the nth partial sums for the following series, i.e. the sequence:

    $$S_1, S_2, S_3, \ldots$$

    If possible, find a formula for S_n and thus evaluate $\lim\limits_{n \to \infty} S_n$.

(i) $1 + \dfrac{1}{2} + \dfrac{1}{2^2} + \dfrac{1}{2^3} + \dfrac{1}{2^4} + \cdots$

(ii) $1 - \dfrac{1}{3} + \dfrac{1}{3^2} - \dfrac{1}{3^3} + \cdots$

(iii) $1 + 3 + 5 + 7 + \cdots$

(iv) $1 + 1 + \dfrac{1}{2!} + \dfrac{1}{3!} + \dfrac{1}{4!} + \cdots$

(v) $1 + \dfrac{1}{2} + \dfrac{1}{3} + \dfrac{1}{4} + \cdots$

16. Obtain a formula for the nth partial sum S_n for the following series and hence investigate their convergence (or otherwise).

(i) $4 + \dfrac{4}{3} + \dfrac{4}{3^2} + \dfrac{4}{3^3} + \cdots$

(ii) $-6 - 2 + 2 + 6 + 10 + 14 + \cdots$

(iii) $1 + 2 + 2^2 + 2^3 + \cdots$

(iv) $1 + \dfrac{x}{2} + \left(\dfrac{x}{2}\right)^2 + \left(\dfrac{x}{2}\right)^3 + \cdots$

17. State, without proof, which of the following series are convergent/divergent. Write down the nth term of each series.

(i) $-1 + 1 - 1 + 1 - 1 + \cdots$

(ii) $1.01 + (1.01)^2 + (1.01)^3 + \cdots$

(iii) $(.99)^2 + (.99)^3 + (.99)^4 + \cdots$

(iv) $\dfrac{1}{50} + \dfrac{1}{51} + \dfrac{1}{52} + \dfrac{1}{53} + \cdots$

(v) $10^6 + \dfrac{10^5}{2} + \dfrac{10^4}{3} + \dfrac{10^3}{4} + \cdots$

(vi) $\dfrac{1}{40} - \dfrac{1}{50} + \dfrac{1}{60} - \dfrac{1}{70} + \cdots$

(vii) $\dfrac{3}{4} + 2^2 \left(\dfrac{3}{4}\right)^2 + 3^2 \left(\dfrac{3}{4}\right)^3 + 4^2 \left(\dfrac{3}{4}\right)^4 + \cdots$

(viii) $\dfrac{1}{2} + \dfrac{2}{3} + \dfrac{3}{4} + \dfrac{4}{5} + \cdots$

(ix) $1 + (.2) + (.2)^2 + (.2)^3 + \cdots$

18. Find the nth term of the following series and test for convergence.

(i) $1 + 1 + 1 + 1 + \cdots$

(ii) $1 - 2 + 3 - 4 + 5 + \cdots$

(iii) $1 + 2(.9) + 3(.9)^2 + 4(.9)^3 + \cdots$

(iv) $1 - \dfrac{1}{\sqrt{2}} + \dfrac{1}{\sqrt{3}} - \cdots$

(v) $1^3 + 2^3(.95) + 3^3(.95)^2 + \cdots$

(vi) $\dfrac{1}{2} + \dfrac{3}{4} + \dfrac{5}{6} + \cdots$

(vii) $\dfrac{1}{2} - \dfrac{3}{4} + \dfrac{5}{6} - \dfrac{7}{8} + \cdots$

(viii) $1 \cdot \dfrac{1}{2} + \dfrac{1}{2} \cdot \dfrac{3}{4} + \dfrac{1}{3} \cdot \dfrac{5}{6} + \dfrac{1}{4} \cdot \dfrac{7}{8} + \cdots$

19. Find the range of values for x for which the following series are convergent.

(i) $x - \dfrac{x^2}{2} + \dfrac{x^3}{3} - \dfrac{x^4}{4} + \cdots$

(ii) $2x + (2x)^2 + (2x)^3 + (2x)^4 + \cdots$

(iii) $1 + x + \dfrac{x^2}{2!} + \dfrac{x^3}{3!} + \cdots$

(iv) $1^5 + 2^5 x + 3^5 x^2 + 4^5 x^3 + 5^5 x^4 + \cdots$

(v) $1^2 + 2^2 x + 3^2 x^2 + 4^2 x^3 + \cdots$

20. For the binomial series $(1 + x)^m$ show that $\left|\dfrac{u_{n+1}}{u_n}\right| = \left|\dfrac{m - n + 1}{n}\right| |x|$ and deduce that the series converges if $|x| < 1$ and diverges if $|x| > 1$ (m not a positive integer).

21. Expand $(1 + 2x)^{\frac{3}{2}}$ as far as the term in x^2. How many terms of the series would be required to give $(1.02)^{\frac{3}{2}}$ correct to three decimal places? For what range of values of x is the series valid?

22. Obtain terms up to x^4 in the Maclaurin's series for the functions

    (i) $\sin 2x$ (ii) e^{x^2} (iii) $\dfrac{1}{1-3x}$

    (iv) $\ln(1+2x)$ (v) $\tan x$ (vi) $\dfrac{1}{(1+x)^2}$

State the values of x for which the series are convergent.

14.13 Applications

1. Since the derivative is defined in terms of a limit, the whole of the theory of differentiation can be based on limits and their properties. While an engineer may not need to know much about the details of this theory, it is useful to have an idea of the basic principles, since limits tell us when we can, for example, approximate a derivative by a simple algebraic difference, for the purposes of numerical methods. This is important in such things as numerical solution of differential equations, which is a common task faced by the engineer.

   Using the limit definition of the derivative, prove the rules of differentiation: sum, product, quotient, and chain rule (see Chapter 8).

2. Show that if h is small then near to a point $x = a$ any differentiable function, $f(x)$, can be approximated by

$$f(x) = f(a+h) \simeq f(a) + hf'(a)$$

This is a **linear (in h) approximation** to $f(x)$ near to $x = a$. It is often used as a basis for numerical differentiation. It essentially replaces the curve $f(x)$ near $x = a$ by a portion of the tangent – Figure 14.10.

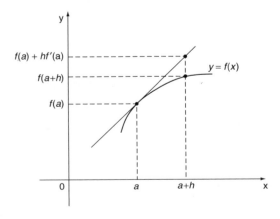

Figure 14.10 Linear approximation to a function.

For $y = f(x) = x^2$ obtain a linear approximation to $f(x)$, and hence evaluate $(1.01)^2$ approximately.

3. You may have heard talk of 'chaos' in recent years. This question, using an iterative process, gives you a simple introduction to this notion. It is based on what is called the 'logistic map', which is an iteration scheme given by

$$x_{n+1} = Ax_n(1 - x_n)$$

where A is some constant. Don't worry where this comes from, just accept that it has a wide applicability in science and engineering and it is important to consider when it yields a convergent solution for x_n – i.e. when the sequence it generates converges to a definite value as we perform the iterations, as we did in Section 14.8. It is found that this depends crucially on the value of A. Experiment with the iterations (15–20 should do, on calculator or computer) for the cases of $A = 2, 1 + \sqrt{5}$ and 4, with a starting value of $x_0 = 0.4$. The case $A = 4$ gives a 'chaotic' result as you will soon discover.

14.14 Answers to reinforcement exercises

1. (i) 3.1, 3.01, 3.001, 3.000001   (ii) 3
2. 2
3. 4/3
4. $\delta s = 32t\delta t + 16(\delta t)^2$ $\dfrac{\delta s}{\delta t} = 32t + 16(\delta t)$

    (i) 67.2   (ii) 65.6   (iii) 64.16   (iv) 64.016   64 m/sec.

5. (i) 16   (ii) 3   (iii) 0   (iv) limit does not exist
    (v) −4   (vi) 1   (vii) 32   (viii) limit does not exist

6. (i) 1   (ii) 3   (iii) α/β (iv) 1
    (v) $(\alpha/\beta)^\gamma$ (vi) 0   (vii) 0

7. (i) continuous for all x (ii) discontinuous at $x = 1$
    (iii) discontinuous at $x = -4$ (iv) discontinuous at $x = 0$
    (v) discontinuous at $x = -2$ (vi) continuous for all x
    (vii) discontinuous at $x = 2$ (viii) discontinuous at $x = 1$

8. $3x^2$, 3

9. (i) 1, 2, 3, 4, (∞)   (ii) $\dfrac{4}{5}, 1, \dfrac{6}{5}, \dfrac{7}{5}, (\infty)$ (iii) 0, −1, 0, 1 (divergent)

    (iv) 0, 1, 0, $\dfrac{1}{2}$, (0)   (v) 1, $\dfrac{1}{\sqrt{2}}, \dfrac{1}{\sqrt{3}}, \dfrac{1}{2}$, (0)   (vi) $\dfrac{2}{3}, \dfrac{2}{5}, \dfrac{2}{7}, \dfrac{4}{9}, \left(\dfrac{1}{2}\right)$

(vii) $a, ar, ar^2, ar^3,$ (divergent if $|r| > 1,$ limit $= a$ if $r = 1,$ convergent to 0 if $|r| < 1$ and oscillating if $r = -1$)

(viii) $-\dfrac{x^3}{3}, \dfrac{x^5}{5}, -\dfrac{x^7}{7}, \dfrac{x^9}{9},$ (limit $= 0$ if $|x| < 1$ and divergent otherwise)

10. (i) $\dfrac{1}{2^n},$ (0)   (ii) $\dfrac{n-1}{n},$ (1)   (iii) $\dfrac{1}{n(n+1)},$ (0)

(iv) $\dfrac{2^n + 1}{2^n},$ (1)   (v) $\dfrac{(-1)^n}{n!},$ (0)

(vi) $\dfrac{nx^n}{(n+1)(n+2)},$ (0 if $|x| = 1,$ divergent otherwise)

(vii) $(-1)^n \dfrac{x^{2n+1}}{2n+1},$ (0 if $|x| = 1,$ divergent otherwise)

(viii) $\dfrac{nx}{n+1},$ (x)

11. (i) 0   (ii) 0   (iii) 0   (iv) 1   (v) 0   (vi) $|x|$

(vii) $\dfrac{a}{1-r}$ if $|r| < 1,$ divergent otherwise

12. (i) 1   (ii) 0   (iii) $\left|\dfrac{x}{2}\right|$

13. -1.55

14. (i) 17.08   (ii) 1.45

15. (i) $S_n = 2 - \left(\dfrac{1}{2}\right)^{n-1},$ $\lim_{n \to \infty} S_n = 2$

(ii) $S_n = \dfrac{1}{4}\left[3 - \left(-\dfrac{1}{3}\right)^{n-1}\right],$ $\lim_{n \to \infty} S_n = \dfrac{3}{4}$

(iii) $S_n = n^2, \lim_{n \to \infty} S_n = \infty$

(iv) No simple expression for $S_n,$ but in fact $\lim_{n \to \infty} S_n = e$

(v) No simple expression for $S_n,$ but in fact $\lim_{n \to \infty} S_n = \infty$

16. (i) $6\left[1 - \left(\tfrac{1}{3}\right)^n\right]$ (C)   (ii) $2n(n-4)$ (D)   (iii) $2^n - 1$ (D)

(iv) $\left[1 - \left(\tfrac{x}{2}\right)^n\right] / \left(1 - \tfrac{x}{2}\right)$ (C for $|x| < 2$)

17. (i) $(-1)^n$ (D)   (ii) $1.01(1.01)^{n-1}$ (D)   (iii) $(.99)^{n+1}$ (C)

(iv) $\dfrac{1}{(49+n)}$ (D)   (v) $\dfrac{10^{7-n}}{n}$ (C)   (vi) $\dfrac{(-1)^{n+1}}{10(n+3)}$ (C)

(vii) $n^2 \left(\dfrac{3}{4}\right)^n$ (C)   (viii) $\dfrac{n}{n+1}$ (D)   (ix) $(.2)^{n-1}$ (C)

18. (i) 1 (D)   (ii) $(-1)^{n+1}n$ (D)   (iii) $n(.9)^{n-1}$ (C)

(iv) $\dfrac{(-1)^{n+1}}{\sqrt{n}}$ (C)   (v) $n^3(.95)^{n-1}$ (C)   (vi) $\dfrac{2n-1}{2n}$ (D)

(vii) $(-1)^{n+1}\dfrac{2n-1}{2n}$ (D)   (viii) $\dfrac{2n-1}{2n^2}$ (D)

19. (i) $|x|<1$ (ii) $|x|<\tfrac{1}{2}$ (iii) all x (iv) $|x|<1$

(v) C for $|x|<1$, D for $|x|>1$, D for $x=\pm 1$

21. $1+3x+\tfrac{3}{2}x^2+\cdots$,   4, $|x|<\tfrac{1}{2}$

22. (i) $2x-\dfrac{4}{3}x^3$ All x (ii) $1+x^2+\dfrac{x^4}{2!}+\cdots$ All x

(iii) $1+3x+9x^2+27x^3+81x^4+\cdots$ $|x|<\dfrac{1}{3}$

(iv) $2x-2x^2+\dfrac{8}{3}x^3-4x^4+\cdots$ $-\dfrac{1}{2}<x<\dfrac{1}{2}$

(v) $x+\dfrac{1}{3}x^3+\cdots$ $x^2<\dfrac{\pi^2}{4}$ (You are not expected to derive this condition, but it is obvious from the graph of $\tan x$!)

(vi) $1-2x+3x^2-4x^3+5x^4+\cdots$ $|x|<1$

15

Ordinary Differential Equations

Ordinary differential equations bring together all the calculus that we have done so far in the book into one of the most powerful and useful tools of engineering mathematics. In this chapter we concentrate on the principles of the key methods, rather than the intricate details of manipulation. It helps to keep in mind the main steps of solving any differential equation:

- identify the **type** of the differential equation by inspecting its form
- choose a **method** of solution appropriate to the type
- **solve** the equation, including any extra conditions
- **check** the solution by substituting back into the differential equation

The last point may seem tedious at times, but not only does it give you greater confidence in the solution, but it gives you essential practice in differentiation and other areas of mathematics.

Prerequisites

It will be helpful if you know something about:

- elementary algebra such as partial fractions (62 ◄)
- the exponential function and its properties (Chapter 4)
- differentiation (Chapter 8)
- integration (Chapter 9)
- sines and cosines and their properties (Chapter 6)
- exponential form of a complex number (361 ◄)
- solving simultaneous equations (391 ◄)

In particular, it will be of greatest benefit if you know and fully understand the following results:

If $y = Ae^{kx}$, where A and k are constants then

$$\frac{dy}{dx} = kAe^{kx} = ky$$

If $y = A \cos kx$ or $A \sin kx$ then

$$\frac{d^2y}{dx^2} = -k^2 y$$

These two results are fundamental to differential equations and represent protype equations of first and second order to which we already know the solutions.

Objectives

In this chapter you will find:

- definitions and terminology for differential equations
- initial and boundary conditions
- direct integration and separation of variables of first order equations
- linear equations and integrating factors
- second order linear homogeneous equations – auxiliary equation
- second order linear inhomogeneous equations – complementary function and particular integral

Motivation

You may need the material of this chapter for:

- modelling the motion of particles in mechanics
- modelling the time behaviour of electrical and electronic circuits
- the study of fluid flow
- modelling of chemical reactions
- modelling economic and financial systems and manufacturing processes

15.1 Introduction

Differential equations are often introduced by talking about rates at which radioactive substances decay. All very well, but how many of us have a nice handy sample of plutonium to play with? Now, bacterial growth – plenty of that found on mouldy bread in the average student kitchen. If you can contain your queasiness long enough to take a detached view of it, that grey, dusty, inedible square of once-bread can in fact provide a ready made mini-laboratory for introducing differential equations.

If you keep a record of the growth of the mould with sufficient accuracy (or, better, let somebody else do it) you will find that the rate at which the bacteria multiply at a given time is roughly proportional (14 ◄) to the number present at that time. Mathematically: if n is the number of bacteria present at time t, then n is a function of t, which we write $n = n(t)$, and it satisfies

$$\frac{dn}{dt} \propto n$$

or

$$\frac{dn}{dt} = kn \tag{15.1}$$

where k is some constant. This is called a **differential equation** (DE) for the **dependent variable** n in terms of the **independent variable** t. It's a little bit more complicated than it need be, so let's tidy it up.

Dividing by k:

$$\frac{dn}{k\,dt} = \frac{dn}{d(kt)} = n$$

and if we now introduce new variables:

$$x = kt, \, y = n$$

it becomes

$$\frac{dy}{dx} = y \qquad (15.2)$$

This is a good example of mathematical modelling. By changing the mathematical variables we have tidied up the equation and reduced the problem to its simplest mathematical form. We always do this when we can.

The differential equation (15.2) is about as simple as you can get for a non-trivial example. Yet it is an extremely important and basic equation. It occurs throughout the theory of DEs, and it can be used directly to solve many more complicated equations, including some of higher order (see Section 15.5).

Problem 15.1

Solve the equation (15.2) – i.e. find y as a function of x.

The simplest approach is to turn the derivative upside down (235 ◄):

$$\frac{dx}{dy} = \frac{1}{y}$$

Now it is simply integration **with respect to y**

$$x = \int \frac{dx}{dy} dy = \int \frac{1}{y} dy = \ln y + C$$

where C is an arbitrary constant.
So

$$\ln y = x - C$$
$$y = e^{x-C} = e^{-C} e^x = A e^x$$

where A is another (positive) arbitrary constant.

So, a solution to $\dfrac{dy}{dx} = y$ is $y = Ae^x$ with A an arbitary constant.

Points to note are:

- The DE is said to be **first order** because the highest derivative involved is first order.
- The solution contains one arbitrary constant, A, giving us an infinity of solutions.
- Just one arbitrary constant arose precisely because to solve a first order DE we need to integrate just once.

- The solution obtained is, in this case, the **most general solution** – there are no others.
- To fix A we would have to specify an extra condition on y – such as its 'initial value' – its value at $x = 0$. If this is y_0 then:

$$y_0 = Ae^0 = A, \text{ so } A = y_0$$

and the solution is

$$y = y_0 e^x$$

Exercise on 15.1

Solve the more general equation (15.1) in the case when the initial number of bacteria is $n_0 = 4$. If time is measured in seconds, and after 5 seconds it is found that the number of bacteria is 10, what is the value of k?

Answer

$n = 4e^{kt}$, $k = \frac{1}{5}\ln(5/2)$

15.2 Definitions

An **ordinary differential equation** (DE) for a **dependent variable** y in terms of an **independent variable** x is any equation that contains one or more derivatives of y with respect to x, and possibly x and y. The order of the highest order derivative in the equation defines the **order** of the DE.

A DE in which the only power to which y or any of its derivatives occurs is zero or one is called a **linear DE**. Any other DE is said to be **nonlinear**. Except for some special cases nonlinear equations are very difficult to deal with. Often, particularly in engineering, they are solved by approximating them by linear equations, although there are many cases where the full nonlinearity must be confronted. Most of this chapter will be devoted to linear equations.

Problem 15.2

State the order of each DE, and state which are nonlinear.

(i) $\dfrac{dy}{dx} = x^2$ (ii) $\left(\dfrac{dy}{dx}\right)^2 + y = x$

(iii) $\dfrac{d^2y}{dx^2} + 4y = 0$ (iv) $\dfrac{d^2y}{dx^2} + 4y^2 = 0$

(i) is first order and linear
(ii) is first order and nonlinear
(iii) is second order and linear
(iv) is second order and nonlinear, because of the y^2.

A **solution** of a DE for y in terms of x is any function, $y = f(x)$, which when substituted into the equation reduces it to an identity – that is, it satisfies it identically, for all values of x (50 ◀). We then say $f(x)$ **satisfies** the equation. In general a DE can have more than one such solution. Sometimes a single function can be found which incorporates all possible solutions of the DE – this is then referred to as the **general solution**. Such solutions contain one or more arbitrary constants, usually depending on the order of the DE. Any other solution than the general solution is called a **particular solution**. Particular solutions can often be found by guesswork (or 'by inspection'), but general solutions are usually much harder to find.

The arbitrary constant(s) occurring in the general solution can be determined by supplementing the DE by specified conditions in which y, or a sufficient number of its derivatives, is given for some particular value(s) of x. Such conditions are called **initial** or **boundary conditions**.

The difference between initial and boundary conditions can be seen by considering a projectile such as a shell from a gun. Even if you don't know the differential equation that describes its motion under gravity alone, you can perhaps appreciate that you could specify the motion completely in one of (at least) two ways:

- Specify the position and velocity at the initial point of projection – **initial conditions**
- Specify two separate points on the trajectory – say point of projection and the furthest point reached, the landing point – **boundary conditions**

Problem 15.3

Show that $y = \dfrac{x^3}{3} + 1$ is a solution of Problem 15.2(i). Can you find any other solutions?

If $y = \dfrac{x^3}{3} + 1$ then $\dfrac{dy}{dx} = x^2$ and the DE is clearly satisfied. Therefore

$$y = \frac{x^3}{3} + 1 \text{ is a solution}$$

You may realise that in fact **any** function of the form

$$y = \frac{x^3}{3} + C$$

Where C is an arbitrary constant is also a solution, because C is knocked out by the differentiation.

Problem 15.4

Show that the following are solutions of Problem 15.2(iii):

(a) $y = \sin 2x$
(b) $y = 3\cos 2x$
(c) $y = 2\sin 2x - \cos 2x$

Can you suggest any more solutions? Is $2\sin x$ a solution?

(a) Differentiating twice we have

$$\frac{d^2y}{dx^2} = -4\sin 2x = -4y$$

giving the required DE

$$\frac{d^2y}{dx^2} + 4y = 0$$

(b) Similarly:

$$\frac{d^2y}{dx^2} = -12\cos 2x = -4y$$

so

$$\frac{d^2y}{dx^2} + 4y = 0$$

i.e. this function is also a solution.

(c) $\frac{d^2y}{dx^2} = -8\sin 2x + 4\cos 2x$

$$= -4(2\sin 2x - \cos 2x)$$

$$= -4y$$

and again therefore

$$\frac{d^2y}{dx^2} + 4y = 0$$

You may again realise that since both $\sin 2x$ and $\cos 2x$ are solutions of the equation, and **since it is a linear** equation, then any function of the form

$$y = A\cos 2x + B\sin 2x$$

where A and B are arbitrary constants is **also** a solution.

If we try $y = 2\sin x$ we have

$$\frac{d^2y}{dx^2} = -2\sin x = -y$$

so this satisfies

$$\frac{d^2y}{dx^2} + y = 0$$

which is **not** the required equation.

Problem 15.5

Find the general solution of Problem 15.2(i) and find particular solutions satisfying

(a) $y = 1$ at $x = 0$ \hspace{2em} (b) $y = 0$ at $x = 1$

In Problem 15.3 we were asked to check a given solution to the equation $\frac{dy}{dx} = x^2$. Here we wish to actually **find** the most general possible solution. We can in fact directly integrate the equation in this case, remembering to add an arbitrary constant.

From $\frac{dy}{dx} = x^2$ we have

$$y = \int x^2 \, dx + C = \frac{x^3}{3} + C$$

which is the required general solution.

(a) If $y = 1$ when $x = 0$ then we have:

$$y = 1 = 0 + C$$

so $C = 1$ and the required particular solution is $y = \frac{x^3}{3} + 1$

(b) $y = 0$ when $x = 1$ gives

$$0 = \frac{1^3}{3} + C = \frac{1}{3} + C$$

so $C = -\frac{1}{3}$ and the required particular solution is

$$y = \frac{x^3}{3} - \frac{1}{3} = \frac{x^3 - 1}{3}$$

Exercises on 15.2

1. State the order of the following differential equations. Which are nonlinear?

(i) $\frac{dy}{dx} = e^x + 1$

(ii) $\frac{d^2y}{dx^2} - 9y = 0$

(iii) $y\frac{d^2y}{dx^2} + \cos x = 0$

(iv) $\frac{dy}{dx}\frac{d^3y}{dx^3} + 2y^2 = 1$

(v) $\frac{d^2y}{dx^2} - 4\frac{dy}{dx} + 3y = 3x + 2$

2. Verify that the following functions are each solutions of one of the equations in Q1, and match the solution to its equation.

(a) $2e^{3x}$

(b) $e^x + x + 2$

(c) $e^{3x} + x + 2$

3. Find the general solution of Q1(i) and the particular solution that satisfies $y(0) = 1$.

Answers

1. (i) 1 (ii) 2 (iii) 2 (iv) 3
   (v) 2

   (iii) and (iv) are nonlinear.

2. (a) 1 (ii) (b) 1 (i) (c) 1 (v)

3. $y = e^x + x + C$; $y = e^x + x$

15.3 First order equations – direct integration and separation of variables

We will only consider cases where we can solve the given DE to give an equation for $\dfrac{dy}{dx}$ of the form

$$\frac{dy}{dx} = F(x, y)$$

where $F(x, y)$ is a 'well behaved' (i.e. we can do whatever we wish with it) function of x and y.

The ease with which we can solve such a DE will depend on the form of $F(x, y)$. We will build up from the simplest cases.

The simplest case is an equation of the form

$$\frac{dy}{dx} = f(x)$$

which can be easily 'solved' in principle by **direct integration**:

$$y = \int f(x)\, dx + C$$

The only difficulty here lies in actually performing the integration. Such simple equations illustrate many of the key points of DEs in general.

A less trivial variation on this is an equation of the form

$$\frac{dy}{dx} = g(y)$$

We can in fact turn this upside down [**not a trivial matter**, but permissible with care (235 ◀)]:

$$\frac{dx}{dy} = \frac{1}{g(y)}$$

Now this can be integrated directly, with respect to y:

$$x = \int \frac{dy}{g(y)} + C$$

This results in principle in a solution of the form

$$x = G(y) + C$$

where $G(y)$ is some function of y. We may or may not be able to solve this integrated equation for y in terms of x.

If we are also given an initial condition, then we can find the value of C by substituting this in the result.

Problem 15.6

Find the general solution of the DEs:

(i) $\dfrac{dy}{dx} = x \cos x$ (ii) $\dfrac{dy}{dx} = \cos^2 y$

In (ii) find the particular solution satisfying $y = \dfrac{\pi}{4}$ when $x = 0$.
Check your answers by substituting back into the equations.

(i) Direct integration gives

$$y = \int x \cos x \, dx + C$$

and the task is simply to do the integration. Integration by parts (273 ◀) gives

$$y = x \sin x - \int \sin x \, dx + C$$
$$= x \sin x + \cos x + C$$

which is the required solution, as you should check by substituting back into the DE.

(ii) $\dfrac{dy}{dx} = \cos^2 y$

In this case, turn both sides upside down

$$\dfrac{dx}{dy} = \dfrac{1}{\cos^2 y} = \sec^2 y$$

Integrating:

$$x = \int \sec^2 y \, dy + C = \tan y + C$$

This is the implicit form of the solution (91 ◀). For the explicit form we solve for y:

$$\tan y = x - C$$

So, as you can check in the DE, the general solution is

$$y = \tan^{-1}(x - C)$$

If $y = \dfrac{\pi}{4}$ when $x = 0$ then we have

$$y = \dfrac{\pi}{4} = \tan^{-1}(-C)$$

from which $C = -1$ and the required solution is

$$y = \tan^{-1}(x + 1)$$

Note that because of the multi-valued nature of the inverse tan function we have to restrict the region on which the DE is defined in order get a unique solution.

The previous cases dealt with either

$$\frac{dy}{dx} = f(x) \quad \text{or} \quad \frac{dy}{dx} = g(y)$$

It is also quite easy to treat equations of the form

$$\frac{dy}{dx} = f(x)g(y)$$

These are called **variables separable** because the expression for dy/dx can be separated into a product of separate functions of x and y alone.

In this case we literally move all y's to one side and x's to the other:

$$\frac{dy}{g(y)} = f(x)\,dx$$

hence, integrating both sides with respect to their respective variables

$$\int \frac{dy}{g(y)} = \int f(x)\,dx + C$$

Of course the left-hand side is now an integral with respect to y, the right-hand side with respect to x. Note that we only need one arbitrary constant. Again, our only real problem arises in actually doing the integrations.

Problem 15.7

Solve the initial value problems

(i) $\dfrac{dy}{dx} = xy \quad y = 1 \quad \text{when} \quad x = 0$

(ii) $\dfrac{dy}{dx} = e^{2x+y} \quad y = 0 \quad \text{when} \quad x = 0$

(i) For $\dfrac{dy}{dx} = xy$ we separate to give

$$\frac{dy}{y} = x\,dx$$

so

$$\int \frac{dy}{y} = \int x\,dx + C$$

and integrating both sides with respect to their respective variables gives

$$\ln y = \frac{x^2}{2} + C$$

STOP AND THINK!

The next step is one where many beginners slip up. To solve for y we 'take the exponential' of both sides (131 ◀):

$$y = e^{\ln y} = e^{\frac{x^2}{2} + C}$$

NOT

$$\text{`}y = e^{\frac{x^2}{2}} + e^C\text{'}$$

We now use (129 ◀) – **AGAIN BE CAREFUL, MANY BEGINNERS MAKE MISTAKES HERE TOO.**

$$e^{A+B} = e^A e^B$$

to obtain

$$y = e^{\frac{x^2}{2}} e^C = A e^{\frac{x^2}{2}}$$

where we have replaced e^C by A, another (positive) arbitrary constant.

Given that $y = 1$ when $x = 0$ we have

$$1 = Ae^0 = A$$

so the required solution is (again, check it in the DE)

$$y = e^{x^2/2}$$

(ii) $\dfrac{dy}{dx} = e^{2x+y} = e^{2x} e^y$ and so, separating the variables

$$e^{-y} dy = e^{2x} dx$$

Integrating both sides gives

$$\int e^{-y} dy = \int e^{2x} dx + C$$

or

$$-e^{-y} = \frac{e^{2x}}{2} + C$$

We might as well find C now

$$-e^0 = \frac{e^0}{2} + C$$

or since $e^0 = 1$, $C = -\frac{3}{2}$ and the solution can be written

$$e^{-y} = \frac{3}{2} - \frac{e^{2x}}{2}$$

Note that since e^{-y} must be positive, we are here restricted to $e^{2x} \leq 3$.
Solving for y gives

$$-y = \ln\left|\frac{3}{2} - \frac{e^{2x}}{2}\right|$$

or

$$y = -\ln\left|\frac{2}{3} - \frac{e^{2x}}{2}\right|$$

The form of variables separable equations is rather restrictive. Even such a simple function as $F(x, y) = x + y$ wouldn't fit into it. However, there are many types of equation that may be reduced to variables separable by some kind of substitution. Consider, for example the equation

$$\frac{dy}{dx} = \frac{x + y}{x} = F(x, y)$$

where $F(x, y)$ is of the form

$$F(x, y) = f\left(\frac{y}{x}\right)$$

Such an equation is said to be **homogeneous** (**not** to be confused with later use of this term). If we change our variables from x, y to x and $v = \frac{y}{x}$ we have $y = xv$ and so

$$\frac{dy}{dx} = v + x\frac{dv}{dx}$$

and the equation becomes

$$\frac{dy}{dx} = v + x\frac{dv}{dx} = f(v)$$

or

$$x\frac{dv}{dx} = f(v) - v$$

This is **separable** and its solution is

$$\int \frac{dv}{f(v) - v} = \int \frac{dx}{x} + C = \ln x + C$$

After evaluating the integral on the left we can then replace v by y/x to get the solution in terms of x and y.

456

Problem 15.8
By substituting $y = xv$ find the general solution of the DE

$$\frac{dy}{dx} = \frac{x+y}{x}$$

We note that

$$\frac{dy}{dx} = 1 + \frac{y}{x} = \text{a function of } \frac{y}{x}$$

and so this equation is homogeneous. So we substitute $y = xv$ as suggested. We have

$$\frac{dy}{dx} = \frac{d(xv)}{dx} = v + x\frac{dv}{dx}$$

But
$$\frac{dy}{dx} = 1 + \frac{y}{x} = 1 + v$$

So
$$x\frac{dv}{dx} = 1$$

or, separating the variables and integrating

$$\int dv = \int \frac{dx}{x} + C = \ln x + C$$

Hence
$$v = \frac{y}{x} = \ln x + C$$

and therefore

$$y = x \ln x + Cx$$

Remember to multiply the C by x also!

Exercise on 15.3
Solve the differential equations

(i) $y' = \sin x$ (ii) $y' = y^2$
(iii) $y' = x^2 y$ (iv) $xy' = 2x + y$

In (i), (ii), (iii) give the particular solutions satisfying the condition $y(0) = 1$. In (iv) give the solution satisfying $y(1) = 0$.

Answer

(i) $y = C - \cos x$, $y = 2 - \cos x$ (ii) $y = -\frac{1}{x+C}$, $y = \frac{1}{1-x}$

(iii) $y = C \exp\left(\frac{x^3}{3}\right)$, $y = \exp\left(\frac{x^3}{3}\right)$ (iv) $y = 2x \ln x + Cx$, $y = 2x \ln x$

15.4 Linear equations and integrating factors

This class of DEs is absolutely fundamental. Such equations occur throughout science and engineering. A **linear equation of the first order** is one that can be put in the form

$$\frac{dy}{dx} + P(x)y = Q(x) \qquad (15.3)$$

where $P(x)$ and $Q(x)$ are functions of x. It is called 'linear' because the non-derivative part is linear in the dependent variable y:

$$\frac{dy}{dx} = -P(x)y + Q(x) \qquad \text{(cf: } ay + b\text{)}$$

This is key to the method of solution – if any other power of y but 0 or 1 occurred then what we are about to do would not be possible. We notice that the left-hand side of equation (15.3) looks very much like the derivative of a product:

$$\frac{dy}{dx} + Py$$

compared to say

$$\frac{d(uy)}{dx} = u\frac{dy}{dx} + \frac{du}{dx}y$$

The resemblance can be improved if we multiply through by a function of x, $I = I(x)$, yet to be determined (called an **integrating factor** because it enables us to integrate the equation). So, compare

$$I\frac{dy}{dx} + IPy = IQ$$

with

$$\frac{d}{dx}(Iy) = I\frac{dy}{dx} + \frac{dI}{dx}y$$

Now IQ is a function of x alone, with no y. It can therefore be integrated with respect to x. On the other hand, if we now take I to be such that

$$\frac{dI}{dx} = IP$$

then

$$I\frac{dy}{dx} + IPy = I\frac{dy}{dx} + \frac{dI}{dx}y$$

$$= \frac{d}{dx}(Iy)$$

and we have

$$\frac{d(Iy)}{dx} = IQ$$

This can now be solved by direct integration:

$$Iy = \int IQ\,dx + C$$

and we finally get the solution

$$y = \frac{1}{I}\left(\int IQ\,dx + C\right)$$

$$= \frac{1}{I}\int IQ\,dx + \frac{C}{I}$$

Thus, the purpose of multiplying by the integrating factor is to convert the left-hand side to the derivative of a product so that we can integrate the resulting equation directly. But all this depends on finding I from the equation

$$\frac{dI}{dx} = IP$$

This equation is separable:

$$\frac{dI}{I} = P\,dx$$

so

$$\int \frac{dI}{I} = \int P(x)\,dx$$

Note that we needn't bother to introduce an arbitrary constant here – you can if you like, but it will simply cancel out in the end.

Thus, for I we obtain

$$\ln I = \int P(x)\,dx$$

So the integrating factor is given by

$$I = e^{\int P(x)\,dx}$$

I don't encourage you to remember the above results, either for I or the solution for y. Rather, you should try to remember how to reproduce the above arguments to derive solutions directly yourself. This may seem daunting, but with plenty of practice it is not too bad, and is a very useful skill. To help with this the following problems work through the procedure step by step. It also illustrates that you can often short cut the process of finding the integrating factor by rearranging the left-hand side of the original equation to be the derivative of a product 'by inspection'.

Problem 15.9
Convert the DE

$$xy' + y = x^2$$

to standard linear form.

Dividing by x to make the coefficient of the derivative unity gives

$$\frac{dy}{dx} + \frac{1}{x}y = x$$

Problem 15.10
Write down the DE satisfied by the integrating factor.

Multiplying through by an, as yet unknown, integrating factor, I, gives

$$I\frac{dy}{dx} + \frac{I}{x}y = xI$$

Now

$$\frac{d(Iy)}{dx} = I\frac{dy}{dx} + \frac{I}{x}y$$

will be satisfied if

$$\frac{dI}{dx} = \frac{I}{x}$$

which is the required equation for the integrating factor. Note that it is separable.

Problem 15.11
Determine the integrating factor I.

Separating the variables of the last equation we have

$$\frac{dI}{I} = \frac{dx}{x}$$

or, integrating through

$$\int \frac{dI}{I} = \int \frac{dx}{x}$$

Performing the integrations gives

$$\ln I = \ln x$$

So in this case we have

$$I = x$$

for the integrating factor. Note that we don't need to include an arbitrary constant at this stage.

Problem 15.12
Solve the original equation.

Multiplying the linear form by the integrating factor x we now have

$$x\frac{dy}{dx} + y = x^2$$

(back where we started, but bear with me on this – we'll come back to it later). We know that the left-hand side must now be the derivative of a product, namely Iy, because that is precisely why we chose the integrating factor:

$$\frac{d}{dx}(Iy) = \frac{d}{dx}(xy) = x^2$$

Integrating both sides finally gives

$$xy = \frac{x^3}{3} + C$$

(**Now** we introduce the arbitrary constant), so the general solution is

$$y = \frac{x^2}{3} + \frac{C}{x}$$

Problem 15.13
Repeat the solution but dispense with the integrating factor and solve the original equation directly.

We ended up, in Problem 15.12, after finding the integrating factor, with the original equation:

$$x\frac{dy}{dx} + y = x^2$$

In fact, in this case, the IF is really using a sledgehammer to crack a nut. Simply notice that by reversing the product rule

$$x\frac{dy}{dx} + y \equiv \frac{d}{dx}(xy)$$

and we have

$$\frac{d}{dx}(xy) = x^2$$

which is where we got to after finding the integrating factor. We are therefore led directly to the result

$$xy = \frac{x^3}{3} + C$$

In general of course it may not be quite so easy to spot the required derivative of a product, and we may need to go through the full procedure of finding the IF, but try to avoid it when you can. You can in any case see that the better developed your skills of differentiation, the easier it will be to spot such short cuts.

Exercise on 15.4
Find integrating factors for the following equations and hence obtain the general solution

(i) $xy' + y = x$

(ii) $xy' - 2y = x^3 + 2$

Can you dispense with the integrating factor, by finding a derivative of a product?

Answer

(i) $y = \dfrac{x}{2} + \dfrac{C}{x}$ (ii) $y = x^3 + Cx^2 - 1$

15.5 Second order linear homogeneous differential equations

If you found the previous section difficult then you may be approaching this section with some trepidation – surely second order will be worse than first order? Relax. We only consider a simple, but extremely important, type of second order equation and it turns out that this is relatively straightforward to solve. The particular type of linear second order equation that we will consider is one of the form

$$a\frac{d^2y}{dx^2} + b\frac{dy}{dx} + cy = f(x) \tag{15.4}$$

where a, b, c are constants (and $a \neq 0$, otherwise it wouldn't be second order!). For such equations the basic tools that you need are simply solution of quadratics (including complex roots) and solution of simultaneous linear equations. Equation (15.4) occurs everywhere in science and engineering, most notably in the modelling of vibrating springs in a resisting medium, and in electrical circuits.

We need some terminology. The equation:

$$a\frac{d^2y}{dx^2} + b\frac{dy}{dx} + cy = 0 \tag{15.5}$$

is the associated **homogeneous equation**, while equation (15.4), with $f(x) \neq 0$, is the **inhomogeneous** form. The **general solution** to the **homogeneous equation** is called the **complementary function**, while **any** solution to equation (15.4) is called a **particular integral**. The **general solution** of equation (15.4) is **the sum of the complementary function and a particular integral**.

For example the differential equation

$$\frac{d^2y}{dx^2} + 3\frac{dy}{dx} + 2y = 2e^x$$

is **inhomogeneous**, and the corresponding **homogeneous form** is

$$\frac{d^2y}{dx^2} + 3\frac{dy}{dx} + 2y = 0$$

The general solution of this homogeneous equation is the **complementary function of the first, inhomogeneous, equation**. In this case it happens to be (see below)

$$y_c = Ae^{-x} + Be^{-2x}$$

where A and B are two arbitrary constants.

Any particular solution of the **inhomogeneous** equation is a **particular integral** of the equation. By substituting it in the equation you should check that a particular integral in this case is

$$y_p = \tfrac{1}{3}e^x$$

The **general solution of the inhomogeneous equation** is thus

$$y = y_c + y_p = Ae^{-x} + Be^{-2x} + \tfrac{1}{3}e^x$$

We therefore have two jobs – to find the complementary function and to find a particular integral. We first concentrate on finding the complementary function and look for general solutions of equations of the form of equation (15.5). There is an additional result which helps us here:

If y_1 and y_2 are two solutions of the homogeneous linear equation

$$a\frac{d^2y}{dx^2} + b\frac{dy}{dx} + cy = 0$$

then $y_1 + y_2$ is also a solution. Thus, the sum (or any other linear combination) of two solutions is also a solution. To see this important result in its most general form let y_1, y_2 be two solutions of the DE and consider the **linear combination** $y = \alpha y_1 + \beta y_2$. Substituting this in the DE gives

$$a\frac{d^2y}{dx^2} + b\frac{dy}{dx} + cy = a\frac{d^2(\alpha y_1 + \beta y_2)}{dx^2} + b\frac{d(\alpha y_1 + \beta y_2)}{dx} + c(\alpha y_1 + \beta y_2)$$

$$= \alpha\left(a\frac{d^2y_1}{dx^2} + b\frac{dy_1}{dx} + cy_1\right) + \beta\left(a\frac{d^2y_2}{dx^2} + b\frac{dy_2}{dx} + cy_2\right)$$

$$= \alpha(0) + \beta(0) = 0$$

So the general linear combination $y = \alpha y_1 + \beta y_2$ is also a solution.

In fact, finding the complementary function, i.e. the general solution to equation (15.5), is not as difficult as might be thought. Note that if we put $a = 0$ then we are back to the simple equation

$$b\frac{dy}{dx} + cy = 0$$

and we know that this has an exponential solution (447 ◂). This encourages us to try a similar exponential function for the second order equation.

We therefore take a trial solution

$$y = e^{\lambda x}$$

where λ is some constant parameter to be determined.

Substituting into the equation:

$$y' = \lambda e^{\lambda x}, \ y'' = \lambda^2 e^{\lambda x}$$

gives

$$(a\lambda^2 + b\lambda + c)e^{\lambda x} = 0$$

Since $e^{\lambda x} \neq 0$ we have

$$a\lambda^2 + b\lambda + c = 0$$

So λ satisfies a quadratic equation with the same coefficients as the DE itself, $ay'' + by' + cy$. This equation in λ is called the **auxiliary** or **characteristic** equation (AE). As with all quadratics with real coefficients (66 ◀) there are three distinct types of solution:

- Real distinct roots
- Real equal roots
- Complex conjugate roots

Each leads to a different type of solution to the DE. In each case we get two distinct types of solution, and the general solution is formed from these. Below we summarise the forms of these solutions. In each case you should verify the stated solutions by substituting in the equations. The problems which follow will confirm the results in particular cases.

Roots of AE real and distinct α_1, α_2

This gives two solutions

$$e^{\alpha_1 x}, e^{\alpha_2 x}$$

and the general solution is then

$$y = Ae^{\alpha_1 x} + Be^{\alpha_2 x}$$

Roots real and equal, α

In this case two distinct solutions can be found:

$$e^{\alpha x}, xe^{\alpha x}$$

and the general solution is then

$$y = (Ax + B)e^{\alpha x}$$

Roots complex, $\alpha \pm j\beta$

This gives two solutions

$$e^{(\alpha + j\beta)x}, e^{(\alpha - j\beta)x}$$

giving a general solution

$$y = Ae^{(\alpha + j\beta)x} + Be^{(\alpha - j\beta)x}$$
$$= e^{\alpha x}(Ae^{j\beta x} + Be^{-j\beta x})$$

The imaginary j is not always welcome here, so we use **Euler's formula** (361 ◀) to put the solution into real form

$$e^{j\beta x} = \cos \beta x + j \sin \beta x$$
$$e^{-j\beta x} = \cos \beta x - j \sin \beta x$$
$$y = e^{\alpha x}((A + B) \cos \beta x + (A - jB) \sin \beta x)$$

or with $\quad C = A + B$ and $D = A - jB$

$$y = e^{\alpha x}(C \cos \beta x + D \sin \beta x)$$

Note: Although we usually prefer such a real form of the solution, particularly in actual physical applications, there are times when the complex exponential form is in fact most convenient – for example when we are using phasors in alternating current electrical work.

We now work through a number of problems to illustrate the above results. As always, you will benefit greatly by checking the solutions obtained by substituting back into the DE.

Problem 15.14

Find values of λ for which $y = e^{\lambda x}$ satisfies the DE

$$y'' + 3y' + 2y = 0$$

and hence determine its general solution. Find the particular solution satisfying the initial conditions $y(0) = 0$, $y'(0) = 1$.

We have

$$y' = \lambda e^{\lambda x}, \, y'' = \lambda^2 e^{\lambda x}$$

Substituting into the equation gives

$$(\lambda^2 + 3\lambda + 2)e^{\lambda x} = 0$$

or the auxiliary equation:

$$\lambda^2 + 3\lambda + 2 = 0$$
$$(\lambda + 1)(\lambda + 2) = 0$$

So

$$\lambda = -1, -2$$

We therefore have two solutions to the DE

$$e^{-x}, e^{-2x}$$

The general solution is then

$$y = Ae^{-x} + Be^{-2x}$$

containing two arbitrary constants A, B, as we might expect for a second order DE. To find them we apply the given initial conditions (remember that $e^0 = 1$ and $\dfrac{d(e^{ax})}{dx} = ae^{ax}$):

$$y(0) = A + B = 0$$
$$y'(0) = -A - 2B = 1$$

Solving these two equations gives $A = 1$, $B = -1$ and the particular solution satisfying the initial conditions is

$$y = e^{-x} - e^{-2x}$$

Problem 15.15

Find the general solution of the DE

$$y'' - 6y' + 9y = 0$$

Substituting $y = e^{\lambda x}$ gives, in this case, the AE

$$\lambda^2 - 6\lambda + 9 = 0$$

or

$$(\lambda - 3)^2 = 0$$

We therefore have two equal roots, $\lambda = 3$ (rather than a single root of $\lambda = 3$!). One solution of the DE is therefore obviously

$$y = e^{3x}$$

But we expect two. On such occasions we try to guess other, similar, types of solution. While there is a full justification for our guesswork in the advanced theory of differential equations here I ask you to simply accept and confirm for yourself that the simplest guess that works is

$$y = xe^{3x}$$

This gives

$$y' = (3x + 1)e^{3x}$$
$$y'' = 3e^{3x} + 3(3x + 1)e^{3x}$$
$$= (9x + 6)e^{3x}$$

Substituting in the equation gives

$$(9x + 6)e^{3x} - 6(3x + 1)e^{3x} + 9xe^{3x} \equiv 0$$

so this is indeed a solution. The general solution in this case is therefore

$$y = Axe^{3x} + Be^{3x}$$
$$= (Ax + B)e^{3x}$$

Problem 15.16

Find the general solution of the equation

$$y'' + 2y' + 2y = 0$$

The AE is

$$\lambda^2 + 2\lambda + 2 = 0$$

which has the complex roots

$$\lambda = -1 \pm j$$

This gives the two solutions

$$e^{(-1+j)x}, e^{(-1-j)x}$$

and the general solution

$$y = e^{-x}(Ae^{jx} + Be^{-jx})$$

Now, using Euler's formula (361 ◄)

$$e^{\pm jx} = \cos x \pm j \sin x$$

we get

$$y = e^{-x}(A(\cos x + j \sin x) + B(\cos x - j \sin x))$$
$$= e^{-x}((A+B)\cos x + j(A-B)\sin x)$$
$$= e^{-x}(C \cos x + D \sin x)$$

on writing $C = A + B$ and $D = j(A - B)$.

In this 'real form' the j is hidden and for physical problems where y represents a real quantity, such as distance or current, never resurfaces. However, as noted previously the complex exponential form is sometimes useful in applications such as alternating current theory.

In the above list of solutions we essentially have just three distinct types of functions: $e^{\alpha x}$, $xe^{\alpha x}$, and $e^{\alpha x} \sin \beta x$. Each of these functions has a particular sort of behaviour which typifies the class to which it belongs, and the type of physical system it represents.

1. $e^{\alpha x}$ gives an increasing ($\alpha > 0$) or decaying ($\alpha < 0$) exponential function:

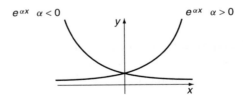

Figure 15.1 The function $e^{\alpha x}$.

2. $xe^{\alpha x}$ gives a similar type of function, but with a kink in it, and it also passes through the origin. Examples of this function were sketched in Chapter 10.

3. $e^{\alpha x} \sin \beta x$ gives either a simple oscillating wave ($\alpha = 0$), or a sinusoidal wave with amplitude that decreases ($\alpha < 0$) or increases ($\alpha > 0$) as

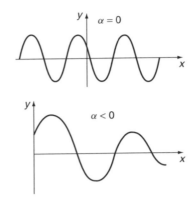

Figure 15.2 Undamped and damped oscillations.

x increases. This could represent an oscillating system in a resisting medium for example.

Exercises on 15.5

1. Solve the following initial value problems:

   (i) $y'' - y' - 6y = 0$ $y(0) = 1$ $y'(0) = 0$

   (ii) $2y'' + y' - 10y = 0$ $y(0) = 0$ $y'(0) = 1$

2. Solve the following boundary value problems:

   (i) $y'' + 4y' + 13y = 0$ $y(0) = 0$ $y\left(\dfrac{\pi}{2}\right) = 1$

   (ii) $y'' - 4y' + 4y = 0$ $y(0) = 0$ $y(1) = 1$

Answers

1. (i) $\dfrac{2}{5}e^{3x} + \dfrac{3}{5}e^{-2x}$ (ii) $\dfrac{2}{9}e^{2x} - \dfrac{2}{9}e^{-5x/2}$

2. (i) $-e^{\pi - 2x} \sin 3x$ (ii) $xe^{2(x-1)}$

15.6 The inhomogeneous equation

We now turn to the solution of the full inhomogeneous equation

$$a\frac{d^2 y}{dx^2} + b\frac{dy}{dx} + cy = f(x)$$

We now know the general form of the CF, so we just need to find particular integrals (PI). These depend on the form of $f(x)$. In fact, in many cases the form of the PI mirrors precisely that of the function $f(x)$ itself. The general approach is to make a trial solution which has the same form as $f(x)$. This does not always work, but is a good start, and we will come back to exceptional cases later. This approach is called the method of **undetermined coefficients**. It in fact works for equations of any order, provided the coefficients are constant, as we are assuming here.

Table 15.1

$f(x)$	Trial solution
Polynomial of degree n: $f(x) = a_n x^n + a_{n-1} x^{n-1} + \cdots$	Polynomial of degree n: $y_p = L x^n + M x^{n-1} + \cdots$
Exponential: $f(x) = e^{kx}$	Similar exponential $y_p = L e^{kx}$
Sinusoidal function: $f(x) = \sin \omega x$ or $\cos \omega x$	Linear combination of similar sinusoidals: $y_p = L \cos \omega x + M \sin \omega x$
'Damped' sinusoidal: $f(x) = e^{kx} \cos \omega x$ or $e^{kx} \sin \omega x$	Linear combination of similar damped sinusoidals $y_p = L e^{kx} \cos \omega x + M e^{kx} \sin \omega x$

The 'forcing functions' $f(x)$ given in Table 15.1 cover most cases of physical interest at the elementary level. They can in fact be used for much more complicated functions, $f(x)$, using linearity and the techniques of Fourier analysis (➤ 517) for example, so they are in fact more general than you might think. We will give an example of each of them before discussing the complications of exceptional cases.

Problem 15.17
Find a particular integral for the equation

$$y'' + 2y' + y = x + 2$$

Hence determine the general solution. Find the particular solution satisfying the initial conditions $y(0) = 0$, $y'(0) = 0$.

For reasons that become clear later, we will always find the CF first. In this case you can treat it as an exercise to check that

$$y_c = (Ax + B)e^{-x}$$

Now for the PI we note that $f(x)$ is in this case a first degree polynomial – i.e. a linear function. We therefore try a solution of the same form:

$$y_p = Lx + M, \text{ so } y'_p = L, y''_p = 0$$

Substituting in the equation gives

$$0 + 2L + Lx + M = Lx + (2L + M) \equiv x + 2$$

Solving this identity (50 ◄) gives

$$L = 1 \quad 2L + M = 2$$

so $M = 0$. Thus, the PI is $y_p = x$, as you can (should!) check by substituting back into the equation:

$$y'' + 2y' + y = 0 + 2(1) + x = x + 2$$

The GS to the inhomogeneous equation is therefore

$$y = (Ax + B)e^{-x} + x$$

Applying the initial conditions we have (remember $e^0 = 1$)

$$y(0) = B = 0$$
$$y'(0) = A + 1 = 0$$

so $A = -1$, $B = 0$ and therefore

$$y = x - xe^{-x} = x(1 - e^{-x})$$

NB: When applying the initial or boundary conditions to solutions of inhomogeneous equations you must **apply them to the full general solution – complementary function plus particular integral**, as we have done here. It is a common mistake for beginners to apply such conditions simply to the complementary function.

Problem 15.18
Find a particular integral for the equation

$$y'' + 3y' + 2y = 3e^{2x}$$

Hence determine the general solution.

The CF is $y_c = Ae^{-x} + Be^{-2x}$ (see Problem 15.14). For the PI we try a solution similar to the right-hand side:

$$y_p = Le^{2x}, \quad y_p' = 2Le^{2x}, \quad y_p'' = 4Le^{2x}$$

Substituting in the equation gives

$$(4L + 6L + 2L)e^{2x} \equiv 12Le^{2x} = 3e^{2x}$$

so

$$L = \tfrac{1}{4} \text{ and the PI is } y_p = \tfrac{1}{4}e^{2x}$$

(check this back in the equation).
The full GS is therefore

$$y = Ae^{-x} + Be^{-2x} + \tfrac{1}{4}e^{2x}$$

Problem 15.19
Find the general solution to the equation

$$y'' + 2y' + 2y = 2\cos 3x$$

The CF is $y_c = e^{-x}(A\cos x + B\sin x)$, from Problem 15.16.

For the PI you might be tempted to try $y_p = L\cos 3x$, but in fact we need both $\cos 3x$ and $\sin 3x$ terms because of the first order derivative on the left-hand side. We therefore try

$$y_p = L\cos 3x + M\sin 3x$$
$$y_p' = -3L\sin 3x + 3M\cos 3x$$
$$y_p'' = -9L\cos 3x - 9M\sin 3x$$

substituting in the DE gives, on collecting cos and sin terms:

$$(-7L + 6M)\cos 3x + (-6L - 7M)\sin 3x \equiv 2\cos 3x$$

So
$$-6L - 7M = 0$$
$$-7L + 6M = 2$$

whence
$$L = -\frac{14}{85}, \quad M = \frac{12}{85}$$

and the PI is
$$y_p = -\frac{14}{85}\cos 3x + \frac{12}{85}\sin 3x$$

and the GS is
$$y = e^{-x}(A\cos x + B\sin x) - \frac{14}{85}\cos 3x + \frac{12}{85}\sin 3x$$

(again, check back in the equation).

Problem 15.20
Find the general solution to the equation

$$y'' + 4y = e^{-x}\cos x$$

The CF is
$$y_c = A\cos 2x + B\sin 2x$$

For the PI choose
$$y_p = Le^{-x}\cos x + Me^{-x}\sin x$$

Then, as an exercise in differentiation by the product rule (234 ◄), you can check that

$$y_p'' = e^{-x}(2L\sin x - 2M\cos x)$$

Substituting in the equation gives, after simplification

$$e^{-x}[(4L - 2M)\cos x + (2L + 4M)\sin x] \equiv e^{-x}\cos x$$

471

Or, cancelling e^{-x} and equating coefficients of $\cos x$ and $\sin x$

$$4L - 2M = 1$$
$$2L + 4M = 0$$

These give
$$L = \frac{1}{5}, \quad M = -\frac{1}{10}$$

So finally, the GS is $y = A\cos 2x + B\sin 2x + \frac{1}{5}e^{-x}\cos x - \frac{1}{10}e^{-x}\sin x$

Two points to note:

- The above method of undetermined coefficients is not the only method – there are others that are more powerful in some cases – D-operators, complex variable methods, Laplace transform, etc. But it is systematic and routine.
- It doesn't always work in such a straightforward way.

To see how the method can break down, and what to do about it, try the following problem.

Problem 15.21
Investigate the particular integral for the following equation

$$y'' - y' - 2y = e^{-x}$$

Trying $y = Le^{-x}$ for a PI, as we might be tempted, gives

$$'0 = e^{-x}'$$

and clearly does not work. In fact in this case the reason a trial function similar to the right-hand side won't work is that it is already a solution of the corresponding homogeneous equation – that is, **it is part of the complementary function**. This is when the method breaks down – when the right-hand side is part of the CF. And that's why we always evaluate the CF first, to check whether it does contain the right-hand side. In this case the CF is the GS of

$$y'' - y' - 2y = 0$$

which you can check is $y = Ae^{-x} + Be^{2x}$. The right-hand side is indeed included in the CF.

So, what do we do in such cases? As usual in mathematics we do as little as possible – that is we look for alternatives that are as close to what we have as possible. Remember that in the case of equal roots for the auxiliary equation, when looking at the CF, we tried – successfully – replacing $e^{\alpha x}$ by $xe^{\alpha x}$ (466 ◀)? Perhaps a similar strategy works this time?

Problem 15.22
Try $y_p = Lxe^{-x}$ in the differential equation of Problem 15.21, where L is to be determined.

We have

$$y_p'' - y_p' - 2y_p = L(x-2)e^{-x} - L(1-x)e^{-x} - 2Lxe^{-x}$$
$$= -3Le^{-x} = \text{the RHS} = e^{-x}$$

So Lxe^{-x} is indeed a solution of the DE if $L = -\frac{1}{3}$ and therefore a particular solution is $y_p = -\frac{1}{3}xe^{-x}$.

Now, you may feel uncomfortable with all this guesswork. Don't worry, it can all be put on a formal basis and derived rigorously, but we do not have the space to go into it here. So, we will simply collect all the necessary results together and state them without proof, for your reference. The method to be described actually works for arbitrary order linear equations with constant coefficients and so we will formulate it in quite general terms. To specialise to the second order case we have been considering so far just take $n = 2$. However, you may meet the general case of higher order n in, for example, control systems. So, consider the general nth order linear differential equation with constant coefficients ($y^{(r)}(x)$ denotes the rth order derivative $\dfrac{d^r y}{dx^r}$):

$$a_n y^{(n)}(x) + a_{n-1} y^{(n-1)}(x) + \cdots + a_1 y^{(1)}(x) + a_0 y(x) = f(x)$$

Then in the following cases:

(1) $f(x)$ an mth *degree polynomial*

(a) If $a_0 \neq 0$ – i.e. the equation contains a non-derivative term, then we choose a particular solution of the form

$$y_p = L_m x^m + L_{m-1} x^{m-1} + \cdots + L_1 x + L_0$$

(b) If $a_0 = 0$ so that there is no non-derivative term in the differential equation and if the lowest order of derivative is r, then choose

$$y_p = L_m x^{m+r} + L_{m-1} x^{m+r-1} + \cdots + L_1 x^{r+1} + L_0 x^r$$

(2) $f(x)$ of the form $\sin \omega x$ or $\cos \omega x$

(a) If $\sin \omega x$ or $\cos \omega x$ do not occur in the complementary function, then take a particular solution of the form

$$y_p = L \cos \omega x + M \sin \omega x$$

(b) If the complementary function contains terms of the form $x^r \sin \omega x$ or $x^r \cos \omega x$ for $r = 0, 1, \ldots, m$, take y_p to be the form

$$L x^{m+1} \cos \omega x + M x^{m+1} \sin \omega x$$

(3) $f(x)$ an *exponential function*, $f(x) = e^{kx}$

(a) If e^{kx} is not contained in the complementary function, take $y_p(x)$ to be of the form

$$y_p = L e^{kx}$$

(b) If the complementary function contains terms $e^{kx}, xe^{kx}, \ldots,$ $x^m e^{kx}$, take $y_p(x)$ to be of the form

$$y_p = Lx^{m+1}e^{kx}$$

(4) $f(x)$ of the form $e^{kx}\cos\omega x$ or $e^{kx}\sin\omega x$

(a) If these terms are not contained in the complementary function, then simply take $y_p(x)$ to be of the form

$$y_p = Le^{kx}\cos\omega x + Me^{kx}\sin\omega x$$

(b) If terms of the form $x^r e^{kx}\cos\omega x$ or $x^r e^{kx}\sin\omega x$, for $r = 0, 1, \ldots, m$, are contained in the complementary functions, then include terms of the form

$$Lx^{m+1}e^{kx}\cos\omega x + Mx^{m+1}e^{kx}\sin\omega x$$

in y_p.

Some obvious patterns may be seen in the above particular integrals. Essentially, if the complementary function does not contain terms identical to the inhomogeneous term, then the particular integral looks much the same as the inhomogeneous term. However, if the inhomogeneous term occurs in the complementary function, then we must modify the particular integral and this is done by multiplying by an appropriate power of x.

In general, by superposition of terms we see that we can treat particular integrals for expressions of the form

$$f(x) = P(x)e^{kx}\cos\omega x + Q(x)e^{kx}\sin\omega x$$

where $P(x)$, $Q(x)$ are polynomials in x. This essentially comes about because any order derivative of such an expression results in an expression of the same form.

Having found the particular integral, we can find the general solution by adding it to the complementary function. Note again that when applying boundary or initial conditions these must always be applied to the full solution – complementary function plus particular integral – and not simply to the complementary function alone.

Exercises on 15.6

1. Find the solutions to each of the following second order equations, with the specified conditions. **Remember to apply the conditions to the full solution – CF + PI.**

   (i) $y'' + 4y' + 3y = 2e^x$ $y(0) = 0$ $y'(0) = 1$
   (ii) $y'' + 4y = x + 1$ $y(0) = 0$ $y\left(\dfrac{\pi}{4}\right) = \dfrac{1}{4}$
   (iii) $y'' + y = \sin 2x$ $y(0) = 0$ $y'(0) = 0$

2. Solve the initial value problem

   $y'' - 4y' + 3y = 3x$ $y(0) = 0$ $y'(0) = 0$

Answers

1. (i) $\dfrac{1}{4}(e^x - e^{-3x})$ (ii) $-\dfrac{1}{4}\cos 2x - \dfrac{\pi}{16}\sin 2x + \dfrac{1}{4}x + \dfrac{1}{4}$

   (iii) $\dfrac{2}{3}\sin x - \dfrac{1}{3}\sin 2x$

2. $\dfrac{1}{6}e^{3x} - \dfrac{3}{2}e^x + x + \dfrac{4}{3}$

15.7 Reinforcement

In all these exercises check your results by substituting back to the equation – the practice will do you good!

1. Radioactive material decays at a rate proportional to the amount present. Construct and solve a mathematical model giving the amount of material remaining after a given time.

2. Solve the following differential equations subject to the conditions given:

   (i) $y' = x$ $\qquad\qquad$ $y(0) = 1$
   (ii) $y' = \cos x$ $\qquad$ $y(\pi) = 0$
   (iii) $xy' = x^2 + 1$ $\qquad$ $y(1) = 0$
   (iv) $y'' = 4$ $\qquad\qquad$ $y(0) = 1$ $\qquad$ $y'(0) = 2$
   (v) $y'' = x^2 - 1$ $\qquad$ $y(0) = 0$ $\qquad$ $y(2) = 1$
   (vi) $y'' = \cos x$ $\qquad$ $y(0) = 0$ $\qquad$ $y(\pi) = 1$
   (vii) $y' = 3y^2$ $\qquad$ $y(0) = 1$
   (viii) $y' = \sec y$ $\qquad$ $y(0) = \pi$

3. Find the general solution of the differential equations $y' = f(x, y)$ where $f(x, y)$ is given by:

   (i) xy^2 $\qquad\qquad$ (ii) $\dfrac{y}{x}$ $\qquad\qquad$ (iii) $x \sec y$
   (iv) e^{x+y} $\qquad\qquad$ (v) $10 - 2y$ $\qquad$ (vi) e^{2x-3y}
   (vii) $\dfrac{2x}{5 - \sin y}$ $\qquad$ (viii) $y \ln x$ $\qquad$ (ix) $(x - y)/x$
   (x) $y(x + 2y)/[x(2x + y)]$ $\qquad$ (xi) $3(y^2 - 3y + 2)$

4. Solve the equations:

   (i) $y' = e^{2x} - y$ $\qquad$ (ii) $y' + 2y = 3e^x$ $\qquad$ (iii) $y' + xy = x^3$
   (iv) $xy' = 3y - 2x$ $\qquad$ (v) $(x^2 - 1)y' + 2y = 0$
   (vi) $(x - 1)y' = 3x^2 - y$ $\qquad$ (vii) $xy' - 2y = x^3 e^{-2x}$

5. Solve the following initial value problems

   (i) $y' - 3y = e^{5x}$ $y(0) = 0$
   (ii) $xy' - 2y = x^2$ $y(1) = 2$
   (iii) $xy' + 2y = x^2$ $y(1) = 0$
   (iv) $xy' + 3y = \dfrac{\sin x}{x^2}$ $y(\pi) = 0$

6. Solve the following second order equations

   (i) $y'' + y' - 2y = 0$ (ii) $y'' - 4y' + 4y = 0$
   (iii) $y'' + 4y' + 5y = 0$ (iv) $y'' + 4y = 0$
   (v) $y'' - 9y = 0$ (vi) $y'' + y = 0$

7. Obtain the general solution of the inhomogeneous equations formed by adding the following right-hand sides to each of the equations of Q6.

   (a) 2   (b) $x + 1$ (c) e^{-2x}
   (d) $2 \sin x$

8. Solve the initial value problem $y(0) = 0$, $y'(0) = 0$ for each of the equations solved in Q7(a). If you feel really keen press on with the other questions – you can check your answers by substituting into the equation.

9. Solve the boundary value problem $y(0) = y(1) = 0$ for each of the equations solved in Q7(a). Again press on with the rest of Q7 if you need more practice.

15.8 Applications

There are a number of well known applications of first order equations which provide classic prototypes for mathematical modelling. These mainly rely on the interpretation of dy/dt as a rate of change of a function y with respect to time t. In everyday life there are many examples of the importance of rates of change – speed of moving particles, growth and decay of populations and materials, heat flow, fluid flow, and so on. In each case we can construct models of varying degrees of sophistication to describe given situations.

1. The rate at which a radioactive substance decays is found to be proportional to the mass $m(t)$ present at time t. Write down and solve the differential equation expressing this relation, given that the initial mass is $m(0) = m_0$ and when $t = 1$, $m = \dfrac{m_0}{2}$.

2. **Newton's law of cooling** states that the rate at which an object cools is proportional to the difference between the temperature at the surface of the body, and the ambient air temperature. Thus, if T is the surface temperature at time t and T_a is the ambient temperature, then

$$\frac{dT}{dt} = -\lambda(T - T_a) \qquad T(0) = T_0$$

where $\lambda > 0$ is some experimentally determined constant of proportionality, and T_0 is the initial temperature. Solve this to give the temperature at $t > 0$.

3. In Applications, Chapter 2, we saw how the differential equation

$$\frac{dx}{dt} = k(x-a)(x-b)(x-c)$$

typical of models of chemical reactions, describes the reaction between three gases A, B, C in a vapour deposition process. Find the general solution of the equation for a number of examples of values of k, a, b, c, starting with a simple case such as $k = a = b = 1$, $c = 0$ and building up to more complicated examples. Also consider various initial conditions. Study the nature of the solutions for the various values of the parameters k, a, b, c.

4. **The** classic application of differential equations is, of course, in Newtonian mechanics. The differential equations arise from either the **kinematic** relation $dx/dt = v$, the velocity, or the **dynamical** relation (Newton's second law) $mdv/dt = md^2x/dt^2 =$ mass × acceleration = force. In general we have

$$\frac{d^2x}{dt^2} = f(x, y, \frac{dy}{dt})$$

This reduces to a first order equation in a number of cases. For example, if f is independent of x we have

$$\frac{dv}{dt} = f(t, v)$$

or if it is independent of t we can write

$$\frac{dv}{dt} = \frac{dx}{dt}\frac{dv}{dx} = v\frac{dv}{dx} = f(x, v)$$

With sufficiently complex force laws, we can obtain a variety of differential equations sufficient to keep the most ardent mathematician satisfied – indeed, much of the modern theory of nonlinear differential equation **is** dynamics, and you will find books on 'dynamical systems' shoulder to shoulder with textbooks on differential equations on the library bookshelves.

A particle falling vertically under gravity, subject to a resistance proportional to its velocity, v, satisfies the equation of motion

$$m\frac{dv}{dt} = mg - kv$$

where m is its mass, g the acceleration due to gravity and k a positive constant. Solve this equation and interpret the motion. Assume the initial velocity is $v = v_0$ at $t = 0$.

5. Many complicated physical, biological, and commercial situations can be modelled by a system of interconnected units or 'compartments' between which some quantity flows or is communicated. Examples include the distribution of drugs in various parts

of the body, water flow between connected reservoirs, or financial transactions in a commercial environment. This exercise illustrates the basic ideas in this topic.

Take as our compartment a tank containing 100L of water, into which a brine solution flows at a rate of 5L/min and out of which solution flows at 5L/min. The concentration of the incoming brine solution is 1 kg/L. Construct a model, based on reasonable assumptions, which will enable you to determine the concentration of salt in the tank at any time after inflow commences.

6. There are many areas of science and engineering where second order linear differential equations provide useful models. Rather than become enmeshed in the technical details of specific applications, we will look at generic models which have utility across a wide range of applications.

The general second order inhomogeneous linear equation with constant coefficients

$$a\ddot{x} + b\dot{x} + cx = f(t) \tag{15.6}$$

arises naturally in dynamics as a consequence of Newton's second law

$$m\ddot{x} = F(t, x, \dot{x})$$

in which the force F arises from a particular physical set up. For example if we are talking about the motion of a particle of mass m attached to a spring, oscillating in a resisting medium, and subject to an additional time dependent force $f(t)$, then we might model the forces acting as follows, taking x as the displacement from equilibrium:

(i) spring restoring force, $-\alpha x \quad \alpha > 0$
(ii) resistance force proportional to velocity, $-\beta\dot{x} \quad \beta > 0$
(iii) forcing term $f(t)$

Newton's second law then gives

$$m\ddot{x} + \beta\dot{x} + \alpha x = f(t)$$

A similar equation was mentioned in Chapter 2 (79 ◀) for the capacitor voltage in a source free electrical circuit containing an inductance L, resistance R, capacitance C. With an appropriate source term $f(0)$ this becomes

$$\frac{d^2V}{dt^2} + \frac{R}{L}\frac{dV}{dt} + \frac{1}{LC}V = f(t)$$

which is, apart from the physical constants, the same as the Newton's law equation above. The same sort of equation describes many other systems of widely varying types – this is mathematical technology transfer!

Such equations are used to describe some type of **oscillatory** behaviour, with some degree of **damping**. We will concentrate our attention on the general properties of such behaviour, and adopt the mechanical notation of the first equation above.

The case $\beta = 0 = f(t)$ is the simplest to deal with, yielding unforced, undamped simple harmonic motion:

$$m\ddot{x} + \alpha x = 0$$

Show that the solution in this case is

$$x = A \sin(\omega t + \phi)$$

where $\omega^2 = \alpha/m$.
In the presence of damping, but no forcing terms, the equation becomes

$$m\ddot{x} + \beta\dot{x} + \alpha x = 0$$

Find the general solution in this case, and consider the cases of

(i) **overdamping**, $\beta^2 > 4\alpha$
(ii) **critical damping**, $\beta^2 = 4\alpha$
(iii) **underdamping**, $\beta^2 < 4\alpha$

If $f(t) \neq 0$ in equation (15.6) then we have an additional forcing term, impelling the particle to respond not just to the spring tension and the resistance, but to some additional force (such as gravity for example). One of the most interesting cases occurs when the forcing term is periodic itself, resulting in an equation of the form

$$m\ddot{x} + \beta\dot{x} + \alpha x = f_0 \cos \upsilon t$$

Show that the general solution in this case is

$$x(t) = Ce^{-\beta t/2} \sin\left(\frac{\sqrt{4\alpha - \beta^2}}{2}t + \phi\right) + \frac{f_0}{\sqrt{(\alpha - \upsilon^2)^2 + \beta^2\upsilon^2}} \sin(\upsilon t + \delta)$$

where $\tan \delta = (\alpha - \upsilon^2)/(\beta\upsilon)$. The first term represents damped oscillatory motion and will eventually die out if $\beta > 0$. For this reason it is called the **transient solution**. The second term, on the other hand, originating from the forcing term, persists so long as the forcing term does, with the same frequency, but a modified amplitude and phase (this is a characteristic of **linear systems** – any sinusoidal input is output with the same frequency but modified amplitude and phase). This second term is therefore referred to as the **steady state** solution – it is what is left after the transients decay away. Note that the amplitude of the steady state solution

$$\frac{f_0}{\sqrt{(\alpha - \upsilon^2)^2 + \beta^2\upsilon^2}}$$

depends on the relation between the damping β, the forcing frequency υ and the natural frequency α. Show that if $\beta^2 < 4\alpha$ (i.e. the system is underdamped), then the amplitude of forced motion has a maximum at

$$\upsilon = \upsilon_r = \sqrt{\alpha - \frac{\beta^2}{2}}$$

At this value of υ we say the system is at **resonance** and $\upsilon_r/2\pi$ is called the **resonant frequency**.

15.9 Answers to reinforcement exercises

1. $m = m_o e^{-\lambda t}$

2. (i) $\dfrac{x^2}{2} + 1$ (ii) $\sin x$ (iii) $\dfrac{1}{2}(x^2 1) + \ln x$

   (iv) $2x^2 + 2x + 1$ (v) $\dfrac{x^4}{12} - \dfrac{x^2}{2} + \dfrac{5}{6}x$

   (vi) $-\cos x - \dfrac{1}{\pi}x + 1$ (vii) $\dfrac{1}{1-3x}$

   (viii) $\sin^{-1} x$

3. (i) $-\dfrac{2}{x^2 + C}$ (ii) Cx (iii) $\sin^{-1}\left(\dfrac{x^2}{2} + C\right)$

   (iv) $-\ln|C - e^x|$ (v) $\dfrac{1}{2}(10 - Ce^{-2x})$ (vi) $\dfrac{1}{3}\ln|C + \dfrac{3}{2}e^{2x}|$

   (vii) $5y + \cos y = x^2 + C$ (viii) $Cx^x e^{-x}$ (ix) $\dfrac{x}{2}(1 - Cx)$

   (x) $(y - x)^3 = Cx^2 y^2$ (xi) $\dfrac{y-2}{y-1} = Ce^{3x}$

4. (i) $Ce^{-x} + \dfrac{1}{3}e^{2x}$ (ii) $e^x + Ce^{-2x}$

   (iii) $x^2 - 2 + C\exp(-x^2/2)$ (iv) $x + Cx^3$

   (v) $C\left(\dfrac{x-1}{x+1}\right)$ (vi) $\dfrac{x^3 + C}{x-1}$

   (vii) $Cx^2 - \dfrac{x^2}{2}e^{-2x}$

5. (i) $\dfrac{1}{2}(e^{5x} - e^{3x})$ (ii) $x^2(\ln x + 2)$

   (iii) $\dfrac{1}{4}\left(x^2 - \dfrac{1}{x^2}\right)$ (iv) $-\dfrac{1}{x^3}(\cos x + 1)$

6. (i) $Ae^x + Be^{-2x}$ (ii) $(Ax + B)e^{2x}$
   (iii) $e^{-2x}(A\cos x + B\sin x)$ (iv) $A\cos 2x + B\sin 2x$
   (v) $Ae^{3x} + Be^{-3x}$ (vi) $A\cos x + B\sin x$

7. (a) (i) $Ae^x + Be^{-2x} - 1$ (ii) $(Ax + B)e^{2x} + \dfrac{1}{2}$

   (iii) $e^{-2x}(A\cos x + B\sin x) + \dfrac{2}{5}$ (iv) $A\cos 2x + B\sin 2x + \dfrac{1}{2}$

   (v) $Ae^{3x} + Be^{-3x} - \dfrac{2}{9}$ (vi) $A\cos x + B\sin x + 2$

(b) (i) $Ae^x + Be^{-2x} - \frac{1}{2}x - \frac{3}{4}$
(ii) $(Ax + B)e^{2x} + \frac{1}{4}x + \frac{1}{2}$

(iii) $e^{-2x}(A\cos x + B\sin x) + \frac{1}{5}x + \frac{1}{25}$

(iv) $A\cos 2x + B\sin 2x + \frac{1}{4}x + \frac{1}{4}$
(v) $Ae^{3x} + Be^{-3x} - \frac{1}{9}x - \frac{1}{9}$

(vi) $A\cos x + B\sin x + x + 1$

(c) (i) $Ae^x + Be^{-2x} - \frac{1}{3}xe^{-2x}$
(ii) $(Ax + B)e^{2x} + \frac{1}{16}e^{-2x}$

(iii) $e^{-2x}(A\cos x + B\sin x) + e^{-2x}$

(iv) $A\cos 2x + B\sin 2x + \frac{1}{8}e^{-2x}$

(v) $Ae^{3x} + Be^{-3x} - \frac{1}{5}e^{-2x}$
(vi) $A\cos x + B\sin x + \frac{1}{5}e^{-2x}$

(d) (i) $Ae^x + Be^{-2x} - \frac{1}{5}\cos x - \frac{3}{5}\sin x$

(ii) $(Ax + B)e^{2x} + \frac{8}{25}\cos x + \frac{6}{25}\sin x$

(iii) $e^{-2x}(A\cos x + B\sin x) - \frac{1}{4}\cos x + \frac{1}{4}\sin x$

(iv) $A\cos 2x + B\sin 2x + \frac{2}{3}\sin x$

(v) $Ae^{3x} + Be^{-3x} - \frac{1}{5}\sin x$

(vi) $A\cos x + B\sin x - x\cos x$

8. (a) (i) $\frac{2}{3}e^x + \frac{1}{3}e^{-2x} - 1$

(ii) $\left(x - \frac{1}{2}\right)e^{2x} + \frac{1}{2}$

(iii) $-e^{-2x}\left(\frac{2}{5}\cos x + \frac{4}{5}\sin x\right) + \frac{2}{5}$
(iv) $\frac{1}{2}(1 - \cos 2x)$

(v) $\frac{1}{9}(e^{3x} + e^{-3x} - 2)$
(vi) $2(1 - \cos x)$

9. (a) (i) $\frac{e+1}{e^2+e+1}e^x + \frac{e^2}{e^2+e+1}e^{-2x} - 1$

(ii) $\left(\frac{1}{2}(1 - e^{-2})x - \frac{1}{2}\right)e^{2x} + \frac{1}{2}$

(iii) $e^{-2x}\left(-\frac{2}{5}\cos x + \frac{2}{5}\left(\frac{\cos 1 - e^2}{\sin 1}\right)\sin x\right) + \frac{2}{5}$

(iv) $-\dfrac{1}{2}\cos 2x + \dfrac{1}{2}\left(\dfrac{\cos 2 - 1}{\sin 2}\right)\sin 2x + \dfrac{1}{2}$

(v) $\dfrac{2}{9}\left[\dfrac{e^3-1}{e^6-1}\right]e^{3x} + \dfrac{2}{9}e^3\left[\dfrac{e^3-1}{e^6-1}\right]e^{-3x} - \dfrac{2}{9} = \dfrac{2}{9(e^2+1)}e^{3x} + \dfrac{2e^3}{9(e^3+1)}e^{-3x} - \dfrac{2}{9}$

(vi) $-2\cos x + \dfrac{2(\cos 1 - 1)}{\sin 1}\sin x + 2$

16

Functions of More than One Variable – Partial Differentiation

This chapter deals with functions of more than one variable and the rates of change of such functions as the different variables change. It involves some visualisation in three dimensions, but on the other hand the manipulation and methods involved are usually quite straightforward – essentially, there is little more to it than ordinary differentiation.

Prerequisites

It will be helpful if you know something about

- function notation (90 ◄)
- limits (412 ◄)
- rules of differentiation (234 ◄)
- three dimensional coordinates and graphs (323 ◄)
- parametric differentiation (240 ◄)
- implicit differentiation (238 ◄)

Objectives

In this chapter you will find

- functions of two variables
- definition of partial differentiation
- rules of partial differentiation
- higher order partial derivatives
- the total differential and total derivative

Motivation

You may need the material of this chapter for

- thermodynamics
- partial differential equations
- vector calculus
- electromagnetic field theory
- fluid and solid mechanics
- least squares in statistics

16.1 Introduction

Often, topics in engineering mathematics can be presented most clearly by means of diagrams rather than symbols. This topic is perhaps the opposite to this – the pictures used in illustrating calculus of more than one variable are sometimes not easy to visualise – but don't worry, in this case the symbols are much easier to handle.

Previously we have considered mainly functions of a **single** variable, e.g.

$$y = f(x)$$

In practice, in engineering and science, we usually have to deal with functions of **many** variables.

For example the pressure P of a perfect gas depends on its volume and temperature:

$$P = \frac{RT}{V} = P(V, T)$$

R being the **gas constant**. We want to consider such questions as, how does P change if V and T vary at given rates?

In general we will restrict consideration to functions of **two variables**, because these can be portrayed graphically. A function of two variables can be represented by a surface in 3-dimensional space. However, most of the ideas we cover are easily extended to functions of more than two variables, even if the pictures are not.

Exercises on 16.1

1. In the relation between P, V, and T, how does P vary with (i) T, (ii) V? How does V vary with (iii) T, (iv) P?

2. If V increases by 10% while T remains constant, by what percentage, approximately, does P change?

Answers

1. (i) In proportion; (ii) In inverse proportion; (iii) In proportion; (iv) In inverse proportion.

2. Decreases by approximately 10%.

16.2 Function of two variables

Let $z = f(x, y)$ be a function of two variables. If we take a 3-dimensional Cartesian coordinate system as described in Section 11.4 then any point in 3-dimensional space can be represented by a point referred to this system, $P(x, y, z)$. For each point (x, y) of the xy plane, we can calculate the value of z from $z = f(x, y)$ and plot this at an appropriate z-level fixed by the z-axis. In this way, the function $f(x, y)$ gives a **surface** in 3 dimensions, as shown in Figure 16.1.

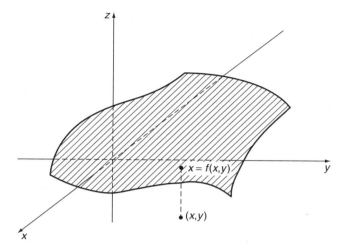

Figure 16.1 Surface defined by $z = f(x, y)$.

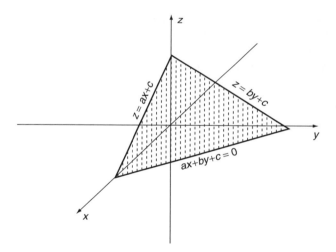

Figure 16.2 The linear function $z = ax + by + c$.

For example, the **linear function**:

$$z = ax + by + c$$

represents a **plane** in 3-dimensional space, as shown in Figure 16.2 (only the portion in the first octant is sketched).

In the theory of single variable we measure the rate of increase or slope of a curve by the slope of the tangent to the curve at a given point. In the case of functions with two variables, where we have to deal with a **surface**, we measure the **rate of increase** or **slope of a surface** by the **orientation** or **slope** of a **tangent plane** to the surface. Thus at any point on the surface we can define a **unique** plane, the tangent plane, which touches the surface at just that one point, **locally**. The orientation of this plane clearly gives a measure

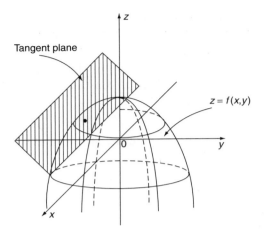

Figure 16.3 The tangent plane.

of the slope and rate of change of the surface at the point of tangency (Figure 16.3 – think of the mortar-board you may one day be wearing!).

Any line in the tangent plane is also tangent to the surface. In particular, the lines parallel to the xz and yz planes – the slopes of these lines give the slope of the surface in the x-direction and y-direction respectively.

Exercise 16.2

Sketch the surface representing the function

$$z = x^2 + y^2$$

Answer

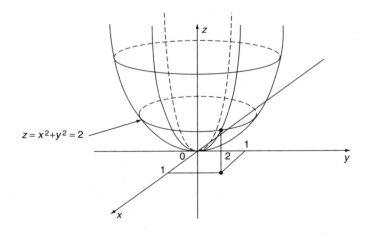

Figure 16.4

16.3 Partial differentiation

The lines in the tangent plane parallel to the xz-plane, give the slope of the surface in the x-direction at each point. When considering rate of change in the x-direction alone, the y-coordinate in $z = f(x, y)$ can be regarded as constant. Then for this fixed valued of y

$$z = f(x, y)$$

is a function of x, a curve parallel to the xz plane. The slope of this curve can be obtained in the usual way by differentiating:

$$\frac{dz}{dx} = \frac{df}{dx}$$

However, this notation is not appropriate because f is now a function of two variables, so we use:

$$\frac{\partial z}{\partial x} = \frac{\partial f}{\partial x}(x, y)$$

read 'curly dee dee ex'.

This is called the **partial derivative of z with respect to x**, and represents the rate of change of z with respect to x assuming y is constant. We will also denote it by f_x. It gives the rate of change in z in the x-direction, or the slope of the surface in the x-direction (see Figure 16.5). We can define the partial derivative rigorously in terms of a limit as follows:

$$\frac{\partial f}{\partial x} = \lim_{h \to 0}\left[\frac{f(x+h, y) - f(x, y)}{h}\right]$$

Similarly, we can define the partial derivative with respect to y:

$$\frac{\partial f}{\partial y} = \lim_{k \to 0}\left[\frac{f(x, y+k) - f(x, y)}{k}\right]$$

which is the derivative of f with respect to y assuming x constant. We will sometimes use f_y. It gives the rate of change of f in the y-direction, or the slope of the surface in the y-direction.

The rules of partial differentiation are the same as those in ordinary differentiation, so long as we remember which of the variables to hold constant. Specifically (234 ◄):

$$\frac{\partial}{\partial x}(f \pm g) = \frac{\partial f}{\partial x} \pm \frac{\partial g}{\partial x}$$

$$\frac{\partial}{\partial x}(fg) = f\frac{\partial g}{\partial x} + \frac{\partial f}{\partial x}g$$

$$\frac{\partial}{\partial x}\left(\frac{f}{g}\right) = \frac{g\dfrac{\partial f}{\partial x} - f\dfrac{\partial g}{\partial x}}{(g)^2}$$

$$\frac{\partial f}{\partial x}(g(x), y) = \frac{\partial g}{\partial x}\frac{\partial f}{\partial g}(g, y)$$

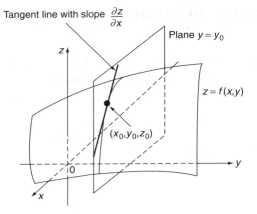

Figure 16.5 Definition of the partial derivative $\dfrac{\partial z}{\partial x}$.

The last rule is the generalisation of the function of a function rule (235 ◀), which may look a little strange in the new notation – again it helps to think of the undifferentiated variable (y in this case) as a constant.

Similar results apply for the partial derivatives with respect to y.

Problem 16.1

Evaluate the partial derivatives

(i) $z = \dfrac{y}{x} + \dfrac{x}{y}$ (ii) $z = y \sin(2x + 3y)$

(i) Treating y as a constant and differentiating with respect to x gives

$$\frac{\partial z}{\partial x} = -\frac{y}{x^2} + \frac{1}{y}$$

Similarly, keeping x constant and differentiating with respect to y gives

$$\frac{\partial z}{\partial y} = \frac{1}{x} - \frac{x}{y^2}$$

(ii) $\dfrac{\partial z}{\partial x} = 2y \cos(2x + 3y)$ and $\dfrac{\partial z}{\partial y} = \sin(2x + 3y) + 3y \cos(2x + 3y)$

Exercise on 16.3

Find f_x, f_y for the following functions

(i) $f(x, y) = \dfrac{xy}{x + y}$ (ii) $f(x, y) = e^{3x + \cos(xy)}$

Answer

(i) $f_x = \dfrac{y^2}{(x + y)^2}$, $f_y = \dfrac{x^2}{(x + y)^2}$

(ii) $f_x = e^{3x + \cos(xy)}(3 - y \sin(xy))$, $f_y = -x e^{3x + \cos(xy)} \sin(xy)$

16.4 Higher order derivatives

If $z = f(x, y)$ is a function of x and y, then $\dfrac{\partial z}{\partial x}$ is also a function of x and y and so may itself be partially differentiated with respect to x, $\dfrac{\partial}{\partial x}\left(\dfrac{\partial z}{\partial x}\right)$ or with respect to y, $\dfrac{\partial}{\partial y}\left(\dfrac{\partial z}{\partial x}\right)$. We write such **second order derivates** as:

$$\dfrac{\partial^2 z}{\partial x^2}, \dfrac{\partial^2 z}{\partial x \partial y}, \dfrac{\partial^2 z}{\partial y \partial x}, \dfrac{\partial^2 z}{\partial y^2}$$

which we will sometimes write as $z_{xx}, z_{xy}, z_{yx}, z_{yy}$, respectively.
For all functions we are interested in we may assume that

$$\dfrac{\partial^2 z}{\partial x \partial y} = \dfrac{\partial^2 z}{\partial y \partial x} \quad \text{or} \quad z_{xy} = z_{yx}$$

Similarly, higher order derivates may be defined, such as $\dfrac{\partial^3 z}{\partial x^2 \partial y} = z_{xxy}$, etc.

Problem 16.2
Evaluate all second order derivatives of the function

$$f(x, y) = x^3 + y^3 + 2xy + x^2 y$$

and verify that

$$\dfrac{\partial^2 f}{\partial x\, \partial y} = \dfrac{\partial^2 f}{\partial y\, \partial x}$$

The differentiation of $f(x, y) = x^3 + y^3 + 2xy + x^2 y$ is routine and we obtain

$$f_x = 3x^2 + 2y + 2xy \qquad f_{xx} = 6x + 2y$$
$$f_y = 3y^2 + 2x + x^2 \qquad f_{yy} = 6y$$
$$f_{xy} = (f_x)_y = (3x^2 + 2y + 2xy)_y = 2 + 2x$$
$$f_{yx} = (f_y)_x = (3y^2 + 2x + x^2)_x = 2 + 2x$$
$$= f_{xy}$$

Exercises on 16.4
1. Find all first and second order partial derivatives of the following functions, $f(x, y)$, checking the equality of the mixed derivatives

   (i) $f(x, y) = x^3 y^2 + 4xy^4$ (ii) $f(x, y) = e^{xy} \cos(x + y)$

2. Show that $f(x, y) = \ln(x^2 + y^2)$ satisfies the **partial differential equation:**

$$\frac{\partial^2 f}{\partial x^2} + \frac{\partial^2 f}{\partial y^2} = 0$$

This is called the **Laplace equation** in two-dimensional rectangular coordinates. It is very important in fluid mechanics, electromagnetism, and many other areas of science and engineering, as well as being a key equation in pure mathematics.

Answers

1. (i) $f_x = 3x^2y^2 + 4y^4$, $f_y = 2x^3y + 16xy^3$,
    $f_{xx} = 6xy^2$, $f_{yy} = 2x^3 + 48xy^2$, $f_{xy} = 6x^2y + 16y^3$
   (ii) $f_x = e^{xy}(y\cos(x+y) - \sin(x+y))$,
    $f_y = e^{xy}(x\cos(x+y) - \sin(x+y))$,
    $f_{xx} = e^{xy}((y^2 - 1)\cos(x+y) - 2y\sin(x+y))$,
    $f_{yy} = e^{xy}((x^2 - 1)\cos(x+y) - 2x\sin(x+y))$,
    $f_{xy} = e^{xy}(xy\cos(x+y) - (x+y)\sin(x+y))$

16.5 The total differential

$\dfrac{\partial z}{\partial x}$, $\dfrac{\partial z}{\partial y}$ give the rates of increase of $z = f(x, y)$ in the x and y directions respectively – what about the increase in a general direction? I.e. if x, y increase by δx, δy, by how much does z increase? From Figure 16.6 we can see that the increase in z, δz, is given approximately by

$$\delta z \simeq \frac{\delta z}{\delta x}\delta x + \frac{\delta z}{\delta y}\delta y$$

Such diagrams are not to everyone's taste, so a few examples might help.

Problem 16.3
Obtain expressions for δz in the cases
(i) $z = x + y$ (ii) $z = xy$ (iii) $z = x^2 + y^2$

What happens if δx, δy are so small that products of them can be neglected?

(i) is not so bad:

$$\delta z = x + \delta x + y + \delta y - (x + y) = \delta x + \delta y$$

This result is exact. There are no products of δx, δy.

(ii) is a little more complicated:

$$\delta z = (x + \delta x)(y + \delta y) - xy$$
$$= y\delta x + x\delta y + \delta x \delta y$$

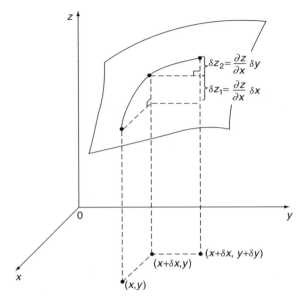

Figure 16.6 The total derivative.

This is exact. However, if we neglect $\delta x \delta y$ then we get an **approximation**:

$$\delta z \simeq y \delta x + x \delta y$$

Notice that since $z_x = y$ and $z_y = x$ this may be written

$$\delta z \simeq z_x \delta x + z_y \delta y$$

(iii) $\delta z = (x + \delta x)^2 + (y + \delta y)^2 - x^2 - y^2$

$$= 2x\delta x + 2y\delta y + (\delta x)^2 + (\delta y)^2$$
$$\simeq 2x\delta x + 2y\delta y$$

if we neglect the δ products. Again, notice that this is

$$\delta z \simeq z_x \delta x + z_y \delta y$$

These examples illustrate the result

$$\delta z \simeq z_x \delta x + z_y \delta y$$

stated above.

Note that this is an **approximate** formula between **increments** δx, δy, δz. It is useful to **define differentials** dx, dy, dz which satisfy:

$$dz = \frac{\partial z}{\partial x} dx + \frac{\partial z}{\partial y} dy$$

dz is called the **total differential** of z. The dx, dy, dz are not actually numerical quantities, but simply symbolise quantities that can be taken as small as we wish, but never zero. They really only serve to define derivatives. Thus, if x, y, are both functions of a parameter t (e.g. in dynamics $t =$ time) we can formally 'divide by dt' and write:

$$\frac{dz}{dt} = \frac{\partial z}{\partial x}\frac{dx}{dt} + \frac{\partial z}{\partial y}\frac{dy}{dt}$$

and this gives the total rate of change of z with t given the rate of change of x and y with t. dz/dt is called the **total derivative of z with respect to t**. Both the formula for dz and dz/dt extend in an obvious way to functions of greater than two variables, although the geometrical significance is not so easy to visualize.

Problem 16.4
Find dz at the point (1, 2, 5) for $z = x^2 + y^2$.

Essentially, all we need are the first partial derivatives. We have

$$\frac{\partial z}{\partial x} = 2x, \quad \frac{\partial z}{\partial y} = 2y$$

The total differential is therefore given by

$$dz = \frac{\partial z}{\partial x} dx + \frac{\partial z}{\partial y} dy$$
$$= 2x\, dx + 2y\, dy$$

So at (1, 2, 5) this gives

$$dz = 2(1)\, dx + 2(2)\, dy = 2\, dx + 4\, dy$$

This is a formal relation between dx, dy, dz, which can for example be used to evaluate the total derivative of z at the point (1, 2, 5) as

$$\frac{dz}{dt} = 2\frac{dx}{dt} + 4\frac{dy}{dt}$$

This gives us the rate of change of z at (1, 2, 5) in terms of the rates of change of x and y at that point. However, the above relation between differentials can also be regarded as an approximate relation between small increments $\delta x, \delta y, \delta z$ to give

$$\delta z \simeq 2\delta x + 4\delta y$$

This gives us the approximate change in z at (1, 2, 5) if x and y are changed by small amounts $\delta x, \delta y$ respectively.

As an example of the use of the total differential in approximations consider the problem of finding the percentage error in functions of the form

$$z = x^\alpha y^\beta$$

due to given percentage errors in x and y.

We have (treating dx, dy, etc. as small increments here)

$$dz = \alpha x^{\alpha-1} y^\beta dx + \beta x^\alpha y^{\beta-1} dy$$

so the relative error in z due to 'errors' dx, dy in x, y respectively is

$$\frac{dz}{z} = \frac{\alpha x^{\alpha-1} y^\beta}{z} dx + \frac{\beta x^\alpha y^{\beta-1}}{z} dy$$

$$= \alpha \frac{x^{\alpha-1} y^\beta}{x^\alpha y^\beta} dx + \frac{\beta x^\alpha y^{\beta-1}}{x^\alpha y^\beta} dy$$

$$= \alpha \frac{dx}{x} + \beta \frac{dy}{y}$$

So the percentage error in z is

$$\frac{dz}{z} \times 100 = \left(\alpha \frac{dx}{x} + \beta \frac{dy}{y}\right) 100$$

$$= \alpha \left(\frac{dx}{x} \times 100\right) + b \left(\frac{dy}{y} \times 100\right)$$

$$= \alpha \text{ percentage error in } x + \beta \text{ percentage error in } y$$

Notice the pattern in which the exponents in the original expression become the linear weightings in the expression for the percentage error.

An easier derivation is to first take logs

$$\ln z = \ln(x^\alpha y^\beta) = \alpha \ln x + \beta \ln y$$

Now differentiate through

$$\frac{dz}{z} = \alpha \frac{dx}{x} + \beta \frac{dy}{y}$$

and then multiplying through by 100 gives the required relation between percentage errors. Extension to functions of more than two variables is obvious.

Problem 16.5

Find the percentage error in the volume of a rectangular box in terms of the percentage errors in the sides of the box.

With sides a, b, c the volume of the box is

$$V = abc = a^1 b^1 c^1$$

So the percentage error in V is

$$\frac{dV}{V} \times 100 = 1.\frac{da}{a} \times 100 + 1.\frac{db}{b} \times 100 + 1.\frac{dc}{c} \times 100$$

$$= \text{percentage error in } a + \text{percentage error in } b$$
$$+ \text{percentage error in } c$$

Note also that these error problems can be taken over directly to problems of small increases – e.g. the expansion of a rectangular box due to heating, etc. (using coefficients of **linear** expansions).

In fact, we have already met the total differential disguised as **implicit differentiation** (238 ◀).

Problem 16.6

If $z = x^2 + 5x^2y + 2xy^2 - y^3 = 4$ find $\dfrac{dy}{dx}$.

We find $\dfrac{dy}{dx}$ by noting that $dz = 0$, using the total differential, and solving the resulting equation to obtain a relation between dx and dy which we then solve for $\dfrac{dy}{dx}$. Thus

$$dz = 2x\,dx + 10xy\,dx + 5x^2\,dy + 2y^2\,dx + 4xy\,dy - 3y^2\,dy = 0$$

Dividing by dx gives

$$2x + 10xy + 5x^2\frac{dy}{dx} + 2y^2 + 4xy\frac{dy}{dx} - 3y^2\frac{dy}{dx} = 0$$

and solving for $\dfrac{dy}{dx}$ gives

$$\frac{dy}{dx} = \frac{2x + 10xy + 2y^2}{3y^2 - 5x^2 - 4xy}$$

Exercises on 16.5

1. Find the total differential dz when

   (i) $z = \ln(\cos(xy))$ (ii) $z = \exp\left(\dfrac{x}{y}\right)$

2. If $z = e^{2x+3y}$ and $x = \ln t$, $y = t^2$, calculate $\dfrac{dz}{dt}$ from the total derivative formula and show that it agrees with the result obtained by substitution for x and y before differentiating.

Answers

1. (i) $-\tan(xy)(y\,dx + x\,dy)$ (ii) $\dfrac{1}{y^2}\exp\left(\dfrac{x}{y}\right)(y\,dx - x\,dy)$

2. $2t(3t^2 + 1)\exp(3t^2)$

16.6 Reinforcement

1. Find the values of the following functions at the points given:

   (i) $f(x, y) = 2xy^3 + 3x^2y$ at the point $(2, 1)$
   (ii) $g(x, y) = (x + y)e^x \sin y$ at the point $(0, \pi/2)$

(iii) $h(x, y, z) = \sqrt{x^2 + y^2 + z^2}$ at the points $(-1, 2, 2)$ and $(3, 2, 4)$

(iv) $l(x, y, z) = e^{x^2} y^4 \cos z$ at $\left(0, -2, \dfrac{\pi}{3}\right)$

2. Sketch the surfaces represented by $z = f(x, y)$ where

(i) $z = 1 - 3y$ (ii) $x^2 + y^2 + z^2 = 9$

3. Determine $\partial z/\partial x$, $\partial z/\partial y$ in each case

(i) $z = x^2 + y^2$ (ii) $z = \dfrac{x}{y}$ (iii) $z = x^3 + x^2 y + y^4$

(iv) $z = \dfrac{1}{\sqrt{x^2 + y^2}}$ (v) $z = e^{xy} \cos(3y^2)$ (vi) $z = \ln(1 + xy)$

(vii) $z = e^{-xy}(2 + 3xy)$ (viii) $z = \dfrac{x^2 + y^2}{\sqrt{1 + y}}$ (ix) $z = x^3 \tan^{-1}\left(\dfrac{x}{y}\right)$

4. For each of the functions in Q3 evaluate $\dfrac{\partial z}{\partial x}(0, 0)$, $\dfrac{\partial z}{\partial y}(1, 2)$ whenever possible.

5. Determine all first order partial derivatives

(i) $w = x^2 + 2y^2 + 3z^2$ (ii) $w = \dfrac{1}{\sqrt{1 - x^2 - y^2 - z^2}}$

(iii) $w = xyz$ (iv) $w = x \cos(x + yz)$

(v) $w = e^{xy} \ln(x + y + z)$

6. Determine all second order partial derivatives for the functions in Q3.

7. Determine all second order partial derivatives for Q5(i), (iii), (iv).

8. Show that $T(x, t) = ae^{-b^2 t} \cos bx$, where a and b are arbitrary constants, satisfies the equation

$$\dfrac{\partial T}{\partial t} = \dfrac{\partial^2 T}{\partial x^2}$$

9. Determine dz for the functions

(i) $z = x^2 - 3y^2$ (ii) $z = 3x^2 y^3$

(iii) $z = \ln(x^2 + y^2)$ (iv) $z = \cos(x + y)$

(v) $z = x^2 e^{-xy}$

10. If $z = 3x^2 + 2xy - y^2$ and x and y vary with time t according to $x = 1 + \sin t$ and $y = 3 \cos t - 1$ evaluate $\dfrac{dz}{dt}$ directly and by using the total derivative (chain) rule.

16.7 Applications

1. Partial differential equations are equations containing partial derivatives – analogous to ordinary differential equations of Chapter 15. Such equations occur frequently in

science and engineering. An important example is Laplace's equation $\dfrac{\partial^2 f}{\partial x^2} + \dfrac{\partial^2 f}{\partial y^2} = 0$ which arises in electromagnetic field theory and fluid flow, for example. Verify that each of the following functions satisfies Laplace's equation

(i) $f(x, y) = \ln(x^2 + y^2)$

(ii) $f(x, y) = e^{-3y} \cos 3x$

(iii) $f(x, y) = e^x(x \cos y - y \sin y)$

(iv) $f(x, y) = x^2 - y^2$

2. An important application of the total differential occurs in estimating the change in a function of a number of variables resulting from changes in the variables. The following give a number of examples of this.

(i) From the ideal gas law $PV = nRT$, where nR is constant, estimate the percentage change in pressure, P, if the temperature, T, is increased by 3% and the volume, V, is increased by 4%.

(ii) Given that time, T, of oscillation of a simple pendulum of length l is

$$T = 2\pi \sqrt{\dfrac{l}{g}}$$

determine the total differential dT in terms of l and the gravitational and constant g. Estimate the percentage error in the period of oscillation if l is taken 0.1% too large and g 0.05% too small.

(iii) The total resistance of three resistors R_1, R_2, R_3, in parallel is given by R, where

$$\dfrac{1}{R} = \dfrac{1}{R_1} + \dfrac{1}{R_2} + \dfrac{1}{R_3}$$

If R_1, R_2, R_3, are measured as 6, 8, 12 Ω respectively with respective maximum tolerances of ± 0.1, ± 0.03, ± 0.15 Ω, estimate the maximum possible error in R.

3. The total derivative is used to determine the rate of change of a function of a number of variables in terms of the rates of change of the variables. The following examples illustrate this.

(i) The radius of a cylinder decreases at a rate 0.02 ms^{-1} while its height increases at a rate 0.01 ms^{-1}. Find the rate of change of the volume at the instant when $r = 0.05$ m and $h = 0.2$ m.

(ii) Find the rate of increase of the diagonal of a rectangular solid with sides 3, 4, 5 m, if the sides increase at $\tfrac{1}{3}$, $\tfrac{1}{4}$, $\tfrac{1}{5}$ ms^{-1} respectively.

16.8 Answers to reinforcement exercises

1. (i) 16 (ii) $\dfrac{\pi}{2}$ (iii) 3, $\sqrt{29}$ (iv) 8

2. (i)

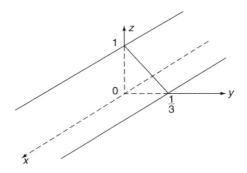

(ii)

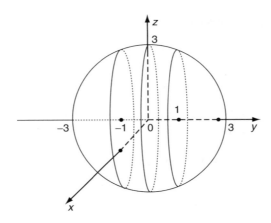

3. (i) $2x, 2y$ (ii) $\dfrac{1}{y}, -\dfrac{x}{y^2}$

(iii) $3x^2 + 2xy, x^2 + 4y^3$ (iv) $-\dfrac{x}{(x^2+y^2)^{3/2}}, -\dfrac{y}{(x^2+y^2)^{3/2}}$

(v) $ye^{xy}\cos(3y^2), e^{xy}(x\cos(3y^2) - 6y\sin(3y^2))$ (vi) $\dfrac{y}{1+xy}, \dfrac{x}{1+xy}$

(vii) $e^{-xy}(y - 3xy^2), e^{-xy}(x - 3x^2y)$ (viii) $\dfrac{2x}{\sqrt{1+y}}, \dfrac{3y^2+4y-x^2}{2(1+y)^{3/2}}$

(ix) $3x^2\tan^{-1}\left(\dfrac{x}{y}\right) + \dfrac{yx^3}{x^2+y^2}, -\dfrac{x^4}{x^2+y^2}$

4. (i) $0, 6$ (ii) Does not exist, $-\dfrac{1}{4}$ (iii) $0, 33$

(iv) Does not exist, $-\dfrac{2}{5\sqrt{5}}$ (v) $0, e^2(\cos(12) - 12\sin(12))$

(vi) $0, \dfrac{1}{3}$ (vii) $0, -5e^{-2}$ (viii) $0, \dfrac{19}{6\sqrt{3}}$

(ix) Doe not exist, $-\dfrac{1}{5}$

5. Answers listed in the order z_x, z_y, z_z

(i) $2x, 4y, 6z$

(ii) $\dfrac{x}{d}, \dfrac{y}{d}, \dfrac{z}{d}$ where $d = (1 - x^2 - y^2 - z^2)^{3/2}$

(iii) yz, xz, xy

(iv) $\cos(x + yz) - x\sin(x + yz), -xz\sin(x + yz), -xy\sin(x + yz)$

(v) $e^{xy}\left(y\ln(x + y + z) + \dfrac{1}{x + y + z}\right), e^{xy}\left(x\ln(x + y + z) + \dfrac{1}{x + y + z}\right),$

$\dfrac{e^{xy}}{x + y + z}$

6. Answers are given in the order w_{xx}, w_{yy}, w_{xy}

(i) $2, 2, 0$

(ii) $0, \dfrac{2x}{y^3}, -\dfrac{1}{y^2}$

(iii) $6x + 2y, 12y^2, 2x$

(iv) $\dfrac{2x^2 - y^2}{(x^2 + y^2)^{5/2}}, \dfrac{2y^2 - x^2}{(x^2 + y^2)^{5/2}}, \dfrac{3xy}{(x^2 + y^2)^{5/2}}$

(v) $y^2 e^{xy}\cos(3y^2), e^{xy}(x^2\cos(3y^2) - 12xy\sin(3y^2) - 6\sin(3y^2)$
$-36y^2\cos(3y^2)), e^{xy}(\cos(3y^2) + xy\cos(3y^2) - 6y^2\sin(3y^2))$

(vi) $-\dfrac{y^2}{(1 + xy)^2}, -\dfrac{x^2}{(1 + xy)^2}, \dfrac{1}{(1 + xy)^2}$

(vii) $e^{-xy}(3xy^3 - 4y^2), e^{-xy}(3x^3y - 4x^2), e^{-xy}(3x^2y^2 - 7xy + 1)$

(viii) $\dfrac{2}{\sqrt{1 + y}}, \dfrac{3y^2 + 8y + 3x^2 + 8}{4(1 + y)^{5/2}}, -\dfrac{x}{(1 + y)^{3/2}}$

(ix) $6x\tan^{-1}\left(\dfrac{x}{y}\right) + \dfrac{2x^4y + 6x^2y^3}{(x^2 + y^2)^2}, \dfrac{2x^4y}{(x^2 + y^2)^2}, \dfrac{-2x^5 - 4x^3y^2}{(x^2 + y^2)^2},$

7. Answers in the order $w_{xx}, w_{yy}, w_{zz}, w_{xy}, w_{xz}, w_{yz}$

(i) $2, 4, 6, 0, 0, 0$

(iii) $0, 0, 0, z, y, x$

(iv) $-2\sin(x + yz) - x\cos(x + yz), -xz^2\cos(x + yz), -xy^2\cos(x + yz),$
$-z\sin(x + yz) - xz\cos(x + yz),$
$-y\sin(x + yz) - xy\cos(x + yz), -x\sin(x + yz) - xyz\cos(x + yz)$

8. Determine dz for the functions

   (i) $2x\,dx - 6y\,dy$

   (ii) $6xy^3\,dx + 9x^2y^2\,dy$

   (iii) $\dfrac{2x}{x^2+y^2}dx + \dfrac{2y}{x^2+y^2}dy$

   (iv) $-\sin(x+y)\,dx - \sin(x+y)\,dy$

   (v) $xe^{-xy}((2-xy)\,dx - x^2\,dy)$

9. $\dfrac{dz}{dt} = 6\cos 2t + 12\sin 2t - 12\sin t + 4\cos t$

17

An Appreciation of Transform Methods

17.1 Introduction

We have already met situations in which changing the variable in a mathematical problem works wonders – as for example in substitution in integration. In fact, this idea of **transforming** variables to a new set is a useful one throughout applied mathematics. In this chapter we are going to look at two particular types of transforms that are virtually essential in engineering and science:

- Laplace transform
- Fourier series (transform)

Both serve useful roles in two key engineering topics:

- the study of initial value problems in control theory, where one is interested in stability properties of a system
- harmonic analysis of signals in which a periodic input to a system is decomposed into sinusoidal components which may be individually processed and the results recombined to produce the output of the system

Although I will try to explain how the transforms arise, the topic may still require something of a leap of faith on your part, particularly in the definitions of the Laplace transform and the Fourier series. This is a place where accepting the results blindly and understanding later is not necessarily a bad learning strategy! Also, we will not prove very much in detail. The priority will be to equip you with the main tools and give you an appreciation of the concepts and methods. This topic is also useful in that it brings together so much of the basic mathematical material that we have covered in this book, perhaps justifying all the hard work put in!

> *Prerequisites*
>
> It will be helpful if you know something about
>
> - the exponential function (Chapter 4)
> - integration, particularly integration by parts (Chapter 9)
> - trig functions (Chapter 6)
> - integration of products of sines and cosines (270 ◄)
> - limits at infinity (417 ◄)
> - function notation (90 ◄)

- continuous and discontinuous functions (418 ◄)
- partial fractions (60 ◄)
- completing the square (60 ◄)
- differential equations (Chapter 15)
- sigma notation (102 ◄)
- infinite series (428 ◄)

Objectives

In this chapter you will find:

- definition of the Laplace transform
- Laplace transforms of elementary functions
- properties of Laplace transforms
- solution of initial value problems by Laplace transform
- the inverse Laplace transform
- linear systems and superposition
- definition of Fourier series
- orthogonality of trig functions
- determination of the coefficients in a Fourier series

Motivation

You may need the material of this chapter for:

- solving differential equations in electrical circuits
- analysing the behaviour of control systems
- signal analysis

17.2 The Laplace transform

Problem 17.1

Integrate the following, where s and a are constants

$$\int_0^a t e^{-st}\, dt$$

Let $a \to \infty$ in the result, assuming that $s > 0$.

This integral has a number of features that we need to look at to get us into the mood for Laplace transforms. Expressions that contain a number of symbols, some constant and some variable, often appear daunting to the most experienced of us. Just take it steady and pick the bones out of the thing. Firstly, since a and s are to be regarded as constant in doing the integral they play no role in the actual integration (other than getting in the way!), which is basically an integration with respect to the variable t. If it helps, think of a and s simply as particular constants, say 3 and 4 respectively, while doing the integration. Then you can concentrate on the actual job of integration and effectively the problem is to integrate something like te^{-4t}, for example. This is a classic case of integration by parts (273 ◄) which you really should revise if you have the slightest doubts. We obtain, on integrating the exponential first,

$$\int te^{-4t} dt = \frac{te^{-4t}}{-4} - \left(-\frac{1}{4}\right) \int e^{-4t} dt$$

$$= -\frac{1}{4} te^{-4t} - \frac{1}{16} e^{-4t}$$

Count the signs carefully here! If we had s instead of 4 then you should now recognise more easily that

$$\int te^{-st} dt = \frac{te^{-st}}{-s} - \left(-\frac{1}{s}\right) \int e^{-st} dt$$

$$= -\frac{1}{s} te^{-st} - \frac{1}{s^2} e^{-st}$$

Now we must put the limits in (279 ◄) to get

$$\int_0^a te^{-st} dt = \left[\frac{te^{-st}}{-s}\right]_0^a + \frac{1}{s} \int_0^a e^{-st} dt$$

$$= \left[\frac{te^{-st}}{-s}\right]_0^a - \frac{1}{s^2} \left[e^{-st}\right]_0^a$$

$$= -\frac{1}{s} ae^{-sa} - \frac{1}{s^2} e^{-sa} + \frac{1}{s^2}$$

Agreed, this is somewhat messy. For the next step, taking the limit of a as it tends to infinity, it again helps to keep thinking of s as some constant, but it is important to remember that it is a **positive** constant. Then we need the limit result (418 ◄)

$$x^n e^{-x} \to 0 \quad \text{as} \quad x \to \infty$$

for any non-negative number n. Then, applying this, since s is positive, we get

$$e^{-sa} \to 0 \quad \text{and} \quad ae^{-sa} \to 0 \quad \text{as} \quad a \to \infty$$

and so as $a \to \infty$ $\quad \int_0^a te^{-st} dt \to \frac{1}{s^2}.$

We write this as

$$\int_0^\infty te^{-st} dt = \frac{1}{s^2}$$

Don't let anyone tell you all this is easy! There is a great deal of sophisticated mathematics to deal with in this problem, and I have spelt it out carefully now because you will need to do a lot of it later – I wanted to get the main difficulties out of the way up front so that we are not too distracted later on. We will now get down to the main business of defining the Laplace transform.

In a similar way to how we transform a **variable** to simplify some mathematical problem such as integration, we can often similarly transform a **function** to facilitate the solution to a differential equation. The most common way to do this is by means of an integral. Thus, if $f(t)$ is a function of t (it is usual to use t as the independent variable, since in applications it usually refers to time), then we transform to a new function $\tilde{f}(s)$ of a new variable s by an expression of the form:

$$\tilde{f}(s) = \int_a^b k(s,t) f(t)\, dt$$

The notation 'f twiddle', $\tilde{f}(s)$, is a standard mathematical way of reminding us that the new function of s comes from f, but is completely different to f, which is a function of t. In Problem 17.1 we had $f(t) = t$, $k(s,t) = e^{-st}$, $a = 0$ and $b = \infty$, and the integral, which would correspond to what we have called $\tilde{f}(s)$ here, came out to be $\dfrac{1}{s^2}$. The above integral expression is called an **integral transform**. Essentially, it replaces the function $f(t)$ by a different function $\tilde{f}(s)$, which may be easier to deal with in certain circumstances.

There are many such integral transforms, usually going under the name of some famous mathematician – Laplace, Fourier, Mellin, Hilbert. The Laplace and Fourier transforms are particularly useful, in scientific and engineering applications such as control theory and signal analysis, and in statistical and commercial applications. I would certainly not waste your time by introducing such complicated objects if they were not so useful! It is worth noting that such transforms were often first invented and developed by scientists and engineers, long before mathematicians got their hands on them.

When we consider a transform such as the Laplace transform we are interested in such questions as:

- What are the Laplace transforms of the elementary functions?
- What general properties does the transform have?
- How does the Laplace transform relate to other mathematical operations (such as differentiation, integration, etc.)?
- How can the transform be 'inverted', i.e. given the transform $\tilde{f}(s)$ of a function $f(t)$, how do we reverse the transform and find the original function $f(t)$?
- What applications does the Laplace transform have?

We now give the formal definition of the Laplace transform. We have already looked at the technical details of integration needed to deal with it, in Problem 17.1, so it should be less of a shock! Be assured that, strange though it might appear, it is absolutely invaluable in engineering mathematics.

Suppose $f(t)$ is a function defined for $t \geq 0$. Then the integral

$$\tilde{f}(s) = \int_0^\infty f(t) e^{-st}\, dt$$

Understanding Engineering Mathematics

is called the **Laplace transform of** $f(t)$. Other notations used are $F(s)$ (lower case goes to upper case when we transform) or $\mathcal{L}[f(t)]$, the square brackets denoting that the Laplace transform acts on a **function**. The appearance of the exponential function here should perhaps come as no surprise when one considers the predominant role that this function plays in, for example, differential equations.

Note that improper infinite integrals of the type occurring in the definition above are defined by:

$$\int_0^\infty g(t)\,dt = \lim_{a \to \infty} \int_0^a g(t)\,dt$$

and this shows how they are evaluated in practice – i.e. do the integral first and then take the limit, as we did in Problem 17.1.

Exercises on 17.2

1. Write down the values of the following limits:

   (i) $\lim_{t \to \infty} \dfrac{e^{-st}}{s}$ $s > 0$ (ii) $\lim_{t \to \infty} \dfrac{e^{-st}}{s^2}$ $s > 0$

   (iii) $\lim_{t \to \infty} \dfrac{te^{-st}}{s}$ $s > 0$ (iv) $\lim_{t \to \infty} t^n e^{-st}$ $s > 0, n$ positive

2. Find the Laplace transform of $f(t) = 3$.

Answers

1. (i) 0   (ii) 0   (iii) 0   (iv) 0   2. $\dfrac{3}{s}$

17.3 Laplace transforms of the elementary functions

Assembling the Laplace transforms of the elementary functions is simply a matter of integration. We flagged up the main points in Problem 17.1, and now you can try developing a table of Laplace transforms for yourself.

Problem 17.2
Find the Laplace transform of the constant function $f(t) = 1$.

From the definition we have

$$\mathcal{L}[1] = \int_0^\infty 1 e^{-st}\,dt$$

$$= \lim_{a \to \infty} \left[\frac{e^{-st}}{-s}\right]_0^a = \frac{1}{s} \text{ provided } s > 0$$

Problem 17.3
Evaluate $\mathcal{L}[t]$, $\mathcal{L}[t^2]$, $\mathcal{L}[t^3]$ and look for a pattern for $\mathcal{L}[t^n]$.

We have already found $\mathcal{L}[t]$ in Problem 17.1, and as a reminder:

$$\mathcal{L}[t] = \int_0^\infty t e^{-st} dt = \left[\frac{te^{-st}}{-s}\right]_0^\infty + \frac{1}{s}\int_0^\infty e^{-st} dt \text{ (by parts)} = \frac{1}{s^2}$$

For $\mathcal{L}[t^2]$ the integration is more lengthy, but follows the same pattern – first integrate with the limits 0, a then let $a \to \infty$.

$$\mathcal{L}[t^2] = \int_0^\infty t^2 e^{-st} dt$$

$$= \frac{2}{s^3} \text{ by integrating by parts twice}$$

$\mathcal{L}[t^3]$ is even more of a slog, but with care and patience you should find that

$$\mathcal{L}[t^3] = \frac{6}{s^4} = \frac{3!}{s^4}$$

So, summarising, we have

$$\mathcal{L}[1] = \frac{0!}{s}, \quad \mathcal{L}[t] = \frac{1!}{s^2}, \quad \mathcal{L}[t^2] = \frac{2!}{s^3}, \quad \mathcal{L}[t^3] = \frac{3!}{s^4}$$

where we have used the convention $0! = 1$ (16 ◄), which we also used in the binomial theorem. These results lead us to **suspect** that in general

$$\mathcal{L}[t^n] = \frac{n!}{s^{n+1}}$$

for n a positive integer. In fact this is a correct generalisation, although we will not prove it.

Proceeding as in the above, exercising your integration, you can if you wish verify the results given in Table 17.1 for the Laplace transforms of the elementary functions (see Exercises on this section and page 285). There are a number of points to note:

- the restrictions on s, which we have assumed to be real
- results involving the exponential function, which clearly has the effect of replacing s by $s - a$ – i.e. of **shifting** or translating s
- the results for $t \sin \omega t$, $t \cos \omega t$ can be obtained by differentiation of those for $\sin \omega t$, $\cos \omega t$ with respect to ω.

Note an important implication of the fact that the Laplace transform is defined by an integral. Specifically, recall that in any mathematical operation, such as differentiation for example, there are precise conditions under which the operations are allowed – for example a function must be continuous at a point if it is to be differentiable there. On the other hand an **integrable** function does not have to be continuous in order to integrate it, meaning that we can apply the Laplace transform to a wider range of functions than continuous ones. A large and important class of functions to which we can apply the Laplace transform includes the class of piecewise continuous functions. A **piecewise continuous function** $f(t)$ on an interval $a \leq t \leq b$ is a function which consists of continuous sections separated by a finite number of isolated points at which the function may not be continuous, but the

Table 17.1

$f(t)$	$\tilde{f}(s) = F(s) = \mathcal{L}[f(t)]$
1	$\dfrac{1}{s}$ $(s > 0)$
t	$\dfrac{1}{s^2}$ $(s > 0)$
t^n (n a positive integer)	$\dfrac{n!}{s^{n+1}}$ $(s > 0)$
e^{at}	$\dfrac{1}{s-a}$ $(s > a)$
$t^n e^{at}$	$\dfrac{n!}{(s-a)^{n+1}}$ $(s > a)$
$\sin \omega t$	$\dfrac{\omega}{s^2 + \omega^2}$ $(s > 0)$
$\cos \omega t$	$\dfrac{s}{s^2 + \omega^2}$ $(s > 0)$
$t \sin \omega t$	$\dfrac{2\omega s}{(s^2 + \omega^2)^2}$ $(s > 0)$
$t \cos \omega t$	$\dfrac{s^2 - \omega^2}{(s^2 + \omega^2)^2}$ $(s > 0)$
$e^{at} \sin \omega t$	$\dfrac{\omega}{(s-a)^2 + \omega^2}$ $(s > a)$
$e^{at} \cos \omega t$	$\dfrac{s-a}{(s-a)^2 + \omega^2}$ $(s > a)$

discontinuity must be finite. We can integrate such a function by integrating the continuous sections separately.

A well known example of a piecewise continuous function is the **unit (or Heaviside) step function**, shown in Figure 17.1, defined by

$$H(t) = 0 \quad t < 0$$
$$ = 1 \quad t > 0$$

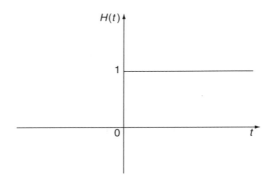

Figure 17.1 The step function.

This function is discontinuous at $t = 0$ and yet has a continuous Laplace transform, namely $1/s$. Integration effectively smoothes things out.

Another well known piecewise continuous function is the square wave, shown in Figure 17.2 and defined by

$$f(t) = -1 \quad -1 < t < 0$$
$$= 1 \quad 0 < t < 1$$
$$f(t+2) = f(t)$$

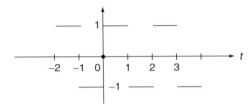

Figure 17.2 The square wave.

This is discontinuous at $t = 1, 2, 3$, etc. and as the name suggests is a periodic function. The ideal tool for dealing with such functions is Fourier series, which again involves an integration that smoothes things out.

Integral transforms can also deal with functions that may be continuous but have discontinuous slope, such as the **saw-tooth wave** shown in Figure 17.3. In this case the derivative fails to exist at the 'corners', and yet it still has a Fourier series. Such functions are called **piecewise smooth**.

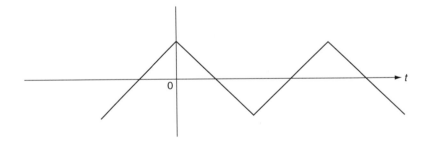

Figure 17.3 The sawtooth wave.

Note that, as mentioned above, the discontinuities must be **finite** jumps. Thus, $1/t$ is **not** piecewise continuous – see Figure 17.4.

Problem 17.4

Evaluate the Laplace transform of the function

$$f(t) = 0 \quad 0 < t < 1$$
$$= 1 \quad 1 < t < 2$$
$$= 0 \quad 2 < t$$

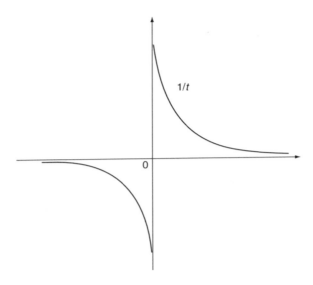

Figure 17.4 An infinite discontinuity.

Note that the value of the integrals involved is unaffected by the absence of the isolated points such as $t = 1, 2$. The Laplace transform of $f(t)$ is

$$\mathcal{L}[f(t)] = \int_0^\infty f(t)e^{-st}\,dt = \int_1^2 e^{-st}\,dt = \left[\frac{e^{-st}}{-s}\right]_1^2 = \frac{e^{-s} - e^{-2s}}{s}$$

This result again illustrates the useful property of the Laplace (and other integral) transforms, that the transform of a discontinuous function may be continuous. Taking the transform **improves** the behaviour of the function.

Exercises on 17.3

1. Derive each of the Laplace transforms in Table 17.1. The results for sin and cos are obtained by integration by parts and from these differentiations with respect to ω will give the results for $t\sin\omega t$ and $t\cos\omega t$.

2. Find the Laplace transform of the piecewise continuous function

$$f(t) = -1 \quad 0 < t < 2$$
$$= t \quad 2 < t < 3$$
$$= 0 \quad 3 < t$$

Answers

2. $\left(\dfrac{1}{s^2} + \dfrac{3}{s}\right)(e^{-2s} - e^{-3s}) - \dfrac{1}{s}$

17.4 Properties of the Laplace transform

Having seen how to obtain elementary Laplace transforms we now turn to the general properties of the transform. These enable us to calculate more complicated transforms and also to apply the Laplace transform to such things as differential equations. We will not be very rigorous about the proofs, the emphasis being on getting across the key ideas.

1. The Laplace transform is linear

The Laplace transform is a **linear operation**. That is, if $f(t)$, $g(t)$ are two functions with Laplace transforms $\tilde{f}(s)$, $\tilde{g}(s)$ and a, b are constants then

$$\mathcal{L}[af(t) + bg(t)] = a\mathcal{L}[f(t)] + b\mathcal{L}[g(t)]$$
$$= a\tilde{f}(s) + b\tilde{g}(s)$$

Or, briefly:

Laplace transform of sum = sum of Laplace transforms.

2. The first shift theorem

The exponential function e^{at} is a mainstay of applied mathematics, and it plays a particularly special role in the theory of the Laplace transform, embodied in the **first shift theorem** (in case you are wondering, there is a 'second shift theorem' but we won't be needing it in this book).

Suppose $\mathcal{L}[f(t)] = \tilde{f}(s)$ for $s > s_0$, and let a be any real number. Then

$$\mathcal{L}[e^{at} f(t)] = \tilde{f}(s - a) \quad \text{for} \quad s > s_0 + a$$

This can be proved by coupling the e^{at} in the Laplace transform with the e^{st} and noting that this effectively replaces s by $s - a$.

Note that the results for $e^{at}t^n$, $e^{at}\sin \omega t$, $e^{at}\cos \omega t$ given in Table 17.1 can be obtained using the shift theorem.

3. The Laplace transform of the derivative

This is the crucial property for the application of Laplace transforms to the solution of differential equations. Suppose that $f(t)$ and its derivative $f'(t)$ have Laplace transforms. Then

$$\mathcal{L}[f'(t)] = s\mathcal{L}[f(t)] - f(0)$$

This follows on integrating by parts:

$$\mathcal{L}[f'(t)] = \int_0^\infty f'(t)e^{-st}dt$$
$$= [f(t)e^{-st}]_0^\infty + s\int_0^\infty f(t)e^{-st}dt$$
$$= -f(0) + s\mathcal{L}[f(t)]$$

on assuming that the limit at infinity vanishes.

Note how differentiation with respect to t in 't-space' becomes multiplication by s in 's-space'. Also, note the appearance of the initial value, $f(0)$, of $f(t)$. This makes the Laplace transform particularly suitable for solving initial value problems. We transform a differential equation from t-space to s-space, thereby obtaining an algebraic equation for the Laplace transform of the solution. We solve this algebraic equation to get the Laplace transform of the solution of the differential equation, which we then invert, going back to t-space to get the solution as a function of t. This methodology is characteristic of 'transform methods' in mathematics, whatever the particular type of transform.

We can now derive the Laplace transform of the second order derivative.

Problem 17.5
Show that

$$\mathcal{L}[f''(t)] = s^2 \mathcal{L}[f(t)] - sf(0) - f'(0)$$

We only have to apply the derivative rule twice in succession:

$$\begin{aligned}\mathcal{L}[f''(t)] &= s\mathcal{L}[f'(t)] - f'(0) \\ &= s(s\mathcal{L}[f(t)] - f(0)) - f'(0) \\ &= s^2 \mathcal{L}[f(t)] - sf(0) - f'(0)\end{aligned}$$

The general result for the Laplace transform of the nth derivative is

$$\mathcal{L}[f^{(n)}(t)] = s^n \mathcal{L}[f(t)] - s^{n-1}f(0) - s^{n-2}f'(0) - \cdots - sf^{(n-2)}(0) - f^{(n-1)}(0)$$

Note again the occurrence of the initial values.

To see how we can use the Laplace transform to solve initial value problems, consider the following problem.

Problem 17.6
Solve the initial value problem

$$y' + y = 1 \qquad y(0) = 0$$

as follows: take the Laplace transform of the equation and obtain an equation for the Laplace transform $\tilde{y}(s)$ of y. Solve this equation. By taking the Laplace transform of the solution found by, say, separation of variables, confirm that $\tilde{y}(s)$ is the Laplace transform of the solution $y(t)$.

Taking the Laplace transform of the equation gives, using linearity and the derivative rule

$$\mathcal{L}[y' + y] = \mathcal{L}[y'] + \mathcal{L}[y] = \mathcal{L}[1]$$

or

$$s\tilde{y}(s) - y(0) + \tilde{y}(s) = \frac{1}{s}$$

Since $y(0) = 0$ this becomes

$$(s+1)\tilde{y}(s) = \frac{1}{s}$$

Solving for $\tilde{y}(s)$ now gives

$$\tilde{y}(s) = \frac{1}{s(s+1)}$$

Now by separation of variables (454 ◀), or integrating factor (458 ◀), the solution of the initial value problem is

$$y(t) = 1 - e^{-t}$$

Taking the Laplace transform of this:

$$\mathcal{L}[y(t)] = \mathcal{L}[1 - e^{-t}]$$

$$= \frac{1}{s} - \frac{1}{s+1}$$

$$= \frac{1}{s(s+1)} = \tilde{y}(s)$$

which is indeed the Laplace transform of the solution as obtained above. So the Laplace transform of the equation did lead us to the Laplace transform of the solution. In fact, in this case it is not too difficult to find the **inverse Laplace transform** of $\tilde{y}(s)$ directly. Denoting the inverse Laplace transform by $\mathcal{L}^{-1}$

$$y(t) = \mathcal{L}^{-1}[\tilde{y}(s)] = \mathcal{L}^{-1}\left[\frac{1}{s(s+1)}\right]$$

$$= \mathcal{L}^{-1}\left[\frac{1}{s} - \frac{1}{s+1}\right]$$

on splitting into partial fractions (60 ◀)

$$= \mathcal{L}^{-1}\left[\frac{1}{s}\right] - \mathcal{L}^{-1}\left[\frac{1}{s+1}\right]$$

Now read Table 17.1 backwards to give

$$\mathcal{L}^{-1}\left[\frac{1}{s}\right] = 1, \qquad \mathcal{L}^{-1}\left[\frac{1}{s+1}\right] = e^{-t}$$

so

$$y(t) = 1 - e^{-t}$$

The above example contains all the essential elements of using the Laplace transform to solve initial value problems. The main new feature is the occurrence of the inverse Laplace transform. Also note that in this method the initial values are actually incorporated in the

Exercises on 17.4

1. Evaluate the Laplace transforms of

    (i) $3t^2 + \cos 2t$

    (ii) $2t + t^3 + 4te^t$

2. Solve the following initial value problem by using the Laplace transform

    $y' + 2y = 3 \qquad y(0) = 0$

Answers

1. (i) $\dfrac{6}{s^3} + \dfrac{s}{s^2 + 4}$ (ii) $\dfrac{2}{s^2} + \dfrac{6}{s^4} + \dfrac{4}{(s-1)^2}$

2. $\frac{3}{2}(1 - e^{-2t})$

17.5 The inverse Laplace transform

If $\tilde{f}(s)$ is some function of the Laplace transform variable s, then that function $f(t)$ whose Laplace transform is $\tilde{f}(s)$ is called the **inverse Laplace transform** of $\tilde{f}(s)$ and is denoted:

$$f(t) = \mathcal{L}^{-1}[\tilde{f}(s)]$$

Crudely, we may think of $\mathcal{L}^{-1}$ as 'undoing' the Laplace transform operation.

Table 17.1 gives a number of important inverse transforms, simply by reading it from right to left.

Problem 17.7

Find the inverse transforms of $3/s^2$, $1/(s-4)$, $4/(s^2+9)$, $(s^2-9)/(s^2+9)^2$, $(s-1)/[(s-1)^2+1]$, from Table 17.1. Check your results by taking their transform.

Many more inverse transforms may be obtained by algebraic and mathematical manipulation on the transform to put it into a suitable form for inversion by already known inverses. Like the Laplace transform itself, the inverse Laplace transform is a linear operator, and this alone accounts for a large number of inverses. Also completing the square (66 ◀) can be useful as for example in

$$\mathcal{L}^{-1}\left[\frac{s-1}{s^2 - 2s + 2}\right] = \mathcal{L}^{-1}\left[\frac{s-1}{(s-1)^2 + 1}\right]$$

$$= e^{-t}\cos t$$

Exercise on 17.5

Find the inverse Laplace transform in each case and check by taking the Laplace transform of your results.

(i) $\dfrac{2s^2 - 3s + 1}{2s^3}$ (ii) $\dfrac{1}{s+1}$ (iii) $\dfrac{1}{s-3}$ (iv) $\dfrac{4}{2s-1}$

(v) $\dfrac{5}{s^2+4}$ (vi) $\dfrac{3s}{s^2+9}$ (vii) $\dfrac{s+1}{s^2-5s+6}$

(viii) $\dfrac{1}{(s+1)(s+2)(s+3)}$

Answer

(i) $1 - \tfrac{3}{2}t + \tfrac{1}{4}t^2$ (ii) e^{-t} (iii) e^{3t} (iv) $2e^{t/2}$

(v) $\tfrac{5}{2}\sin 2t$ (vi) $3\cos 3t$ (vii) $4e^{3t} - 3e^{2t}$

(vi) $\tfrac{1}{2}e^{-t} - e^{-2t} + \tfrac{1}{2}e^{-3t}$

17.6 Solution of initial value problems by Laplace transform

The inverse Laplace transform enables us to solve linear initial value problems. With what you now know about the inverse transform, try this on your own.

Problem 17.8
Solve the initial value problem

$$y' + y = 1 \qquad y(0) = 1$$

by Laplace transform.

Taking the Laplace transform of the equation:

$$\mathcal{L}[y' + y] = \mathcal{L}[y'] + \mathcal{L}[y]$$
$$= s\tilde{y}(s) - y(0) + \tilde{y}(s)$$
$$= (s+1)\tilde{y}(s) - 1 = \mathcal{L}[1] = \dfrac{1}{s}$$

Hence

$$\tilde{y}(s) = \dfrac{1}{s}$$

so the solution is

$$y(t) = 1$$

which you can check by direct solution of the equation.

The above example illustrates the general approach to solving linear initial value problems by Laplace transform:

1. Take the Laplace transform of the equation, inserting the initial values.
2. Solve for the transform of the solution, obtaining

$$\tilde{y}(s) = f(s)$$

3. Invert this transform to obtain the solution

$$y(t) = \mathcal{L}^{-1}[f(s)]$$

Note an important feature of the Laplace transform method concerning initial conditions. In most elementary methods for solving differential equations we first find the general solution, involving the appropriate number of arbitrary constants. These constants are then found by applying the initial or boundary values, producing a number of equations which are solved for the constants. In the Laplace (and other) transform method however the initial conditions are automatically included, and there is no need to find the general solution.

The extension of this method to second and higher order initial value problems is conceptually simple and merely requires more complicated manipulation and inversion.

Problem 17.9
Solve the initial value problem

$$y'' + y = 0 \qquad y(0) = 1, \quad y'(0) = 0$$

by Laplace transform.

Taking the Laplace transform through the equation and applying the initial conditions gives

$$\mathcal{L}[y''] + \mathcal{L}[y] = (s^2 + 1)\tilde{y} - s = 0$$

Hence

$$\tilde{y}(s) = \frac{s}{s^2 + 1}$$

so

$$y(t) = \mathcal{L}^{-1}\left[\frac{s}{s^2 + 1}\right] = \cos t$$

Exercises on 17.6
1. Solve the initial value problems

   (i) $y' + 3y = 2 \qquad y(0) = 4$ \qquad (ii) $y' - y = t \qquad y(0) = 0$

   Check by using alternative solutions, or substituting back in the equations.

2. Solve the initial value problem

$$y'' + 3y' + 2y = 20e^{-3t} \qquad\qquad y(0) = y'(0) = 0$$

   Verify that your solution satisfies the equation and the initial conditions.

Answers

1. (i) $\dfrac{10}{3}e^{-3t} + \dfrac{2}{3}$ (ii) $e^t - 1 - t$

2. $10e^{-t} - 20e^{-2t} + 10e^{-3t}$

17.7 Linear systems and the principle of superposition

Before we move on to a different type of transform, we will look briefly at the sort of systems for which it is used. Consider a system whose response to a time dependent input $x(t)$ is an output $y(t)$. The system is called **linear** if the output of the sum of two inputs is the sum of the separate outputs to the two inputs. That is, if $x_1(t) \to y_1(t)$ and $y_2(t) \to y_2(t)$ implies $x_1(t) + x_2(t) \to y_1(t) + y_2(t)$.

Many, but not all, systems behave in this way – any that do not are called **non-linear** and are very difficult to analyse. This additivity of outputs of linear systems is called the **principle of superposition**.

If a complicated signal can be split up into a linear combination of simpler signals then the effect of a linear processing system on the total signal can be analysed by adding up the separate effects on the component input signals. This is the philosophy behind **Fourier analysis**. A **Fourier series** splits a periodic input (such as a square wave) into a sum of sinusoidal components of different amplitude and frequency. The effect of a linear system on each separate sinusoid is easy to determine in general and the separate outputs can be added up to synthesize the total output corresponding to the original periodic interval. This is illustrated in Figure 17.5.

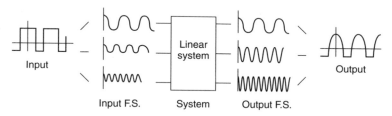

Figure 17.5 Fourier decomposition of input and output.

We will now summarise some terminology for a general sinusoidal function:

$$f(t) = A \sin(\omega t + \alpha)$$

A is called the **amplitude**, ω the **angular frequency** (radians per second) and α is the **phase** relative to $A \sin \omega t$. The period of such a sinusoid is $T = 2\pi/\omega$. The inverse of T, $\omega/2\pi$, is the **frequency** in cycles per second, or Hertz (Hz). See Figure 17.6.

Linear operations such as differentiation, integration, addition, etc. change the amplitude and phase of such a sinusoid, **but not its frequency**. For example

$$\frac{d}{dt}[A \sin(\omega t + \alpha)] = A\omega \cos(\omega t + \alpha)$$
$$= A\omega \sin(\omega t + \alpha + \pi/2)$$

which has the same frequency as the original, but a modified amplitude and phase.

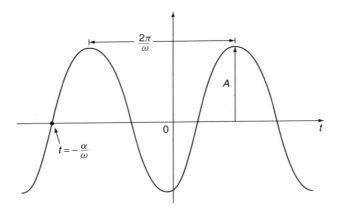

Figure 17.6 Sinusoidal function $A\sin(\omega t + \alpha)$.

Exercise on 17.7

Show that (i) the sum of two sinusoids with the same frequency.

$$A\sin(\omega t) + B\sin(\omega t + \alpha)$$

and (ii) the integral of a sinusoid

$$\int a\sin(\omega t + \phi)\,dt$$

each have the same frequency as the original sinusoids, and determine the amplitude and phase of the result in each case.

Answer

(i) Amplitude is $\sqrt{(A + B\cos\alpha)^2 + B^2\sin^2\alpha}$

Phase is β where $\tan\beta = \dfrac{B\sin\alpha}{A + B\cos\alpha}$

(ii) Amplitude is $\dfrac{a}{\omega}$. Phase is shifted by $\dfrac{\pi}{2}$

17.8 Orthogonality relations for trigonometric functions

In the process of expressing a general waveform of a given period in terms of sinusoids of given frequency – the object of Fourier analysis – we will need certain integral formulae. These are so important that we devote this whole section to them. Make sure that you fully understand what they mean, and can derive them (285 ◀). We will only consider integrals and sinusoids defined over a period of 2π, usually taken to be $-\pi$ to π. This is no real restriction – if we have a region of different length then we can simply scale it to 2π. Our integrals concern products of sines and cosines over the range $-\pi$ to π.

If m, n are any integers, then:

$$\int_{-\pi}^{\pi} \sin mt \sin nt \, dt = 0 \quad \text{if} \quad m \neq n$$

$$\int_{-\pi}^{\pi} \sin^2 nt \, dt = \pi$$

$$\int_{-\pi}^{\pi} \cos mt \cos nt \, dt = 0 \quad \text{if} \quad m \neq n$$

$$\int_{-\pi}^{\pi} \cos^2 nt \, dt = \pi$$

$$\int_{-\pi}^{\pi} \sin mt \cos nt \, dt = 0 \quad \text{for all} \quad m, n$$

$$\int_{-\pi}^{\pi} \sin mt \, dt = \int_{-\pi}^{\pi} \cos mt \, dt = 0$$

These are called the **orthogonality relations for sine and cosine.** The limits $-\pi$, π on the integrals may in fact be replaced by **any** integral of length 2π, or integer multiple of 2π. The orthogonality relations should not be thought of as simply specific integral relations – many other functions satisfy similar orthogonality relations (often called **special**, or **orthogonal functions**) and can be used in a similar way to expand in a 'generalised Fourier series'.

Exercise on 17.8

Prove the orthogonality relations.

17.9 The Fourier series expansion

A Fourier series is essentially a means of expressing any periodic function $f(t)$ as a sum (possibly infinite) of sines or cosines of different frequencies. The point is of course that sine and cosine are relatively simple functions to deal with.

Let $f(t)$ be a function with period 2π. It may be a square wave, or triangular wave, for example. If we compare it with a sine or cosine wave with the same period there may be little resemblance, on the face of it. However, it turns out that if we combine a large enough number of sine and/or cosine waves of appropriate amplitudes and frequencies then it is possible, under certain conditions, to approximate to any function of period 2π. This is certainly not obvious, but is mathematically well established and so here we will simply **assume** that $f(t)$ can be expanded in a series of sines and cosines in the form

$$f(t) = \frac{a_0}{2} + \sum_{n=1}^{\infty} a_n \cos nt + \sum_{n=1}^{\infty} b_n \sin nt$$

This is called a **Fourier (series) expansion** for $f(t)$. Note that each term on the right-hand side has period 2π, like $f(t)$, but as n increases $\cos nt$ and $\sin nt$ oscillate an increasing number of times in this period. The above series is therefore adding together a (possibly infinite) number of different oscillatory functions, with different amplitudes or weighting. The hope is that by suitably choosing these weightings, one can approximate any other function of the same period, 2π. The problem is of course to find the appropriate 'weightings' or coefficients, a_0, a_n, b_n for any given function $f(t)$.

The notation in the above series is conventional, as is the factor $\frac{1}{2}$ with a_0, which simplifies later results. a_0, a_n, b_n are constants which will depend upon $f(t)$ and which are to be determined.

Particularly in the case of time dependent periodic functions, it is usual to speak of the $n = 1$ term in the Fourier decomposition as the **'fundamental component'** (it is often the dominant one) and the nth term ($n > 1$) as the **nth harmonic**.

As an example of a Fourier series, consider the square wave period of 2π and amplitude A, shown in Figure 17.7.

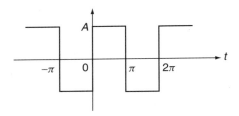

Figure 17.7

This can be described by the function

$$f(t) = \begin{array}{ll} -A & -\pi < t < 0 \\ A & 0 < t < \pi \end{array} \qquad f(t + 2\pi) = f(t)$$

Mathematically this is a difficult functional form to handle and to process – for example we cannot differentiate it, and if it appeared as a 'forcing function' (469 ◄) in a differential equation then there is not a lot we could do with it as it is. However, the fact that it is periodic with period 2π does mean that we can express it as a superposition of sinusoids of the same period – i.e. as a Fourier series. We will show in Section 17.10 that the corresponding Fourier series is in fact

$$f(t) = \frac{4A}{\pi}\left(\sin t + \frac{1}{3}\sin 3t + \frac{1}{5}\sin 5t + \ldots\right)$$

The amplitude of the nth harmonic is $\dfrac{4A}{n\pi}$. Now although this series looks nothing like the original functional form of the square wave it is in fact equivalent to it. If you have a graphical calculator you might like to try plotting each of the harmonics and the sum of the first few. You will see that the result starts to look like a humpy square wave, and gets more and more like it as you take more and more terms in the series. The whole point of course is that the sinusoidal functions to which we have converted the wave are continuous and far easier to handle than the original discontinuous form. There are, it is true, now an infinite number of them – but as with all such series we can get as good an approximation as we please by taking enough terms in the series.

There are a number of points that are worth noting, although we do not want to be too picky at this stage, so treat these as refinements for (much) later assimilation.

(i) At a discontinuity (e.g. at $t = 0$ in the above example) the Fourier series gives the 'average value':

$$\tfrac{1}{2}[f_-(t_0) + f_+(t_0)]$$

where $f_-(t_0)$ denotes the limiting value of $f(t)$ as t approaches the discontinuity at $t = t_0$ from **below**, $f_+(t_0)$ denotes the limit as t approaches from **above**.

(ii) The Fourier series of an **even** function can contain only the **constant** and the **cosine** terms, because cosine is an even function. The Fourier series of an **odd** function can contain only the **sine** terms, since sine is odd.

(iii) Interesting results for infinite series can be deduced from Fourier series. In the above example, putting $t = \pi/2$ gives:

$$f\left(\frac{\pi}{2}\right) = A = \frac{4A}{\pi}\left(1 - \frac{1}{3} + \frac{1}{5} - \cdots\right)$$

$$\text{or } 1 - \frac{1}{3} + \frac{1}{5} - \frac{1}{7} + \cdots = \frac{\pi}{4}$$

(iv) While the Fourier series represents a periodic function, it can be used to represent a non-periodic function just over one period if required. For example the square wave considered above can be represented by a Fourier series over just one period $-\pi < t < \pi$.

Exercise on 17.9

Sketch the triangular wave

$$f(t) = t \qquad 0 < t < \pi$$
$$= -t \qquad -\pi < t < 0$$
$$f(t) = (t + 2\pi)$$

for $-5\pi < t < 5\pi$.

Is the function odd or even? The corresponding Fourier series is

$$f(t) = \frac{\pi}{2} - \frac{4}{\pi}\left(\cos t + \frac{1}{3^2}\cos 3t + \cdots + \frac{1}{(2r+1)^2}\cos(2r+1)t + \cdots\right)$$

(you will be asked to derive this in Section 17.10). What is the average value of the function over all time (consider the average value of any sinusoid over a complete period)? What is the fundamental component? Write down the amplitude of the nth harmonic.

Deduce from the series that

$$1 + \frac{1}{3^2} + \frac{1}{5^2} + \cdots + \frac{1}{(2r+1)^2} + \cdots = \frac{\pi^2}{8}$$

Answer

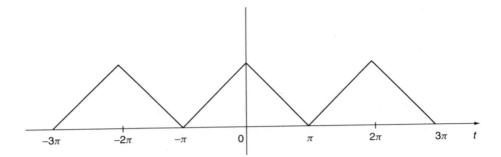

Even; $\dfrac{\pi}{2}$; $-\dfrac{4}{\pi}$; 0 if n is even, and $-\dfrac{4}{\pi(2r+1)^2}$ if $n = 2r+1$ is odd

17.10 The Fourier coefficients

There are, in general, an infinite number of the coefficients a_n, b_n of the Fourier series. So we cannot expect to determine them 'all at once'. We had a similar situation with power series in Section 14.11 (435 ◂). There we found the coefficients one at a time by successive differentiation to remove all terms except the one that we wanted to determine at each stage. Such an approach won't work in this case because no matter how many times you differentiate a sinusoidal function, it won't go away! However, the integral formulae or 'orthogonality relations' given in Section 17.8 can serve a similar purpose in enabling us to eliminate all coefficients except the one we want, one at a time. We do this by multiplying through by $\cos nt$ or $\sin nt$, integrating over a single period and using the orthogonality relations to remove all but the desired coefficient. An example will illustrate this more clearly.

Suppose we want to find a_3 in the series

$$f(t) = \frac{a_0}{2} + \sum_{n=1}^{\infty} a_n \cos nt + \sum_{n=1}^{\infty} b_n \sin nt$$

Multiply through by $\cos 3t$, the sinusoid that goes with a_3, and integrate over $(-\pi, \pi)$ (any interval of length 2π will do). By the orthogonality relations of Section 17.8 every resulting integral on the right-hand side of the series will vanish **except** that corresponding to $\cos 3t$ itself, and we will obtain

$$\int_{-\pi}^{\pi} f(t) \cos(3t)\, dt = a_3 \int_{-\pi}^{\pi} \cos^2(3t)\, dt$$

$$= \pi a_3 \quad \text{from Section 17.8}$$

So

$$a_3 = \frac{1}{\pi} \int_{\pi}^{-\pi} f(t) \cos(3t)\, dt$$

In general, we have

$$a_n = \frac{1}{\pi} \int_{-\pi}^{\pi} f(t) \cos nt\, dt \quad n = 0, 1, 2, \ldots$$

$$b_n = \frac{1}{\pi} \int_{-\pi}^{\pi} f(t) \sin nt\, dt \quad n = 1, 2, \ldots$$

Problem 17.10
Verify the series for the square wave given in Section 17.9.

We have

$$f(t) = -A \quad -\pi < t < 0$$
$$ = A \quad 0 < t < \pi$$

Since the function has period 2π, we can take its series to be:

$$f(t) = \frac{a_0}{2} + \sum_{n=1}^{\infty} a_n \cos nt + \sum_{n=1}^{\infty} b_n \sin nt$$

Also, since the function is **odd** it can't contain any even terms in the series, so we can take all $a_n = 0$, and write the series as

$$f(t) = \sum_{n=1}^{\infty} b_n \sin nt$$

We therefore have to find the b_n.

Multiplying through by $\sin mt$ and integrating over $[-\pi, \pi]$ we have

$$\int_{-\pi}^{\pi} f(t) \sin mt\, dt = \int_{-\pi}^{\pi} \sum_{n=1}^{\infty} b_n \sin nt \sin mt\, dt$$

$$= \sum_{n=1}^{\infty} b_n \int_{-\pi}^{\pi} \sin nt \sin mt\, dt$$

$$= b_m \int_{-\pi}^{\pi} \sin^2 mt\, dt$$

since

$$\int_{-\pi}^{\pi} \sin nt \sin mt\, dt = 0 \text{ if } m \neq n$$

Now

$$\int_{-\pi}^{\pi} \sin^2 mt\, dt = \frac{1}{2} \int_{-\pi}^{\pi} (1 - \cos 2mt)\, dt$$

$$= \pi$$

So we get

$$\pi b_m = \int_{-\pi}^{\pi} f(t) \sin mt \, dt$$

$$= \int_{-\pi}^{0} (-A) \sin mt \, dt + \int_{0}^{\pi} A \sin mt \, dt$$

$$= A\left[\frac{\cos mt}{m}\right]_{-\pi}^{0} - A\left[\frac{\cos mt}{m}\right]_{0}^{\pi}$$

$$= \frac{A}{m}(1 - (-1)^m) - \frac{A}{m}((-1)^m - 1)$$

$(\cos m\pi = (-1)^m)$

$$= \frac{2A}{m}(1 - (-1)^m)$$

Note that since $f(t)$ is odd we could have used (284 ◀)

$$\int_{-\pi}^{\pi} f(t) \sin mt \, dt = 2\int_{0}^{\pi} f(t) \sin mt \, dt$$

So finally

$$b_m = \frac{2A}{m\pi}(1 - (-1)^m)$$

$$= 0 \quad \text{if } m \text{ even}$$

$$= \frac{4A}{m\pi} \quad \text{if } m \text{ odd}$$

or, reverting to n

$$b_n = \frac{2A}{n\pi}(1 - (-1)^n)$$

The series is thus

$$f(t) = \sum_{n=1}^{\infty} b_n \sin nt = \sum_{n=1}^{\infty} \frac{2A}{n\pi}(1 - (-1)^n) \sin nt$$

$$= \frac{4A}{\pi} \sin t + \frac{4A}{3\pi} \sin 3t + \frac{4A}{5\pi} \sin 5t + \cdots$$

$$= \frac{4A}{\pi}\left(\sin t + \frac{1}{3}\sin 3t + \frac{1}{5}\sin 5t + \cdots\right)$$

Exercise on 17.10

Verify the series given in Exercise on 17.9 for the triangular wave

$$f(t) = t \quad 0 < t < \pi$$

$$ -t \quad -\pi < t < 0$$

$$f(t) = f(t + 2\pi)$$

17.11 Reinforcement

1. Find the Laplace transforms of

(i) $2 + t + 3t^2$

(ii) $2\sin 3t + e^{-2t}$

(iii) $e^t \cos(2t)$

(iv) $\sin^2 t$

2. Write down the inverse Laplace transforms of

(i) $\dfrac{1}{s^4} + \dfrac{8}{s^3}$

(ii) $\dfrac{s^2 + 5s + 7}{s^4}$

(iii) $\dfrac{1}{s-4}$

(iv) $\dfrac{1}{s+4}$

(v) $\dfrac{2}{4s-3}$

(vi) $\dfrac{c}{as+b}$

(vii) $\dfrac{s}{s^2+9}$

(viii) $\dfrac{6}{s^2+9}$

(ix) $\dfrac{5s+4}{s^2+9}$

(x) $\dfrac{as+6}{s^2+c^2}$

where a, b, c are constants.

3. Find the inverses of the following Laplace transforms:

(i) $\dfrac{s+3}{s(s+2)}$

(ii) $\dfrac{1}{(s+1)(s-3)}$

(iii) $\dfrac{s^2+2}{(s^2+1)(s+1)}$

(iv) $\dfrac{1}{s^2(s-4)}$

(v) $\dfrac{1}{(s-1)(s+2)^2}$

(vi) $\dfrac{1}{(s^2+4)(s^2+9)}$

4. Solve the following initial value problems using the Laplace transform, and the results of Question 3.

(i) $y' + 2y = 3$ $y(0) = 1$

(ii) $y' - 3y = e^{-t}$ $y(0) = 0$

(iii) $y' + y = 2\sin t$ $y(0) = 2$

(iv) $y' + 4y = t$ $y(0) = 0$

(v) $y' - y = te^{-2t}$ $y(0) = 0$

(vi) $y'' + 9y = 3\sin 2t$ $y(0) = y'(0) = 0$

In each case check your result by solution by another means (e.g. undetermined coefficients).

5. State which of the following functions of t are periodic and give the period when they are.

   (i) $\tan 2t$
   (ii) $\cos\left(\dfrac{3\pi t}{2}\right)$
   (iii) $\cos(\sqrt{t})$
   (iv) $\sin t + \cos 2t$
   (v) $\sin |t|$
   (vi) $|\cos t|$
   (vii) $\sin\left(\dfrac{2\pi}{L}t\right)$
   (viii) $\cos(4\omega t)$
   (ix) $4\cos 2t + 3\sin 4t$
   (x) $t^2 \cos t$.

6. What is the period of

$$f(t) = \frac{1}{2}a_0 + \sum_{n=1}^{\infty}(a_n \cos n\omega t + b_n \sin n\omega t)?$$

7. Obtain the Fourier series for the following functions defined on $-\pi < t < \pi$:

   (i) $2|t| \quad -\pi < t < \pi$
   (ii) $t \quad -\pi < t \le \pi$
   (iii) $t^2 \quad -\pi < t < \pi$
   (iv) $f(t) = 0 \quad -\pi < t < 0$
   $ = t \quad 0 < t < \pi$
   (v) $f(t) = -t^2 \quad -\pi < t < 0$
   $ = t^2 \quad 0 < t < \pi$
   (vi) $f(t) = 0 \quad -\pi < t < 0$
   $ = 1 \quad 0 < t < \pi$
   (vii) $f(t) = \begin{cases} \dfrac{1}{2} & -\pi < t < -\dfrac{\pi}{2} \\ 1 & -\dfrac{\pi}{2} < t < \dfrac{\pi}{2} \\ 0 & \dfrac{\pi}{2} < t < \pi \end{cases}$

17.12 Applications

Transform methods form a vast area of engineering mathematics, and this chapter only scratches the surface. In these applications we just flag up some of the key fundamental ideas which may come up in your engineering subjects.

1. One of the most important applications of Laplace transforms in engineering is in the solution of initial value problems of the sort discussed in Chapter 15. Particularly important are the sorts of engineering models described by inhomogeneous second order differential equations of the type covered in Chapter 15, Applications, question 6:

$$m\ddot{x} + \beta\dot{x} + \alpha x = f(t)$$

In Chapter 15 we solved this type of equation by the auxiliary equation and the method of undetermined coefficients. In the case when we have initial conditions, specifying x and $\dot{x}$ at $t = 0$, the Laplace transform provides a powerful tool for solving such equations and studying the properties of the solutions. This question is a significant project in which you are asked to repeat as much of the Chapter 15, Applications, Question 6 as you can using Laplace transform methods.

Show that with zero initial conditions (i.e. a system that is initially at rest) the Laplace transform of $x(t)$ can be written

$$\mathcal{L}[x(t)] = \tilde{x}(s) = \frac{\tilde{f}(s)}{ms^2 + \beta s + \alpha}$$

where $\tilde{f}(s)$ is the Laplace transform of the right-hand side $f(t)$.

Find the solutions $x(t)$ for the constant forcing function

$$f(t) = F_0 H(t) \qquad (H(t) \text{ the unit stepfunction})$$

for various values of m, β, α, considering cases of real, equal and complex roots of the quadratic.

Consider also the sinusoidal forcing function

$$f(t) = f_0 \cos \upsilon t$$

and compare your solutions with the results of Chapter 15, Applications, Question 6.

2. The sorts of equations considered in question 1 are often used to describe a **control system** in which $x(t)$ represents the **response** or output of the system to an **input** $f(t)$. In this case the Laplace transform of the equation plays an important role in control theory, and the connection between the Laplace transform of the input to that of the output is called the **transfer function** of the system, denoted $F(s)$. We write:

$$\tilde{x}(s) = F(s)\tilde{f}(s)$$

So, for example the transfer function of the system described in question 1 would take the form

$$F(s) = \frac{1}{ms^2 + \beta s + \alpha}$$

The **stability** of the control system depends crucially on the **poles** of the transfer function – that is the roots of the denominator, $ms^2 + \beta s + \alpha$, in this case. Control systems can obviously be of great complexity, and a particularly important feature is that of **feedback**. A control system has feedback when the output or some other signal is fed back to the input as a means of influencing the behaviour of the control system. A simple example is shown in Figure 17.8. Here, the input $x_i(t)$ is modified by subtraction of a feedback $x_f(t)$ to form $x_i(t) - x_f(t)$ which is input to the control system which has transfer function $G(s)$. The feedback $x_f(t)$ is formed by a modifying control system that converts the output $x_o(t)$ to $x_f(t)$ with a transfer function $F(s)$, thus forming a feedback loop. If $\tilde{x}_i(s)$ and $\tilde{x}_o(s)$ are respectively the Laplace transforms of the input and output, show that the overall transfer function of the feedback system is

$$\frac{G(s)}{1 + F(s)G(s)}$$

i.e.
$$\tilde{x}_o(s) = \frac{G(s)}{1 + F(s)G(s)} \tilde{x}_i(s)$$

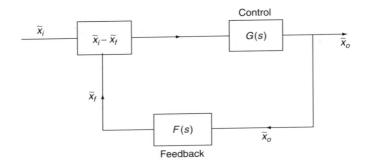

Figure 17.8

By such means we can derive the transfer functions for ever more complex control systems. Many linear control systems have a transfer function that is a rational function. Show that if $F(s)$ and $G(s)$ are rational functions then so is the overall transfer function derived above. Discuss the poles of the overall transfer function in terms of those of the $F(s)$ and $G(s)$. Essentially, the study of the stability of a control system reduces to the study of the poles of such transfer functions – this is where all the work we did on rational functions in Chapter 2 comes in handy.

3. We have already emphasised the use of Fourier series in, for example, breaking up periodic signals into sinusoidal components that are more easily analysed and then recombined in a linear system. No doubt you will see more of this in your engineering subjects, whether it be in the analysis of coupled dynamical systems, heat transfer in solid bodies, or analysis of electronic and optical signalling systems. Here we look at a use of Fourier series in providing a succinct expression for the power in a periodic signal.

The average power in a periodic signal $x(t)$ (assumed to be of period 2π) is defined by

$$P = \frac{1}{2\pi} \int_{-\pi}^{\pi} x^2(t)\, dt$$

If the signal can be expressed as a Fourier series in the form

$$x(t) = \frac{a_0}{2} + \sum_{n=1}^{\infty} a_n \cos nt + \sum_{n=1}^{\infty} b_n \sin nt$$

show that the average power can be written as

$$P = \frac{a_0^2}{4} + \frac{1}{2} \sum_{n=1}^{\infty} (a_n^2 + b_n^2)$$

This is a particular case of a famous mathematical result called **Parseval's theorem**, and the general principle it expresses is a key principle of the study of any sort of periodic motion – the energy/power in any periodic phenomenon is proportional to the sum of the squares of the amplitudes of the sinusoidal components. If this reminds you of Pythagoras' theorem, well …

17.13 Answers to reinforcement exercises

1. (i) $\dfrac{2s^2 + s + 2}{s^3}$ (ii) $\dfrac{s^2 + 6s + 21}{(s+2)(s^2+9)}$

   (iii) $\dfrac{s-1}{s^2 - 2s + 5}$ (iv) $\dfrac{2}{s^2 + 4}$

2. (i) $\dfrac{t^2}{6}(t + 24)$ (ii) $t + \dfrac{5}{2}t^2 + \dfrac{7}{6}t^3$

   (iii) e^{4t} (iv) e^{-4t}

   (v) $\dfrac{1}{2}e^{3t/4}$ (vi) $\dfrac{c}{a}e^{-bt/a}$

   (vii) $\cos 3t$ (viii) $2\sin 3t$

   (ix) $5\cos 3t + \dfrac{4}{3}\sin 3t$ (x) $a\cos ct + \dfrac{b}{c}\sin ct$

3. (i) $\dfrac{3}{2} - \dfrac{1}{2}e^{-2t}$ (ii) $\dfrac{1}{4}(e^{3t} - e^{-t})$

   (iii) $\dfrac{3}{2}e^{-t} - \dfrac{1}{2}\cos t + \dfrac{1}{2}\sin t$ (iv) $\dfrac{1}{16}e^{-4t} - \dfrac{1}{16} + \dfrac{1}{4}t$

   (v) $\dfrac{1}{9}e^t - \dfrac{1}{9}e^{-2t} - \dfrac{1}{3}te^{-2t}$ (vi) $\dfrac{1}{10}\sin 2t - \dfrac{1}{15}\sin 3t$

4. (i) $\dfrac{3}{2} - \dfrac{1}{2}e^{-2t}$ (ii) $\dfrac{1}{4}(e^{3t} - e^{-t})$

   (iii) $3e^{-t} - \cos t + \sin t$ (iv) $\dfrac{1}{16}e^{-4t} - \dfrac{1}{16} + \dfrac{1}{4}t$

   (v) $\dfrac{1}{9}e^t - \dfrac{1}{9}e^{-2t} - \dfrac{1}{3}te^{-2t}$ (vi) $\dfrac{3}{5}\sin 2t - \dfrac{2}{5}\sin 3t$

5. (i) Period $\dfrac{\pi}{2}$ (ii) Period $\dfrac{4}{3}$ (iii) Not periodic

   (iv) Period 2π (v) Not periodic (vi) Period π

   (vii) Period L (viii) Period $\dfrac{\pi}{2\omega}$ (ix) Period π

   (x) Not periodic

6. $\dfrac{2\pi}{\omega}$

7. (i) $\pi - \dfrac{8}{\pi} \sum\limits_{n=1}^{\infty} \dfrac{\cos(2n-1)t}{(2n-1)^2}$

(ii) $2 \sum\limits_{n=1}^{\infty} \dfrac{(-1)^n}{n} \sin nt$

(iii) $\dfrac{\pi^2}{3} + 4 \sum\limits_{n=1}^{\infty} \dfrac{(-1)^n}{n^2} \cos nt$

(iv) $\dfrac{\pi}{4} + \dfrac{1}{\pi} \sum\limits_{n=1}^{\infty} \dfrac{((-1)^n - 1)}{n^2} \cos nt + \sum\limits_{n=1}^{\infty} \dfrac{(-1)^n}{n} \sin nt$

(v) $\sum\limits_{n=1}^{\infty} \left[\dfrac{4}{\pi n^2}((-1)^n - 1) - \dfrac{2\pi}{n}(-1)^n \right] \sin nt$

(vi) $\dfrac{1}{2} + \sum\limits_{n=1}^{\infty} \dfrac{(1-(-1)^n)}{n\pi} \sin nt$

(vii) $\dfrac{\pi}{2} + \dfrac{1}{8} + \sum\limits_{n=1}^{\infty} \dfrac{3}{2n\pi} \sin\left(\dfrac{n\pi}{2}\right) \cos nt + \sum\limits_{n=1}^{\infty} \dfrac{1}{2n\pi} \left(\cos n\pi - \cos\left(\dfrac{n\pi}{2}\right) \right) \sin nt$

Index

Absolute convergence 433
Acute angles 149
Addition of integrals 257
Addition of matrices 383
Adjoint matrix 393
Algebra 40
Algebra of complex numbers 353
Algebraic division 60
Algebraic equation 40
Algebraic expression 40
Alternate angles 149
Alternating series 431
Amplitude 182
Analytical geometry 205
Angle 148
Angle between two lines 327
Angle bisector theorem 153
Anti-derivative 253
Arc of a circle 156
Area under a curve 305
Argand plane 355
Argument 90
Argument of a complex number 355
Arithmetic series 104
Associative rule 41
Asymptotes 300
Auxiliary equation 464
Axioms 147

Base 18, 122
Base of natural logarithms 124
Basis vectors 328
Binomial 43, 72
Binomial theorem 72
BODMAS 11
Boundary conditions 449

Calculators 2
Cartesian coordinate system 91, 205
Centre of mass 312
Centroid 312

Chain rule 235
Chaos 442
Choosing integration methods 276
Chord 157
Closed interval 98
Coefficients 43
Cofactor 388
Complementary angles 149, 177
Common ratio 103
Commutative rule 41
Comparison test 430
Complementary function 462
Completing the square 66, 265
Complex number 353
Complex root 44
Components of a vector 328
Composition of functions 97
Compound angle formulae 187
Congruent triangles 152
Constant of integration 254
Constant 40
Continuity 418
Convergence 426, 428
Coordinate geometry 205
Coordinates 91, 205
Corresponding angles 149
Cosec 175
Cosine 175
Cosine rule 179
Cotangent 175
Cramer's rule 391
Curve sketching 299
Cyclic quadrilateral 159

D'Alembert's ratio test 432
De Moivre's theorem 362, 365
Definite integral 278
Degree of polynomial 43
Degrees 148
Denominator 12, 55
Dependent variable 90

529

Index

Derivative   231
Derivative of a vector   340
Determinant   388
Difference   10
Differential coefficient   231
Differential equation   246, 446
Differentiation   231
Differentiation from first principles   230
Direction cosine   325
Direction ratio   326
Discontinuity   414, 419
Discriminant   66
Distance between two points   208, 324
Distributive rule   41, 45
Divergence   426, 428
Division of a line   147
Domain   90
Dot product   334
Double angle formulae   188
Dummy index   102

Eigenvalue equation   398
Eigenvalue   398
Eigenvector   398
Electrical circuits   13, 31, 247, 371, 403
Equation   52
Equation of a circle   217
Equation of a line   212
Equilateral triangle   150
Equivalent fractions   12
Estimation   25
Euler's formula   361, 464
Even function   94
Exponent   24
Exponential form of a complex number   361
Exponential function   122, 126
Expression   52
Exterior angle   151

Factor   8, 45, 52
Factor theorem   52
Factorial   16
Factorising   8, 45
First order equations   452
First shift theorem   509
Forced damped oscillations   477
Forcing functions   469
Formulae   93
Fourier coefficients   520
Fourier expansion   517
Fourier series   285, 515
Fraction   6, 12
Function   90

Function of a function   97
Function of a function rule   235
Function of two variables   484
Fundamental component   518
Fundamental theorem of algebra   44

General solution   449, 463
Geometric progression   103
Gradient of a curve   230, 292
Gradient of a line   210
Graph   91

Half angle formulae   188
Heron's formula   178
Higher order derivatives   241
Higher order partial derivatives   489
Highest common factor (HCF)   9
Homogeneous equations   49
Homogeneous first order equation   456
Homgeneous system   397
Hyperbolic functions   138

Ideal gas law   94
Identity   50
Image   90
Imaginary axis   355
Imaginary number   19
Imaginary part   353
Implicit differentiation   238, 494
Implicit function   91
Improper fraction   12
Improper integral   279
Incompatible equations   49
Indefinite integral   253
Independent variable   90
Indeterminate form   412
Index   18, 70
Index of summation   102
Inequality   7, 97
Infinite arithmetic series   429
Infinite binomial series   106
Infinite geometric series   428
Infinite powers series   106, 423, 434
Infinite sequence   424
Infinite series   423, 428
Infinity   6
Inhomogeneous equation   462, 468
Initial conditions   449
Initial value problems   513
Integers   5
Integral transform   503
Integrand   253
Integrating factor   458

Index

Integrating rational functions   265
Integration   253
Integration by parts   273
Intercept of a line   212
Intercept theorem   153
Intersecting lines   216
Interval   98
Inverse Laplace transform   512
Inverse matrix   381, 395
Inverse of a function   100
Inverse trig functions   184
Irrational number   6, 411
Isosceles triangle   150
Iteration   426

Kinematics   311

Laplace equation   490
Laplace transform   285, 504
Laplace transform of the derivative   509
Law of natural growth   127
Limit   125, 412
Line   147
Line segment   147
Linear approximation   441
Linear combination   463
Linear equation   48, 65
Linear expression   40
Linear first order equation   458
Linear inequality   99
Linear programmming   223
Linear substitution   260
Linear superpositions   515
Linear systems   515
Logarithm   130
Lowest common multiple (LCM)   9

Maclaurin's series   434
Magnitude of a vector   320, 331
Mantissa   24
Mapping   90
Matrix   378
Maximum   294
Mean   110, 314
Method of least squares   312
Midpoint of a line   209
Minimum   294
Mixed fraction   12
Modulus   8
Modulus function   96
Modulus of a complex number   355
Moment of a force   347
Moment of inertia   313

Monomial   43
Multiplication by a scalar   322, 331
Multiplication of matrices   383

Natural exponential function   124
Natural logarithm   130
Natural numbers   5
Negative numbers   5
Newton's laws   245, 477
Newton's method   427
Normal to a curve   230, 293
nth harmonic   518
nth term   105
Numerator   12, 55

Obtuse angles   149
Odd function   95
Open interval   98
Opposite angles   149
Order of differential equation   448
Ordinary differential equation   448
Orthogonality relations   517
Orthonormal basis   329
Orthonormal set   329

Parallel lines   149, 214
Parallelogram law   321
Parametric differentiation   240
Parametric representation   91
Parametric representation of curves   219
Partial differential equation   490
Parseval's theorem   526
Partial derivative   487
Partial fractions   62, 265
Partial sum   428
Particular integral   462, 469
Particular solution   449
Percentage   13
Period   182
Periodic functions   180
Permutation   16
Perpendicular lines   214
Phasors   198, 371
Piecewise continuous function   505
Plane   485
Plane geometry   149
Plotting a graph   92
Point   147
Point of inflection   294
Polar coordinates   206
Polar form of a complex number   355
Polynomial   43, 52
Polynomial equation   43, 52

531

Position vector   319
Power   18, 70
Prime number   8
Principal value   184
Product rule   234
Proof   155
Proof by contradiction   411
Proper fraction   12
Proportion   14
Pythagoras' theorem   154
Pythagorean identity   185

Quadratic equation   65
Quadratic expression   41
Quadratic inequality   99
Quotient   10
Quotient rule   235

Radians   148, 173
Radius of convergence   436
Range   90
Rate of change   292
Ratio   14
Rational function   54
Rational number   6, 12, 410
Rationalisation   20
Rationalisation of complex numbers   355
Real axis   355
Real line   6
Real numbers   6
Real part   353
Reciprocal   13
Reciprocal function   90
Rectangular coordinate system   205
Reflex angles   149
Remainder theorem   61
Restrictions   300
Right angle   149
Right-angled triangle   150
Right-handed set   323
Rigour   88
Root mean square   314
Roots   44
Rounding   24
Rules of algebra   41
Rules of arithmetic   10
Rules of differentiation   234
Rules of indices   18

S(ketch) GRAPH   299
Scalar   318
Scalar product   334
Scalene triangle   150

Scientific notation   24
Secant   157
Second order equations   462
Sequence   102
Series   102
Sigma notation   102
Significant figures   23
Similar triangles   152
Simple harmonic oscillator   404
Simultaneous equations   48
Sine   175
Sine rule   178
Sine wave   182
Sinusoidal function   182, 515
Sketching a graph   92
Slope of a curve   230, 292, 421
Slope of a line   210
Solution   52
Solution of triangles   180
Solving inequalities   98
Square root   19
Stability   525
Standard derivatives   232
Standard integral   255
Stationary points   294
Step function   419, 506
Subscript   43
Substitution   260
Sum   10
Sum to infinity   105
Supplementary angles   148
Surd   20
Surface   485
Symbols   40
Symmetry   299

Tangent   157, 175
Tangent plane   486
Tangent to a curve   230, 293
Taylor series   434
Tensor   318
Torque   347
Total derivative   492
Total differential   492
Transfer function   525
Transpose   381
Transposing a formula   94
Transversal   149
Trap-door principle   9
Triangle   150
Triangle law   321
Triangular wave   507, 519
Trig identities in integration   269

Trig substitutions in integration 272
Trigonometric equations 191
Trigonometric functions 175
Trigonometric ratios 174
Trivial solution 397
Turning point 294

Undetermined coefficients 469
Unit matrix 381
Unit vector 331

Variables 40
Variables separable 454

Variance 110
Vector 318
Vector function 339
Vector product 336
Volume of revolution 308
Vulgar fraction 12

Work done 334

Zero matrix 381
Zero vector 319
Zeros 6, 44